that experiencing this evolution of thought very much helped to clarify the development of my own ideas. Thinking that this might be equally beneficial to readers of this book, I have included a considerable discussion of the studies in the older literature as well as of those of recent origin. Because of the quantity of material available, I have had to restrict the coverage of the literature to those studies that either led up to the use of, or actually dealt with, thermochemically definable, growth-process systems. This has meant considering only batch cultures in which microorganisms are capable of nonlimited growth for most of the duration of an experiment. Consequently, I have purposely ignored many excellent studies dealing with continuous culture systems, in which the growth of microorganisms is rate-limited. From the thermodynamic point of view, batch cultures provide the simplest possible systems to investigate.

For those who have not used thermodynamics recently, the first chapter summarizes the comparatively simple thermodynamics, involving little more than addition and subtraction, which is all that is needed to understand the thermodynamic concepts developed in the rest of the book. This is followed by chapters intended to carry readers through the historical development of thought on how energy and the production of heat is related to microbial growth. The middle chapters describe the evolution of ideas on microbial growth-process systems that can be used to study thermodynamic changes. The development of modern theories is then covered in the last chapter, which is devoted entirely to comparatively easy methods for calculating the free-energy, enthalpy, and entropy changes accompanying the growth of a microorganism on a known quantity of substrate and how these changes can be interpreted.

I hope that readers will feel the same fascination that I did in observing how many different kinds of scientists have tried to understand this subject and how most of them did, in fact, contribute a piece to fill in the puzzle of what constitutes the energetics of microbial growth.

EDWIN H. BATTLEY

Stony Brook, New York
November 1986

PREFACE

The most basic relationship in microbial ecology is undoubtedly that between the quantity of matter and energy available to a microorganism from its environment and the quantity of growth that can be obtained from it. It is therefore all the more astonishing that until recently so little effort has been made to learn more about the thermodynamics, or energetics, of microbial growth. This has probably been a result of the complexity of the subject and the necessity of dealing with both thermochemical and biological concepts. Even in the 1980s it is certainly true that many biological processes do indeed seem complex. However, this does not mean that they are not susceptible to thermodynamic treatment. Every mass, including that of a living cell, has thermodynamic properties. Using appropriate thermodynamic systems it ought to be possible to determine the thermodynamic properties of growing, microbial cells; to calculate the energy exchanges that occur between these cells and their environment; and to learn how these energy exchanges are related to their increase in biomass. This book was conceived and written in an effort to stimulate more interest in the energetics of microbial growth. Studies on this subject are extremely important. They are of interest not only to microbial ecologists, but also to other scientists. The energetics of cellular growth is basic to far more than microbial ecology; it is the most basic of all general, biological phenomena. It just so happens that the subject is most conveniently studied through the use of microorganisms.

While reviewing the literature on the energetics of microbial growth, I noticed the thin threads of comprehension that linked the studies of many different groups of scientists as far back as the time of Pasteur. The early ideas of the relation of heat to growth were more simple than later ones, and I found

To Susan

***Library of Congress Cataloging in Publication Data*:**

Battley, Edwin H.
Energetics of microbial growth.

"A Wiley-Interscience publication."
Includes bibliographies and index.
1. Microbial growth. 2. Bioenergetics.
3. Thermodynamics. I. Title.
QR86.B37 1987 576′.131 86-15937
ISBN 0-471-08492-1

Printed in the United States of America

10 9 8 7 6 5 4 3 2 1

ENERGETICS OF MICROBIAL GROWTH

Edwin H. Battley
State University of New York, Stony Brook

A Wiley-Interscience Publication
JOHN WILEY & SONS
New York · Chichester · Brisbane · Toronto · Singapore

ACKNOWLEDGMENTS

During the writing of this book many people at the State University of New York at Stony Brook have been exceedingly helpful. Reviewing the literature became a pleasure because of the fine facilities of the Biological Sciences Library and the helpfulness of Doris Williams, its Head Librarian, in facilitating interlibrary loans when these were necessary. Linda Josephs prepared the manuscript, Joyce Schirmer and Diane Chandler drew the original illustrations, and Joseph Lofreddo did the photography. Chapter 1 and Section 10.3 were read by Dr. Nelson Johnson, who was most helpful in his criticism. The entire manuscript was reviewed by Professor William J. Payne of the University of Georgia, who made several editorial and other suggestions. I am grateful for the care, scientific objectivity, and thoughtfulness with which he did this. Special thanks go to Dr. Mary M. Conway, whose encouragement and skill as Life Sciences Editor for John Wiley & Sons led to the publication of this book.

Many of my colleagues and friends have offered encouragement for this endeavor. But the most encouraging of all was my wife, Susan, who endured for three years, with characteristic patience and love, the preoccupations of an author-husband. The writing of this book would have been far more difficult without the support she gave me.

E.H.B.

CONTENTS

APPENDIXES

ENERGETICS OF MICROBIAL GROWTH

1

A BRIEF INTRODUCTION TO CERTAIN ASPECTS OF CHEMICAL THERMODYNAMICS

In any book describing the energetics of microbial growth there quite reasonably ought to be a chapter devoted to the subject of energy, the various forms it can take, and the various ways in which it can be transformed. Chapter 1 is not intended for this purpose. Unfortunately, one chapter is not enough to cover what is otherwise described in complete books. Because of this, what is presented here is largely in the form of statements of fact devoid of formal derivation. Only that material which is directly applicable to an understanding of the energy transformations that accompany microbial growth will be described. Thus, this chapter is truly only an introduction. It should not be expected to replace the many books on chemical thermodynamics, some of which are listed in the bibliography at the end of this chapter. On the other hand, even those who are already familiar with energy changes as they occur in chemical reaction systems may find this a useful "refresher" chapter.

1.1. WHAT IS MEANT BY CHEMICAL THERMODYNAMICS?

The word "energy" as used in physics represents the capacity for performing work. Energy and energy transformations are the objects of study in thermodynamics, the laws of which have been derived by physicists. Energy can be regarded as occurring in two major forms: thermal and nonthermal. In addition, nonthermal energy occurs in several alternate forms, these being chemical, electrical, kinetic, mechanical, potential, and radiant energies. All forms of energy are interconvertible. However, although thermal energy cannot be quantitatively converted into nonthermal, all forms of nonthermal energy can be quantitatively converted into thermal and for this reason quantities of energy have been historically expressed in thermal units. Possibly because of this the word "thermodynamics" became accepted as representing the study of nonthermal–thermal energy exchanges, and the distinction arose between nonthermal and thermal energy as being two general forms of energy. But even though the energy changes accompanying a chemical or a physical event can usually be measured in terms of a change in a quantity of heat, such changes are actually manifestations of possibly several forms of energy.

The subject of chemical thermodynamics is the nonthermal–thermal energy exchanges accompanying chemical events. The events that take place around or within cells are almost exclusively of a chemical nature, the major exceptions being diffusion, the absorption of radiant energy by pigments and its emission during fluorescence, osmotic work, and the performance of mechanical work by contractile proteins, all of which have more of a physical nature.

What the study of chemical thermodynamics seeks to determine is the identification of the kinds of energy and the quantification of the energy changes that take place, relative to a chemical event. From this information it seeks also to determine what causes the event to take place and the extent to which it can proceed. Conversely, it is not in the nature of thermodynamics, chemical or otherwise, to provide information as to the rate at which or the mechanism by which a chemical event proceeds.

1.2. THERMODYNAMIC SYSTEMS

A thermodynamic system is a bounded space containing mass and energy that, for purposes of study and experimentation, can be intellectually and physically separated from and related to the environment that surrounds it. The physical state (condition) of such a system can be defined in terms of the temperature at which it exists, the physical pressure that is exerted upon it by the environment, and the material composition of the mass that comprises it. There are four kinds of systems that can be intellectually conceived and practically obtained. These are (1) the isolated system; (2) the closed system; (3) the adiabatic system; and (4) the open system. In an *isolated system* there can be no exchange of either matter or energy with the environment, and the system is thus completely insulated from the environment. A *closed system* is one that is materially self-contained, but there can be a free exchange of energy with the environment. In an *adiabatic system* matter, but not energy, may be exchanged with the environment. In an *open system* there may be a free exchange of both matter and energy with the environment. Biological examples of all four systems can be found in nature, depending on how these examples are arbitrarily defined. They may also be created arbitrarily in the laboratory through an appropriate selection of physical and chemical conditions with which to set up a system experimentally. Because of their intellectual and practical convenience it is usually laboratory-created biological systems that are chosen for thermodynamic study, the major consideration for biologists being that, unless there is some interest in studying abnormal or "unnatural" biological events, the arbitrarily created system should model as closely as possible what is imagined to be the natural system relative to the purpose of the study. Thus, for example, unless there is a specific reason for doing otherwise the temperature at which a biological or biochemical event should be studied is the physiological temperature at which it normally occurs in nature, if this can be established.

Most biological events can be considered to take place at a constant temperature. This is because of the capacity for absorbing heat of the water which forms most of the composition of all living cells and which is usually the most abundant substance in their immediate environment. In nature this temperature constancy applies over time intervals of minutes to days in poikilothermic organisms, and indefinitely in homoiothermic organisms. It can also be made to apply indefinitely under arbitrarily chosen laboratory conditions, when it becomes possible to control the temperature more accurately than it is controlled under natural conditions. Also, because most biological events occur in a restricted environment, they may generally be considered to take place under a constant atmospheric or hydrostatic pressure.

A great many, perhaps most, biological assemblages of matter and energy can be intellectually conceived in such a way that they may be considered closed systems, and this type of system is also the one most amenable to convenient study and experimentation. A closed system that does not possess the capability of free exchange of energy with its environment is said to be in a state, or condition, of stable thermodynamic equilibrium with its environment, and in this condition no energy changes will take place and no events or processes will occur. However, under the condition of an unstable thermodynamic equilibrium the potentiality for a free exchange of energy between a closed system and its environment does exist, and some event or process will take place.

The *initial state* of a system is that which exists before any change in the matter or energy of the system has occurred, and when a maximum potentiality for change is present. The *final state* of a system is that existing after all possible changes in the matter or energy of the system have occurred, and when no further possibility for change is present under the then-existing physical conditions.

The matter and energy changes that occur as a closed system undergoes a transition from an initial, defined state of unstable thermodynamic equilibrium to a final, defined state of stable thermodynamic equilibrium can be determined, and because these states can be defined in terms of the temperature, pressure, and material composition of the system, if the temperature and pressure are kept constant, any energy changes can be related directly to changes in the material composition. Conversely, under constant temperature (isothermal), constant pressure (isobaric) conditions, if the change in the composition of a system as it undergoes a transition from an initial to a final state is known, it is possible to relate this change directly to a change in the quantity and the distribution of energy within the system.

The transition of a system from an initial to a final state is called a *reaction*, and in chemical thermodynamics the compositions of the initial and final states can be represented in the construction of a *chemical reaction equation,* usually simply called a *chemical equation,* which is an expression of the Law of the Conservation of Mass. The molecules or substances that constitute the initial state of a system are called the *reactants,* and those constituting the final state are called the *products.* All chemical reactions, no matter what form they take, are subject to the laws of the conservation of mass and of energy, and to the three laws of thermodynamics. Historically, all forms of energy were evaluated on the basis of their equivalence to heat. However, since thermodynamics actually deals with all forms of energy and not heat alone, there is a present tendency on the part of biologists and biochemists to replace the name "thermodynamics" by the name "energetics." The thermodynamics of biological systems thus becomes "biological energetics."

1.3. QUANTITIES, DIMENSIONS, AND UNITS

Quantities are portions of things. A specified area represents a quantity of surface. A given volume represents a quantity of space. A velocity represents a quantity of distance traversed within a given time. But in order to compare one quantity with another, suitable dimensions having appropriate units have to be used with which to describe the magnitudes of these quantities. In biology and chemistry the system (not to be confused with a thermodynamic system) that is most commonly used in this respect is the centimeter-gram-second (c.g.s.) system, in which the centimeter is the unit of the quantity "length"; the gram is the unit of the quantity "mass"; and the second is the unit of the quantity "time." Although not specifically included as a part of the c.g.s. system, the degree is the unit of the quantity "temperature." These dimensions of length, mass, time, and temperature are fundamental in that other physical quantities can be defined in terms of them, for example, area, volume, velocity, acceleration, force, pressure, density, and energy, this last being of great importance in energetics. The relationships between a number of quantities and the dimensions and units that can be used to describe them are given in Table 1.1.

Most American biologists are familiar with the data in Table 1.1 since they were probably educated in the use of the c.g.s. system. However, it sometimes seems that no sooner do we become accustomed to one particular system than a different one is proposed to supplant it. This is the case with the c.g.s. system, which will sooner or later be replaced by the *Systeme International* (SI) *d'Unites.* This is the most recent form of the meter-kilogram-second (m.k.s) system which was internationally accepted in 1960. The two systems are, of course, related, this relationship being given in Table 1.2. It will be seen that the SI is simpler than the c.g.s. system, and this may have some merit associated with it. On the other hand, biologists seldom deal with such large units as meters and kilograms, and many of us do not presently use the SI. Two major differences between the two systems are the use of the newton as the unit of

TABLE 1.1.

Quantity	Dimensions	c.g.s. Primary Units	Secondary Units
Length	$[L]$	cm	1 in = 2.54 cm 1 μ = 10^{-4} cm 1 Ångström = 10^{-8} cm
Area	$[L^2]$	cm^2	1 in^2 = 6.46 cm^2
Volume	$[L^3]$	cm^3	1 liter = 1000.028 cm^3 ∴ 1 ml ≅ 1 cm^3
Mass	$[M]$	g	
Time	$[t]$	sec	1 min = 60 sec 1 hr = 60 min 1 day = 24 hr
Velocity	$[Lt^{-1}]$	cm sec^{-1}	
Acceleration	$[Lt^{-2}]$	cm sec^{-2}	
Force or weight	$[MLt^{-2}]$	dyne (= g cm sec^{-2})	1 g wt = 980.7 dyne (average)
Pressure	$[ML^{-1}t^{-2}]$	dyne cm^{-2}	1 g wt cm^{-2} = 980.7 dyne cm^{-2} 1 mm Hg at 0°C = 1 torr = 1,333 dyne cm^{-2} 1 atmosphere = 760 mm Hg at 0°C 1 bar = 10^6 dyne cm^{-2}
Density	$[ML^{-3}]$	g cm^{-3}	
Work or energy	$[ML^2t^{-2}]$	erg (= dyne cm = g cm^2 sec^{-2})	1 absolute joule = 10^7 ergs 1 defined calorie = 4.184 abs. joules 1 abs. volt Faraday = 96,487 abs. joules

Source: Reprinted with permission from J. G. Morris, *A Biologist's Physical Chemistry*, 2nd ed., "Contemporary Biology Series" published by Edward Arnold (Publishers) Limited, London.

force in the SI and the joule as the unit of work or energy. The extent to which the SI will actually be put into practical use by biologists will probably be a function of the resistance to change of those who have been educated in the c.g.s. system and who have found it more convenient to use.

1.4. THERMOMETRIC MEASUREMENTS

A thermometer is a device which makes it possible to determine whether two bodies are at the same temperature, or if they are not, to provide some objective numerical measurement of their relative temperatures. The first

TABLE 1.2.

Quantity	SI Unit	c.g.s. Unit	To Convert c.g.s. Unit to SI Unit Multiply by
Length	meter (m)	centimeter (cm)*	10^{-2}
Area	square meter (m^2)	square centimeter (cm^2)*	10^{-4}
Volume	cubic meter (m^3)	cubic centimeter (cm^3)*	10^{-6}
		liter (l)**	10^{-3}
Mass	kilogram (kg)	gram (g)*	10^{-3}
Force	newton (N)	dyne	10^{-5}
Work or energy	joule (J)	erg	10^{-7}
		calorie (thermochemical)	4.184
		liter atmosphere	101.328
Pressure	newton/square meter (N/m^2)	dyne cm^{-2}	10^{-1}
		mm Hg	133.322
		atmosphere	101.325×10^3

* These are "acceptable" submultiples of the SI unit (though use of the cm is deprecated).
**This is an "acceptable" practical unit within the SI system.
Other useful conversion factors include the following:

Quantity in non-SI units	Same quantity in SI units
1 inch	25.4 mm
1 foot	0.3048 m
1 g/l	1 kg/m^3
1 lbf/sq in	6894.76 N/m^2 (i.e. 6.89476 kN/in^2)
t°C	$(t + 273.15)$°K
Gas constant (R) = 1.987 cal $mole^{-1}$ $degree^{-1}$	8.3144 J mol^{-1} $°K^{-1}$

Source: Reprinted with permission from J. G. Morris, *A Biologist's Physical Chemistry*, 2nd ed., "Contemporary Biology Series" published by Edward Arnold (Publishers) Limited, London.

thermometer appears to have been invented by Galileo about 1592. It consisted of an air-filled glass bulb drawn out into a tube partly filled with wine which led down into an open reservoir. As the volume of the air increased or decreased with rising or falling temperature, the change in volume was indicated by the level of the wine in the tube. A scale was marked on the tube at intervals which corresponded roughly to $\frac{1}{1000}$ of the volume of air in the bulb. In 1643 Galileo's former assistant, E. Torricelli, invented the mercury barometer (Torricelli tube) and discovered atmospheric pressure, and it became evident that Galileo's air thermoscope was sensitive to changes both in air pressure and in temperature. A short time later, about 1660, Grand Duke Ferdinand of Tuscany designed the first hermetically sealed liquid-in-glass thermometer. This was very similar to a modern alcohol thermometer and consisted of a glass bulb connected to a capillary tube, or stem. The bulb and part of the stem were filled with wine or alcohol, after which the end of the stem was sealed off. Such thermometers were used by scientists at the famous Academia del Cimento, in Florence, which was founded by Duke Ferdinand. These came to be known as Florentine thermometers. Some were used for weather observations, and were standardized by making marks on the stem corresponding to the position of the liquid boundary on the hottest and coldest days during the year. Intermediate temperatures were judged as "extremely cold," "moderately cold," and so on, according to a scale subdivided between these extreme "fixed points" of reference. It was these Florentine scientists who first used a thermometer for the purpose of measuring body temperature. The fixed points used by the Florentines were not very reliable, and in 1701 Isaac Newton used a linseed oil-filled, liquid-in-glass thermometer to establish a numerical temperature scale. Assigning the numerical value of 0 to the temperature of melting, packed snow, and the value of 12 to the temperature of the human body, he determined the boiling point of water to be 34 and the melting points of metals such as tin to be 72, bismuth 81, lead 96, and so on. D. G. Fahrenheit made several improvements in the construction of liquid-in-glass thermometers and in 1724 introduced the mercury thermometer. During the course of experiments with freezing mixtures (mixtures of salts and ice, similar to those that can be used for making "homemade" ice cream) he discovered reproducible fixed temperatures considerably below that of melting ice. In 1714 he proposed the original Fahrenheit scale, in which the number 0 was assigned to the lowest temperature he was able to achieve at that time with a freezing mixture of ice and ammonium chloride. The number 12 was first assigned to human body temperature, but this was later changed to 96, presumably because this provided a standard interval that could be subdivided into a large number of rational fractions (e.g., halves, fourths, eighths, sixteenths, and thirty-seconds). It became obvious that human body temperature is not a satisfactory fixed point, and in 1742 A. Celsius proposed the assignment of the number 0 to the steam point and the number 100 to the ice point. Later it was suggested by his colleague, M. Stromer, that it would be more convenient if the scale were reversed (since the numbers would then rise with the temperature), resulting in the modern centigrade, or Celsius scale of temperature. The Fahrenheit scale is based in the same two fixed points, but it has been redefined since Fahrenheit's time such that the ice point is assigned a value of 32 and the steam point a value of 212, these values being in close agreement with the approximate values for the same points on Fahrenheit's original scale. The accepted value for normal, human body temperature is now 98.6°F. Other temperature scales have been proposed, but have largely dropped out of use. It is the Celsius or the Absolute (to be described below) temperature scales that are almost universally used at the present time in scientific work, and the degree is presently defined as one-hundredth of the temperature difference between the freezing and the boiling points of water at atmospheric pressure.

1.5. THERMAL ENERGY—ENERGETICS

Thermal energy, or heat, is a form of kinetic energy (energy of motion) possessed by the atoms and molecules of which substances are made. It exists as a function of the translational, rotational, vibrational, and internal rotational movements of these atoms and molecules. According to kinetic theory these movements increase as the temperature of the substance rises. If two substances are brought in contact, heat can be transmitted from the one having the higher temperature to the one having the lower temperature by the processes of conduction or convection. Effectively, the atoms or molecules of the hotter substance, being in a state of more violent motion, collide with those of the cooler substance by reason of the contact of the two substances and transmit this kinetic energy to the cooler substance until the two substances are in thermal equilibrium. The measurement of heat involves the dimension of temperature, which is sensed by some thermometric device having a sensing element that assumes the temperature of the substance with which it is in contact.

1.6. UNITS OF ENERGY USED IN THERMODYNAMIC STUDIES

Because all forms of energy are interconvertible, it is possible to use the same unit for all of them. However, the unit which has found the most practical use in thermodynamic studies has in general been the *gram calorie* (usually referred to simply as the "calorie"). As originally defined this unit represented the amount of heat which is necessary to raise the temperature of 1 g of water by 1°C. Subsequently, it was found that this quantity of heat differed, depending on the mean temperature at which it was measured. Thus, the amount of heat required to raise the temperature of 1 g of water from 14.5 to 15.5°C (the 15° calorie) is slightly different from that required to raise the temperature of the same mass of water from 16.5 to 17.5°C (the 17° calorie). For many years it was the 15° *practical calorie* which was used as the standard unit of thermal energy. However, with the adoption of the *joule* as the international standard unit of energy of all kinds (i.e., the SI unit of energy) it became necessary to adopt a standard conversion factor to enable a conversion to be made from calories to joules. For this reason a *defined calorie* was introduced, the magnitude of which is independent of temperature. Thus, as seen in Tables 1.1 and 1.2, one defined calorie is equal to 4.184 joules. The defined calorie is slightly different from the 15° practical calorie in that it is more nearly equal to the 17° gram calorie. For most purposes, especially in biological work in which an experimenter is lucky to obtain an accuracy of $\pm 2\%$, the difference between the practical calorie and the defined calorie is not significant. Formally, it is the defined calorie rather than the 15° practical calorie which has become the standard unit of thermal energy.

It is worth considering whether the use of the calorie or the joule is preferable in thermochemical studies. In the strict sense we are mandated by the adoption of the *Systeme International d'Unites* to use the joule. All forms of energy can be expressed at this upper level of abstraction without regard to the actual form of energy being evaluated. However, it is exactly this which makes the use of the joule so unappealing. The word "joule" does not even tell us whether the energy is thermal or nonthermal. The calorie is the historic unit of measurement of thermal energy. If reactions are being studied in which chemical energy becomes converted into thermal energy, or can be conceived to do so, the visualization and conceptualization of this conversion is most easily made if the calorie is used as the unit of energy. Although calories and joules are interconvertible by means of the appropriate conversion factor, what is the point in expressing heat (as 17° gram calories) in terms of work (as joules)? For this reason use of the calorie as the unit of thermochemical energy will be retained in this book. For a more complete discussion of this controversy the reader is referred to the short article by M. Kleiber.[2]

1.7. CONSERVATION OF MASS DURING A CHEMICAL REACTION

The principle that the heat released by chemical reactions is in no way associated with sensible changes in the total mass of the system undergoing reaction was first established by A. L. Lavoisier. It was later confirmed by the classical work of E. W. Morley on the synthesis of water from oxygen and hydrogen, and the series of measurements made by H. Landolt on several different reactions carried out in sealed vessels. Since then hundreds of workers in the field of quantitative chemistry have abundantly confirmed this principle, which came to be known as the *Law of the Conservation of Mass.* Exceptions to this principle do occur in reactions involving changes in atomic nuclei during which there is a net conversion of mass to energy. However, the net conversions of mass to energy in ordinary chemical reactions, in which only the electrons in the outer orbits of the atoms are involved, are insignificant by comparison and cannot be detected by chemical analytical means. Under these latter conditions the Law of the Conservation of Mass states that "Matter can neither be created nor destroyed." Thus, in an ordinary isolated, or a closed thermodynamic system the total mass remains constant, although within the system it may undergo a change or rearrangement in composition.

1.8. CONSERVATION OF ENERGY DURING A CHEMICAL REACTION. THE FIRST LAW OF THERMODYNAMICS

Largely through the classic experiments of J. P. Joule during the years 1843 to 1850 it became possible to conclude that heat may be regarded as a form of energy. Thus, whenever a certain amount of energy becomes lost, or "dissipated" from a real mechanical or electrical process (through mechanical friction or electrical resistance), without exception some effect is produced which is exactly proportional to the mechanical or electrical energy which is lost. Energy is conserved in that the quantities of nonthermal and thermal energy which can be accounted for at the termination of such processes are exactly equal to the quantity of nonthermal energy available at the beginning. Conversely, energy is also conserved in a heat engine, in the conversion of thermal energy from a high-temperature source into mechanical energy. In addition, thermal energy is itself conserved in processes of thermal interchange, such as the mixing of two volumes of liquid at different temperatures, heat conduction, thermal radiation, and so on. The idea that energy might be conserved appears to have been conjectured in 1842 by R. J. Mayer, who made an estimate of the mechanical equivalent of heat based on the thermal properties of gases. He is not generally credited with formulating the principle because his reasoning did not have the experimental foundation provided a few years later through the work of J. P. Joule. The general principle of the conservation of energy was recognized by H. Helmholtz, who first extensively employed it in 1847. This principle subsequently became known as the *Law of the Conservation of Energy,* or as the *First Law of Thermodynamics.* As with the conservation of mass, exceptions to the conservation of energy occur in reactions that involve changes in atomic nuclei during which there is a net conversion of mass to energy. In ordinary chemical reactions, however, involving only electrons in the outer atomic orbits, the net conversions of mass to energy are insignificant and cannot be detected by calorimetric means. Under these latter conditions the First Law of Thermodynamics states that "Energy can neither be created nor destroyed." Thus, in an ordinary, isolated, chemical thermodynamic system, but not in a closed system by itself, the total amount of energy remains constant, although within that system it may undergo a change in form and distribution. In a closed system the conservation of energy exists in the sum of the energy within the system plus that exchanged with the environment.

1.9. HEAT AND WORK

Energy was previously defined as "the capacity for performing work," with no definition being given as to the meaning of the word "work." This latter can best be defined in relation to the meaning of the word "heat."

During the historical development of the concept of the conservation of energy it came to be generally recognized that the conservation of mechanical energy in conservative mechanical systems, and the conservation of heat in processes which are purely thermal, are both special cases of a more general conservation phenomenon which includes thermodynamic processes involving interchanges of mechanical and thermal energy. Following the establishment on a chemical basis of the molecular nature of matter two different, though related, approaches were made toward correlating the phenomenon of heat with ordinary mechanical energy. One of these was the kinetic approach developed through the work of R. J. E. Clausius, L. Boltzmann, and J. Clerk Maxwell, and the other was the statistical mechanical approach of J. Willard Gibbs. In the theories advanced by both of these schools, thermal energy is considered to be a random molecular motion which is different from ordinary mechanical motion only in that the measuring instruments used in thermodynamic investigations are not sufficiently sensitive to enable random molecular motion to be perceived. Indirect evidence for this motion exists in the form of the Brownian motion of particles of colloidal size. Similarly, the random motion of the molecules comprising an otherwise motionless mass can be ordinarily only indirectly perceived through the sensing of temperature. Heat is thus considered as the kinetic and potential energy of random molecular motion, and the conversion of mechanical energy into heat is simply the randomization of the energy of bulk masses of molecules that had been previously associated with an average motion in a particular direction, or an average location in a particular spatial position.

Energy, to be capable of being measured, must produce some sort of effect which can be detected, or sensed. The effect invariably produced by thermal energy is the heating of a colder by a hotter mass. The effect produced by nonthermal energy is to cause one mass to act upon another in such a manner that something happens of a chemical, electrical, kinetic, mechanical, or radiant nature. Energy can therefore be considered as something "in transition" that is capable of being exchanged, as a result of which some effect is caused. Thus, *heat is thermal energy that is exchanged between a hotter and a colder mass because of a difference in temperature between them,* and *work is nonthermal energy that is exchanged between two masses because of a force which is exerted between them.* It is important to realize that the concept of work, although ultimately capable of being described in mechanical terms, embraces *more* than mechanical effects and can include other nonthermal effects as well. Among these are chemical effects. Also, inherent in the concepts of either heat or work is that one mass acts upon another, either to heat it (in which case the mass having the greater amount of thermal energy loses some of it to the mass having the lesser amount, which thereby gains thermal energy), or to perform work upon it (in which case the mass having the greater amount of nonthermal energy loses some of it to the mass having the lesser amount, which thereby gains nonthermal energy).

1.10. INTERNAL ENERGY

Because energy is conserved, in an isolated, thermodynamic system (one which is completely insulated from the environment) the total quantity of energy within the system must remain constant, although within that system it may change its form. However, *energy is also conserved in the exchange of energy between a closed material system and its environment if the closed system PLUS its environment are regarded as an isolated system with respect to energy.* Regarded in this way the system can still be considered closed relative to the exchange of matter. The total energy contained within a system is designated as its *internal energy,* represented by the symbol U, to differentiate it from the energy that is external to the system. The internal energy is also often called the *intrinsic energy.* It is a function of a system that depends only on the physical state, or condition, of the system at the time it is being considered, and is independent of its previous history. Because of this the internal energy is said to be a "function of state." The actual amount of internal energy that is associated with any given state of a closed system is impossible to measure. It is thus a variable which does not have a unique value. However, it *is* frequently possible to measure the *difference* in the amounts of energy which exist with respect to two states of the same closed system. This difference is a *quantity* of energy represented by the symbol ΔU.

Any closed system that performs work on its environment can do so only through the expenditure of an equivalent amount of its internal energy. In accord with the definition of "work" which was given previously, the work done is represented by the nonthermal energy that is exchanged between two masses, one of which is contained within the system and the other within the environment. The quantity of work can be determined as a function of a measurable difference in something associated with the state of the system before and after the work was performed. For example, if a closed, isothermal, isobaric system were to undergo an increase in volume by an amount ΔV, the amount of work performed on the environment would be $P\,\Delta V$, where $V = \text{cm}^3$, $P = \text{dyne cm}^{-2}$; and $P\,\Delta V = \text{dyne cm}^{-2}\,\text{cm}^3 = \text{dyne cm} = \text{ergs}$. In this case, the pressure–volume work being performed can be considered to be that of compressing the environment. Because the system is performing work *on* the environment, its internal energy must decrease by an amount $(-P\,\Delta V)$ of nonthermal, pressure–volume energy equal to the amount $(P\,\Delta V)$ of nonthermal, pressure–volume work gained by the environment. Conventionally, energy changes are only considered relative to the system that is undergoing change, and not relative to the environment. Thus, a loss of energy from a closed system is indicated by a minus sign.

A closed system may perform various kinds of work on the environment or on another system, depending on the form of nonthermal energy that is exchanged (e.g., chemical, electrical, kinetic, mechanical, or radiant energy), but all of these kinds of work may be represented by the single symbol w. The total nonthermal energy expended by a system in the performance of work is then represented by $-w$, to indicate that energy is lost from the system. Conversely, if work were performed *on* the system w would be positive. There can also be an exchange of thermal energy between a closed system and the environment. This is represented by the symbol q if thermal energy is absorbed by the system from the environment, and by $-q$ if thermal energy is transferred from the system to the environment. A change in the internal energy of a system can thus be a function of an exchange of nonthermal energy, thermal energy, or both with the environment. Since any change in the internal energy of a closed system can only occur as a result of some change in the system from an initial state having an internal energy U_1 to a final state having an internal energy U_2. a change in the internal energy of a system can be represented by means of the relation

$$\Delta U = U_2 - U_1 = (q - w)$$

Because U_1 and U_2 are functions of state, depending only on the temperature, pressure, and composition of the system, if the temperature and pressure are kept constant, the value of $(q - w)$ will be the same no matter how devious the route along which the system may pass as it changes from state 1 to state 2.

1.11. THE RELATIONSHIP BETWEEN THE HEAT OF REACTION AT CONSTANT VOLUME AND AT CONSTANT PRESSURE

For any closed, isothermal system that is characterized by the performance of no work other than that associated with the expansion or contraction of the system,

$$\Delta U = (q - w) = q_v \tag{1.1}$$

where q_v represents the thermal energy absorbed under conditions such that the system remains at a constant volume, and

$$\Delta U = (q - w) = q_p - P\,\Delta V \tag{1.2}$$

where q_p represents the thermal energy absorbed under conditions such that the system remains at a constant pressure, P the pressure, V the volume, and $P\,\Delta V$ the pressure–volume work performed as a result of the change in the volume of the system. From Eqs. 1.1 and 1.2 it is evident that q_p and q_v do not have the same value for the same closed system, and that

$$q_p = q_v + P\,\Delta V \tag{1.3}$$

1.12. THE MEASUREMENT OF HEAT EXCHANGES IN CHEMICAL REACTIONS

Except under unusual circumstances a chemical reaction will be associated with either a liberation of thermal energy to, or an absorption of thermal energy from, the environment of a chemical thermodynamic closed system. The measurement of the quantity of heat exchanged in a chemical reaction is made in an instrument called a *calorimeter,* of which several kinds are in use. One of these, called a *bomb calorimeter,* is used for the determination of the heat of combustion of a substance. Because the substance to be analyzed is burned to completion in an atmosphere of pure oxygen inside the calorimeter vessel, there is a transient production of a high pressure inside the calorimeter vessel associated with the production of thermal energy during the reaction, before this energy becomes transferred to the mass of the vessel and thence to the mass of water in the water jacket surrounding the vessel and to the mass of the water jacket container. For this reason the calorimeter vessel, or "bomb," must be thick walled, made of steel, and sealed to prevent the escape of hot gases. The bomb calorimeter is thus a constant volume system. During the course of measurements made with a conventional bomb calorimeter (usually only a few minutes), a small amount of heat may be lost by conduction to the environment because of the difference in temperature between the calorimetric system and the insulated shield which surrounds it, from which it is separated by a layer of air. This heat loss can be minimized through the use of a much more elaborate instrument called an *adiabatic* bomb calorimeter, in which the environment of the calorimeter is heated at the same rate and to the same extent as the calorimeter itself is heated by the reactions that take place within it. Because there is never any appreciable temperature difference between the calorimetric system and its environment, no heat can be exchanged with the environment and highly accurate measurements can be made. An adiabatic, bomb calorimeter is thus a real example of an isolated, constant volume, thermodynamic system. The temperature at which the combustion reaction is considered to take place is the average of the temperatures of the initial and final states of the system. The actual thermodynamic processes that take place in a combustion bomb, with oxygen under pressure serving as the oxidant, are by no means as simple as has been implied here. Because combustion reactions outside of bomb calorimeters do not take place at constant volume, it is necessary to correct the observed heat of combustion at constant volume to what it would be if the reaction were to take place at constant pressure. For this it is necessary to have some idea of the nature of the Gas Laws.

1.13. THE GAS LAWS

The Gas Laws are a set of mathematical relationships involving the temperatures, pressures, and volumes of gases. They are fundamental to any study of energetics and it is therefore worth devoting some attention to them.

1.13.1. The Relationship Between Gas Volume and Pressure at Constant Temperature (Boyle's Law)

There are three fundamental states of matter, represented by gases, liquids, and solids. All matter occupies space, the quantity of space that is occupied being its volume. The concept of a "volume" necessarily implies that some limiting boundary is present to contain it. This limiting boundary is represented by a surface or a set of surfaces. These surfaces are usually themselves the limiting boundaries of some material which is of a different phase from the material which is being contained and which occupies the volume. For example, a gas might be contained in a rubber balloon or a glass flask which might also contain an aqueous culture medium designed for the growth of microorganisms. Here, the limiting boundaries would be partly the gas–solid (glass) interface and partly the gas–liquid (water) interface within the flask. The shape of the container is, of course, unimportant when considering gases.

The gaseous state, according to the Kinetic Theory of Gases, consists of particles of molecular size that are in constant motion during the course of which they individually collide with one another in a completely random manner. They also collide with and bounce off the walls of whatever contains the gas, the summation of the forces of these collisions over a unit area of container wall surface being measurable as the *pressure* of the gas. Each gas molecule is considered to be completely elastic, and to rebound after a collision with the same energy it possessed when entering the collision. Because of the independent, random motion of each gas molecule a gas occupies fully any volume that encloses it, unlike liquids and solids, which need not. For the same reason, if a space occupied by a gas is increased while the temperature is held constant, the gas will still come to occupy fully the new volume. However, the concentration of gas molecules per unit volume and thus the density of the gas will have decreased, and the pressure will have become lowered. This tendency for gas molecules to move spontaneously from a condition of higher concentration to a condition of lower concentration is called the force of diffusion. The opposite situation, in which the space occupied by a gas becomes decreased in volume, does not occur spontaneously at constant temperature. To accomplish this a force must be applied to cause the gas molecules to become closer together such that their density is increased with a concomitant increase in pressure on the walls of the container. This force is called the force of compression. There is thus an inverse relationship between the volume and pressure of a gas which can be expressed by the following equation:

$$V \propto \frac{1}{P} \tag{1.4}$$

where V = volume, usually expressed in liters, and P = pressure, usually expressed in atmospheres with 1 atm being equal to the mass of a column of liquid mercury 760 mm high and 1 cm^2 in cross section. From Eq. 1.4 it follows that

$$PV = K$$

where K is a constant having the dimensions of liter-atmospheres. This relationship was discovered by R. Boyle in 1662 and is known as Boyle's Law.

Since, with respect to a given gas at a constant temperature, K would be expected to have the same value for any combination of P and V, then $P_1V_1 = K$ and $P_2V_2 = K$. From this it also follows that

$$P_1V_1 = P_2V_2$$

Thus, if three of these values are known, it becomes possible to calculate the fourth, even if the value of K is not known.

1.13.2. The Relationship Between Gas Volume and Temperature at Constant Pressure (Gay Lussac's Law, Also Known as Charles' Law)

According to the Kinetic Theory of Gases, the effect of heating a given volume of gas such that its temperature rises is to cause the kinetic energy of its molecules to become greater so that they strike one another and the walls of a containing vessel with greater force. This has two consequences. If the volume of the gas is held constant the pressure of the gas will rise. If the pressure of the gas is maintained constant the effect of the increased kinetic energy will be to cause the gas molecules to have a tendency to maintain a greater average distance from one another, and the volume will increase. This average distance can actually be determined, and is called the "mean free path" of the gas molecules. Conversely, if the temperature of the gas is lowered, the pressure of the gas will decrease at constant volume, and the volume will decrease at constant pressure. The quantitative aspects of this relationship were first observed by J. A. C. Charles about 1787 (but apparently never published) and by J. L. Gay-Lussac in 1802. They found that at constant pressure, a given mass of gas increases its volume by the same relative amount for each degree rise in temperature. This can be expressed by the following equation in which, at constant pressure

$$V_t = V_0 + (\alpha_v V_0)t \tag{1.5}$$

where V_t = the volume of a given mass of any gas at temperature t°C, V_0 = the volume of the same quantity of gas at 0°C, α_v = the coefficient of expansion (which should be the same for all ideal gases), and t = the temperature in degrees Celsius. This is a general equation representing what is usually called Gay-Lussac's Law. Both Charles and Gay-Lussac determined the coefficient of expansion to have a value of $(1/273)V_0$°C^{-1} at atmospheric pressure. Inserting this in Eq. 1.5,

$$V_t = V_0 + \left(\frac{V_0}{273}\right)t$$

or

$$V_t = V_0\left(1 + \frac{t}{273}\right) \tag{1.6}$$

If the value of -273 is substituted for t in Eq. 1.6 the value within the parentheses becomes reduced to 0. From this it would appear that at this temperature the gas would occupy no volume at all. This is not likely, since considerably before this temperature is reached almost all gases would have become first liquefied and then solidified, and would certainly have a finite volume. Thus, Gay-Lussac's Law tends to become less valid as very low temperatures are reached. Nevertheless, the idea that there is an absolute minimum temperature of $-273°C$ (actually, more recent calculations have indicated that this value is more accurately $-273.15°C$) has been corroborated by evidence from other gas properties. Experimentally, $-273.15°C$ has been closely approached but never achieved. The apparent existence of an absolute minimum temperature was the basis of a different temperature scale proposed by William Thompson (Lord Kelvin) in 1852. On the Kelvin scale this absolute minimum temperature is assigned a numerical value of 0°K, and since it developed that 1°K was exactly equivalent to 1°C, the ice point acquired a value of 273.15°K. The use of the Kelvin scale is more direct than the use of the Celsius scale, although any temperature scale for which conversion factors are known can in principle be used. For example, at amospheric pressure 274 ml of gas at 1°C would occupy a volume of 275 ml at 2°C. Here the temperature has doubled, but the volume has not. It is easy to make the conversion, since all that is necessary is to add the constant 273 to the Celsius temperature. Thus, the Celsius temperatures plus this constant become 274 and 275°K, respectively, these temperatures being directly proportional to the corresponding volumes. This can be expressed mathematically as follows:

If

$$V_1 = V_0 \left(1 + \frac{t_1}{273}\right)$$

and

$$V_2 = V_0 \left(1 + \frac{t_2}{273}\right)$$

then

$$\frac{V_1}{V_2} = \frac{(273 + t_1)}{(273 + t_2)}$$

Combining $(273 + t_1)$ into T_1 and $(273 + t_2)$ into T_2 results in the following equation.

$$\frac{V_1}{V_2} = \frac{T_1}{T_2} \tag{1.7}$$

where T_1 and T_2 are the temperatures in degrees Kelvin. Although the Celsius scale could be used, it is obviously easier to use Eq. 1.7, from which it is also apparent that

$$V \propto T \tag{1.8}$$

and that

$$\frac{V}{T} = K$$

where K is a constant having the dimensions of liters per degree. Since, with respect to a given gas at a constant pressure K would be expected to have the same value for any combination of V and T, then $V_1T_1^{-1} = K$ and $V_2T_2^{-1} = K$, from which $V_1T_1^{-1} = V_2T_2^{-1}$. By exchanging T_1 and V_2 one arrives at Eq. 1.7. Thus, even if the value of K is not known, any value in Eq. 1.7 can be calculated if the other three are known.

1.13.3. The Volume Change Accompanying Changes in Both Temperature and Pressure

Using Eqs. 1.4 and 1.8, if

$$V \propto \frac{1}{P} \quad \text{at constant temperature} \tag{1.4}$$

and

$$V \propto T \quad \text{at constant pressure} \tag{1.8}$$

then

$$V \propto \frac{T}{P}$$

and

$$\frac{PV}{T} = K \tag{1.9}$$

where K is still another constant having the dimensions of liter-atmospheres per degree. Equation 1.9 is one of a class of equations called equations of state. These equations relate variables, called variables of state, which define the condition, or state, of a given substance. In Eq. 1.9 the physical state of a gas is defined solely in terms of its pressure, volume, and temperature.

Since with respect to a given gas K would be expected to have the same value for any combination of P, V, and T, then $P_1V_1T_1^{-1} = K$ and $P_2V_2T_2^{-1} = K$, from which

$$\frac{P_1V_1}{T_1} = \frac{P_2V_2}{T_2} \tag{1.10}$$

and if five of these values are known, the sixth may be calculated even if a value for K is not known.

1.13.4. The Molar Gas Constant, *R*

Equation 1.9 contains a constant that cannot be specifically evaluated since it applies only to particular quantities of gas, and not to the behavior of gases in general. To make such an evaluation it is necessary to standardize the conditions under which the evaluation is being made. The volume of a gas can be used to define its mass, but only if its temperature and pressure are also defined. Thus, gas volumes are recorded at a Standard Temperature and Pressure (abbreviated STP), these being 0°C (273.15°K) and 1 atm (760 mm Hg), respectively. Gas volumes do not, of course, have to be measured under these standard conditions, and it is seldom convenient to do so. It is only necessary to record the temperature and pressure at the time of measurement and to convert this measured value to what the volume would be were it to be determined at STP, by means of Eq. 1.10. If the composition of a given substance is known, its mass is most conveniently expressed in terms of its gram-molecular weight, or mole. Since it was determined by Avogadro that 1 mole of any substance contains 6.023×10^{23} molecules, and since it has also been determined that at STP 1 mole of any gas will occupy a volume of 22.414 liters, it is evident that at STP 22.414 liters of any gas will contain 6.023×10^{23} molecules. But, PVT^{-1} is constant for any given quantity of gas and therefore, at the same temperature and pressure (not necessarily STP) equal volumes of all gases will contain the same number of molecules. This last conclusion is known as Avogadro's Law. Because of this, for 1 mole of gas PVT^{-1} must have a unique value and becomes a universal gas constant. This relationship is conventionally expressed in the following manner:

$$PV = nRT \tag{1.11}$$

where n represents the number of moles in the quantity of gas and R represents the gas constant, replacing the unspecified K in Eq. 1.9. For 1 mole of gas,

$$R = \frac{PV}{nT}$$

and at STP, $P = 1$ atm, $V = 22.414$ liters, $T = 0°C = 273.15°K$, and therefore

$$R = \frac{1 \text{ atm} \times 22.414 \text{ liters}}{1 \text{ mole} \times 273.15 \text{ deg}}$$

$$= 0.082 \text{ liter atm mole}^{-1} \text{ deg}^{-1}$$

(for whatever reason the product PV is usually expressed as liter atm, rather than atm liter)

The gas constant R is actually equivalent to units of energy per mole per degree, and its value will depend on the purpose for which a calculation making use of this constant is intended. To make this more clear, l atm are really secondary units representing the quantities of volume and pressure. Volume has the dimensions of (length)^3, or L^3, and pressure has the dimensions of force (in dynes) per unit area, or $MLt^{-2}L^{-2}$. Thus, l atm $= (MLt^{-2}L^{-2}) \times L^3 = ML^2t^{-2}$ = ergs = work energy, and R has the dimensions of energy mole^{-1} deg^{-1}. Since all of the energy secondary units are equivalent, R can equal 8.314 joules, 8.314×10^7 ergs, 1.987 defined calories, 0.082 liter atm, or 8.56×10^{-5} volt Faradays, all per mole per degree, respectively.

Equation 1.11 is known as the *Ideal Gas Equation*, representing the *Ideal Gas Law*. This law was formulated on the basis of the generalized behavior of gases and not on any one gas in particular. The gases to which the law applies are thus, in the strict sense, "ideal" gases. Actually, ideal gases do not exist, and real gases vary to a greater or lesser extent in their behavior from that predicted by the ideal gas law. However, at low pressures the departure from ideality is sufficiently small that the ideal gas equation is applicable for all practical purposes.

1.13.5. Partial Pressures in Mixtures of Gases (Dalton's Law)

The ideal gas law applies both to pure gases and to mixtures of gases, and it is these latter which are normally found in nature. It was postulated by J. Dalton in 1801 that each component of a gas mixture occupying a given volume exerts a pressure that is the same as that which it would exert if it alone occupied that same volume, and that the total pressure of the mixture is the sum of the partial pressures of the component gases. This postulate has since become known as Dalton's Law. It is evident from this and the Ideal Gas Law that the ratio of the partial pressure of a given component in a gas mixture to that of the mixture as a whole is the same as the ratio of the quantity in moles of the component gas to the quantity in moles of all gases in the mixture. Thus,

$$\frac{p_c}{P_m} = \frac{n_c}{N_m}$$

where p_c represents the partial pressure of a given component of a gas mixture, P_m the total pressure of the mixture, n_c the number of moles of the component gas, and N_m the total number of moles of gas in the mixture, respectively. The ratio $n_c N_m^{-1}$ is the "mole fraction" of the component gas with respect to the whole mixture.

1.14. THE DIFFERENCE BETWEEN THE VALUE OF THE HEAT OF COMBUSTION AT CONSTANT VOLUME AND AT CONSTANT PRESSURE

Many microorganisms can utilize acetic acid (CH_3COOH) and ethyl alcohol (CH_3CH_2OH) as sources of carbon and energy, provided that these substrates are not present in too high a concentration. One thermodynamic measure of the amount of energy available from the total oxidation of a substrate is its heat of combustion under the normal, physiological conditions for growth of the specific organism. Lacking information to the contrary, these conditions are often taken as a constant temperature of 25°C and a pressure of 1 atm. However, the most convenient way of measuring the heat of combustion is with a bomb calorimeter, which is a constant volume, rather than a constant pressure system. Because of this, the heat of combustion at constant volume has to be corrected to take into account the pressure–volume work that would have been done if the determination of the heat of combustion could have been made under the condition of constant pressure. For example, when acetic acid is burned in a bomb calorimeter, the heat of combustion is found to be -209.0 kcal mole^{-1} at 25°C. The reaction can be represented by the following equation:

$$CH_3COOH(l) + 2O_2(g) \longrightarrow 2CO_2(g) + 2H_2O(l)$$

Because 2 moles of CO_2 gas are formed and 2 moles of O_2 gas consumed, the system will not undergo any significant change in volume at constant pressure. Using Eq. 1.3,

$$q_p = q_v + P\,\Delta V \tag{1.3}$$

but since $\Delta V = 0$,

$$q_p = -209.0 \text{ kcal mole}^{-1} + 0 = -209.0 \text{ kcal mole}^{-1}$$

Thus, in this example the heats of combustion at constant volume and at constant pressure are the same. On the other hand, when 1 mole of ethyl alcohol is burned in a bomb calorimeter, the heat of combustion is found to be -326.1 kcal mole^{-1} at 25°C. The reaction can be represented as follows:

$$CH_3CH_2OH(l) + 3O_2(g) \longrightarrow 2CO_2(g) + 3H_2O(l) \tag{1.12}$$

In this example 2 moles of CO_2 gas are formed, but 3 moles of O_2 gas are consumed, and there will be a negative change in the volume of the system at constant pressure (i.e., the system will contract). The relationships between the pressure, volume, and temperature of an ideal gas are represented by the ideal gas equation

$$PV = nRT \tag{1.11}$$

According to Eq. 1.12 there will be a change in volume (ΔV) which is a function of the change in the number of moles (Δn) of gas from that of the initial state to that of the final state (i.e., $\Delta n = 2 - 3 = -1$). Equation 1.11 then becomes

$$P\,\Delta V = \Delta nRT$$

Substituting for Δn, R, and T,

$$\begin{aligned} P\,\Delta V &= -1 \text{ mole} \times 1.987 \text{ cal mole}^{-1} \text{ deg}^{-1} \times 298.15 \text{ deg} \\ &= -592 \text{ cal} \\ &= -0.6 \text{ kcal} \end{aligned}$$

and

$$\begin{aligned} q_p &= q_v + P\,\Delta V \\ &= -326.1 \text{ kcal} + (-0.6 \text{ kcal}) \qquad (1.3) \\ &= -326.7 \text{ kcal} \end{aligned}$$

From these examples of the combustion of acetic acid and ethanol it is evident that the heat of combustion at constant pressure can be significantly different from that at constant volume, but depending on the particular reaction this difference may be of no practical importance. Thus, in the latter case the pressure–volume work represents only 0.18% of the total thermal exchange. Because the value for the heat of reaction is a function of the way in which it is measured (i.e., under conditions of constant volume or of constant pressure), it can only be used qualitatively to characterize the energy changes in a reaction. For a quantitative characterization of these changes it is necessary to inquire into the nature of the origins of the heat of reaction, and to relate the quantity of heat exchanged between an isothermal, isobaric, closed system, and its environment to the total change in the amount of internal energy in the system, and in its distribution. For this purpose three functions of state of isothermal, isobaric, closed thermodynamic systems have been defined. These are the ENTHALPY, the ENTROPY, and the FREE ENERGY.

1.15. ENTHALPY

From Eq. 1.2 it is evident that in an isobaric, closed system which performs no work on the environment other than that associated with a change in volume,

$$q_p = \Delta U + P\,\Delta V \quad (1.13)$$

Thus, the thermal energy that can be absorbed by such a system is *greater* than the change in internal energy by an amount that is determined by the pressure and the volume of the system. This function of the state of the system is called its *enthalpy* and is represented by the symbol H. Its relationship to the internal energy, pressure, and volume of the system is described by the following equation:

$$H = U + PV$$

Just as with the internal energy, U, the enthalpy, H, and all other functions of state can be defined as variables that do not in themselves have unique values. It is only changes in the values of these functions which can be measured and expressed as quantities. A change in enthalpy, ΔH, is defined as the quantity of heat that is absorbed from its environment by an isothermal, isobaric, closed system which undergoes a change in state during which no work is performed on the environment except that associated with expansion. The quantity ΔH is thus equivalent to the quantity q_p and may be substituted for it in Eq. 1.13.

$$\Delta H = \Delta U + P\,\Delta V \quad (1.14)$$

The change in value of any function of state is invariant for any given change of an isothermal, isobaric, closed system. It depends only on the initial and final states of the system, and is completely independent of the manner in which, or the path by which, the change in state is accomplished. This was first discovered in principle in 1840, when G. H. Hess proposed that the net heat of a chemical reaction was equal to the algebraic sum of the heats of any intermediate reactions by means of which the same net reaction could be brought about. Hess's proposal subsequently became known as Hess's Law of Constant Heat Summation, although the principle has come to be recognized as applying to changes in all functions of state associated with changes in isothermal, isobaric, closed systems. As applied to changes in enthalpy, the use of Hess's Law often makes it possible to calculate an otherwise unobtainable value for ΔH, if the values for ΔH for all other reactions in a given set of reactions are known. For example, if

$$\begin{array}{l} A \longrightarrow B, \quad \Delta H = a \\ B \longrightarrow C, \quad \Delta H = b \\ C \longrightarrow D, \quad \Delta H = c \\ \hline A \longrightarrow D, \quad \Delta H = d \end{array}$$

then,

$$d = a + b + c$$

and if the values for a, b, and d are known, the value of c can be calculated algebraically. Thus,

$$c = d - (a + b)$$

Because changes in value of the functions of state are conventionally referred to a system and not to its environment, a negative enthalpy change, $-\Delta H$, conventionally represents a quantity of thermal energy that becomes transferred from the system to the environment, and thus becomes lost to the system. Chemical reactions that are accompanied by a decrease in enthalpy are said to be *exothermic* (ΔH is negative). Conversely, chemical reactions that are accompanied by an increase in enthalpy are said to be *endothermic* (ΔH is positive).

According to Eq. 1.14 the value of ΔH will not differ significantly from that of ΔU unless the system undergoes an appreciable change in volume. This will happen *only* if gases are produced or consumed because of changes in the system as it proceeds from some initial to some final state, such that there is a new increase or decrease in the total volume of the system, respectively. In the case of chemical reactions that take place entirely in dilute aqueous solution, as is characteristic of most, if not all biochemical reactions, ΔH and ΔU will have the same value. It thus matters little whether considerations of the energetics of microbiological reactions are made in terms of ΔH or ΔU, since all microorganisms can be regarded as living in an aqueous environment. However, there is probably less chance of error if all biological heat exchanges are related to the constant pressure of 1 atm, and ΔH is the term most generally used.

1.16. STANDARD MOLAR ENTHALPIES OF FORMATION

The standard molar enthalpy of formation, ΔH_f°, represents the amount of thermal energy which becomes exchanged with the environment during the formation of 1 mole or 1 g at wt of a substance from its constituent elements at a constant temperature and pressure. The standard temperature is 298.15°K and the standard pressure is 1 atm. There is thought to be no heat change associated with the formation of an element from itself under any conditions of constant temperature and pressure and for this reason, by convention, the ΔH_f° of all elements in their standard states is taken arbitrarily as 0.000 kcal mole^{-1}. This is not true for substances other than the elements, since the reactions whereby these are formed from their constituent elements are usually, although not always, accompanied by the release of heat to the environment. Thus, values of ΔH_f° for substances other than the elements are usually negative. If only substances containing C, H, O, and N are considered, the reactions by which these are formed from the elements can be generalized by the following equation:

$$cC + hH_2 + oO_2 + nN_2 \longrightarrow C_cH_{2h}O_{2o}N_{2n} \tag{1.15}$$

where c, h, o, and n represent the appropriate coefficients for C, H, O, and N, respectively. If the reaction represented by Eq. 1.15 is carried out at 298.15°K and 1 atm, the physical phases of the elements become defined, and the reaction is better represented as follows:

$$cC(s) + hH_2(g) + oO_2(g) + nN_2(g) \longrightarrow C_cH_{2h}O_{2o}N_{2n}(s, l, g) \tag{1.16}$$

The enthalpy change accompanying the reaction represented by Eq. 1.16 is the difference between the enthalpies of formation of the product and the reactants:

$$\Delta H^\circ = \Delta H_f^\circ(\text{product}) - \Sigma\ \Delta H_f^\circ(\text{reactants}) \tag{1.17}$$

However, since $\Sigma\ \Delta H_f^\circ(\text{reactants})$ is 0.000 (the reactants are all elements), $\Delta H^\circ = \Delta H_f^\circ(\text{product})$.

Values for ΔH_f° are determined directly or indirectly by means of bomb calorimetry at 298.15°K (25°C). As an example, the combustion of C(graphite) to $CO_2(g)$ is represented by the following equation:

$$C(s) + O_2(g) \longrightarrow CO_2(g) \tag{1.18}$$

When this reaction is carried out in a bomb calorimeter the heat of reaction at constant temperature and constant volume q_v, is equal to that at constant temperature and constant pressure, q_p, because there has been no volume change in the system as it passes from its initial to its final state, and ΔH° is found directly to be -94.051 kcal. However, since this reaction also represents the heat of formation of $CO_2(g)$ from the elements, $\Delta H^\circ = \Delta H_f^\circ\ CO_2(g)$. Similarly, the combustion of $H_2(g)$ to $H_2O(l)$ is represented by the following equation:

$$H_2(g) + 0.5O_2(g) \longrightarrow H_2O(l) \tag{1.19}$$

When this reaction is carried out in a bomb calorimeter the heat of reaction at constant temperature and constant volume is different from that which it would be at constant pressure because of a considerable change in the volume of the system as it proceeds from its initial to its final state. After this correction has been made $\Delta H°$ is found to be -68.315 kcal. Again, since this reaction also represents the heat of formation of $H_2O(l)$ from the elements, $\Delta H° = \Delta H_f^\circ H_2O(l)$.

The combustion at 298.15°K and 1 atm of crystalline glucose, $C_6H_{12}O_6(s)$, which is an organic compound, is represented by the following equation:

$$C_6H_{12}O_6(s) + 6O_2(g) \longrightarrow 6CO_2(g) + 6H_2O(l) \tag{1.20}$$

Since there is no volume change in this system between the initial and the final states, just as with the reaction represented by Eq. 1.18, $q_v = q_p = \Delta H°$, which is found to be -669.556 kcal mole^{-1}. However, although this is the heat of combustion it is *not* the heat of formation. This latter can be calculated by means of an equation analogous to Eq. 1.17.

$$\Delta H° = \Sigma\ \Delta H_f^\circ \text{ (products)} - \Sigma\ \Delta H_f^\circ \text{ (reactants)} \tag{1.21}$$

In the case of the oxidation of glucose at 298.15°K and 1 atm, the values for $\Delta H_f^\circ CO_2(g)$ and $\Delta H_f^\circ H_2O(l)$ have been determined from the reactions represented by Eqs. 1.18 and 1.19, and $\Delta H°$ for the combustion of glucose has been determined from the reaction represented by Eq. 1.20, all by bomb calorimetry. $\Delta H_f^\circ O_2(g)$ has a value of 0.000 kcal mole^{-1} by convention. Thus,

$$\Delta H° = \begin{array}{l} 6\ \Delta \mathrm{H}_f^\circ CO_2(g) + 6\ \Delta H_f^\circ H_2O(1) \\ -\ \Delta H_f^\circ C_6H_{12}O_6(s) - 6\ \Delta H_f^\circ O_2(g) \end{array}$$

from which

$$\begin{aligned} -669.556 &= 6(-94.051) + 6(-68.315) \\ &\quad - \Delta H_f^\circ C_6H_{12}O_6(s) - 6(0.000) \end{aligned}$$

and

$$\begin{aligned} \Delta H_f^\circ C_6H_{12}O_6(s) &= -564.306 + (-409.890) + 669.556 \\ &= -304.640 \text{ kcal mole}^{-1} \end{aligned}$$

The values of ΔH_f° for most inorganic and many organic compounds are available in various compilations which are found in the literature. Some of these which are of biological importance are listed in Appendix 1. Such values of ΔH_f° can be used to calculate by difference the *standard enthalpy change*, represented by the symbol ΔH°, by means of Eq. 1.21.

1.17. STANDARD MOLAL (OR MOLAR) ENTHALPIES OF FORMATION OF SUBSTANCES IN AQUEOUS SOLUTION

When dealing with chemical reactions that are of biological significance it is important to realize that such reactions always take place in an aqueous environment (for example, enzyme-catalyzed reactions within cytoplasm) and most frequently in aqueous solution. For this reason the conventional standard states of solids, liquids, and gases are not the appropriate ones to use when determining the enthalpy changes that accompany biological reactions, and it is the value of the enthalpy function relative to an arbitrarily defined *aqueous standard state* that should be used instead as a reference. The aqueous standard state with respect to the enthalpy function is defined as a hypothetical one molal solution which corresponds to the limiting condition implied by Henry's Law, that is, a 1 mole quantity of solute dissolved in 1000 g of water to give a solution in which the solute behaves as if it were present at infinite dilution. Such solutions are more conceptual than real, since the solutes in most solutions do not obey Henry's Law except at very low concentrations (~0.001 m or less). The standard molal enthalpy of formation in aqueous solution can be designated either as $\Delta H_f^{\circ\prime}$ or as ΔH_f° plus some other indication that the substance referred to is in aqueous solution under standard conditions. Conventionally, the superscript (′) indicates that the substance is at a pH of 7 (and therefore aqueous). $\Delta H_f^{\circ\prime}$ is equal to the standard enthalpy of formation, ΔH_f°, plus the heat exchange with the environment that accompanies the solution of 1 mole of the substance in an infinite volume of water. Heats of solution may be either positive or negative, depending on the physical nature of the substance being dissolved. Values of heats of solution are determined in solution calorimeters, the construction and operation of which are beyond the scope of this book. However, the data in Table 1.3 will give some idea of what is meant by the concept of the heat of solution at infinite dilution. The ΔH_f° for ethanol is -66.37 kcal mole^{-1} at 298.15°K and 1 atm pressure. If 1 mole of ethanol is dissolved in 1 mole of water at the same temperature and pressure the heat of solution is found to be -0.217 kcal. The physical reaction that takes place can be represented by the following equation.

$$CH_3CH_2OH(\text{l, 298.15°K, 1 atm}) \longrightarrow CH_3CH_2OH\left(\begin{matrix}\text{aq, in 1 mole } H_2O\\ \text{298.15°K, 1 atm}\end{matrix}\right)$$

TABLE 1.3. Values for $\Delta H_f'$ of 1 mole of pure ethanol dissolved in successively larger quantities of water at 298.15°K and 1 atm

Dilution	Molality	$\Delta H_f'$, kcal mole^{-1}
Standard state $m = 1$		−68.9
In 0.25 mole H_2O	222.024	−66.438
In 0.50 mole H_2O	111.012	−66.477
In 1.0 mole H_2O	55.506	−66.587
In 5.0 moles H_2O	11.099	−67.465
In 10.0 moles H_2O	5.555	−68.063
In 50.0 moles H_2O	1.110	−68.740
In 100.0 moles H_2O	0.555	−68.836
In 200.0 moles H_2O	0.2775	−68.888
In 500.0 moles H_2O	0.1110	−68.893
In 1000.0 moles H_2O	0.0555	−68.897
In ∞ H_2O	0.0000	−68.9

Source: Reference 3.

from which

$$\Delta H' = \Delta H_f'\begin{pmatrix}\text{aq, in 1 mole } H_2O\\ 298.15°K,\ 1\ \text{atm}\end{pmatrix} - \Delta H_f^\circ$$

and

$$\begin{aligned}\Delta H_f'\begin{pmatrix}\text{aq, in 1 mole } H_2O\\ 298.15°K,\ 1\ \text{atm}\end{pmatrix} &= \Delta H_f^\circ + \Delta H'\\ &= -66.37 + (-0.217)\\ &= -66.587\ \text{kcal mole}^{-1}\ \text{of}\\ &\quad\ \text{dissolved ethanol}\end{aligned}$$

As shown in Table 1.3, if 1 mole of ethanol is dissolved in successively greater quantities of water, the differential heats of solution will become less and less as the molecules of ethanol become more and more separated from one another and surrounded by water molecules. Eventually, further dilution will cause no further heat exchange, thus establishing a limiting condition with respect to the heat of solution. This is the limiting condition which is implied by Henry's Law and which is defined as being the standard state with respect to aqueous solutions. The enthalpy of formation of 1 mole of substance in this standard state can therefore be calculated in the following manner.

$$\Delta H_f^{\circ\prime} = \Delta H_f^\circ + \Delta H'\begin{pmatrix}\text{of solution at}\\ \text{infinite dilution}\end{pmatrix}$$

In the case of ethanol this limiting condition is closely approached when 1 mole (44.052 g) of ethanol is dissolved in 1000 moles (18,016 g) of water to provide a concentration of 0.055 *m*, for which the heat of solution has been determined to be -2.527 kcal mole^{-1}. The heat of solution at infinite dilution has been determined to be -2.530 kcal mole^{-1}, or only -3 cal greater than that at a dilution of 1000 moles of water. The enthalpy of formation of 1 mole of ethanol in aqueous solution at infinite dilution is therefore

$$\Delta H_f^{\circ\prime} \text{ ethanol} = \Delta H_f^\circ + \Delta H' \begin{pmatrix} \text{of solution at} \\ \text{infinite dilution} \end{pmatrix}$$

$$= -66.37 + (-2.530)$$

$$= -68.9 \text{ kcal mole}^{-1}$$

It is apparent that the $\Delta H_f^{\circ\prime}$ value of -68.897 kcal mole^{-1} for the solution of 1 mole of ethanol in 1000 moles of water is only 0.004% less than that at infinite dilution, and for all practical purposes may be considered the same. Thus, the theoretical standard state of infinite dilution may be very closely approached at low, but finite, concentrations of solutes which are well within the range of concentrations encountered by microorganisms grown in a natural environment. In fact, 0.001 *m* (or *M*) concentrations of solutes are quite practical for experimental studies on microbial growth, and such *real* solutions may well be regarded as infinitely dilute at least for the purpose of calculating $\Delta H^{\circ\prime}$ values. Values of $\Delta H_f^{\circ\prime}$ for a number of inorganic and organic substances are given in Appendix 1.

1.18. NONSTANDARD ENTHALPIES OF FORMATION

Nonstandard enthalpies of formation are those that are not relative to a particular, arbitrarily chosen standard condition, such as that involving 1 mole of substance at 298.15°K and 1 atm. Since the conditions are not standard, the superscript ° is left off the ΔH_f symbol. Thus, the enthalpy of formation of 1 g of substance in solid form (such as a 1-g mass of dried bacterial cells) would be represented by ΔH_f. Similarly, the enthalpy of formation of 1 g of a given concentration of protein in aqueous solution at pH 7.0 would be represented by $\Delta H_f'$. When using values represented by these symbols the quantity of substance present and the physical conditions under which the thermal measurement was made should be stated. Otherwise, such values are meaningless. The advantage of using standard states is that the quantity of substance and the physical conditions under which the measurement was made are understood.

It should be recognized that standard states are arbitrarily chosen. If there is a good reason for adopting a standard state that is different from those conventionally used this can be done, provided that correct values can be found for the thermodynamic quantities relative to substances in the new standard state. For example, this could be done with respect to $\Delta H_f^{\circ\prime}$ values. The usual standard reference state is a hypothetical solution in which the solute is present at a 1 *m* concentration but is supposed to behave physically as if it were at infinite dilution. However, it should be possible to have a real standard solution in which the substance is present at a 0.001 *m* concentration which actually does behave physically very nearly as if it were at infinite dilution. There would be no difference between the two standard states if we dealt with the solute as if it were a 1 mole *quantity* of substance at infinite dilution, rather than as if it were at a 1 *m concentration*.

Although the standard enthalpy of formation in aqueous solution is defined in terms of molal rather than molar concentrations, it is apparent that at infinite dilution $\Delta H_f'$ values for the two are the same. $\Delta H_f'$ values for substances in real solutions at concentrations as high as 0.002 *m* and 0.002 *M* are also so close that for all practical purposes they may be considered the same.

1.19. METHODS OF OBTAINING VALUES FOR ENTHALPY CHANGES

There are several ways of obtaining values for enthalpy changes, including ΔH, ΔH°, $\Delta H'$, and $\Delta H^{\circ\prime}$. These are:

1. From solution calorimetry at constant pressure. An example of this is the measurement of the heat produced during an enzymatic reaction occurring in aqueous solution, or during the growth of a microorganism in aqueous suspension. In this case $\Delta H' = q_p$, and a value for $\Delta H'$ can be obtained directly. If the reaction is carried out at 298.15°K and 1 atm, and if the reactants and products are present at concentrations of 0.002 *m* (or *M*) or less, then $\Delta H'$ will closely approximate $\Delta H^{\circ\prime}$.
2. From calorimetry at constant volume in a bomb calorimeter. An example of this is the determination of the heat of combustion of a given weight of an organic substance such as glucose, or dried microbial cells. In this case a value of q_v is obtained from which q_p and thus ΔH can be calculated, provided that the changes in the volume of the calorimetric system are known. If the reaction is carried out at 298.15°K and corrected to 1 atm, $\Delta H = \Delta H^\circ$.
3. From the heats of combustion of the components of a reaction.
4. From the use of Hess's Law (one example of which is 3 above).
5. By calculation using the values of ΔH_f, ΔH_f°, $\Delta H_f'$, and $\Delta H_f^{\circ\prime}$ for the reactants and products and Eq. 1.21.
6. From the way in which the equilibrium constant with respect to a reversible reaction taking place in aqueous solution changes with temperature (not discussed here).

Quite certainly, method 5 (above) is the most convenient for biologists provided that the appropriate enthalpy of formation values are available in the literature.

The values for ΔH° and $\Delta H^{\circ\prime}$ for a given reaction are often not greatly different from one another. Thus, if 1 mole of glucose were to be completely oxidized to carbon dioxide and water, each reactant and product being in its standard state, the enthalpy change for the reaction is calculated to be -669.556 kcal. The same reaction carried out in aqueous solution, each reactant and product being in its aqueous standard state, is accompanied by an enthalpy change which is calculated to be -684.04 kcal mole^{-1}. The former value is therefore less negative than the latter by 2.11%, which may not seem like a large discrepancy. Nevertheless, in principle it is always ΔH values relative to the aqueous standard state which should be employed if the reactions being considered are of a biological nature.

Except for an understanding of its role in energy transformations in biological systems, a knowledge of the enthalpy change accompanying a chemical reaction of microbiological importance is only of limited value. It can be useful in making calculations of free energy and entropy changes, and can be of practical importance in calculating the amount of heat dissipated in industrial fermentations and in sewage treatment plants. Otherwise, all microorganisms are poikilothermic (do not maintain their body heat constant), and are so small that any heat that is dissipated from microbial cells in a natural environment is quickly and to all practical purposes completely absorbed by the environment.

1.20. ENTROPY

The entropy function is not an easy one to describe briefly. Its definition can be approached intellectually either on the basis of strictly thermal considerations or on the basis of statistical mechanics. However, the former approach is probably a better one to use in order to relate entropy to biological processes, most of which are less readily characterized than the thermal processes studied by chemists and physicists.

The entropy of a closed system is a thermodynamic function of state, represented by the symbol S which relates the *total* quantity of thermal energy, Q, absorbed by a system, to the temperature of the system. This relationship is described by the following equation:

$$\text{entropy} = S = \frac{Q}{T}$$

As with the other functions of state, the entropy cannot be evaluated directly, and only changes in entropy can be experimentally determined *unless* one special condition can be found that enables the entropy to be defined absolutely.

A "reversible" process by which a system passes from one thermodynamic state to another is defined as being always in equilibrium with any tendency that may exist to oppose it, such that if this tendency is infinitesimally increased the process will be reversed, and vice versa. It is considered to be one from which all extraneous causes of energy dissipation such as frictional and electrical resistance have been eliminated, and it can be represented by the summation of a continuous series of infinitesimal increments of change that pass through a continuous series of equilibrium conditions. Actually, a truly reversible process is not a natural one, and only exists as a concept in the minds of thermodynamicists. Nevertheless, it is a useful concept in that a reversible process can be regarded as the unattainable limit of a practically realizable process. A few practically realizable processes are very nearly reversible. One of these is the voltaic cell, the electromotive force of which can be opposed by an external electromotive force acting in the opposite direction. Here a condition can be realized such that, in favorable cases, a reversal of current flow can be caused to take place in the voltaic cell by external changes of as little as $\pm 1 \times 10^{-6}$ volt.

The absorption of thermal energy by matter is considered to be a "reversible" process. In making this consideration a given quantity of matter is regarded as a closed system, and by a reversible process is meant that if the temperature of the environment of the system is increased or decreased infinitesimally, there will be a corresponding infinitesimal increase or decrease in the temperature of the system such that there is no sensible departure from the condition of temperature equilibrium between the system and its environment. Under these conditions the quantity of heat absorbed can be represented by dQ_{rev}, and the change in entropy that accompanies it by dS. Accordingly,

$$dS = \frac{dQ_{\text{rev}}}{T} \tag{1.22}$$

For a finite, or large change in the quantity of heat absorbed by a system that is the result of the summation of a continuous sequence of reversible changes, the entropy change can be expressed in terms of finite units. Thus, for a finite, but reversible change in entropy that takes place during a change in the state of a system from A to B,

$$\Delta S = S_B - S_A = \int_A^B \frac{dQ_{\text{rev}}}{T} \tag{1.23}$$

where Q is expressed in calories, T in degrees Kelvin, and ΔS has the dimensions of cal deg^{-1}. For a finite, but irreversible entropy change,

$$\Delta S = S_B - S_A = \frac{\Delta Q}{T} \tag{1.24}$$

Because it is a system containing a quantity of matter that undergoes an entropy change, for actual calculations a dimension representing this quantity is introduced along with the value for the integral of dQ_{rev} or of ΔQ, which is then usually expressed in cal mole^{-1}, or in cal g^{-1} if the composition of the system is not known. The value for ΔS then becomes expressed in cal mole^{-1} deg^{-1}, or cal g^{-1} deg^{-1}.

1.20.1. Entropy Changes Accompanying Fusion and Vaporization at Constant Pressure

The process of fusion is considered to be reversible. The system in this case is a bounded space containing a quantity of matter at its melting point. At this temperature and at a constant pressure (usually stated to be 1 atm) an equilibrium exists between the solid and liquid phases such that if the temperature of the environment of the system is infinitesimally increased some of the solid will melt, and if it is infinitesimally decreased some of the liquid will freeze. Exactly the same considerations apply to the process of vaporization, except that in this case the system is a bounded space containing a quantity of matter at its boiling point. Under isothermal, isobaric conditions, the entropy change accompanying the change in phase from ice to water can be represented as follows:

$$\Delta S = S_{\text{water}} - S_{\text{ice}} = \int_O^Q \frac{dQ_{\text{rev}}}{T} = \frac{1}{T}\int_O^Q dQ_{\text{rev}} = \frac{\Delta Q_{\text{rev}}}{T}$$

An analogous equation represents the entropy change accompanying the change in phase from water to water vapor.

The equations representing the change in phase of crystalline ice to water at 0°C and 1 atm pressure and of water to water vapor at 100°C and 1 atm pressure are given below.

$$\begin{gathered} \mathrm{H_2O(s,\ 0°C,\ 1\ atm)} \longrightarrow \mathrm{H_2O(l,\ 0°C,\ 1\ atm)} \\ \Delta H = L_{sl} = 1436.3 \text{ cal mole}^{-1} \end{gathered} \tag{1.25}$$

$$\begin{gathered} \mathrm{H_2O(l,\ 100°C,\ 1\ atm)} \longrightarrow \mathrm{H_2O(g,\ 100°C,\ 1\ atm)} \\ \Delta H = L_{lg} = 9717.5 \text{ cal mole}^{-1} \end{gathered} \tag{1.26}$$

where L_{sl} and L_{lg} represent the values for the molal latent heats of fusion and of vaporization, respectively. The adjective "latent" refers to the fact that in these systems quantities of heat can be added without this being reflected by a rise in temperature.

Calculations of the entropy changes accompanying the change in phase of ice into water, and of water into water vapor are as follows:

From Eq. 1.25

$$L_{sl} = 1436.3 \text{ cal mole}^{-1}$$

From Eq. 1.22

$$\Delta S = \frac{\Delta Q_{\text{rev}}}{T} = \frac{L_{sl}}{T} = \frac{1436.3 \text{ cal mole}^{-1}}{273.15°K} = 5.26 \text{ cal mole}^{-1} \text{ deg}^{-1}$$

From Eq. 1.26

$$L_{lg} = 9717.5 \text{ cal mole}^{-1}$$

From Eq. 1.22

$$\Delta S = \frac{\Delta Q_{\text{rev}}}{T} = \frac{L_{lg}}{T} = \frac{9717.5 \text{ cal mole}^{-1}}{373.15°\text{K}} = 26.04 \text{ cal mole}^{-1} \text{ deg}^{-1}$$

1.20.2. Heat Capacity

Calculations of nonisothermal entropy changes at constant pressure involve an understanding of the concept of heat capacity.

The heat capacity of a system, represented by the symbol C, is defined as the ratio of the quantity of thermal energy supplied to a system to the corresponding increase in temperature that it undergoes. If the temperature increase of a system is infinitesimal in response to an infinitesimal increase of the temperature of its environment such that there is no sensible departure from a condition of thermal equilibrium between the system and its environment, the process of temperature increase can be said to be reversible. Given the condition of reversibility the quantity of thermal energy supplied to the system can be expressed as dQ and its corresponding temperature increase by dT. Heat capacity can then be described by the following equation:

$$\text{heat capacity} = C = \frac{dQ_{\text{rev}}}{dT} \tag{1.27}$$

For a finite, or large change in the temperature of a system, whether brought about reversibly or irreversibly, the heat capacity can be expressed in finite units, so that

$$\text{heat capacity} = C = \frac{\Delta Q}{\Delta T}$$

where Q is in calories, T is in degrees Kelvin, and C has the dimensions of cal deg^{-1}.

The heat capacity of a closed system is determined most accurately with an adiabatic calorimeter. To accomplish this a known amount of substance is placed in an adiabatic calorimeter which has been calibrated electrically between the temperatures T_1 and T_2, the temperature range being kept as small as is experimentally practicable. Usually this range is from one to a few degrees. The amount of thermal energy required for the calibration of the empty calorimeter (ΔQ_{cal}), subtracted from the amount of thermal energy which must be added to the calorimetric system in order to raise it through the same temperature range as that used for the calibration of the calorimeter (ΔQ_{cal} + substance), divided by the change in temperature which the system undergoes while passing from the initial to the final state (ΔT), equals the heat capacity of the substance.

$$C = \frac{(\Delta Q_{\text{cal + substance}} - \Delta Q_{\text{cal}})}{(T_2 - T_1)} = \frac{\Delta Q_{\text{substance}}}{\Delta T}$$

Because a substance has mass, the heat capacity must be expressed in terms of a given amount of substance and is usually evaluated in cal mole^{-1} deg^{-1}, or in cal g^{-1} deg^{-1}. If the heat capacity determination is carried out in a completely closed container such that the calorimetric system undergoes no change in volume, the heat capacity is represented by the symbol C_v. If the determination is carried out in a calorimetric system that is kept under constant pressure, the heat capacity is then represented by the symbol C_p. Although the temperature range in the determination of heat capacity may be a few degrees, the actual temperature of the determination is taken as the average of T_1 plus T_2, the assumption being that the heat capacity does not change significantly within a small temperature range provided that no phase changes occur within that range.

Related to the heat capacity is the *specific heat*, represented by the symbol c, which is the quantity of heat required to raise the temperature of 1 g of a substance 1°C, the standard temperature interval usually being specified as that between 16.5 and 17.5°C. Thus, water has a specific heat of 1.000.

1.20.3. Nonisothermal Entropy Changes at Constant Pressure

From Eq. 1.23 the entropy change accompanying a reversible change in the temperature of a system can be represented by the following equation:

$$\Delta S = S_{T_2} - S_{T_1} = \int_{T_1}^{T_2} \frac{dQ_{\text{rev}}}{T} \tag{1.28}$$

where T_1 and T_2 represent the initial and final temperatures of the system, respectively.

Rearranging Eq. 1.27,

$$dQ_{\text{rev}} = C\,dT$$

and if the heat capacity is determined when the system is under a constant pressure, $C = C_p$, and

$$dQ_{\text{rev}} = C_p\,dT$$

Substituting $C_p\, dT$ for dQ_{rev} in Eq. 1.28,

$$\Delta S = S_{T_2} - S_{T_1} = \int_{T_1}^{T_2} \frac{C_p\, dT}{T} = \int_{T_1}^{T_2} C_p\, d \ln T \qquad (1.29)$$

If C_p happens to be constant over the temperature range T_1 to T_2, then

$$\Delta S = S_{T_2} - S_{T_1} = C_p \ln \frac{T_2}{T_1}$$

Usually, however, the value of C_p *does* change as a function of temperature. In this case, in order to evaluate the integral in Eq. 1.29 an empirical relationship must be established between C_p and T which can be expressed in the form of an equation, or alternatively, a graphical method of integration can be used. In the application of the graphical method C_p is experimentally determined and plotted as a function of ln T, after which the area enclosed by the curve between T_1 and T_2 on the graph can be evaluated. It is important to realize that Eq. 1.29 can be applied to determine the difference in the entropy of *any* substance between T_1 and T_2, whether or not the substance happens to be pure or of known composition. It is only necessary to know the specific heats or the heat capacities of the substance over the temperature range being considered, and the temperatures of and the heats absorbed during any phase changes that may occur within this temperature range. For example, to obtain the difference between the entropy of a substance such as water at 298.15°K (25°C) and in the solid phase at 0°K, the following equation would apply:

$$\Delta S = S_{298.15} - S_0 = \int_0^{T_{sl}} C_p\, d \ln T + \frac{L_{sl}}{T_{sl}} + \int_{T_{sl}}^{298.15} C_p\, d \ln T \qquad (1.30)$$

where T_{sl} represents the melting point of ice (273.15°K), L_{sl} the latent heat of fusion of water (1436.3 cal mole^{-1} deg^{-1}), and T the temperature difference represented by an integral. The integral from 0 to T_{sl} represents the increase in the entropy of 1 mole of water in the solid state (ice) while the temperature is being raised from 0 to 273.15°K (0°C), at which temperature the ice melts. The isothermal increase in the entropy of water as it passes from the solid to the liquid phase is represented by the expression L_{sl}/T_{sl}. The integral from T_{sl} to 298.15°K represents the increase in the entropy of 1 mole of water as the temperature is being raised from the melting point of ice to 298.15°K (standard temperature). Depending on the substance and the temperature for which ΔS is determined there may be more than one or no phase changes. The values for C_p over the whole temperature range for L_{sl} (or L_{lg}) can all be measured experimentally, usually being expressed in moles of substance. However, if the substance is not pure or if its composition has not been determined, thus making it impossible to calculate molal values, the values for C_p, L_{sl}, L_{lg}, and for other phase transitions can be expressed in grams of substance.

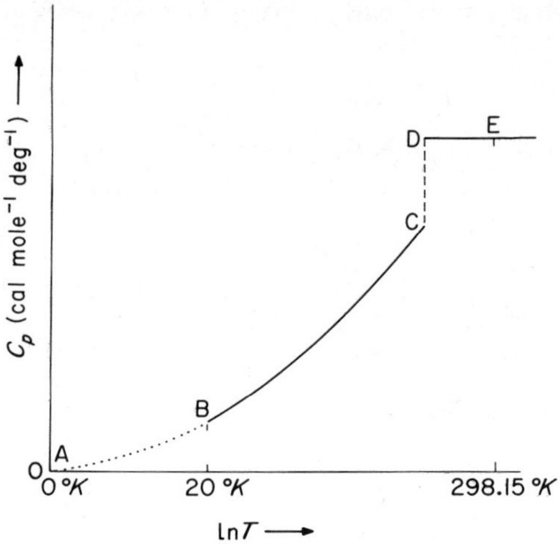

FIGURE 1.1. Heat capacity of water in relation to temperature.

The various integrals in an equation such as Eq. 1.30 can be evaluated graphically. The manner in which this is done is illustrated in Fig. 1.1, in which C_p in cal mole^{-1} deg^{-1} is plotted as a function of ln T, and in which the solid lines represent curves that must be integrated and the dashed line the isothermal change in entropy as the result of a phase change. By evaluating the areas under the curves enclosed between lines representing two temperatures such as A and B, or B and C, and adding to these values those of any entropy changes due to phase changes that occur within the total temperature range being considered, the total entropy change accompanying a change in the temperature of a closed system over any experimentally obtainable temperature range can be calculated, using nothing but thermal data. The calculation of the entropy change that occurs when 1 mole of water is raised from 0 to 298.15°K (25.00°C) is given in Table 1.4.

The expression that represents $S_T - S_0$, as in Eq. 1.30, can be used to evaluate the entropy *difference* of a substance between 0°K and any required temperature. However, it does not enable the total entropy of a substance, S_T, at temperature T to be calculated since the entropy of the substance at 0°K cannot be experimentally measured. Thus,

$$\Delta S = S_T - S_0 = \int_0^T C_p \, d \ln T$$

or,

$$\Delta S - S_T = \int_0^T C_p \, d \ln T + S_0 \qquad (1.31)$$

where S_0 becomes a constant of integration which cannot be evaluated.

TABLE 1.4. Calculation of the entropy change that occurs when 1 mole of water is raised from 0 to 298.15°K (25.00°C)

1. $H_2O(s, 0°K) \rightarrow H_2O(s, 273.15°K)$

$$\Delta S_1 = S_{273.15} - S_0 = \int_{T_0}^{T_{273.15}} C_p \, d \ln T = 9.88 \text{ cal mole}^{-1} \text{ deg}^{-1}$$

2. $H_2O(s, 273.15°K) \rightarrow H_2O(l, 273.15°K)$

$$\Delta S_2 = S_{\text{water}} - S_{\text{ice}} = \frac{\Delta Q_{\text{rev}}}{T} = \frac{L_{\text{cl}}}{T} = \frac{1436.3}{273.15}$$

$$= 5.26 \text{ cal mole}^{-1} \text{ deg}^{-1}$$

3. $H_2O(l, 273.15°K) \rightarrow H_2O(l, 298.15°K)$

$$\Delta S_3 = S_{298.15} - S_{273.15} = \int_{T_{273.15}}^{T_{298.15}} C_p \, d \ln T$$

$$= C_p \ln \frac{T_2}{T_1} = C_p \, 2.303 \log \frac{298.15}{273.15}$$

$$= 18.016 \text{ cal mole}^{-1} \text{ deg}^{-1} \times 2.303 \log 1.091$$

$$= 1.57 \text{ cal mole}^{-1} \text{ deg}^{-1}$$

Adding: $H_2O(s, 0°K \rightarrow H_2O(l, 298.15°K)$

$$\Delta S° = S_{298.15} - S_0 = \Delta S_1 + \Delta S_2 + \Delta S_3$$

$$= 9.88 + 5.26 + 1.57 = 16.71 \text{ cal mole}^{-1} \text{ deg}^{-1}$$

1.21. FREE ENERGY

At this point it is necessary to introduce the concept of free energy. The free energy of a thermodynamic system is the maximum amount of energy that is potentially available for accomplishing something as the system proceeds from a given initial to a given final state. With respect to the process of microbial growth, the available free energy is the nonthermal, chemical energy that is incorporated within the substances used as substrates by microorganisms. Just as with H and S, a detailed treatment of free energy is beyond the scope of this book, and the reader is directed toward the bibliography at the end of this chapter. Two forms of free energy are recognized. The Gibbs free energy (named after W. Gibbs) is designated by the symbol G. As is the case with all thermodynamic functions of state, G is defined as a variable that does not in itself have a unique value. It is only a change in G which can be measured and expressed as a quantity. An irreversible change in the Gibbs free energy, ΔG, represents the change in the free energy of a closed system which occurs as the result of an isothermal reaction at constant pressure. The Helmholtz free energy (named after H. von Helmholtz, who first called this function of state the "free energy") is designated by the symbol A (from German *Arbeit*, work; historically from the identification of A as the "work function"). An irreversible change in the Helmholtz free energy, ΔA, represents the change in free energy of a closed system which results from an isothermal reaction at constant volume. Because no pressure–volume work is expended by a closed system undergoing an isothermal reaction at constant volume, ΔA represents the maximum potential within a closed system for performing all kinds of work (total work) as opposed to ΔG, which represents the potential for performing only that work ("useful work") which is not associated with expansion. The difference between the two is similar to that between ΔU and ΔH in that it is attributable to the volume change of a closed system which is kept isothermal and at constant pressure. In addition, just as with ΔU and ΔH, if a reaction takes place entirely in dilute aqueous solution the pressure and volume remain constant, and $\Delta G = \Delta A$. With a few possible exceptions all biochemical reactions and certainly all microbial "growth processes" (a complex of reactions comprising the process of growth) do take place in an aqueous environment, and it does not matter whether energetic considerations of microbiological phenomena are made in terms of ΔG or ΔA. However, since it is usually more convenient to carry out chemical reactions at constant temperature and pressure, and to grow microorganisms under the same conditions, it is usually the change in Gibbs free energy, ΔG, which is taken into consideration. At constant temperature and pressure ΔG is related to ΔH and ΔS through the Gibbs free-energy equation, as follows:

$$\Delta G = \Delta H - T\,\Delta S \tag{1.32}$$

Under standard conditions Eq. 1.32 becomes

$$\Delta G^\circ = \Delta H^\circ - T\,\Delta S^\circ \tag{1.33}$$

where ΔG° and ΔH° have the dimensions of cal mole^{-1}, and ΔS° has the dimensions of cal mole^{-1} deg^{-1}.

1.22. THE SECOND LAW OF THERMODYNAMICS

A spontaneous reaction is one that "goes by itself," without any external energetic input from the environment. The Second Law of Thermodynamics states that "Spontaneous reactions are those that, when carried out under suitable conditions, can be made to perform work." (Morris,[1] p. 172). In addition, spontaneous reactions within closed systems are irreversible, and if one considers the system *plus* the environment, are always accompanied by an increase in entropy. Because changes in the values of the thermodynamic functions of state are conventionally referred to the system and not to its environment, a negative free-energy change, $-\Delta G$, conventionally represents a quantity of free energy which is potentially available for expenditure, and which potentially can thus be transferred to the environment and lost to the

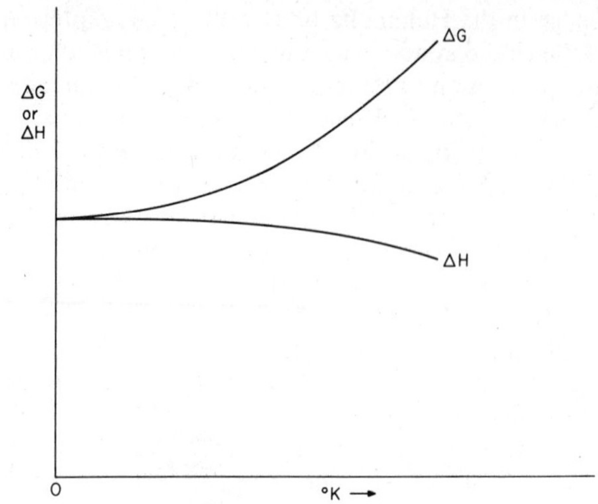

FIGURE 1.2 Hypothetical limiting approach of ΔG and ΔH as T approaches 0°K. The difference between ΔG and ΔH at any given temperature is $-T\Delta S$.

reaction system. Under suitable conditions this energy *could* be made to do work, and it is therefore a negative sign before a ΔG which indicates whether a reaction is able to perform work. A positive sign before a ΔG indicates that the reaction is not by itself capable of performing work, and that if it is to proceed there must be a concomitant Gibbs free-energy input into the system from the environment. A reaction having a negative value of ΔG is called an *exergonic reaction* because free energy passes out of and is lost from the reaction system as it proceeds spontaneously from its initial to its final state. A reaction having a positive value of ΔG is called an *endergonic* reaction because it cannot proceed spontaneously, and free energy must be put into the reaction system in order for it to proceed from its initial to its final state. It is important to realize that although the sign of ΔG gives information as to the spontaneity of a reaction system, no information is given as to the length of time required for the reaction to proceed to completion.

1.23. THE THIRD LAW OF THERMODYNAMICS. STANDARD MOLAR ENTROPIES

In 1903 T. W. Richards[4] published the results of a comprehensive series of investigations of the free-energy and heat exchanges accompanying reactions taking place in a number of galvanic cells. These results indicated that ΔG and ΔH approach one another more and more closely in the low-temperature range, as generalized in Fig. 1.2. According to curves such as this which Richards constructed, $d(\Delta G)/dT$ and $d(\Delta H)/dT$ are opposite in sign and approach zero in value and each other tangentially as the temperature approaches 0°K. Thus,

$$\lim_{T \to 0} (\Delta G - \Delta H) = 0$$

However, because $d(\Delta G)/dT = -\Delta S$ and $d(\Delta H)/dT = \Delta C_p$ (see Lewis and Randall,[5] p. 437), it follows that $-\Delta S$ and ΔC_p for these galvanic cell reactions also approach zero as a limit as the temperature approaches 0°K. As pointed out by Lewis and Randall, "except for the fact that these cells involve solutions, the curves presented by Richards very nearly imply the generalizations which were later to be embodied in the third law of thermodynamics." (Lewis and Randall,[5] p. 437.)

In a paper published in 1906 W. Nernst[6] attempted to calculate chemical equilibria from thermal measurements. Using the results of Richards, he postulated *as a general principle* that in *any* chemical reaction between solid or liquid substances, $d(\Delta G)/dT$ and $d(\Delta H)/dT$ approach zero (tangentially) as the temperature approaches 0°K, and thus that for any such chemical reaction system $-\Delta S$ and ΔC_p also approach zero as the temperature approaches 0°K. In order to explain the vanishing of ΔC_p Nernst made the assumption that the heat capacity of solid and liquid substances approaches the same *finite value, per atom of substance*, at absolute zero. In 1907 A. Einstein[7] made a prediction, based upon a mathematical treatment of a hypothetical solid, that heat capacities would approach not a finite value, but *zero* with decreasing temperature in the low-temperature range. Subsequent experiments by several investigators, however, showed that actual substances do not possess quantitatively the properties of Einstein's hypothetical solid. Nevertheless, although the experimental values subsequently obtained were not the same as those suggested by Einstein's simple equation, in all cases so far investigated it has been shown that the value for the heat capacity at constant pressure diminishes rapidly with decreasing temperature in the low-temperature range, so that not only C_p/T, but S approaches zero as a limit at the absolute zero of temperature.

In his postulate, which came to be known as the Nernst Heat Theorem and which served as the basis for the development of the Third Law of Thermodynamics, Nernst made no statement as to the value of the entropy of substances at absolute zero, although it must obviously be finite or zero if ΔS at 0°K is to be finite for reactions involving condensed phases, that is, all solid and liquid substances. Nernst's postulate was extended in 1911 by M. Planck,[8] who made the additional postulate that the value of the entropy of a *pure solid* or a *pure liquid* approaches zero as a limit at 0°K. Thus,

$$\lim_{T \to 0} S = 0$$

Planck's modification of the Nernst Heat Theorem was based upon his reasoning that the entropies of solutions, which are impure solid or liquid substances, cannot be assigned the value of zero at 0°K because there would still be a residual positive entropy associated with the solution which is equal to the entropy of mixing of the dispersing and the dispersed phases of the solution. For this reason he considered the Nernst Heat Theorem to be applicable only to pure solids and liquids. The positive entropy of solutions at 0°K was also emphasized in 1920 by Lewis and Gibson,[9] who considered the changes in heat capacity with decreasing temperatures in the low-temperature range of a number of solutions, including mixtures of organic isomers and of metal isotopes, and concluded that the entropy change that results from mixing remains essentially constant down to absolute zero. Lewis and Gibson also discussed entropy from the point of view of the randomness of the system being considered, as follows:

> When we turn to the consideration of crystalline solids we not only find numerous cases in which [what had come to be known as] the third law has been verified experimentally, but we also see some *a priori* reasons for the existence of such a law, even though these reasons cannot at present be stated with all the lucidity that might be desired.
>
> Since entropy is a quantity defined for the purpose of giving mathematical expression to the second law of thermodynamics, and since that law, from the standpoint of statistical mechanics, is essentially a consequence of the law of probability, it is natural to look for a relation between the entropy of a substance in a given state and the probability of the particular arrangement of molecules which determines that state. Essays in this direction by Boltzmann and more recently by Planck, important and suggestive as they have been, have not sufficed to show us what this relationship is. Nevertheless, it seems to us that if the entropy of a given state be regarded as in some sense a measure of the randomness of that state, the condition of a perfect crystal of a pure substance at the absolute zero is unique. In a solution there is a random distribution of several types of molecules. Even in a pure liquid or glass there is some randomness of arrangement. In any substance at a finite temperature there is a random distribution of energy among the individual molecules. But in the pure crystal at the absolute zero no randomness remains, for when the positions and properties of a few molecules are fixed, the positions and properties of all other molecules are completely determined; thus, when a few elements of the crystalline structure are known we may build up the whole crystal by a process of repetition.
>
> It is this lack of any sort of randomness that we believe to be the theoretical basis for the conclusion that the entropy of a perfect crystal of a pure substance vanishes at the absolute zero.
>
> We, therefore, state the third law of thermodynamics as follows: *if the entropy of each element in some crystalline form be taken as zero at the absolute zero, the entropy of any pure crystal at the absolute zero is zero, and the entropy of any other substance is greater than zero.*
>
> It seems likely, however, that the difference between the entropy of a pure substance in a crystalline state and in an amorphous state may, in many cases, prove to be very small. [(Lewis and Gibson,[9] pp. 1532–1533.) Reprinted with

permission from G. N. Lewis and G. E. Gibson, The third law of thermodynamics and the entropy of solutions and liquids., *J. Am. Chem. Soc.* **42,** 1529–1533 (1920). Copyright 1920 American Chemical Society.]

The extent to which the above difference between the entropy of a pure substance in a crystalline state and in an amorphous state actually proves to be very small has not been thoroughly investigated. Obviously, there are a great many substances for which a perfect, crystalline state has never been demonstrated under any conditions. Among these substances are solid solutions, high polymers such as polyethylene, polypropylene, teflon, and so on, biological macromolecules such as nucleic acids, polysaccharides, lipoproteins, proteins, and so on (although quite a number of proteins have, in fact, been crystallized), and complex lipids. Indeed, it may well be precisely because the physical and thermodynamic states of most systems of biological interest cannot be exactly defined that the thermodynamics of these systems has been so little studied. It is possible that some substances for which a perfect, crystalline state at 0°K cannot be demonstrated are physically incapable of forming crystals, and even if the temperature of these substances were to be lowered sufficiently slowly that it were certain that a condition of thermodynamic reversibility actually existed (this might require an indefinitely long period of time), they might simply pass into a state of minimum, but positive, entropy at 0°K and never into a perfect, crystalline state in which the entropy can theoretically be considered as zero. Or, it is also possible that other substances, when subjected to the same slow, reversible condition of temperature decrease, might eventually undergo a slow rearrangement of their structures such that given enough time, a perfect, crystalline structure would exist at 0°K, even though this did not occur if the same substances were to be cooled more rapidly to 0°K, under which conditions a residual entropy would become "frozen in." In either case, the actual experiments are impractical to carry out, and the question then arises as to whether there is any rationale by which we can evaluate the entropy at 0°K of substances that do not form perfect crystals.

Three years after the publication of the forgoing statement of Lewis and Gibson, Lewis and Randall in 1923 incorporated the idea of entropy as a positive entity into the definition of the third law, and also made provision for the fact that some pure substances may have more than one crystalline form. Their "final" statement of the third law is as follows:

> If the entropy of each element in some crystalline state be taken as zero at the absolute zero of temperature: *every substance has a finite positive entropy, but at the absolute zero of temperature the entropy may become zero, and does become so in the case of perfect crystalline substances.* [Lewis and Randall,[10] p. 448.]

It is this statement of the third law that is usually accepted as the most convenient with respect to making free-energy calculations.

Since 1923 when Lewis and Randall made their "final" statement of the third law, a great many investigations have shown that the Debye equation adequately expresses the behavior of the heat capacities of crystallizable substances below 20°K. Graphical methods of integrating heat capacities can be used down to temperatures in the range of 20°K, but it is difficult and expensive to acquire data below this temperature. This is because isotopes of helium must be used to achieve these low temperatures and because of the inherent difficulties of extremely low-temperature calorimetry. The Debye equation is as follows:

$$C_p \cong C_v = 464.5 \frac{T^3}{O^3} \text{ cal mole}^{-1} \text{ deg}^{-1}$$

where the only constant, O, which is called the characteristic Debye temperature, can be determined from one value of C_p at approximately 20°K. It is the integral of this equation which is actually used to calculate the entropy difference between 0°K and T. Thus,

$$S_T^\circ = \int_0^T \frac{kT^3}{T} \, dT = \int_0^T dT^2 \, dT \qquad (1.34)$$

The form of the curve represented by Eq. 1.34 does not show any positive excursions of C_p between approximately 20 and 0°K which would lead us to believe that the entropy changes in a nonuniform manner in this temperature range. Even if it did, as a result of electronic degeneracies in the atoms of a substance within a few degrees of 0°K, it is unlikely that the magnitude and extent of such changes would contribute appreciably to the absolute entropy at 298.15°K. For this reason, if the Debye equation can be said to apply adequately for crystallizable substances at temperatures between 0 and 20°K, the assumption can be made that it also applies for noncrystallizable substances, and that no marked C_p deviations occur within this temperature range. The final statement of the third law as made by Lewis and Randall, in which perfect, crystalline substances can be regarded as having zero entropy at 0°K, is theoretically sound and has held without serious dispute since the time it was stated. However, as a practical, working assumption, Nernst's original suggestion, that the entropies of all condensed phases (including solutions) are zero at 0°K is worth reconsideration. From the point of view of statistical mechanics, one is led to wonder in exactly what mechanical form a residual entropy of solution could exist. Another practical working assumption, to the same effect, is to believe, with Lewis and Gibson,[9] that the difference between the entropy of a pure substance in a crystalline state and in an amorphous state is "likely" to be small enough to be considered negligible relative to the entropy at the working temperature of a biochemical or microbial reaction. It is only by making such assumptions that investigators will be encouraged to look into the thermodynamic behavior of biological systems. We assume, for example, that cells are highly organized, but at the present time have very little idea of the magnitude of the entropy of an individual cell, or that of a gram of cells, or something equivalent to a "mole" of cells. Surely there is more to be learned in this respect.

In the practical application of the third law, the assumption that the entropy of each element in some crystalline state may be taken as zero at 0°K is taken as fact. Then, if, as Nernst postulated as a general principle, the entropy change which accompanies the transformation of elements into compounds in a chemical reaction at 0°K is also taken as zero, the entropy must be zero for every perfect crystalline compound (substance) which is formed *from* the elements of 0°K. Thus, for the reaction,

$$\begin{array}{ccc}\text{elements (reactants)} & \longrightarrow & \text{compound (product)} \\ \text{crystalline, 0°K} & & \text{crystalline, 0°K}\end{array}$$

$$\Delta S_0 = (\text{sum of } S_0 \text{ values of elements}) - (S_0 \text{ value of compound})$$

and if both ΔS_0 and the sum of the S_0 values of the elements are taken as zero, so also is the S_0 value of the compound. For this reason, at least for perfect crystalline compounds, the integration constant, S_0, in Eq. 1.31 can be set equal to zero, and the entropy of any perfect crystalline substance (elements or compounds) at any specified temperature can be represented as follows:

$$\Delta S = S_T = S_T^\circ = \int_0^T C_p \, d \ln T$$

It becomes apparent, then, that the entropy change that accompanies the elevation in temperature of a perfect crystalline substance from 0°K to any specified temperature can be evaluated simply by integrating the heat capacity data between 0°K and the specified temperature. By adding to this the entropy change accompanying a change in state of the substance from the perfect crystalline state to any other state, the entropy change that accompanies the transformation of a perfect crystalline substance at 0°K to the same substance in any other state at any specified temperature can be evaluated. The entropy changes obtained in this manner are usually called "absolute" entropies. Because of the convenience of having them available for reference and calculation, tables of the "standard" molar entropies of commonly occurring substances have been compiled in which the values given are those relative to 1 mole of the substance in a specified standard state (i.e., s, l, g, or aq) at a temperature of 298.15°K (25°C) and a pressure of 1 atm. These standard values are represented by the symbol S°, the temperature and pressure being understood, and are either expressed in cal mole^{-1} deg^{-1}, or in "entropy units" (eu) mole^{-1}. The entropy change represented by S° might seem to be equivalent to an "entropy of formation from the elements" in the same sense as that implied by the symbol ΔH_f°. This is true, of course, provided that one agrees with Nernst. On the other hand, it is usually found that the sum of the atomic S° values for a compound is *not* the same as the empirical S° value for the assembled compound at 298.15°K and 1 atm, thus implying that a finite entropy change is associated with the formation of a compound from the elements at this pressure and at this higher temperature. Thus, the postulate of Nernst does not seem to apply with respect to these latter conditions, under which $S°$ is not equal to $\Delta S_f°$. The $S°$ values for many substances of biological interest are given in Appendix 1. It should be noticed that although the $\Delta G_f°$ and $\Delta H_f°$ values for the elements in their particular allotropic forms stable at 298.16°K and 1 atm pressure are taken as zero, by convention, the $S°$ values are *all positive*. Also, the entropy values for ions in aqueous solution have significance, by convention, only relative to balanced equations for complete ionic reactions, the $S°$ value for H^+(aq) being taken conventionally as zero.

There is another method of evaluating the absolute entropy of a substance, and this is through the application of statistical mechanics (statistical molecular theory) to a given thermodynamic system, whereby entropy is identified with the state of maximum thermodynamic probability of the system. The fundamental equation of statistical mechanics is that formulated by L. Boltzmann in 1896, as follows:

$$S = k \ln W + \text{constant} \tag{1.35}$$

This equation relates the entropy of a system to its thermodynamic probability, W, where k is the Boltzmann constant (i.e., the gas constant, R, per single molecule) having the dimensions of erg deg^{-1} or cal deg^{-1}. By thermodynamic probability was originally meant the ratio of the probability of an actual state of a system to the probability of a state of the system in which the molecules are completely ordered, the total energy and volume being the same for each probability state. It was suggested in 1912 by Planck[11] that the undetermined constant in Eq. 1.35 should be zero, so that the Boltzmann equation then took the following form:

$$S = k \ln W \tag{1.36}$$

Thus, in the completely ordered arrangement of a perfect, crystalline substance at 0°K, W would be unity and S would be zero. In modern statistical mechanics Eq. 1.36 is retained in form, but the symbol W has taken on a somewhat modified form which will not be discussed in detail here. By means of statistical mechanics the absolute entropy can be evaluated in a manner completely independent of thermal measurements. This evaluation involves complex and ingenious mathematics, and a consideration of the translational, vibrational, rotational, and internal rotational energies of the molecules of the substance, the absolute entropy of which is required. The assumption is made that these energies behave independently of each other and can thus be separately calculated, their sum at 298.15°K and at 1 atm pressure being equal to the standard entropy. The agreement of standard entropy values obtained by the application of statistical molecular theory with those obtained by the integration of heat capacity data is in most cases excellent, and where the agreement has not been good it has usually been found that the thermal data were in error. For example, hydrogen was found to have a considerably lower standard entropy as calculated statistically than it had as determined from thermal data. The discrepancy was found to be due to the presence of ortho and para hydrogen in the liquid and solid hydrogen substance that was used for the determination of the heat capacity data at low temperatures. Thus, hydrogen was not actually a single substance, although the existence of these two forms of hydrogen was not apparent from the thermal data. In this example as well as in several others, the statistical method of evaluating standard entropies has served to complement the thermal method, the two methods being used together to arrive at what seems to be the more correct data. The values for standard entropies such as those appearing in Appendix 1 represent the results of both thermal measurements and statistical calculations, where these are possible, the latter being used to correct the former where this was thought necessary.

According to M. A. Paul,[12] the development of statistical molecular theory is "one of the most brilliant intellectual achievements in the entire realm of scientific thought" (Paul,[12] p. 604), and because of the general excellent agreement of the entropy values obtained by statistical theory with those obtained by means of the third law assumptions, the concept of entropy has been made very much clearer. However, the identification of entropy with probability has often been the source of considerable confusion on the part of students in biology with respect to the "information" which cells or molecules (especially DNA molecules) are supposed to contain. There seems to be the philosophically satisfying correlation in that through statistical mechanics a theoretical treatment is provided which accounts for the empirically observable fact that biological entities such as organisms, cells, or macromolecules *do* appear to possess a higher degree of organization. On the other hand, there are both fundamental and operational difficulties associated with the use of statistical mechanics. The fundamental difficulty is that probability is dimensionless. It can be applied to almost any number of seemingly unrelated situations. Thus, to identify probability with any given situation it is necessary to introduce the relevant dimensions. This can be done in statistical mechanics by introducing the Boltzmann constant, which gives the expression on the right of Eq. 1.36 the dimensions of entropy. In this manner probability is equated with a thermodynamic property. From this example it can easily be seen that, with the proper statistical manipulations and with the appropriate constants, probability can be equated with almost anything. Not only this, but through the medium of probability a thermodynamic property such as entropy, which has a definite, physical meaning, can be equated with something as indefinite as "information." Exactly this has been done by N. Wiener, who stated that a certain "... amount of information is the negative of the quantity usually defined as entropy in similar situations." (Wiener,[13] p. 76.) However, as Brillouin[14] has pointed out, if we adopt Wiener's extension of the notion of entropy, we cannot avoid extending it to all types of intelligence and intellectual thinking. When one considers the general acceptance of "information theory," it appears that entropy has taken on a generalized meaning by which it can be identified with many things. The fact still remains, however, that physical (structural) entropy is not the same as informational entropy, and in this respect entropy does not have a generalized meaning. A hand of cards, for example, might or might not contain a great deal of information which is of value to a card player, depending on a number of factors relative to the game and his ability as a player, and yet the same hand of cards would be expected to have the same physical entropy no matter how the cards were used. There is also an operational difficulty to the use of statistical mechanics relative to biological systems. Although it seems to be readily applicable to small molecules for which probability states can be imagined, it would seem to be virtually impossible to apply practically (even though in theory one can do anything) to large molecules, to high polymers, to cells, or to organisms. Thus, the use of statistical mechanics to calculate biological entropies seems foredoomed to failure. What, for example, can be imagined to be the probability states of proteins or nucleic acids? For this reason, the physical approach to evaluating the entropies of biological systems appears to be by far the most practical.

1.24. STANDARD MOLAR ENTROPIES OF FORMATION

In the practical use of the standard entropy values listed in Appendix 1, the entropy change accompanying a chemical reaction at 298.15°K and 1 atm pressure is calculated by first calculating the entropies of formation of the reactants and products. Unlike the values for the ΔH_f° of the elements, the corresponding ΔS_f° values for the elements are not zero but all have positive values. Thus, for the elements, $\Delta S_f^\circ = S^\circ$. For compound substances containing C, H, O, and N, ΔS_f° values can be calculated through the use of the following equation,

$$\Delta S_f^\circ = S^\circ - 1.372n\text{C} - 15.604n\text{H} - 24.501n\text{O} - 22.885n\text{N} \quad (1.37)$$

where S° represents the standard (absolute) entropy of the compound; n represents the numbers of carbon, hydrogen, oxygen, and nitrogen atoms, respectively, in the compound; the constant 1.372 is the entropy of solid graphite; and the constants 15.604, 24.501, and 22.885 are one-half the entropies of $H_2(g)$, $O_2(g)$, and $N_2(g)$, respectively, all at 298.15°K and 1 atm. As an example, the standard entropy of ethanol(l) is 38.49 cal mole^{-1} deg^{-1}. The formula for ethanol indicates that there are two carbon atoms, six hydrogen atoms, and one oxygen atom. Using Eq. 1.37,

$$\begin{aligned}\Delta S_f^\circ \text{ ethanol} &= 38.49 - (1.372 \times 2) \\ &\quad - (15.604 \times 6) - (24.502 \times 1) \\ &= -82.38 \text{ cal mole}^{-1} \text{ deg}^{-1}\end{aligned}$$

ΔS_f° can also be calculated using the Gibbs free-energy equation, provided that the respective standard molar enthalpy of formation, ΔH_f° (see Section 1.16), and the standard molar free energy of formation, ΔG_f° (see Section 1.28), are known. Thus,

$$\Delta G_f^\circ = \Delta H_f^\circ - T\,\Delta S_f^\circ \tag{1.33}$$

from which

$$\Delta S_f^\circ = \frac{\Delta H_f^\circ - \Delta G_f^\circ}{T} \tag{1.38}$$

1.25. STANDARD MOLAL (OR MOLAR) ENTROPIES OF FORMATION OF SUBSTANCES IN AQUEOUS SOLUTION

The aqueous standard state with respect to the entropy function is defined as a hypothetical 1 *m* solution corresponding to the limiting condition implied by Henry's Law, and in this respect is in general similar to that with respect to the enthalpy function (see Section 1.17) in that this limiting condition is approached at infinite dilution. However, there *may* also be some solutes that exhibit this limiting condition at finite, not zero, concentrations at which the activity (see Section 1.29.7) of the solute is not unity. The standard molar entropy of formation in aqueous solution can be designated either as $\Delta S_f^{\circ\prime}$ or as ΔS_f° plus some other indication that the substance referred to is in aqueous solution under standard conditions. $\Delta S_f^{\circ\prime}$ values can be calculated from $S^{\circ\prime}$ values for substances in aqueous solution at 298.15°K and 1 atm pressure, using an equation that is analogous to Eq. 1.37, as follows:

$$\Delta S_f^{\circ\prime} = S^{\circ\prime} - 1.372n\mathrm{C} - 15.604n\mathrm{H} - 24.501n\mathrm{O} - 22.885n\mathrm{N} \tag{1.39}$$

where $\Delta S_f^{\circ\prime}$ represents the standard entropy of formation of a substance in aqueous solution, $\Delta S^{\circ\prime}$ represents the absolute entropy of the same substance in aqueous solution, and where all the other terms and constants have been previously defined for Eq. 1.37 or above. $\Delta S^{\circ\prime}$ values are in general available from the literature more for inorganic than for organic substances. $\Delta S_f^{\circ\prime}$ values for organic substances can be calculated using the Gibbs free-energy equation, provided that the respective standard molal enthalpies of formation in aqueous solution, $\Delta H_f^{\circ\prime}$ (see Section 1.17), and the standard molal free energies of formation in aqueous solution, $\Delta G_f^{\circ\prime}$ (see Section 1.29), are known. Thus,

$$\Delta G_f^{\circ\prime} = \Delta H_f^{\circ\prime} - T\,\Delta S_f^{\circ\prime} \tag{1.40}$$

from which

$$\Delta S_f^{\circ\prime} = \frac{\Delta H_f^{\circ\prime} - \Delta G_f^{\circ\prime}}{T} \tag{1.41}$$

It should be noted that in the case of compounds and except for some ions, whereas $S°$ and $S°'$ are positive, ΔS_f° and $\Delta S_f^{\circ\prime}$ are usually negative.

1.26. NONSTANDARD ENTROPIES OF FORMATION

If the absolute entropy of a substance is determined with respect to temperatures and pressures other than 298.15°K and 1 atm, or if the empirical composition or molecular formula for a substance is not known, its absolute entropy value should be represented by the symbol S, or by the symbol S' if the substance is in aqueous solution. As with nonstandard enthalpy changes, if the quantity of the substance or the physical conditions under which the thermal measurements were made are not stated when the value is used, such values have no meaning.

1.27. METHODS OF CALCULATING ENTROPY CHANGES

All biochemical reactions that are not in a state of equilibrium are accompanied by a change in entropy which is represented by the symbols ΔS, $\Delta S°$, $\Delta S'$, and $\Delta S°'$ when these changes are nonstandard, standard, nonstandard in aqueous solution, and standard in aqueous solution. There are two practical methods of making these calculations. These are:

1. From known values for ΔS_f, ΔS_f°, $\Delta S_f'$, or $\Delta S_f^{\circ\prime}$, using the following equations:
   a. For nonstandard but otherwise specified conditions.

   $$\Delta S = \Sigma\, \Delta S_f(\text{products}) - \Sigma\, \Delta S_f(\text{reactants})$$

   b. For conditions involving the reactants and products in their standard states at 298.15°K and 1 atm pressure.

   $$\Delta S° = \Sigma\, \Delta S_f^\circ(\text{products}) - \Sigma\, \Delta S_f^\circ(\text{reactants})$$

   c. For nonstandard but otherwise specified conditions involving aqueous solutions.

   $$\Delta S' = \Sigma\, \Delta S_f'(\text{products}) - \Sigma\, \Delta S_f'(\text{reactants})$$

   d. For conditions involving the reactants and products in their aqueous standard states at 298.15°K and 1 atm pressure.

   $$\Delta S°' = \Sigma\, \Delta S_f^{\circ\prime}(\text{products}) - \Sigma\, \Delta S_f^{\circ\prime}(\text{reactants})$$

2. From known values for ΔG, ΔG°, $\Delta G'$, $\Delta G^{\circ\prime}$, ΔH, ΔH°, $\Delta H'$, and $\Delta H^{\circ\prime}$ using the following forms of the Gibbs free-energy equation:
   a. For nonstandard but otherwise specified conditions.

$$\Delta S = \frac{\Delta H - \Delta G}{T}$$

   b. For conditions involving the reactants and products in their standard states at 298.15°K and 1 atm pressure.

$$\Delta S^\circ = \frac{\Delta H^\circ - \Delta G^\circ}{T}$$

   c. For nonstandard but otherwise specified conditions involving aqueous solutions.

$$\Delta S' = \frac{\Delta H' - \Delta G'}{T}$$

   d. For conditions involving the reactants and products in their standard states in aqueous solution at 298.15°K and 1 atm pressure.

$$\Delta S^{\circ\prime} = \frac{\Delta H^{\circ\prime} - \Delta G^{\circ\prime}}{T}$$

Obviously, if any of the terms in these equations represent substances that are not in acceptable standard states, the entropy change reverts to being nonstandard.

1.28. STANDARD MOLAR FREE-ENERGIES OF FORMATION

The standard molar free energy of formation, ΔG_f°, represents the amount of nonthermal energy which becomes converted into heat and entropy during the formation of 1 mole or of 1 g at wt of a substance from its constituent elements at a constant temperature and pressure. The standard temperature is 298.15°K and the standard pressure is 1 atm. Just as with standard molar enthalpies of formation (see Section 1.16) there is thought to be no free-energy change associated with the formation of an element from itself under any conditions of constant temperature and pressure, and ΔG_f° for the elements is also taken as 0.000 kcal mole^{-1}, by convention. Similarly, the ΔG_f° values for substances other than the elements are almost always negative. Whereas ΔH_f° and ΔS_f° values can be obtained by means of thermal or other measurements, values for ΔG_f° can be obtained only by calculation. One method is through the use of Eq. 1.33. As an example, taking the value for ΔH_f° of ethanol from Appendix 1 as -66.37 kcal mole^{-1} and that for ΔS_f° of ethanol as -82.38 cal mole^{-1} deg^{-1}, the value for ΔG_f° of ethanol is calculated to be

$$\begin{aligned}\Delta G_f^\circ &= \Delta H_f^\circ - T\,\Delta S_f^\circ \\ &= -66.37 \text{ kcal mole}^{-1} - (298.15°\text{K} \times -82.38 \text{ cal mole}^{-1} \text{ deg}^{-1}) \\ &= -41.80 \text{ kcal mole}^{-1}\end{aligned}$$

1.29. STANDARD MOLAL (OR MOLAR) FREE-ENERGIES OF FORMATION OF SUBSTANCES IN AQUEOUS SOLUTION

When a substance is dissolved in water, the fact that there is usually an enthalpy change (heat of solution) associated with this phenomenon is not difficult to appreciate, nor is the idea that this enthalpy change can be added to that representing the standard enthalpy of formation to give a standard enthalpy of formation in aqueous solution of a given substance. The same general idea as applied to free energy is a good deal more complex because it is necessary to consider aspects of solution chemistry which at first may seem considerably removed from the concept of free energy. The necessity of dealing with the free energies of substances in aqueous solution is simply that all cells live in an aqueous environment and take their nutrition ultimately in the form of substances that are dissolved in water.

1.29.1. What Is an Aqueous Solution?

Any single-phase, homogeneous mixture of two or more substances is called a solution, although the phase can be solid, liquid, or gaseous. The component of the solution which determines the phase and which is usually present in the highest concentration is said to be the *solvent*, whereas the components which are present in lesser concentrations and which do not determine the phase are called *solutes*. A solute is said to be *dissolved* in a solvent.

In biological systems at physiological temperatures, the solvent is water, a liquid having markedly different physical properties from most other solvents. The heat capacity, the heat of vaporization, the surface tension, and the boiling point of water are higher than would be predicted for similar, but less polar substances, and the freezing point is lower. Water is an exceptionally good solvent, capable of dissolving ionic substances, inorganic polar substances, and many organic substances having polar groups. It is most dense at 4°C, rather than at its freezing point of 0°C. These unusual properties make water a substance that is peculiarly suited as a solvent in solutions that form the environment of living cells. When solutes become dissolved in water the process of solution involves, in general, an interaction of the solvent molecules with the solute molecules such that the latter become physically separated from one another as individual molecules and move about randomly within the solvent as a consequence of their kinetic energy. In this respect solute molecules in aqueous solution behave similarly to solute molecules in gaseous solutions except that the latter move about in a void whereas the former move about within the matrix of the water solvent. The rate of diffusion of a gas in aqueous solution is therefore very much less than that of the same solute in gaseous solution at the same temperature and pressure.

1.29.2. The Colligative Properties of Dilute Aqueous Solutions

All systems have physical properties of one kind or another, and many of these latter can be placed in three broad categories. These are *additive*, *constitutive*, and *colligative* properties. Additive properties are those that represent a sum of the corresponding properties of individual components of a system. Molecular weight is an additive property in that it represents the sum of the atomic weights of the different kinds of atoms which are found in a molecule multiplied by the number of each which are present in the same molecule. When two or more gases are mixed at constant temperature and volume their partial pressures are additive, and when they are mixed at constant temperature and pressure their volumes are additive (Dalton's Law). A constitutive property is one that depends on the arrangement of atoms in molecules, and to some extent on the nature and number of these atoms. Freezing and boiling points of pure liquids are examples of these. Colligative properties are those that depend on the number of molecules in a system and not on their nature. Examples of these are given below. Some properties, for example, molar volumes, are partly additive and partly constitutive.

When solute molecules become separated from one another in aqueous solution, each molecule in the case of nonelectrolytes or each ion in the case of electrolytes acts as a separate entity and can be regarded as a single "particle." As a function of the number of these "particles" that become dissolved in it a solution acquires physical properties that are different from those of the pure solvent. These are the colligative properties of solution. They are applicable to all solvents, not only to water. The differences in the physical behavior of an aqueous solution as compared with that of the pure water solvent are as follows:

1. The vapor pressure of water over an aqueous solution at a given temperature and at atmospheric pressure is lower than that of pure water.
2. The boiling point of the water solvent in solution at atmospheric pressure is higher than that of pure water.
3. The freezing point of the water solvent in solution at atmospheric pressure is lower than that of pure water.
4. The aqueous solution has an osmotic pressure not exhibited by pure water.

It has been found that the extent to which the colligative properties are exhibited in very dilute solutions is directly proportional to the *number* of solute "particles" per unit volume of solution, but is completely independent of the physical nature of these particles, for example, their size, shape, chemical composition, electrical charge, and so on.

1.29.3. The Solution of Pure Gases in Liquids (the Bunsen Absorption Coefficient)

Every gas is soluble to a greater or lesser extent in every liquid. Given a container partly filled with liquid, the rest of the volume of the container being filled with gas, some of the gas will sooner or later dissolve in the liquid until an equilibrium condition is reached, at which point the liquid will be saturated with the gas under the then-existing conditions of temperature and pressure. This phenomenon is standardized through the use of the Bunsen absorption coefficient, which is equal to the volume of gas in liters, reduced to standard conditions, that can be absorbed by 1 liter of solution at a given temperature when the gas phase consists of the pure gas at a pressure of 1 atm. The Bunsen coefficient is represented by the letter alpha (α). Tables of these coefficients exist for most gases with respect to their solution in water at atmospheric pressure and at various temperatures.

1.29.4. The Solution of Mixtures of Gases in Liquids (Henry's Law)

Just as the component gases of a mixture behave independently of one another at constant temperature with respect to the partial pressures they exert, so do gases in solution behave independently of one another with respect to the extent to which they dissolve in solution. The relationship between the partial pressure and the solubility of a gas was first expressed in 1803 by W. Henry, who came to the conclusion that the *mass* of any gas that will dissolve in a unit volume of solvent at a given temperature is directly proportional to the partial pressure of the gas in equilibrium with the solution. This can be represented in the following manner:

$$\mathrm{M} \propto p \tag{1.42}$$

where M = the mass of gas dissolved by a unit volume of solvent at the equilibrium partial pressure p. From Eq. 1.42 it follows that

$$\mathrm{M} = Kp$$

where K is a proportionality constant. At constant temperature the quantity M is equivalent to n/N, where n is the number of moles of gas dissolved in a unit volume of solvent, and N is the number of moles of solvent in the same volume. Thus,

$$\frac{n}{N} = Kp$$

In the practical use of Henry's Law the mass of a small volume of gas is much less easy to determine than the volume itself, reduced to standard conditions. Thus, Henry's Law can also be written

$$V = K \alpha p$$

where V = the volume of gas under standard conditions that can be dissolved in a unit volume of solvent at the given temperature, K = a constant, α = the Bunsen absorption coefficient for that gas at the given temperature, and p is the partial pressure of the gas above the solvent. By combining K and α into a single constant, K', and using a derivation similar to that used for Eq. 1.10,

$$\frac{V_1}{p_1} = \frac{V_2}{p_2}$$

If three of these values are known, the fourth may be calculated even if a value for K' is not known.

1.29.5. The Relationship Between the Vapor Pressure of a Solvent and That of a Solution (Raoult's Law)

Wherever there exists a liquid–gas interface, there is a tendency for those molecules of the liquid that have a sufficiently great kinetic energy to escape from the liquid phase and to enter the gaseous phase. This phenomenon is called vaporization. The same tendency exists with respect to solids, but to a much lesser extent. If a container is partly filled with water and then closed, water will evaporate from the liquid phase until the partial pressure of its vapor has increased to a value at which water molecules from the gas phase enter the liquid as fast as they leave it, and an equilibrium condition becomes established. The partial pressure of the water vapor at this equilibrium condition is a function of both the temperature and the external pressure above the liquid, and is called its *vapor pressure*. If the temperature of a quantity of water is increased at atmospheric pressure, the water vapor pressure above the liquid phase will increase exponentially until the temperature of 100°C is reached, at which temperature the vapor pressure of the water equals the pressure of the gas phase above the liquid, and the water will begin to boil. If the external pressure were less than 1 atm the water would boil at a lower, but predictable temperature. The reverse is also true. Similarly, if the temperature of a quantity of water is decreased at atmospheric pressure until a temperature of 0°C is reached, its vapor pressure will decrease exponentially until it equals that over ice at the same temperature and the water will begin to freeze. Although it might be predicted that by raising the external pressure above the water at 0°C the freezing point would be raised, actually it is lowered, a consequence of the fact that water is less dense at 0°C than at 4°C.

In 1887–1888 F. M. Raoult proposed that the relative lowering of the vapor pressure above a solvent at a given temperature and pressure as a consequence of the presence of a solute is directly proportional to the mass of solute per unit mass of solution, that is, to the mole fraction of the solute in the solution. This relationship became known as Raoult's Law, and can be represented by the following equation.

$$\frac{(p - p_s)}{p} = \frac{n}{(N + n)} = \text{mole fraction of solute} \qquad (1.43)$$

where p = the vapor pressure of the pure solvent, p_s = the vapor pressure of solvent above a solution, N = the number of moles of solvent per unit volume of solution, and n = the number of moles of solute in the same volume.

If $n/(N + n)$ in Eq. 1.43 is taken to represent the mole fraction of the solute, then $N/(N + n)$ must represent the mole fraction of the solvent, both per unit volume of solution, since the sum of the mole fractions of the solute and the solvent must equal unity. If the mole fraction of the solute is represented by x_n and the mole fraction of the solvent by x_N, then

$$x_n + x_N = 1$$

and

$$x_n = (1 - x_N)$$

Substituting $(1 - x_N)$ for the mole fraction of the solute in Eq. 1.43,

$$\frac{(p - p_s)}{p} = (1 - x_N)$$

$$\frac{p}{p} - \frac{p_s}{p} = (1 - x_N)$$

$$1 - \frac{p_s}{p} = (1 - x_N)$$

and therefore

$$p_s = p x_N \qquad (1.44)$$

This form of Raoult's Law states that the vapor pressure of the solvent above a solution is directly proportional to the mole fraction of the *solvent*, the proportionality constant (p) being the vapor pressure of the pure solvent at the same temperature and pressure.

Raoult's Law and the Ideal Gas Laws are similar in that just as the Gas Laws apply strictly only to ideal gases, Raoult's Law applies strictly only to ideal solutions. Real solutions do not behave exactly in accord with Raoult's Law, although the more dilute the solution, the more Raoult's Law tends to be obeyed.

Raoult's Law and Henry's Law are similar in that Raoult's Law predicts the ideal behavior of a solvent, whereas Henry's Law predicts the ideal behavior of a solute. Henry's Law has been represented previously by the following equation:

$$\mathrm{M} \propto p \tag{1.42}$$

However, the mass of any gas that will dissolve in a unit volume of solvent at a given temperature is also proportional to its mole fraction with respect to the solution; therefore,

$$\mathrm{M} \propto \frac{n}{(N + n)}$$

and

$$\frac{n}{(N + n)} \propto p$$

Substituting x_n for $n/(N + n)$ and turning the proportionality around gives:

$$p \propto x_n$$

from which

$$p = Kx_n$$

This is an equation identical in form to Eq. 1.44 describing Raoult's Law. Thus, while with Raoult's Law the ideal solvent is considered to be one the vapor pressure of which is directly proportional to its (the solvent's) mole fraction in the solution, with Henry's Law the ideal solute is considered to be one the vapor pressure of which is directly proportional to *its* (the solute's) mole fraction. Henry's Law was originally conceived with respect to its application to gases; however, in principle there is no reason why it should not also apply to any solute that has a detectable vapor pressure above a solution in which it is dissolved. If a solution behaves ideally with respect to Henry's Law, it will also behave ideally with respect to Raoult's Law. The reverse need not be true if the solute does not have a detectable vapor pressure.

1.29.6 The Concept of Infinite Dilution

Both Raoult's Law and Henry's Law describe the behavior of solutions more and more accurately as these latter become more and more dilute with respect to the presence of a solute. A solvent exhibits ideal behavior when it has a mole fraction of unity (i.e., it is a pure solvent), and it is not difficult to imagine a solution in which the solute is sufficiently dilute that the solvent approaches a mole fraction of unity as a limit. It is more difficult to apply the same concept to a solute, since as this becomes infinitely dilute its mole fraction approaches zero as a limit, and there is no solute at all. It has thus become the practice to define a hypothetical, reference solution in which the solute is present at a unit concentration but in which the solute molecules *behave* as if they were present in an infinitely dilute solution. To understand this better, imagine a mole of solute dissolved in an infinite volume of solvent, after which all the "particles" of solute (i.e., individual molecules or ions) together with the properties they exhibit at infinite dilution, are collected together to give a molal solution. Such a hypothetical solution in which the solute is said to be present at *unit activity*, defines the standard state of the solute and is represented by the postscript (aq), as in $C_6H_{12}O_6$(aq). On a molal scale of concentrations such a solution is said to be "hypothetical 1 *m* at infinite dilution." The concept of infinite dilution is extremely important with respect to determining the energies of molecules in solution. Although the properties of solutes could not be practically determined if the solute were infinitely diluted, in real solutions that are highly diluted the behavior of the solute molecules approaches closely that imagined with respect to hypothetical, infinitely diluted solutes.

Another way of attempting to understand the concept of infinite dilution is to imagine 1 mole of solute dissolved in an infinite volume of solvent, the *mass* of this solute being still capable of reacting in a given manner just as if it were in a 1 *m* solution. This latter is, of course, impossible in that if the molecules of a solute were infinitely removed from one another and from other kinds of reactive solutes, no reaction could occur. Nevertheless, the behavior of a solute at, for example, 10^{-3} *m* is sufficiently close to that which it would have at infinite dilution that for all practical purposes it is the same and we may regard it as such.

1.29.7. Activities

It is typical that many things in nature do not behave in exactly the way that theory, which is largely based on hypothesized or "ideal" conditions, predicts. Thus, the physical behavior of real gases may deviate considerably from that predicted from the application of the "ideal" Gas Laws, which would better be considered as generalizations that have not been found to apply accurately to every kind of real gas. The same applies to the physical behavior of solvents as described by Raoult's Law and of solutes as described by Henry's Law. Because of this techniques must be used to bring the application of these laws more into line with the real nature of the substances to which they apply.

The *activity* of a substance is a function that was introduced by G. N. Lewis to aid in the thermodynamic treatment of real systems. The use of this function permits the chemical potential of a substance to be correlated with experimentally determined quantities, such as the concentration of a solute in an aqueous solution, by a relationship that is formally equivalent to that described as having an activity of one, or "unit activity." This is also what is ascribed to the hypothetical, infinitely dilute solution. Through the use of an activity coefficient which can be practically determined for a given substance at a given concentration, the activity of a solute can be equated with its *effective* concentration (i.e., the concentration at which it *appears* to be present on the basis of the colligative properties that are imparted to the solution, rather than the actual quantity of solute per unit volume of solution). The activity of a substance in solution can be defined by the following equation

$$a = \gamma m$$

where a represents the activity, γ the molal activity coefficient, and m the concentration of the solute in moles per 1000 g of solvent. The greater the deviation of the solute from ideal behavior in solution, the greater will be the difference between the values of a and m. On the other hand, the more dilute the solution, the more m will approach a until at hypothetical infinite dilution $\gamma = 1$. Activities can also be expressed in terms of molar concentrations, as follows:

$$a = fM$$

where a represents the activity, f the molar activity coefficient, and M the concentration of solute in moles per liter. In very dilute solutions the difference between f and γ is negligible. It is not within the scope of this book to describe the various methods of determining the activity coefficient, for which the bibliography at the end of this chapter can be consulted. The need for having some idea of the concept of the activity coefficient is to understand how free energies of formation in aqueous solution are determined.

1.29.8. Free-Energy Changes of Reversible Reactions in Aqueous Solution

If a set of reactants for an isothermal, isobaric, reversible reaction are mixed together in aqueous solution, a reaction will take place that will proceed until the reactants have reached a thermodynamic equilibrium with the products, at which time the reaction will cease since the products will be forming the reactants at the same rate as the reactants are forming the products. Under these conditions the ratio of the products to the reactants becomes constant. For the isothermal, isobaric, reversible reaction,

$$aA + bB \rightleftharpoons cC + dD \tag{1.45}$$

this relationship can be expressed as follows:

$$K_{\text{eq}} = \frac{(C)^c(D)^d}{(A)^a(B)^b} \tag{1.46}$$

where K_{eq} represents the true (thermodynamic) equilibrium constant for the reaction at temperature T°K as the reaction proceeds from left to right; (A) and (B) represent the activities of the reactants and (C) and (D) those of the products, respectively; and a, b, c, and d represent the respective coefficients. A relationship exists between the thermodynamic equilibrium constant for an isothermal reaction that proceeds at atmospheric pressure and its standard Gibbs free-energy change, $\Delta G°$, in the manner defined by the following equation:

$$\Delta G° = -RT \ln K_{\text{eq}} \tag{1.47}$$

where $\Delta G°$ represents $\Delta G°$ at T°K, R the gas constant (1.987 cal $\text{mol}^{-1}\,\text{deg}^{-1}$), T the temperature in degrees Kelvin, and K_{eq} the thermodynamic equilibrium constant at T°K.

Another way of expressing the equilibrium constant for a reaction such as that represented by Eq. 1.46 is in terms of concentrations rather than activities, concentrations being a great deal easier to determine than activities. Thus,

$$K_c = \frac{[C]^c[D]^d}{[A]^a[B]^b} \tag{1.48}$$

where K_c represents the apparent equilibrium constant for the reaction at T°K as the reaction proceeds from left to right; $[A]$ and $[B]$ represent the concentrations of the reactants and $[C]$ and $[D]$ those of the products, respectively; and a, b, c, and d represent their respective coefficients. For reactions that take place in very dilute solutions (roughly 0.001 M or less) K_c closely approximates K_{eq}, and under these conditions,

$$\Delta G° = -RT \ln K_c$$

Equations 1.46 and 1.48 are important in that they enable a calculation to be made of the ΔG_f° for a given substance when this calculation cannot be made because of a lack of values for ΔH_f° and/or ΔS_f° for use in Eq. 1.33. Because free-energy changes are additive, the following equation relates the $\Delta G°$ to the ΔG_f° values of the reactants and products.

$$\Delta G^\circ = \Sigma\ \Delta G_f^\circ(\text{products}) - \Sigma\ \Delta G_f^\circ(\text{reactants}) \tag{1.49}$$

Thus, if the values for all but one of the terms in Eq. 1.49 are known, the value for the remaining term can be calculated by difference. As an example, glutamine can be reversibly hydrolyzed in aqueous solution to produce glutamic acid according to the following equation.

$$\text{glutamine} + H_2O \rightleftharpoons NH_4^+ + \text{glutamate}^- \tag{1.50}$$

According to Morris[1] (p. 204), at a temperature of 298.15°K and a pressure of 1 atm K_{eq} for the reaction represented by this equation has a value of 334, from which

$$\begin{aligned}\Delta G^\circ &= -RT \ln K_{eq}\\ &= -(1.987 \text{ cal mole}^{-1} \text{ deg}^{-1} \times 298.15°\text{K} \times \ln 334)\\ &= -3442 \text{ cal mole}^{-1}\\ &= -3.442 \text{ kcal mole}^{-1}\end{aligned}$$

1.29.9. Standard Molal (or Molar) Free-Energies of Formation of Substances in Aqueous Solution

If a substance in its pure, standard state is dissolved in water until the point of saturation, an equilibrium condition will have been established in which the molecules of the substance can pass reversibly between the standard state and the dissolved state of saturation. Under these conditions $\Delta G^\circ = 0.000$, and ΔG_f° for the substance in its pure standard state is equal to ΔG_f° for the same substance in saturated solution. As with any solute in solution, a solute in a saturated solution has an activity which is a product of the concentration of the solute times its activity coefficient at that concentration (see Section 1.29.7). The free energy of formation of a solute in aqueous solution at any concentration other than that at saturation will be different from that of the same substance in its pure, standard state, the difference being the $\Delta G'$ accompanying the process of changing the activity of the solute, by dilution or concentration, to a new activity. This can be represented by the following equations.

$$\Delta G_f^\circ \begin{pmatrix}\text{sat'd sol'n}\\ \text{activity} = a\end{pmatrix} \longrightarrow \Delta G_f' \begin{pmatrix}\text{nonsat'd sol'n}\\ \text{activity} = a'\end{pmatrix} \tag{1.51}$$

from which

$$\Delta G' = \Delta G'_f \begin{pmatrix} \text{nonsat'd sol'n} \\ \text{activity} = a' \end{pmatrix} - \Delta G^\circ_f \begin{pmatrix} \text{sat'd sol'n} \\ \text{activity} = a \end{pmatrix}$$

and

$$\Delta G'_f \begin{pmatrix} \text{nonsat'd sol'n} \\ \text{activity} = a' \end{pmatrix} = \Delta G^\circ_f \begin{pmatrix} \text{sat'd sol'n} \\ \text{activity} = a \end{pmatrix} + \Delta G'$$

The free energy of formation in aqueous solution can be designated by the symbol $\Delta G'_f$ or by the symbol ΔG_f plus some indication that a given substance is dissolved.

The free-energy change accompanying a reaction such as that represented by Eq. 1.51 can be calculated by means of the following equation:

$$\Delta G' = -RT \ln \frac{(\text{activity coeff.} \times \text{conc. in sat'd sol'n})}{(\text{activity coeff.} \times \text{conc. in nonsat'd sol'n})} \quad (1.52)$$

$$= -RT \ln \frac{a}{a'}$$

For the same reasons that it is useful to have standard states with respect to the calculation of enthalpy and entropy changes that accompany reactions occurring in aqueous solution, it is also useful to have a standard state with respect to the calculation of free-energy changes accompanying aqueous reactions. This latter is defined as a hypothetical 1 m solution which corresponds to the limiting condition implied by Henry's Law, in which the solute has an activity of one (unity). However, it is entirely possible that the condition of unit activity may occur at a finite, not zero, concentration of solute. Thus, a solution in which the solute exists at unit activity may not necessarily be infinitely diluted, and the concept of infinite dilution does not usually apply to the standard state with respect to free-energy changes. This is especially true of strong electrolytes. As a somewhat extreme example, at unit activity at 25°C a sulfuric acid solution has a molality of nearly 14. However, with most substances of biological interest a condition of unit activity exists at low, but finite concentrations of solutes, of the order of 10^{-3} m. At these concentrations the difference between molalities and molarities becomes negligible.

Because in the aqueous standard state the activity coefficient is taken as unity and the concentration of the solute as either 1 m or 1 M, the denominator in Eq. 1.52, a', assumes a value of one and a standard free-energy change, $\Delta G^{\circ\prime}$, can be calculated in the following manner.

$$\Delta G^{\circ\prime} = -RT \ln (\text{activity coeff.} \times \text{conc. in sat'd sol'n}) \quad (1.53)$$

$$= -RT \ln (\text{activity of solute in sat'd sol'n})$$

Thus,

$$\Delta G_f^\circ \begin{pmatrix} \text{sat'd sol'n} \\ \text{activity} = a \end{pmatrix} \longrightarrow \Delta G_f^{\circ\prime} \begin{pmatrix} \text{std. 1 } m \text{ sol'n} \\ \text{activity} = 1 \end{pmatrix}$$

from which

$$\Delta G^{\circ\prime} = \Delta G_f^{\circ\prime} \begin{pmatrix} \text{std. 1 } m \text{ sol'n} \\ \text{activity} = 1 \end{pmatrix} - \Delta G_f^\circ \begin{pmatrix} \text{sat'd sol'n} \\ \text{activity} = a \end{pmatrix}$$

and

$$\Delta G_f^{\circ\prime} \begin{pmatrix} \text{std. 1 } m \text{ sol'n} \\ \text{activity} = 1 \end{pmatrix} = \Delta G_f^\circ \begin{pmatrix} \text{sat'd sol'n} \\ \text{activity} = a \end{pmatrix} + \Delta G^{\circ\prime}$$

The term $\Delta G_f^{\circ\prime}$ $\left(\substack{\text{std. 1 } m \text{ sol'n} \\ \text{activity} = 1}\right)$ is far more conveniently written simply as $\Delta G_f^{\circ\prime}$, to indicate that it represents the free energy of formation in the aqueous standard state at a pH of 7.0. When such a term represents a specific value such as the free-energy of formation of glucose in aqueous solution, it would be better written ΔG_f° glucose(aq), where the suffix (aq) designates the standard state of the substance.

Using the data employed by Morris[1] (p. 192), given that glutamic acid species in saturated aqueous solution at 298.15°K and 1 atm have a total concentration which is 0.0595 m, that the acid is dissociated into its constituent ions to the extent of 3.8%, that the ΔG_f° of glutamic acid in its standard state is -174.05 kcal mole^{-1}, that the molal activity coefficient for the undissociated glutamic acid is 0.55, and that the $\Delta G^{\circ\prime}$ for the ionization of glutamic acid is 5.89 kcal mole^{-1}, what are $\Delta G_f^{\circ\prime}$ values for undissociated glutamic acid and for glutamate$^-$ ion in aqueous solution? Using these data, the concentration of undissociated acid in aqueous solution is

$$0.0595\ m - (0.038 \times 0.0595\ m) = 0.0572\ m$$

The activity of the undissociated acid in saturated aqueous solution is equal to (activity coefficient × concentration), or

$$\text{activity of sat'd solute} = 0.55 \times 0.0572\ m = 0.0314\ m$$

The free-energy change accompanying the process of changing the concentration of the acid from that of saturation to that of a 1 m concentration at unit activity can be found using Eq. 1.53.

$$\begin{aligned} \Delta G^{\circ\prime} \text{ solution} &= -(1.987 \text{ cal mole}^{-1} \text{ deg}^{-1} \times 298.15°\text{K} \times \ln 0.0314) \\ &= -(-2050 \text{ cal mole}^{-1}) \\ &= 2.05 \text{ kcal mole}^{-1} \end{aligned}$$

The $\Delta G_f^{\circ\prime}$ for glutamic acid in solution in its standard state is therefore

$$\begin{aligned}\Delta G_f^{\circ\prime} &= \Delta G_f^{\circ}(\text{s}) + \Delta G^{\circ\prime} \text{ solution} \\ &= -174.05 \text{ kcal mole}^{-1} + 2.05 \text{ kcal mole}^{-1} \\ &= -172.00 \text{ kcal mole}^{-1}\end{aligned}$$

The $\Delta G_f^{\circ\prime}$ of glutamate$^-$ ion in aqueous solution will equal the $\Delta G_f^{\circ\prime}$ of undissociated glutamic acid in aqueous solution plus the $\Delta G^{\circ\prime}$ of ionization. Thus,

$$\begin{aligned}\Delta G_f^{\circ\prime} &= -172.00 \text{ kcal mole}^{-1} + 5.89 \text{ kcal mole}^{-1} \\ &= -166.11 \text{ kcal mole}^{-1}\end{aligned}$$

The $\Delta G_f^{\circ\prime}$ of a given substance can also be calculated if appropriate values are known for all of the terms in a reaction equation except that of the $\Delta G_f^{\circ\prime}$ of the given substance. Thus, with respect to the reaction represented by Eq. 1.50, if the ΔG_f° values for H_2O(l), NH_4^+(aq) and glutamate$^-$(aq) are given as -56.687, -18.97, and -166.11 kcal mole^{-1}, respectively, and the $\Delta G^{\circ\prime}$ for this reaction is known to be -3.442 kcal mole^{-1}, the $\Delta G_f^{\circ\prime}$ value for glutamine(aq) can be calculated using Eq. 1.49 as follows:

$$\begin{aligned}\Delta G^{\circ} &= \Delta G_f^{\circ}\, NH_4^+(\text{aq}) + \Delta G_f^{\circ} \text{ glutamate}^-(\text{aq}) \\ &\quad - \Delta G_f^{\circ} \text{ glutamine(aq)} - \Delta G_f^{\circ}\, H_2O(\text{l}) \\ -3.442 &= -18.97 + (-166.11) - \Delta G_f^{\circ} \text{ glutamine(aq)} - (-56.687)\end{aligned}$$

from which

$$\begin{aligned}\Delta G_f^{\circ} \text{ glutamine(aq)} &= 3.442 - 18.97 - 166.11 + 56.687 \\ &= -124.95 \text{ kcal mole}^{-1}\end{aligned}$$

Many biological reactions have hydrogen ions as a reactant or a product, for which the standard state need not (and in most cases can not) involve a concentration that is 1 m or 1 M. In these cases a new standard state is defined in which the hydrogen ion concentration is that at neutrality, which is 10^{-7} m. Free-energy changes accompanying reactions that take place in aqueous solution in which all the reactants and products are present in their aqueous standard states except for hydrogen ion, which is present at a concentration of 10^{-7} m or (M) are designated by the symbol $\Delta G^{\circ\prime}$.

1.29.10. The Calculation of Standard Free-Energy Changes Accompanying Reactions in Aqueous Solution

Provided that the appropriate data are available, probably the most useful method of calculating ΔG° values for reactions in aqueous solution is through the use of Eq. 1.49. Consider as an example, the fermentation of glucose to carbon dioxide and ethyl alcohol which is represented by the following equation.

$$\underset{\text{Glucose}}{C_6H_{12}O_6(aq)} \longrightarrow 2CO_2(aq) + \underset{\text{Ethyl alcohol}}{2CH_3CH_2OH(aq)}$$

Given that the ΔG_f° values for glucose(aq), CO_2(aq), and CH_3CH_2OH(aq) taken from Appendix 1 are -218.58, -92.26, and -43.44 kcal mole^{-1}, respectively, the $\Delta G^{\circ\prime}$ accompanying this reaction can be calculated using Eq. 1.49 as follows:

$$\begin{aligned} \Delta G^{\circ\prime} &= 2(-92.26) + 2(-43.44) - (-218.58) \\ &= -184.52 - 86.88 + 218.58 \\ &= -52.82 \text{ kcal mole}^{-1} \end{aligned} \tag{1.54}$$

Values for $\Delta G^{\circ\prime}$ can be calculated by methods other than by means of Eq. 1.49. If a reaction taking place in aqueous solution also happens to be an oxidation–reduction reaction with respect to which a difference in standard electrode potentials can be measured, $\Delta G^{\circ\prime}$ is related to ΔE° by the following equation.

$$\Delta G^\circ = -nF\,\Delta E^\circ$$

where n = the number of electrons transferred from reductant to oxidant, F = the Faraday constant = 23.062 kcal volt^{-1} equivalent^{-1}, and ΔE° = the electromotive force in volts representing the difference between the standard electrode potentials of the oxidation–reduction couple which is interacting. In addition, if the reaction taking place in aqueous solution is reversible and if the activities of the reactants and products can be determined at equilibrium, then Eq. 1.47 can be used to determine $\Delta G^{\circ\prime}$. Further, if $\Delta H^{\circ\prime}$ and $\Delta S^{\circ\prime}$ for a given reaction can be determined, then $\Delta G^{\circ\prime}$ can be calculated using Eq. 1.33. Unfortunately, all of these methods are probably unsuitable for determining the free-energy change accompanying the growth of a microorganism.

1.29.11. The Unsuitability of the Conventional Aqueous Standard State for Determining Free-Energy Changes Accompanying Microbial Growth

What is inherent in the calculation involving Eq. 1.54 is the concept that the free energies of formation of the reactant and the products are maintained in their standard states throughout the reaction. This is equivalent to stating that at the beginning of the reaction the glucose and the carbon dioxide and alcohol are *already* present in their aqueous standard states, and that as fast as glucose is used up it is replaced by more glucose so that its 1 m (or M) concentration at unit activity is always maintained, and that the carbon dioxide and alcohol are removed from solution as fast as they are produced, so that their 1 *m* (or *M*) concentrations at unit activity are also maintained. This provision is necessary in order that the ΔG_f° values for glucose(aq), CO_2(aq), and alcohol(aq) do not change. Otherwise as the reaction proceeded from its initial to its final state the concentration of glucose would drop from 1 *m* (or *M*) at unit activity to zero as it became converted into CO_2 and alcohol. At the same time, concentrations of these latter substances would rise from zero to 1 *m* (or *M*) at unit activity. Under real conditions the $\Delta G_f'$ values for glucose, CO_2, and alcohol would be constantly changing with their concentrations and activity coefficients at these concentrations, and all these data would have to be determined and integrated with respect to each substance over the time period of the reaction. No one seems particularly to want to do this.

The process of microbial growth can be represented in several ways, one of which is the following.

$$\left.\begin{array}{l} S\mathrm{Cn_CHn_HOn_O(aq)} + O\mathrm{O_2(aq)} \\ + N\mathrm{NH_3(aq)} \end{array}\right\} \longrightarrow \begin{array}{l} Y\mathrm{CH}n_\mathrm{H}\mathrm{O}n_\mathrm{O}\mathrm{N}n_\mathrm{N}\mathrm{(cells)} \\ + C\mathrm{CO_2(aq)} + W\mathrm{H_2O(l)} \end{array} \qquad (1.55)$$

where $\mathrm{Cn_CHn_HOn_O}$(aq) represents the molecular formula of a carbon and energy source for the growth of a microorganism; $\mathrm{CH}n_\mathrm{H}\mathrm{O}n_\mathrm{O}\mathrm{N}n_\mathrm{N}$(cells) represents the empirical, unit-carbon formula (obtained from the ash-free dry weight of the microorganism) for the cells produced during the growth reaction; $\mathrm{n_C}$, $\mathrm{n_H}$, $\mathrm{n_O}$, and n_N represent the relative numbers of C, H, O, and N atoms in the formulas for the carbon source and the cells, respectively; and *S*, *O*, *N*, *Y*, *C*, and *W* represent the coefficients of their respective terms in the equation. Equations such as 1.55, if constructed from accurate data, can represent the actual mass balance that occurs during the process of growth to within 98–99%, exclusive of the components of the ash content of cells. They have been called "elemental balance equations" by Minkevich and Eroshin[15] (also later referred to in the literature as "mass-balance equations"), although such equations were earlier called "growth-reaction equations" by Battley.[16] This last designation seems preferable. All chemical reaction equations are elemental balance equations. Battley's designation is equally concise and considerably more descriptive.

In their natural environments the substances used by microorganisms as sources of carbon and energy are usually present in concentrations far less than 1 *m*, and are almost always completely consumed so that nothing remains. At the same time, the products of their metabolism diffuse away into the environment. Thus, the use of the conventional standard state in aqueous solution (hypothetical 1 *m* at unit activity) for the calculation of energy changes accompanying microbial growth is seriously to be questioned. It would be preferable to use a different standard state for the purpose of studying the energetics of microbial growth in which the substances in solution are present in concentrations similar to those encountered by microorganisms under natural conditions.

1.29.12. The Calculation of Nonstandard Free-Energy Changes Accompanying Reactions in Aqueous Solution

It can be stated without fear of contradiction that almost all biological reactions take place in aqueous solution under conditions that are not standard. Taking again the reversible reaction represented by Eq. 1.45,

$$aA + bB \rightleftharpoons cC + dD$$

where A and B represent the reactants, C and D the products, and a, b, c, and d their respective coefficients. If (A), (B), (C), and (D) represent the nonstandard activities of the reactants and products at 298.15°K and 1 atm pressure which are *maintained* during the reaction in which 1 mole each of A and B produce 1 mole each of C and D, the nonstandard free-energy change, $\Delta G'$, which accompanies this reaction is related to the standard free-energy change, $\Delta G^{\circ\prime}$, by the following equation.

$$\Delta G' = \Delta G^{\circ\prime} + RT \ln \frac{(C)^c(D)^d}{(A)^a(B)^b} \tag{1.56}$$

Using another example from Morris[1] (p. 188), the ΔG° for the reaction

$$\underset{\text{(FDP)}}{\text{fructose diphosphate}} \rightleftharpoons \underset{\text{(G3P)}}{\text{glyceraldehyde phosphate}} + \underset{\text{(DHAP)}}{\text{dihydroxyacetone phosphate}}$$

is 5.52 kcal mole^{-1} (i.e., this reaction is not spontaneous in the direction from left to right). However, if under more physiological conditions than those represented by the standard state the fructose diphosphate is maintained at a concentration of 10^{-2} M and the glyceraldehyde phosphate and the dihydroxyacetone phosphate are maintained at concentrations of 10^{-5} M, when this reaction takes place under these specified, nonstandard conditions and assuming that for purposes of this calculation concentrations equal activities,

$$\Delta G' = \Delta G^{\circ\prime} + RT \ln \frac{[\text{G3P}][\text{DHAP}]}{[\text{FDP}]}$$

from which

$$\begin{aligned}\Delta G' &= \Delta G^{\circ\prime} + 1.987 \text{ cal mole}^{-1} \text{ deg}^{-1} \times 298.15°\text{K} \times \ln \frac{10^{-5} \times 10^{-5}}{10^{-2}} \\ &= 5.520 \text{ kcal mole}^{-1} - 10.912 \text{ kcal mole}^{-1} \\ &= -5.392 \text{ kcal mole}^{-1}\end{aligned}$$

It is apparent from this result that changing the concentrations of the reactants and the products from the standard to a nonstandard state can result in a change from a reaction which is not spontaneous to one that *is* spontaneous, and for which the value for $\Delta G'$ is markedly different from that for $\Delta G^{\circ\prime}$. It is thus values for $\Delta G'$ which should be sought with respect to biological reactions. However, in the case of microbial growth it may not even be possible to use Eq. 1.56, again because the concentrations of the reactants and products are not maintained, and because of the formation of a precipitate (the cells) as a product.

1.30. SUMMARY

The purpose of the foregoing chapter has been to point out some of the ideas and conceptual tools that classical thermodynamics can offer with respect to determining the changes in energy, or more specifically, the changes in the thermodynamic quantities G, H, and S that accompany chemical reactions as they proceed from an initial to a final state. The symbols representing these changes are summarized in Table 1.5. As a generalization it can be stated that with respect to irreversible chemical reactions taking place at constant tem-

TABLE 1.5. Symbols used in this book for making thermodynamic calculations

ΔG	=	Gibbs free-energy change: when the reaction takes place under nonstandard conditions.
ΔG°	=	Standard Gibbs free-energy change: when the reaction takes place under standard conditions, all components being in their standard states.
$\Delta G'$	=	Gibbs free-energy change: when the reaction takes place in aqueous solution under nonstandard conditions. The components are not in their standard states and the H^+ ion activity is taken to be 10^{-7} unless otherwise stated.
$\Delta G^{\circ\prime}$	=	Standard Gibbs free-energy change: when the reaction takes place in solution under standard conditions. All components are in their standard states except for the H^+ ion activity, which is taken to be 10^{-7} unless otherwise stated.
ΔH, ΔH°, $\Delta H'$, $\Delta H^{\circ\prime}$, ΔS, ΔS°, $\Delta S'$, $\Delta S^{\circ\prime}$ represent changes in enthalpy and entropy as described for the Gibbs free-energy changes above. All ΔG, ΔH, and ΔS symbols followed by the subscript f represent changes accompanying formation from the elements.		
T	=	Absolute temperature in degrees Kelvin.
M	=	Molarity in moles/liter^{-1} of solution.
m	=	Molality in moles/kg^{-1} of solvent.
R	=	Gas constant; here, 1.987 cal mole^{-1} deg^{-1}.
K_{eq}	=	Thermodynamic equilibrium constant, calculated using the activities of the reactants and products.
K_c	=	Molal (or molar) equilibrium constant, calculated using molal (or molar) concentrations of reactants and products.
( )	=	Parentheses used to indicate activities.
[ ]	=	Brackets used to indicate concentrations.

perature and pressure there is a conversion of reactants to products such that the Law of the Conservation of Mass is obeyed. At the same time there is a conversion of nonthermal, chemical energy into heat, as represented by the Gibbs free-energy equation.

Reactants	$\longrightarrow$	Products		
ΔG	$\longrightarrow$	ΔH	$-$	$T\,\Delta S$
Initial state nonthermal, chemical energy		Final state thermal energy exchanged with the environment as the heat of reaction		Final state thermal energy absorbed by the reaction system because of the change in the physical nature of the system

It is apparent that many of the classical methods for calculating ΔG values cannot be applied to the overall reaction whereby microbial growth takes place. Enthalpy changes accompanying growth reactions ought to be measurable. Entropy changes, although theoretically capable of being calculated, are not likely to be so since the absolute entropy of a unit mass of microbial cells grown at constant temperature and pressure on any given substrate has never been determined. Nevertheless, it might be possible to determine ΔG, ΔH, and ΔS for the growth of a given microorganism on a known quantity of a thermochemically defined substrate provided that appropriate methods of growing cells or of constructing and manipulating growth-reaction equations can be discovered. Once this is done it ought to be possible to learn to what extent the quantity of microbial growth is related to the quantity of free energy potentially available from a given substrate. It should also be possible to predict the amount of heat exchanged with the environment by a microbial growth reaction. Finally, it may be possible to learn something about the entropy of cells, and the extent to which structural entropy and informational entropy can be correlated.

BIBLIOGRAPHY

Thermodynamics and the Free Energy of Chemical Substances, by G. N. Lewis and M. Randall, McGraw-Hill, New York, 1923, 653 pp.

Textbook of Physical Chemistry, by S. Glasstone, 2nd ed., van Nostrand, New York, 1946, 1320 pp.

Chemical Thermodynamics. Basic Theory and Methods, by I. M. Klotz, Prentice-Hall, New York, 1950, 369 pp.

Principles of Chemical Thermodynamics, by M. A. Paul, McGraw-Hill, New York, 1951, 740 pp.

A Biologist's Physical Chemistry, by J. G. Morris, Edward Arnold (Publishers), Ltd.,

London, 1968, 367 pp. For most purposes this is by far the best book available for the use of persons interested in the biological applications of physical chemistry.

An Introduction to the Study of Chemical Thermodynamics, by D. H. Everett, Longman Group, Ltd., London, 1971, 271 pp.

REFERENCES

1. J. G. Morris, *A Biologist's Physical Chemistry*, Edward Arnold, Ltd., London, 1968.
2. M. Kleiber, Joules vs calories in nutrition, *J. Nutrition* **102,** 309–312 (1972).
3. D. D. Wagman, W. H. Evans, V. B. Barker, I. Halow, S. M. Bailey, and R. H. Schumm, NBS Technical Note 270-3. *Selected Values of Chemical Thermodynamic Properties. Tables for the first thirty-four elements in the standard order of arrangement*, National Bureau of Standards, Washington, D.C., 1968.
4. T. W. Richards, Die Bedeutung der Änderung des Atomvolums. III. Die Beziehung zwischen Änderung der Wärmekapazität und Änderungen der frien Energie, Reaktionswärme, Volumänderung, und chemische Affinität, *Z. Physik Chem.* **42,** 129–154 (1903).
5. G. N. Lewis and M. Randall, *Thermodynamics and the Free Energy of Chemical Substances*, McGraw-Hill, New York, 1923.
6. W. Nernst, Nachr. kgl. Ges. Wiss, Göttingen, *Math.-physik. Klasse*, 1, 1906.
7. A. Einstein, Die Planchsche Theorie der Strahlung und die Theorie der spezifischen Wärme, *Ann. Physik* (4) **22,** 180–190 (1907).
8. Max Planck, *Thermodynamik*, 3rd ed., Veit & Co., Leipzig, 1911, p. 279.
9. G. N. Lewis and G. E. Gibson, The third law of thermodynamics and the entropy of solutions and liquids, *J. Am. Chem. Soc.* **42,** 1529–1533 (1920).
10. G. N. Lewis and M. Randall, *Thermodynamics and the Free Energy of Chemical Substances*, McGraw-Hill, New York, 1023.
11. M. Planck, Über neure thermodynamische Theorien (Nernstsches Wärmetheorem und Quanten Hypothese), *Ber. Dtsch. Chem. Ges.* **45,** 5–23 (1912).
12. M. A. Paul, *Principles of Chemical Thermodynamics*, McGraw-Hill, New York, 1951.
13. N. Wiener, *Cybernetics, or Control and Communication in the Animal and Machine*, Wiley, New York, 1948.
14. L. Brillouin, Life, thermodynamics, and cybernetics, *Am. Sci.* **37,** 554–568 (1949).
15. I. G. Minkevich and V. K. Eroshin, Oxygen and water expense during the growth of microorganisms. (in Russian), *Izvestiya AN SSSR*, *Ser. Biol.*, **N2,** 245–254 (1972).
16. E. H. Battley, Growth-reaction equations for *Saccharomyces cerevisiae*, *Physiol. Plant.* **13,** 192–203 (1960).

2

EARLY STUDIES ON MICROBIAL THERMOGENESIS RESULTING FROM THE UTILIZATION OF COMPLEX SUBSTRATES, 1864–1931

It has been known for many centuries by brewers and vintners that fermentation is accompanied by an increase in the temperature of the fermentation liquor. Other observations such as the apparently spontaneous heating of manure piles, compost heaps, and ensilage have also been well known. However, it was not until the presence of microorganisms was definitely linked to these otherwise unrelated phenomena that this production of heat acquired any biological significance. Eventually in those cases where heat production was known to be caused by microbial activity, this came to be called "microbial thermogenesis" (sometimes "bacterial thermogenesis"). This chapter describes several early studies involving the demonstration or quantitative measurement of heat produced by known or unknown microorganisms grown on chemically complex substances. In these environmental conditions it is usually not known what the microorganisms are using as sources of carbon and energy, and thus the actual heat source is never established.

2.1. THE POLEMIC BETWEEN PASTEUR AND BERTHELOT CONCERNING THE HEAT OF MICROBIAL GROWTH

Although Dubrunfaut[1] made a quantitative measurement of the heat of fermentation in 1856 (see Section 3.1), L. Pasteur[2] was probably the first to comment on the phenomenon of what we now call thermogenesis. He pointed out in 1864 that fermentations constitute a type of decomposition which takes place with the evolution of heat. In addition, these fermentative decompositions have a certain analogy with explosions (Pasteur did not amplify this statement further), and are all the more remarkable because of their prolonged duration. In 1872 he published his opinion that the heat produced during a fermentation was to the profit of the microorganisms causing it (Pasteur[3]), and in 1875 made the statement that, "[The theory of fermentation] will be established mathematically the day when science is sufficiently advanced to relate the quantity of heat that the life of yeast, in the absence of air, releases during the decomposition of sugar with the quantity of heat furnished by substrates in the presence of oxygen gas when the life of the yeast is carried out under conditions where this gas is furnished in more or less great abundance (Pasteur[4], p. 457)." At first Pasteur's observations were confined to the fact that microbial "life" is accompanied by the production of heat. Later, during the course of a minor polemic[5–17] with M. Berthelot over the results of Claude Bernard's posthumously published research Pasteur[11] indicated that he thought microorganisms had need of the heat produced during the fermentation or oxidation of their substrates. This opinion was severely criticized by Berthelot[16] in an exceedingly clear analysis of the possible energetic alternatives involved in yeast growth.

Berthelot pointed out that, according to rigorous thermochemical reasoning, it was necessary to define alcoholic fermentation as a chemical phenomenon occurring during the transition from an initial to a final state. A certain weight of sugar and yeast constituted the initial state, the final state being constituted by the weights of carbon dioxide, alcohol, and yeast produced during the reaction. The amount of heat that was formed during this reaction was independent of any hypothesis relative to the nature of the actual fermentation reaction (Hess's Law). Berthelot also divided alcoholic fermentation into two concomitant reactions. In one reaction the principal portion of the initial weight of sugar becomes decomposed into carbon dioxide and alcohol. This reaction evolves heat. In the other reaction a minor portion of the sugar becomes converted into yeast in a manner which is little understood. Since there is no way of knowing whether this latter reaction evolves or absorbs heat, there is no way of knowing whether or not heat is actually needed by the yeast.

In reply to Berthelot's criticism Pasteur[17] emphasized that the energy appearing as heat had already been used by the yeast to carry out its life processes. He stated that Berthelot had only concerned himself with the energy required to form the bodies of the yeast cells, and had forgotten that the rest of the energy was expended in the process of growing.

The opinions of Pasteur and of Berthelot cited above illustrate the views held by the vitalists and the mechanists during the latter half of the nineteenth century with respect to the phenomenon of energy exchange and heat evolution as these are seen in biological systems. Since that time a great deal of accumulated data has provided a basis for the belief that the two approaches are not mutually exclusive, but rather that a combination of the two gives a clearer insight into what actually happens energetically.

2.2. WHAT IS MICROBIAL THERMOGENESIS?

The term "thermogenesis" as applied to the production of heat due to the biological activities of microorganisms was first used in 1888 by F. Cohn,[18] although the phenomenon itself had been commonly known for decades before then. Most people who have lived on a farm have had the experience of seeing steaming haystacks or piles of manure on cold winter mornings and of discovering that the temperature of the interior of these masses of material is considerably higher than that of the surrounding environment. The use of "hot beds" consisting of mixtures of earth and manure for the early germination of seeds is a good example of the practical use of thermogenesis. The same heating process can be observed in the utilization of stable manure for the growing of mushrooms. Substances other than manure and moist hay also exhibit thermogenesis, and the phenomenon can be considered a general one which, with some exceptions, takes place when soft plant material is piled or stored in such a manner as to produce a porous mass through which air can diffuse. The heat production is a function both of the respiration of the plant tissues and of the respiration of mixed cultures of microorganisms which are normally found on the surfaces of plant parts. In 1893 Cohn[19] stated that the rise in the temperature of a mass of plant material to 35°C can be attributed to plant cell respiration, and above this temperature to the presence of molds. He also reported the observation of a temperature of 67.2°C in a mass of raw cotton. Following this Suchsland (1894)[20] observed heat production in fermenting tobacco, Behrens (1897, 1901)[21,22] in stored hops, Holdefleiss (1899)[23] in hay, and Hunter (1917)[24] in silage, each attributing all or some of this to the presence of microorganisms. In most instances microbial thermogenesis is either harmless or beneficial to the material being heated, depending on what the effect is. In some cases it can be detrimental. The "scorching" of hays, grains, or fodder resulting in a grey or blackish product definitely lowers their market value. In addition, microbial thermogenesis may lead to "spontaneous combustion." This was first investigated in 1873 by Ranke[25] who studied the possibility of spontaneous ignition in hay but made no connection between the presence of microorganisms and the production of heat. Subsequently, Cottrell (1902),[26] Laupper (1920),[27] Heaton (1921),[28] and Schwartz (1921)[29] reported other cases of spontaneous ignition in plant products. In general, it was recognized that the first stage of heating occurs soon after the thermogenic material is piled or stored, with the temperature of the stored material never rising to more than 80°C. Since this is far below the ignition temperature of these materials, the inference is that volatile, combustible substances are generated as a result of microbial activity and become ignited in some manner unrelated to the presence of the microorganisms.

In all of the above studies the data were accumulated as the result of direct observation, not as the result of experimentation involving known microbial cultures growing on natural products. The first experiments of this latter type were carried out in 1907 by Miehe[30] who constructed a device consisting of concentric, cylindrical, wire mesh chambers which were graduated in size, the outer chambers serving as insulation for the innermost chamber which held the experimental material. All the outer chambers contained packed hay. The innermost chamber was removable and could be filled with chopped hay, "sterilized" with steam for 1 hr, inoculated with an aqueous suspension of the test organisms, and then replaced. The whole device was then put in an incubator held at 40°C and measurements were taken of any rise in temperature which might occur in the inner chamber. Miehe observed that *Bacillus calfactor* which he isolated from heating hay could raise the temperature of sterilized hay to 46°C, whereas *Oidium lactis* and *Aspergillus niger* raised the temperature to 58.5 and 48.0°C, respectively. However, since he demonstrated that pure cultures of *B. calfactor* could grow at temperatures as high as 75°C, he concluded that the heating of haystacks to this temperature could be caused by this microorganism.

2.3. RUBNER'S DEVELOPMENT OF A DIFFERENTIAL MICROCALORIMETER AND HIS EXPERIMENTS ON MICROBIAL THERMOGENESIS

Classical thermodynamics can be said to have begun with Mayer and with Helmholz who in 1842 and 1847, respectively, postulated the law of the conservation of energy. In addition, Mayer also suggested that this law would apply to living systems. The actual proof of this suggestion was provided by M. Rubner, who demonstrated that a dog living 45 days in a respiration calorimeter produced 17,349 cal of heat energy at the same time that it utilized 17,406 cal of nutrient (chemical) energy as calculated from the dog's respiratory metabolism and nitrogen excretion. Rubner[31] summarized the results of his many investigations on animal heat in 1902. Following this he turned his attention to lower organisms and in 1903 (Rubner[32]) published the first description of a microcalorimeter which he devised to measure the heat production of microorganisms. This was a simple, yet sophisticated instrument that was designed to eliminate many of the difficulties encountered by Dubrunfaut,[1] Bouffard,[33] and A. J. Brown[34] (see Sections 3.1–3.3). More complete descriptions were published in 1911 (Rubner[35]) and 1913 (Rubner[36]) and included useful modifications of the basic instrument.

Rubner's calorimeter consisted of a 300–400 ml (although sometimes up to 2 liter) triple-walled, spherical flask opening to the environment through a neck, the walls being separated by a distance of 0.5 cm with the space between the walls being evacuated. It was thus essentially a triple-walled Dewar flask. The double vacuum caused the heat loss by conduction to be extremely low, but did not eliminate it completely. The temperature was read off from a sensitive thermometer graduated to 0.005°C which was passed through a cork in the neck of the flask, thereby closing it off from the environment. Several of these calorimeters could be placed in a cylindrical copper bacteriological incubator built for this purpose, their thermometers passing through the top cover of the incubator so that they could be read with a magnifying glass without opening the cover. In later models the incubator was automatically electrically thermostated and the internal air circulated by means of an electric fan. This enabled all calorimeters to come to a reasonably fast temperature equilibrium with the incubator. At the beginning of an experiment all the thermometers had to have the same reading. One calorimeter filled with uninoculated, sterile nutrient solution served as a control. Diagrams of a calorimeter and of several of these in an incubator are shown in Fig. 2.1. To make the temperature control even better the incubator with its calorimeters was often placed in a special room which was thermostated to ±1–2°C.

In practice one experimental calorimeter plus the control calorimeter together constituted what later came to be known as a "differential" calorimeter. Although the two flasks were not directly connected, the readings of the thermometer in the control flask provided a baseline temperature against which

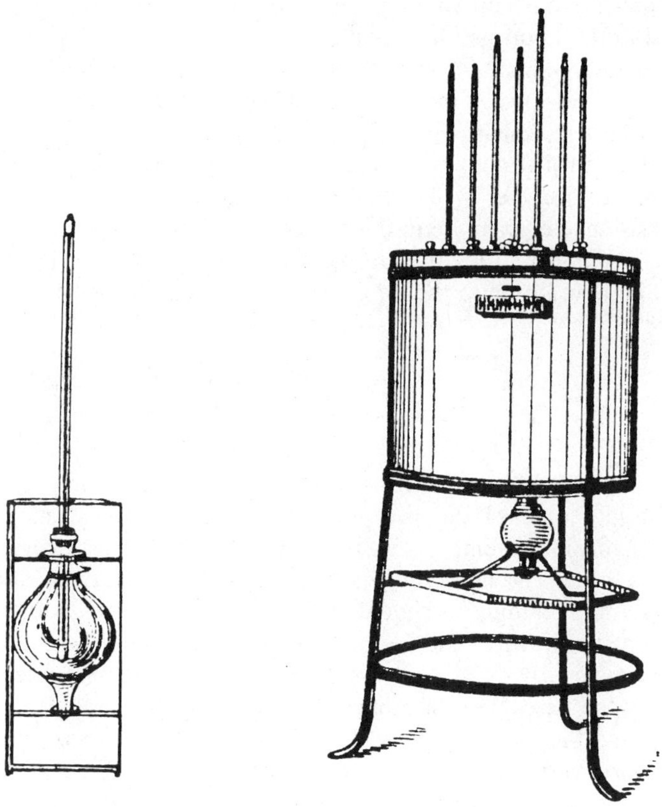

FIGURE 2.1. Rubner's microcalorimeter. Source: Reference 36.

the readings of the thermometer in the experimental flask could be compared, and any effect that slight variations in the incubator temperature might have had could thus be eliminated. Inherent in this practice was the assumption that small changes in the temperature of the incubator affected all the calorimeter flasks equally. This type of calorimetry, which is not automatic, was made automatic by Hill[37] when he connected the experimental and control flasks by means of a thermocouple (see Section 2.5). One great advantage of the use of these glass calorimeters was that they could be easily sterilized so that pure cultures of microorganisms could be used if desired. Another advantage was that the development of microbial cultures could be visually observed. Rubner apparently did not sterilize his nutrient solutions directly in the calorimeter flasks, since he stated that one of the difficulties with the experiments was in introducing the sterile, nutrient solution into the calorimeter flasks such that it would be at nearly the same temperature as the incubator. Otherwise, temperature equilibrium could take a long time. He did not state how he sterilized his thermometers. Possibly this was done by washing them with a solution of $HgCl_2$ and then rinsing well with sterile water. The calorimeter flasks could be sterilized with dry heat or with the use of an autoclave, and the nutrient solutions by autoclaving or by the use of Chamberland or Berkfeld porous filter candles.

Rubner calibrated his calorimeters electrically. A given amount of water or nutrient solution was placed in a calorimeter flask, the flask being then closed with a cork through which passed the thermometer and two wire leads connected to a coiled platinum resistance wire which became immersed in the liquid when the cork was in place, as did the bulb of the thermometer. The electrical resistance of the wire in the coil could be accurately measured, and if a current of known strength were passed through it, a known amount of heat would be liberated into the liquid according to the following formula:

$$\text{cal} = \frac{I^2Rt}{4.184} \tag{2.1}$$

where I = amperes (A), R = resistance (ohms, Ω), t = time (sec), and 4.184 = absolute joules (J) per calorie. By introducing a known, small amount of heat continuously the temperature of the calorimeter rose to a certain value and stabilized there. This latter temperature was a function of the heat capacity of the apparatus (including the liquid in the flask, the inner wall of the calorimeter, the platinum wire, and the bulb of the thermometer) and the rate at which heat was lost by conduction from the flask to the environment of the incubator. In one example of a calibration which Rubner cited $R = 0.25\ \Omega$, $I = 0.76$ A, and the temperature in the calorimeter flask stabilized at 2.22°C above the temperature of the incubator over an experimental period of about 12 hr. Using Eq. 2.1, the number of calories of heat energy introduced into the flask during 1 hr was then

$$\text{cal} = \frac{0.76^2 \text{ A} \times 0.25\ \Omega \times 3600 \text{ sec hr}^{-1}}{4.186 \text{ J cal}^{-1}}$$

$$= \frac{519.84 \text{ J hr}^{-1}}{4.184 \text{ J cal}^{-1}}$$

$$= 124 \text{ cal hr}^{-1}$$

and the amount of heat necessary to raise the temperature of the calorimeter flask by 1°C was

$$\frac{124 \text{ cal hr}^{-1}}{2.22 \text{ deg}^{-1}} = 56 \text{ cal deg}^{-1} \text{ hr}^{-1}$$

Rubner carried out calibrations at several temperatures to get an average rate for each calorimeter.

In practice, during an actual experiment with microorganisms the reading of the thermometer in degrees Celsius above the temperature of the control flask was plotted against time in hours until the thermometer reading again returned to that of the control flask. The average temperature rise for the entire duration of the experiment was then determined and multiplied by the appropriate calibration factor. For example, if the calibration factor were 52 cal deg^{-1} hr^{-1}, the average temperature rise were 0.161°C, and the length of the experiment were 14 days, the amount of heat produced by the microorganism during this time would be

$$0.161°\text{C} \times 52 \text{ cal } °\text{C}^{-1} \text{ hr}^{-1} \times 24 \text{ hr d}^{-1} \times 14 \text{ d} = 2812 \text{ cal}$$

This observed heat production then had to be corrected for the mechanical heat effects of any air passed through the system to aerate the culture, for water loss through evaporation, and for heat loss because of the carbon dioxide passing out of solution into the gas phase. The greatest correction was for the heat of vaporization of some of the water from the culture solution. This could be determined by multiplying the difference between the weight of this solution before and after the experiment by the latent heat of vaporization of water, which Rubner took as 600 cal g^{-1}. If 0.238 g of water were lost from the system during an experiment the correction would be 0.238 g × 600 cal g^{-1} = 142.8 cal that would have to be added to the observed heat production. The return of the temperature of the experimental flask to that of the control flask was asymptotic, and Rubner would frequently terminate his experiments when they were 0.04–0.05°C above the control flask temperature; otherwise they would take much longer. In this case a correction had to be made for the residual heat within the system by multiplying the residual temperature rise (for example, 0.04°C) by the "water value," or heat capacity of the system. This latter was determined by adding the heat capacities of the individual masses comprising the components of the system. Thus, if the specific heat of 1 g of culture solution were 0.986 cal °C and there were 250 g of this solution in the calorimeter, the heat capacity of the solution would be 250 g × 0.986 cal g^{-1} deg^{-1} = 246.5 cal $deg.^{-1}$ In general, the specific heat of a given culture solution had to be determined experimentally at the incubator temperature of an experiment. The heat capacity of the thermometer bulb was determined by measuring the mass of mercury in the bulb and of the glass surrounding it, and multiplying these values by the specific heats of mercury and of glass, respectively, at the experimental temperature. The same was true for the heat capacities of the inner wall of the calorimeter and of the platinum resistance wire used to calibrate the system. Rubner even went so far as to determine the heat capacity of the mass of microorganisms produced during growth, although in this case the heat capacity of a filtrate of the culture medium would have to be determined after the experiment had been terminated, rather than to use the heat capacity value of the original culture solution. With 250 g of culture solution in the calorimeter flask the total heat capacity of the system usually ran about 265 cal deg^{-1}. Were the experiment to be terminated at 0.04°C above the control flask temperature, the correction would be 0.04°C × 265 cal deg^{-1} = 10.6 cal which would then have to be added to the observed heat production.

These descriptions of some of the procedures Rubner used in making his calorimetric measurements are only intended to illustrate the exactitude of his technique and the attention he gave to detail. Although subsequent authors have stated that his measurements were inaccurate, it is hard to believe that this was so, even though he did not provide statistical data on the sensitivity, accuracy, and stability of his instruments. He could detect the production of 0.44 cal of heat per hour (511 μW), which is good even by today's standards. The difficulty therefore would seem to be not so much the accuracy of his measurements but what they represented. We still have this problem today. If the initial and final states of a system are not strictly thermochemically defined or if the exact nature of the reaction is not known, it is impossible to know to what to ascribe the observed heat production.

In 1904 Rubner[38] published the results of his calorimetric experiments on alcoholic fermentation. It is extremely unfortunate that considering the instrument he had at his disposal, the care he took in his technique, and the amount of work he did, his data are not of greater value. Many of his experiments were with various quantities of yeast suspended in sugar solutions. Figure 2.2 is a graph of temperature versus time from one such experiment which has less to do with growth than with the metabolism of the sugar, since in the absence of a nutrient culture solution of minerals and a nitrogen source growth would not be expected to occur to any appreciable extent. Or would it? In the experiment represented by Fig. 2.2, for example, 6 g of yeast were

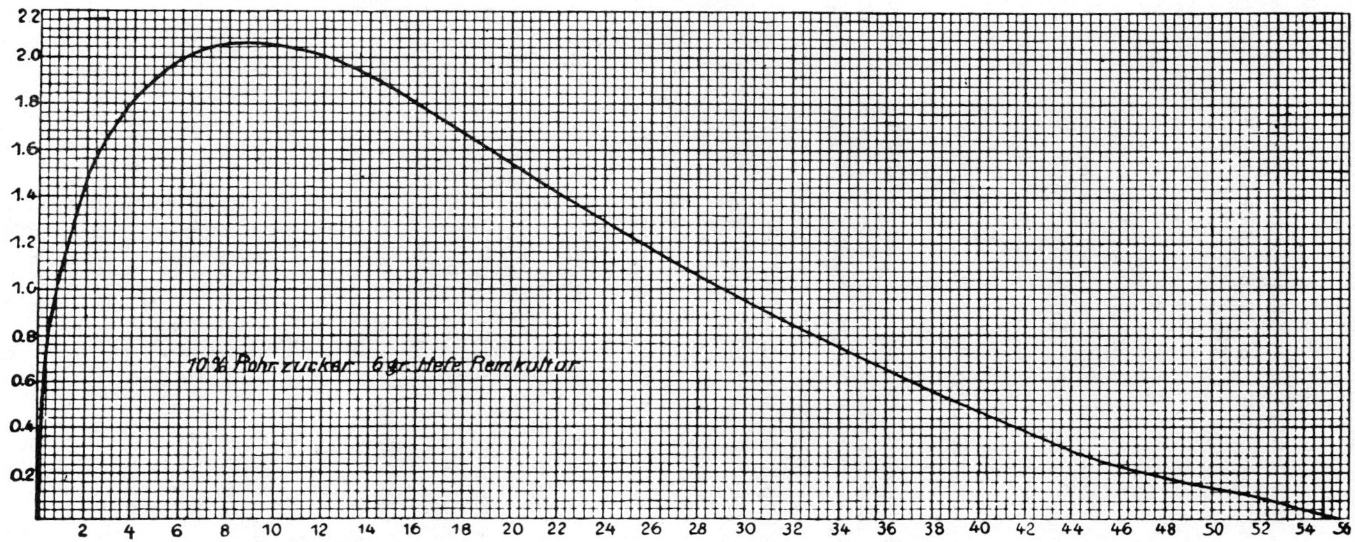

FIGURE 2.2. Rubner's thermogram showing temperature rise in degrees Celsius plotted against time in hours for a suspension of 6 g of yeast in a 10% sugar solution. Source: Reference 35.

used. If the yeast cells had happened to have a source of endogenous nitrogen in the form of an amino acid pool, a certain amount of growth would certainly have taken place, with some of the energy which would otherwise be lost as heat during the alcoholic fermentation becoming incorporated into the substance of the newly grown yeast cells. Thus, the amount of heat produced during the experiment would have been less than that predicted on the basis of the Gay-Lussac equation (see Eq. 3.1) whereby 2 moles of carbon dioxide and of alcohol are produced for every mole of sugar that disappears. In fact, this is what Rubner observed. The only puzzling thing about his data is that no matter what kind of experiment he did on alcoholic fermentation (and there were twelve of these) he always got nearly the same results averaging 149.5 ± 2.35 cal g^{-1} of sugar consumed. He observed this even when growing cells were used. With respect to these latter, however, he seemed not to be aware of the effects of temperature on the growth of yeasts, since he carried out one of the experiments at a temperature of 38°C. At this temperature although the yeast cells did, indeed, ferment and might well have incorporated some of the nitrogenous substances in the beer wort he used as a nutrient medium, they would have been close to the maximum temperature of growth for many yeast species and it is possible that they may not have been actually growing. There is little point in speculating on this even though one or another speculation might account for the astonishing uniformity of the results he obtained. The value he finally arrived at for the heat of fermentation was −25.6 kcal $mole^{-1}$ of glucose (equivalent) fermented. This can be compared with the values of −21.4, −22.7, and −23.5 kcal $mole^{-1}$ for the same fermentation which were obtained previously with growing yeasts by A. J. Brown,[34] Dubrunfaut,[1] and Bouffard,[33] respectively.

In 1904 Rubner[39] extended his techniques to the use of indirect calorimetry as applied to microorganisms. The apparatus required, the extended length of time necessary for accurate temperature measurements, and the number and magnitude of the corrections that had to be applied all tended to make the technique of direct calorimetry extremely inconvenient. A simpler method was indirect calorimetry, in which the heat of combustion of the reactants and products was determined by bomb calorimetry, the difference between the two being the heat of reaction. A solution calorimeter was not required and corrections did not have to be made for heats of solution, hydrolysis, or residual heat. In practice, samples of the nutrient material comprising the culture medium were burned in a bomb, the measured heat of combustion being converted into cal g^{-1} of nutrient. This value multiplied by the quantity in grams of the nutrient used in the culture medium gave the heat of combustion of the medium at the beginning of an experiment (the initial state). When the experiment was over an aliquot of the cell suspension in the residual medium was dried and its heat of combustion also measured. This value multiplied by the number of aliquots making up the whole culture gave its heat of combustion at the end of the experiment (the final state). The difference between them was the heat of reaction. Sometimes the heat of combustion of the bacteria which were grown and that of the culture medium from which they had been filtered were also determined. The major error in indirect calorimetry is that if the heats of solution or suspension of the substances in the initial and final states are not added to the values for the combustion of these substances after they have been dried, the bomb calorimetric values themselves do not indicate accurately the changes in heat content which take place under the actual conditions of the experiment, in which all substances are dissolved or suspended in water.

Rubner[40,41] made many measurements of heat production by bacteria using both direct and indirect calorimetry. Again, not much information can be gained from his data. One outstanding advantage of working with yeasts is that they can ferment sugar, a specific substance, according to a given equation. We largely know what the yeasts are using in their metabolism. The extent to which they actually carry out a straight alcoholic fermentation or form products other than carbon dioxide and alcohol such as glycerine and succinic acid (as Pasteur[42] demonstrated) or even growing cells will be indicated by the observation of a lesser heat production in the latter case than would be expected to be produced in the former case. We are thus able to compare the theoretical maximal amount of heat which they can produce by fermentation with what they actually do produce. What such a comparison means is a matter of interpretation. However, no interpretation is possible without some means of comparison. For example, Rubner[39,40] made measurements of the heat produced when various bacteria grew on such complex substances as peptone, meat extract, beer wort, and so on. In all cases he observed heat production. However, to what to ascribe this heat production cannot be ascertained since it is not known exactly what substances in these media the microorganisms were fermenting or oxidizing. Using indirect calorimetry Rubner compared the amounts of heat produced by several species of bacteria growing separately on similar substrates, and observed that some bacteria produced more heat during a given period of time than others. These data are not of great use, although they did demonstrate that in all cases growing bacteria produced heat and that some produced more heat than others. Rubner then went on to use direct calorimetry to demonstrate that during the growth of bacteria heat was not produced at a constant rate but in general was produced at an increasing rate until it reached a maximum, after which it fell off rapidly to a low level. These changes in rate could easily be demonstrated by plotting the rise in the temperature of the calorimeter vessel above ambient temperature as a function of time. This is illustrated by Fig. 2.3 showing the growth of *Bacterium coli* on peptone over a period of nine days. Looking at this figure with the clarity of hindsight we might suggest that after a lag phase the bacteria passed into a phase of exponential growth which terminated when certain nutrients became exhausted at which time, because of heat conduction to the incubator, the temperature of the calorimeter flask rapidly decreased to a low level when there occurred a slow but more or less constant rate of heat production, for whatever reason. We can suggest this, but we do not know this to be true. Rubner did not determine bacterial numbers as a function of time during this experiment. The rate of heat production might only resemble what we suspect

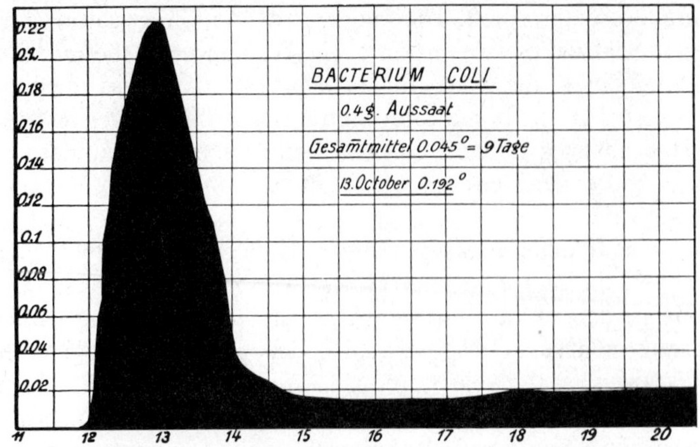

FIGURE 2.3. Rubner's thermogram showing temperature rise in degrees Celsius plotted against time in days for the growth of *Bacterium coli* on 0.4 g peptone in solution. Source: Reference 40.

could have happened. We do not know the reactants or products of whatever reaction took place within the calorimeter except that bacteria were visibly produced. We only know the observed quantity of heat produced (probably very accurately measured), but this, by itself, is not very meaningful at least with respect to this particular experiment.

Rubner called figures such as 2.2 and 2.3 "heat curves" and was the first to publish these. They have since come to be called "thermograms." Such data have proven to be of little use unless it is also known to what to attribute the heat production. Nevertheless, Rubner's studies demonstrated conclusively the universal fact of heat production by growing or metabolizing suspensions of microorganisms, and that in this respect they are similar to other living beings. In 1906 Rubner[41] made a determination of bacterial mass as a function of time during some experiments on the spontaneous heat production which occurs during the souring of milk, and came to the conclusion that the greatest rate of heat production occurred when the bacteria were most actively growing. Rubner did not originate quantitative microbial microcalorimetry. Bouffard[33] did this. Rubner's great contribution consisted of his extensive development of the technique to the point where it came to be regarded as a practical tool.

2.4. THE EXPERIMENTS OF TANGL

It was inevitable that the technique of indirect calorimetry developed by Rubner should be used by other investigators. One of these was F. Tangl[43] who in 1908 published the results of his studies on the consumption of energy during the development of bacterial cultures. His method was to use bomb calorimetry to measure the heat of combustion of the dry substance in 100 ml of nutrient bouillon solution before growth began and after the experiment was terminated. The bacteria that had grown were then filtered from suspension and the heat of combustion of the dry substance in 100 ml of the filtrate was determined. The heat of combustion of the dry substance of the bacteria in 100 ml of suspension could then be determined by difference. From this he could obtain values for the heat of combustion of unit amounts (1 g) of dry bacterial cells, something that had first been done by Nencki[44] in 1884 when he found that the heat of combustion of dried bacteria was about 5 kcal g^{-1}. Tangl found that 1 g each of *Bacillus anthracis, Bacillus suipestifer,* and *Bacillus subtilis* have heats of combustion of 4.25, 4.04, and 4.65 g^{-1}, respectively, and that for the development of 1 g of dry bacterial cells of these species there were required 4.4, 11.9, and 7.9 kcal, respectively. By this was meant that during the synthesis of 1 g of cellular material these latter amounts of chemical energy were converted into heat. What Tangl was measuring for the first time in microorganisms was the amount of "work" (*specifische Entwicklungsarbeit*), expressed as heat, required for the synthesis of 1 g dry weight of bacterial cells. Since *Bacillus sp.* are aerobic, this represents the enthalpy change accompanying the catabolic, respiratory reactions of the bacteria that provide energy for synthesizing cellular material from the various substances present in the bouillon nutrient medium. It seems to be the first time that specific attention was paid to the energy requirements for the synthesis (growth) of microorganisms, rather than simply to the amounts of heat produced or to the rate of heat production from a microbial culture.

The efficiency with which the energy of the medium became incorporated into bacterial cells was found to be $(4.25/8.65) \times 100 = 49\%$ for *B. anthracis,* $(4.04/15.94) \times 100 = 25\%$ for *B. suipestifer,* and $(4.65/12.55) \times 100 = 37\%$ for *B. subtilis*. This seems like quite a spread of values for organisms of the same genus growing aerobically on the same substrate. In order to obtain aliquots of culture medium, culture, and filtrate for analysis Tangl placed 300–400 ml nutrient solution into Erlenmeyer flasks. These flasks were then autoclaved, cooled, inoculated, and placed in an incubator at 37–38°C. No mention is made of methods of stirring or aeration, and since these are aerobic bacteria it is likely that oxygen became limiting within a short period of time. That this may have been a possibility is indicated by the excessive periods of time required for the experiments, which lasted from 7 to 45 days.

Tangl seems to have come to no firm conclusions from his data. The importance of his research lies in his measurement of the thermodynamic efficiency of bacterial growth in terms of heat energy.

2.5. A. V. HILL'S DEVELOPMENT OF AN AUTOMATIC, DIFFERENTIAL MICROCALORIMETER

In 1911–1912 A. V. Hill[37] published a description of a new type of calorimeter that came to be used for many subsequent studies of microbial thermogenesis. Hill's calorimeter was of the "differential" type, which measures the temperature difference between two calorimeter vessels. These consisted of two narrow-necked, silvered Dewar flasks (Hill used thermos bottle replacements), one serving as an experimental vessel and the other as a control. A modified Hill calorimeter is illustrated in Fig. 2.6. For observing heat production during the growth or metabolism of microorganisms a given quantity of nutrient solution or cell suspension was placed in the experimental flask and an appropriate amount of water in the control flask. As microorganisms metabolized in the experimental flask they gave off heat. The *difference* in the temperature of the fluid in the experimental flask from that in the control flask was measured by the deflection of a sensitive mirror galvanometer connected to a copper–constantan thermocouple one junction of which was placed in each flask. Since both of the vessels were positioned next to one another, small temperature changes could be expected to affect both vessels equally and the effect of such temperature changes would therefore be eliminated. There was thus no need to have a complex thermostat for maintaining an exact, constant temperature as would be the case if only a single, experimental calorimeter vessel were used. Just as with Rubner's calorimeters there was some heat loss due to conduction and this was usually different for the experimental and control vessels. Hill devised an ingenious method of equalizing the loss due to heat conduction from each vessel by using different amounts of fluid in the control vessel such that the coefficient of conduction became the same for both. The calorimeter was calibrated electrically in the manner described by Rubner, and the temperature rise of heat production in calories plotted as a thermogram. In one of these Hill represented heat production during the souring of milk. It is immediately apparent that this type of data is of very limited use with respect to understanding heat production by microorganisms, even though it involves a phenomenon (the souring of milk) which is known to be associated with microbial activity. Although we know this is largely a lactic fermentation, we do not know what microorganisms were involved, what substances were being used as sources of carbon (i.e., carbohydrates or other substances in the milk) and in what quantity, and what substances were produced in what quantity as a result of the growth or metabolism of the microorganisms that were present. In short, the initial and the final states of the calorimetric system in this instance are largely unknown. Hill was undoubtedly aware of this, because the intent of his experiment with souring milk was to demonstrate that the amount of heat given off was for the most part in direct proportion to the extent to which whole milk was diluted. The same quantity of "pure" milk diluted by 50% with water gave off half the heat as did pure milk. A similar experiment was performed with fermenting yeasts. A 15% suspension of ordinary baker's yeast was incubated in the experimental vessel of the calorimeter together with 3.2% cane sugar and the heat production measured. In a second experiment a similar suspension of yeast was incubated together with 6.4% cane sugar. As shown in Fig. 2.4 the heat production in both experiments occurred at the same rate. The thermogram shows that there was a sharp change in the slope of the thermogram of the 3.2% sugar experiment at the level of about half the heat production which occurred at the level of a similar

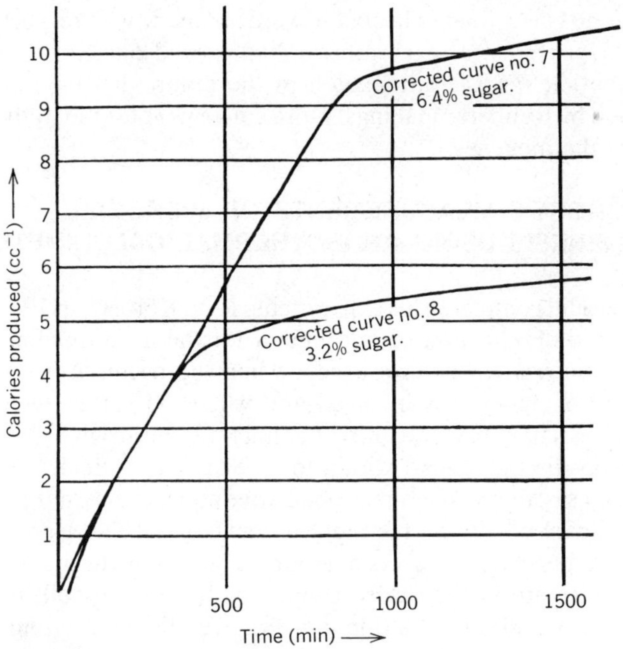

FIGURE 2.4. Hill's thermogram of identical suspensions of yeast fermenting sucrose at two concentrations. Reprinted from Reference 37 with permission.

sharp change in the slope of the thermogram of the 6.4% sugar experiment. Finally, the heat production did not stop at the points of slope change, but continued on at a much slower but more or less constant rate. Although not understood by Hill, in the light of what we know today all these observations are quite meaningful. The initial rates of heat production were the same because the same density of cell suspension was used for both experiments and the concentration of sugar was in excess in both experiments until just before the slopes change. Twice as much heat was produced in the presence of twice as much sugar. Finally, in all probability there was a certain amount of fermentative assimilation going on during the initial, fast phase of heat production which was due to the metabolism of the sugar in solution. This was followed by a slow fermentation of the stored carbohydrate after the sugar in solution was exhausted. Hill has thus provided us with an extremely interesting early demonstration of the physiology of fermenting yeast suspensions. We know what microorganism was carrying out the fermentation, we know what was being fermented and in what quantity, but we still do not know what was being produced, although this could be assumed to be carbon dioxide and alcohol. Also, we have only an observed heat of fermentation, even though Hill did correct for the heat loss due to conduction. What is missing was a better definition of the initial state, a complete definition of the final state, and the corrections for the heats of solution of the reactants and products which then became possible. However, Hill cannot be criticized for not having made these definitions and corrections, since these were not made with respect to microbial calorimetry for some 40 years after he had carried out his experiments. His real contribution was the automation of the differential type of calorimeter first invented by Rubner,[32] making it more convenient for the study of microbial (and other) thermogenesis.

2.6. v. KÖRÖSY'S MEASUREMENTS OF MICROBIAL THERMOGENESIS USING AN ISOTHERMAL CALORIMETER

Shortly after Hill completed his experiments K. v. Körösy[45] in 1913 constructed a completely different form of calorimeter for the measurement of heat production by bacteria. In nature, except when the bacterial decomposition of large masses of material is involved and when the surface/volume ratio of these masses is small, bacteria and other microorganisms live under essentially isothermal conditions at least within the life span of a single microorganism. This is not to say that a given microbial environment may not heat up during the day and cool off during the night, depending on the degree of exposure to the sun. In general, however, microorganisms living in the hydrosphere or the rhizosphere are in such close contact with water or soil, both of which have a high heat capacity, that the temperature of their environment is effectively constant. In the single-vessel microbial calorimeters used before 1913 the amount of heat produced was measured as a function of the rise in temperature of the fluid in the calorimeter, and these calorimeters were therefore not isothermal. It was v. Körösy who constructed the first isothermal calorimeter which could be used for the measurement of bacterial heat production. This was an experimental vessel on the top of which was a flat boiler containing diethyl ether. This latter was connected to three condensing bulbs which emptied into a graduated cylinder. The glassware containing the ether was sealed off from the atmosphere at atmospheric pressure. An evacuated, silvered glass envelope surrounded the experimental vessel. A diagram of the entire apparatus is shown in Fig. 2.5. The calorimeter was placed in a 850-liter thermostat at 35 ± 0.01°C, which is just above the boiling point of ether. Calorimetric measurements were made by introducing a nutrient solution into the calorimeter vessel at the exact temperature of the thermostat and inoculating this with the test microorganism. As the microorganisms grew or metabolized they produced heat, raising the temperature of the experimental vessel slightly. This simultaneously caused some of the ether to evaporate and distill over into the condensing bulbs which were at the slightly lower temperature of the thermostat. Here the ether condensed and dripped into the graduated cylinder, the amount of ether accumulating being directly related to the amount of heat produced by the microbial suspension. The temperature of the experimental vessel was kept constant by the removal of the heat of vaporization of the ether. The whole calorimetric system could be calibrated electrically.

v. Körösy carried out only three experiments with his calorimeter, in which he incubated a suspension of *Bacterium* (now *Pseudomonas*) *stutzeri* anaerobically in the presence of 1.0% KNO_3 and 1.0, 2.0, and 5.0% glucose, re-

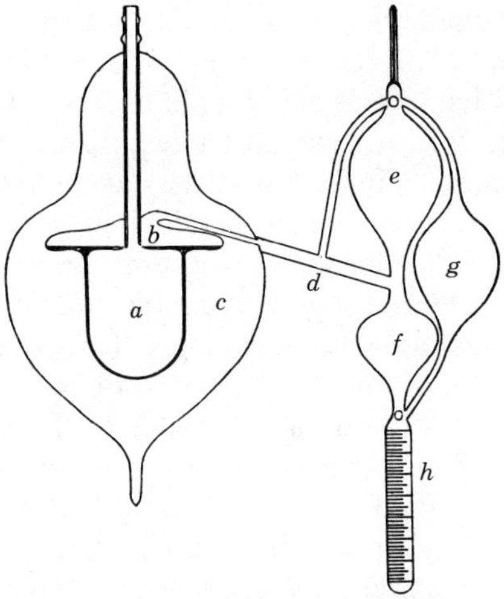

FIGURE 2.5. Diagram of von Körösy's isothermal microcalorimeter. *a,* calorimeter vessel; *b,* evaporator containing diethyl ether; *c,* evacuated chamber having a silvered inner surface; *d,* delivery tube; *e, f, g,* condensing bulbs; *h,* graduated receiving tube. The whole apparatus was made of glass and mounted in a thermostated water bath. Reprinted with permission from K. V. Körösy, Mikrokalorimeter zur Bestimmung der Wärmeproduktion von Bakterien, *Hoppe-Seyler's Zeitschrift für Physiologische Chemie* **LXXXVI,** 391 (1913).

spectively, finding that an average of 1.42 kcal of heat energy was produced per gram of nitrate reduced for all three sugar concentrations. *Pseudomonas stutzeri* is a nominally aerobic bacterium which is capable of anaerobic respiration in the presence of nitrate if suitable organic electron donors are provided. v. Körösy observed the disappearance of nitrate, the nitrogen of which was apparently reduced to some unspecified form. However, van Iterson[46] and van Niel and Allen[47] have stated that dentrification with P. *stutzeri* does not take place with a carbohydrate as the electron donor. Thus, we are not certain of the nature of the microorganism that v. Körösy used in his experiments. Unfortunately, as with the experiments of Hill,[37] not enough is known about the initial and final states of v. Körösy's experiments or even about the bacterium that produced the heat, and there is little of value in his data which can be used to increase our understanding of the energetics of microbial growth. In retrospect it seems unfortunate that v. Körösy's brilliantly conceived and elegantly designed isothermal calorimeter was never again used by him or anyone else. That it was isothermal for only one temperature may have restricted its usefulness and the necessity of a large thermostat may simply have made its use inconvenient.

2.7. SHEARER'S EXPERIMENTS ON HEAT PRODUCTION BY *BACILLUS COLI*

In 1921 C. Shearer[48] used a Hill calorimeter to measure the heat produced when *Bacillus* (now *Escherichia*) *coli* was grown in the presence of free amino acids. The major reasons why Hill[37] developed his design of the differential calorimeter were to avoid the necessity of working with complicated thermostats and to automatically eliminate the effects of variations in environmental temperature. That he was not completely successful in this is suggested by his statement that, ". . . if available, a cellar is the best place in which to conduct these experiments. Firstly it is far less liable to vibration and steadier for the galvanometer: secondly a cellar undergoes no very wide temperature variations." (Hill,[37] p. 276.) Shearer solved this latter problem by putting his calorimeter back into a thermostat. Stirring the contents of the experimental vessel was accomplished by gently shaking it for a short time before each reading and at appropriate intervals, since bubbling air through it, as recommended by Hill, caused considerable frothing of the nutrient solutions which were used. Five experiments were carried out using nutrient solutions containing peptone, glucose, and a broth of casein that had been digested with pancreatic extract for varying periods of time. After inoculating 400 cc of a given medium with 1 g dry-weight equivalent of a living *E. coli* suspension in saline solution, the heat production was measured for 24 hr. Shearer determined that growth of *E. coli* was taking place during these experiments by comparisons of the dry weights of bacteria present before and after heat production measurements were made. However, it is curious that a plot of galvanometer scale readings against time shows that in each experiment for which this was done the temperature rise was arithmetic rather than logarithmic, as it would have been expected to be. The reason for this is not clear, but it does illustrate the difficulty of microbial microcalorimetry. Obviously, something was not properly corrected. Shearer's experiments tell us that *E. coli* produces heat when it is grown on amino acids or on proteinaceous substances, and little else.

2.8. THE STUDIES BY JAMES AND HIS COLLABORATORS ON SPONTANEOUS HEATING

In an attempt to learn more about the nature of the "spontaneous heating" that is exhibited by uncured farm products L. H. James[49] in 1927 devised an apparatus for the measurement of spontaneous heat production. The experiments of previous investigators had been directed toward determining the number of calories, either totally or per unit time, which were produced as the result of the activities of microorganisms that were metabolizing various nutrients in solution aerobically or anaerobically. James was more interested in determining the maximum temperature to which farm products would heat aerobically. His device was a wide-mouthed Dewar Flask embedded in a wooden box filled with powdered cork which served as insolation. The material to be tested was placed in a cylindrical glass receptacle, together with the bulb and part of the stem of the thermometer and a glass tube extending to the bottom of the receptacle through which pure oxygen gas (so as not to remove too much heat) could be slowly passed. The filled receptacle was next stoppered with cotton and placed in the Dewar flask, the top of the receptacle then being further insulated from the environment by shredded asbestos fiber. No effort seems to have been made to correct for heat loss due to conduction. James found that unsterilized commercial cornmeal and cracked yellow field corn could spontaneously heat to 60°C, and that marked heating did not take place in the absence of oxygen.

One year later, in 1928, James, Rettger, and Thom[50] published a continuation of James's work in which they investigated the optimum moisture content of cracked corn for the production of heat, and then attempted to confirm that microbial development does indeed occur concomitantly with heat production. In general they found that as the temperature in the experimental containers rose from 50 to 55°C the number of living bacteria markedly decreased. Experiments were also carried out with sterilized cracked corn that had been inoculated with individual cultures of 12 strains of microorganisms isolated from aerated cracked corn that had spontaneously heated. These included isolates of *Bacillus subtilis, Bacillus vulgatus, Proteus vulgaris,* and *Aspergillus fumigatus.* Suspensions of the test organisms were made in sufficient water to provide a moisture content in the corn of 30–35%. The inoculated corn was then packed into the receptacle of the thermogenesis-measuring device described by James.[49] Four strains of *B. subtilis,* one of *B. vulgatus,* and one of *P. vulgaris* were found to be highly thermogenic, producing temperatures from 49 to 57°C. The other isolates were not markedly thermogenic. Fresh manure, when placed in James's device, produced temperatures of 71.0–73.5°C, although no heating was observed in the absence of oxygen.

An additional study was carried out by Stuart and James[51] in 1928 in which they studied the effect of salting on the microbial heating of alfalfa hay, finding that the addition of salt up to 5% at a moisture content of 30% failed to inhibit heat production, although this was considerably delayed in starting. Bacterial growth was inhibited at 2% salt, but this failed to prevent mold growth.

The work of James and his collaborators, along with that of other investigators at about the same time, is important because it established conclusively that spontaneous heating to at least 60°C was possible due to the growth and metabolic activity of microorganisms.

2.9. THE DETERMINATIONS OF BAYNE-JONES AND RHEES REGARDING HEAT PRODUCTION DURING THE PHASES OF GROWTH OF BACTERIA

One of the difficulties which is apparent with respect to the quantitative measurement of the heat produced by growing and metabolizing microbial cultures is that those investigators who made such measurements were for the most part animal or plant physiologists who had little interest in the niceties of pure-culture technique. The microcalorimeters of Hill[37] and Shearer[48] were therefore not suitable for growing pure cultures of microorganisms or for properly aerating these when necessary. These deficiencies were overcome by S. Bayne-Jones[52] in 1929 when he drew attention to these problems and modified Hill's calorimeter so as to eliminate them. Bayne-Jones found that it was possible to insert a large-diameter, thin-walled glass cylinder serving as a reaction vessel into each of the thermos bottles comprising the differential calorimeter, much as was done by James[49] with his single-vessel device. These could each be removed, filled with suitable quantities of nutrient medium or of water, closed with a rubber stopper through which three tubes were passed, and autoclaved to sterilize the contents of the vessel. Just as Shearer[48] had done, Bayne-Jones found that bubbling air through the medium for the purpose of stirring caused froth to form, and in addition caused excessive evaporation and irregular distribution of bacteria within the liquid. Thus, one of the tubes through the rubber stopper was a thick-walled capillary tube through which passed a fine wire connected to a glass stirring bulb partly filled with mercury which had been sealed into it. After autoclaving, the upper part of the capillary tube was filled with vaseline to exclude the entrance of bacteria and to prevent evaporation. During an experiment a small ring at the upper end of the wire was connected to the arm of an automobile windshield wiper which, when oscillating, raised and lowered the bulb, thus stirring the contents of the vessel. A second tube simply passed through the rubber stopper into the reaction vessel and was sealed off from the environment with a cotton stopper. Inoculation of the nutrient medium could be made through this tube, and samples withdrawn through it by means of a sterile capillary pipette. The third tube was sealed off at the bottom and filled with a small amount of water. This served as a receptacle for one junction of the thermocouple, which could be inserted after the vessel with its contents had been autoclaved. Once each vessel had been placed into its respective thermos bottle, the thermos bottle itself was closed with a second rubber stopper. The calorimeter consisted of two of these units connected by a thermocouple, one unit serving as an experimental vessel and one as a control. A diagram of the calorimeter is given in Fig. 2.6. The whole apparatus was placed in a water bath the temperature of which was maintained within ±0.01°C. The method of ensuring that the thermal conductivity from each side of the calorimetric system is the same was that described by Hill, as was the method of calibration. An accuracy of about ±3% with respect to heat measurements could be obtained with an input of 23 cal hr^{-1}, as compared with Hill's original calorimeter, which had an accuracy of about 9% with an input of 37–52 cal hr^{-1}.

Bayne-Jones's calorimeter was used by Bayne-Jones and Rhees[53] to study the relation of heat production to phases of growth in bacteria. Many authors had previously published thermograms but had not attempted to correlate changes in the slopes of the curves with anything specific. For example, Hill's[37] thermogram representing the souring of milk shows that there was a relatively low rate of heat production during the first 9 to 10 hr followed by a much

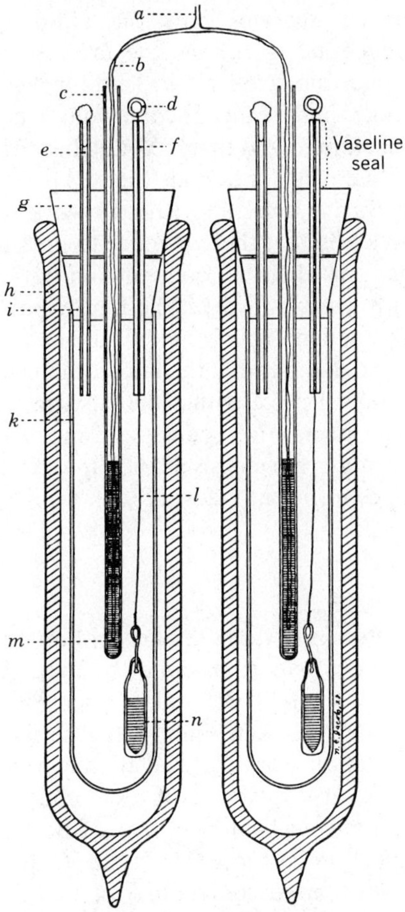

FIGURE 2.6. Bayne-Jones's differential microcalorimeter. Reprinted from Reference 52 with permission.

higher rate during the next 5 hr, after which it slowly declined to zero within the next 5 to 6 hr. Hill made no attempt to measure the concentration of bacteria as a function of time, but an inspection of his curves suggests that the heat production was logarithmic, at least for the first 12 or 13 hr, after which it declined.

The experiments of Bayne-Jones and Rhees consisted of inoculating 1 ml of a 12–18 hr *Bacterium coli* broth culture into an experimental culture vessel containing 100 ml of a solution of 2% peptone alone or supplemented with sodium chloride or glucose. The experiments were carried out at 37.5°C. Samples were withdrawn at regular intervals to determine the number of bacteria per milliliter so that correlations could be made between the phases of growth and the rate of heat production. The data they obtained are most confusing.

An inspection of their thermograms shows that although the increase in bacterial numbers is in general logarithmic the increase in heat production is arithmetic, and this makes no sense at all. Little in their data warrants their conclusion that, "Changes in the rate of heat production are related to changes in the growth rate of bacteria," and that, "Young bacterial cells produce more heat than older ones." (Bayne-Jones and Rhees,[53] p. 140.)

The importance of Bayne-Jones and Rhees's work is as much negative as positive. It illustrates clearly that there is more to measuring heat production by growing microorganisms than the construction of a sensitive, nicely calibrated calorimeter with which to work. The purely physical measurement of heat production is not enough. This is merely the observed heat of reaction. It is also necessary to understand what the microorganisms are doing and what chemical changes are taking place within the calorimetric system which may be extraneous to the production of heat by the microorganisms themselves. The early investigators of thermogenesis did not appreciate this largely because their science was not yet sufficiently advanced.

2.10. SUMMARY

In describing the early studies it was necessary for brevity's sake to select those that seemed to be the most important either because of their chronology or because of the type of work that was done. Many investigators during the period 1856–1931 have been left out simply because they repeated in greater detail or with other substances work that had already been established. Among these are the extensive investigations of Boekhout and DeVries[54–58] and Hildebrandt[59,60] on spontaneous heating in hay, of Boekhout and DeVries[61] in tobacco, and of Surauer[62] in cotton waste. Except for the studies of Calvet and his collaborators which are described in Section 7.5, it is not necessary to deal further with the subject of heat production by microorganisms growing on or in complex substances such as manure, straw, fodder, beer wort, wine must, and peptone. Thermogenesis with respect to these substances has been conclusively demonstrated. Furthermore, not much can be learned concerning the microbial energetics of systems involving such substances because we know so little about their initial and final states. Even though a given mass of manure or straw, for example, does indeed have a free energy, enthalpy, and entropy of formation there would be little point in trying to establish these values. For this reason it is far more convenient to deal with systems for the study of microbial growth which involve a single, thermochemically defined substance as the source of carbon and energy. The early approaches to these studies are described in Chapter 3.

REFERENCES

1. M. Dubrunfaut, Note sur la chaleur et le travail méchanique produit par la fermentation vineuse, *C. R. Hebd. Séances Acad. Sci.* **42,** 945–948 (1856).

2. L. Pasteur, Remarques sur une classe de phénomènes de décomposition s'effectuant avec dégagement de chaleur, *C. R. Hebd. Séances Acad. Sci.* **59,** 689 (1864).
3. L. Pasteur, Faits nouveaux pour servir à la connaissance de la théorie des fermentations proprement dites, *C. R. Hebd. Séances Acad. Sci.* **75,** 784–790 (1872).
4. L. Pasteur, Nouvelles observations sur la nature de la fermentation alcoolique (Réponse à *MM.* Brefeld et Traube), *C. R. Hebd. Séances Acad. Sci.* **80,** 452–457 (1875).
5. L. Pasteur, Sur la théorie de la fermentation, *C. R. Hebd. Séances Acad. Sci.* **87,** 125–128 (1879).
6. M. Berthelot, Réponse à la communication de *M.* Pasteur., *C. R. Hebd. Séances Acad. Sci.* **87,** 128–129 (1878).
7. L. Pasteur, Nouvelle communication au sujet des notes sur la fermentation alcoolique, trouvées dans les papiers de Cl. Bernard, *C. R. Hebd. Séances Acad. Sci.* **87,** 185–188 (1878).
8. M. Berthelot, Observations à la suite de la communication de *M.* Pasteur, *C. R. Hebd. Séances Acad. Sci.* **87,** 188–189 (1878).
9. L. Pasteur, Examen critique d'un écrit posthume de Claude Bernard sur la fermentation alcoolique., *C. R. Hebd. Séances Acad. Sci.* **87,** 813–819 (1878).
10. M. Berthelot, Observations sur la note de *M.* Pasteur, relative à la fermentation alcoolique, *C. R. Hebd. Séances Acad. Sci.* **87,** 949–952 (1878).
11. L. Pasteur, Réponse à *M.* Berthelot, *C. R. Hebd. Séances Acad. Sci.* **87,** 1053–1058 (1878).
12. M. Berthelot, Réponse à *M.* Pasteur, *C. R. Hebd. Séances Acad. Sci.* **88,** 18–20 (1879).
13. L. Pasteur, Deuxième réponse à *M.* Berthelot, *C. R. Hebd. Séances Acad. Sci.* **88,** 58–61 (1879).
14. M. Berthelot, Observations sur la deuxième réponse de *M.* Pasteur, *C. R. Séances Acad. Sci.* **88,** 103–106 (1879).
15. L. Pasteur, Troisième réponse à *M.* Berthelot, *C. R. Hebd. Séances Acad. Sci.* **88,** 133–137 (1879).
16. M. Berthelot, Remarques sur la troisième réponse de *M.* Pasteur, *C. R. Hebd. Séances Acad. Sci.* **88,** 197–201 (1879).
17. L. Pasteur, Quatrième réponse à *M.* Berthelot, *C. R. Hebd. Séances Acad. Sci.* **88,** 255–261 (1879).
18. F. Cohn, Ueber thermogene Wirkung von Pilzen, *Schles. Ges. vaterl. Cultur. Jahr.* **66,** 150–156 (1888).
19. F. Cohn, Ueber thermogene Bakterien, *Ber. dtsch. bot. Ges.* **12,** 66–69 (1893).
20. E. Suchsland, Beobachtung ueber die Selbsterwärming des fermentirenden Tabaks., in Festschr. D. Latina, Univ. Halle-Wittenberg, 1894.
21. J. Behrens, Der Ursprung des Trimethylamins in Hopfen und die Selbsterhitzung desselben, *Inst. Karlsruhe* **1,** 187–200 (1897).
22. J. Behrens, Ueber die oxydierenden Bestandteile und die Fermentation des deutschen Tabaks, *Centralbl. Bakt. Parasit., Abt. II,* **7,** 1–12 (1901).
23. F. Holdefleiss, Weitere Untersuchungen ueber den Einfluss der Gärung auf den Wert des Heues, *Mitt. landwirt. Inst. Konig, Univ. Breslau* **1,** 59–74 and 233–249 (1899).

24. O. W. Hunter, Microorganisms and heat production in silage fermentation, *J. Agric. Res.* **10,** 75–83 (1917).

25. H. Ranke, Experimenteller Beweiss der Möglichkeit der Selbstenzundung des Heues (Grummets), *Ann. Chem.* **167,** 361–368 (1873).

26. H. M. Cottrell, *Spontaneous combustion of alfalfa, Exp. Sta. Bull.* **109,** Kansas Agric. Coll. (1902).

27. G. Laupper, Die neusten Ergebnisse der Neubrandforschung, *Landw. Jahrb. Schweiz* **34,** 1–54 (1920).

28. E. P. Heaton, Spontaneous combustion in hay mows and stacks, Fire Marshal Office, Toronto, Canada, 1921.

29. H. Schwartz, Brandursache unbekannt, *Schweizerischen Feuerwehrzeitung,* 1–16, Bern (1921).

30. H. Miehe, *Die Selbsterhitzung des Heues, Jena,* 1–127 (1907).

31. M. Rubner, *Die Gesetze des Energieverbrauchs bei der Ernährung*, Franz Deuticke, Leipzig und Wein, 1902.

32. M. Rubner, Ueber die Wärmebildung durch Mikroorganismen und über die Methodik einer quantitativen Wärmemessung, *Hyg. Rund.* **13,** 857–864 (1903).

33. A. Bouffard, Détermination de la chaleur dégagée dans la fermentation alcoolique, *C. R. Hebd. Séances Acad. Sci.* **121,** 357–360 (1895).

34. A. J. Brown, The heat of fermentation, *Inst. Brew., London, J.* **7,** 93–103 (1903).

35. M. Rubner, Die Kalorimetrie., in *Handbuch der physiologischen Methodik,* Robert Tigerstedt, ed., S. Hirzel, Leipzig, 1911, pp. 150–228.

36. M. Rubner, *Die Ernährungsphysiologie der Hefezelle bei alkoholischer Gärung,* Veit & Comp., Leipzig, 1913.

37. A. V. Hill, A new form of differential microcalorimeter, for the estimation of heat production in physiological, bacteriological, or ferment actions, *J. Physiol.* **43,** 261–285 (1911–1912).

38. M. Rubner, Die Umsetzungswärme bei der Alkoholgärung, *Arch. Hyg.* **49,** 355–418 (1904).

39. M. Rubner, Energieverbrauch im Leben der Mikroorganismen, *Arch. Hyg.* **48,** 260–311 (1904).

40. M. Rubner, Energieverbrauch im Leben einiger Spaltpilze, *Arch. Hyg.* **57,** 193–242 (1906).

41. M. Rubner, Ueber spontan Wärmebildung in Kuhmilch und die Milksäuregärung, *Arch. Hyg,* **57,** 244–268 (1906).

42. L. Pasteur, Memoire sur la fermentation alcoolique, *Ann. Chem. Phys.* **58,** 323–426 (1860).

43. F. Tangl, Beiträge zur Energetik der Ontogenese. Zweite Mittheilung. Ueber den Verbrauch an Chemischer Energie während der Entwicklung von Bakterienculturen, *Pfluger's Arch. gesamte Physiol. Menschen Teire* **98,** 475–489 (1908).

44. M. Nencki, Über das Eiweiss der Milzbrandbacillen, *Ber. dtsch. chem. Ges.* **17,** 2605–2609 (1884).

45. K. von Körösy, Mikrokalorimeter zur Bestimmung der Wärmeproduktion von Bakterien, *Z. physiol. Chem.* **86,** 383–400 (1913).

46. G. Van Iterson, Jr., Accumulation experiments with dentrifying bacteria, *Proc. Sec. Sci. Koninkl. Akad. Wetenschap. Amsterdam* **5,** 148–162 (1902).

47. C. B. Van Niel and M. B. Allen, A note on *Pseudomonas stutzeri.*, *J. Bact.* **64,** 413–422 (1952).
48. C. Shearer, On the amount of heat liberated by *Bacillus coli* when grown in the presence of free amino acids, *J. Physiol.* **55,** 50–60 (1921).
49. L. H. James, Studies in microbial thermogenesis, I. Apparatus, *Science* (*Wash. D.C.*) **65,** 504–506 (1927).
50. L. H. James, L. F. Rettger, and C. Thom, Microbial thermogenesis. II. Heat production in moist materials with reference to the part played by microorganisms, *J. Bact.* **15,** 117–141 (1928).
51. L. S. Stuart and L. H. James, Studies in microbial thermogenesis. III. The effect of salting on the microbial heating of alfalfa hay, *J. Bact.* **19,** 53–54 (1930).
52. S. Bayne-Jones, Bacterial Calorimetry I. General considerations. Description of differential calorimeter, *J. Bact.* **17,** 105–122 (1929).
53. S. Bayne-Jones and H. S. Rhees, Bacterial Calorimetry II. Relationship of heat production to phases of growth of bacteria, *J. Bact.* **17,** 123–140 (1929).
54. F. W. J. Boekhout and J. J. O. de Vries, Ueber die Selbsterhitzung des Heues, *Zentral. Bakt. Parasitenkd. Infektionskr. Hyg., Zweite naturwiss. Abt. allg. landwirtsch. tech. Mikrobiol.* **12,** 675–681 (1904).
55. F. W. J. Boekhout and J. J. O. de Vries, Ueber de Selbsterhitzung des Heues, *Zentral. Bakt. Parasitenkd. Infektionskr. Hyg., Zweite naturwiss. Abt. allg. landwirtsch. tech. Mikrobiol.* **15,** 568–573 (1905).
56. F. W. J. Boekhout and J. J. O. de Vries, Ueber die Selbsterhitzung des Heues, *Zentral. Bakt. Parasitenkd. Infektionskr. Hyg., Zweite naturwiss. Abt. allg. landwirtsch. tech. Mikrobiol.* **21,** 398–407 (1908).
57. F. W. J. Boekhout and J. J. O. de Vries, Ueber die Selbsterhitzung des Heues, *Zentral. Bakt. Parasitenkd. Infektionskr. Hyg., Zweite naturwiss. Abt. allg. landwirtsch. tech. Mikrobiol.* **23,** 106–108 (1909).
58. F. W. J. Boekhout and J. J. O. de Vries, Ueber die Selbsterhitzung des Heues, *Zentral. Bakt. Parasitenkd. Infektionskr. Hyg., Zweite naturwiss. Abt. allg. landwirtsch. tech. Mikrobiol.* **44,** 290–304 (1915).
59. F. Hildebrandt, Beiträge zur Frage der Selbsterwärming des Heues, *Zentral. Bakt. Parasitenkd. Infektionskr. Hyg., Zweite naturwiss. Abt. allg. landwirtsch. tech. Mikrobiol.* **71,** 440–490 (1927).
60. F. Hildebrandt, Neue Untersuchungen zur Frage der Selbsterhitzung des Heues, *Fortschr. Landwirtsch.* **2,** 414–417 (1927).
61. F. W. J. Boekhout and J. J. O. de Vries, Über Tabaksfermentation, *Zentral. Bakt. Parasitenkd., Sect. II* **24,** 496–511 (1909).
62. H. Surauer, Ueber einem aus baumwollabgefallen isolierte thermophilen Bacillus, *Beitr. Biol. Pflanz.* **16,** 1–38 (1928).

3

EARLY STUDIES ON MICROBIAL THERMOGENESIS RESULTING FROM THE UTILIZATION OF SPECIFIC SUBSTRATES, 1856–1932

Most of the early studies on microbial thermogenesis involved systems in which microorganisms utilized nonspecific sources of carbon and energy during their growth or metabolism. Examples of these have been given in Chapter 2. A few of the early studies dealt with systems which were better defined in that (1) known microorganisms were used; (2) the microorganisms were known to be actively growing; and (3) specific substrates were known to serve as sources of carbon and energy, being sugars which are found as components in complex, largely undefined media such as wine must or beer wort. These experiments were on alcoholic fermentation, and because the environment was anaerobic only the sugars could be used as sources of energy. It was generally thought then that some of the sugar was also used to form the substance of the yeast cells that were grown. Later investigations involved aerobic systems in which a single substance served as a source both of carbon and of energy, or of energy alone. There was thus slow progress toward defining more clearly the initial and the final states of these systems. These developments will be discussed in this chapter.

3.1. DUBRUNFAUT'S MEASUREMENT OF THE HEAT OF FERMENTATION

The first measurement of the amount of heat produced during a fermentation (in this case an alcoholic fermentation) was published by M. Dubrunfaut,[1] who must surely be credited with having carried out the most massive experiment of this sort which has ever been done. The vessel in which he performed the fermentation was a large oak vat in the shape of a truncated cone having an average diameter of 3 m and being of the same depth. The wood itself had a volume of 1.4 m^3 and a weight of 1120 kg. In it he placed 21,400 liters of a molasses solution containing 2559 kg of crystallizable cane sugar to which enough yeast inoculum had been added that the fermentation could be completed in four days. The amount of this latter was not more specifically stated, although it must have been considerable! The fermentation itself was thought to be represented by the Gay-Lussac equation, showing that 4 moles of carbon dioxide and 4 moles of ethanol are produced for every mole of sucrose fermented, as follows:

$$C_{12}H_{22}O_{11} + H_2O \rightarrow 4CO_2 + 4CH_3CH_2OH$$

During the course of the fermentation the temperature of the vat rose from 23.7 to 33.7°C, an average of which was considerably above the ambient temperature, which varied from 12 to 16°C. Dubrunfaut corrected for the loss of heat from the vat due to radiation and convection by measuring the rate at which the vat cooled after fermentation had ceased, and calculating the heat loss by means of Newton's Law of Cooling. Corrections were also made for the heat absorbed by the wood in the vat, for the negative heat of solution of the carbon dioxide which passed from solution in the fermentation liquor into the gas phase, for the heat equivalent of the mechanical work done by the expansion of carbon dioxide gas against the atmosphere, and for the evaporation of water from the vat which occurred during the experiment. Other minor corrections were also made. Dubrunfaut's experiment thus had considerable sophistication considering the year in which it was done. In retrospect we can now make the observation that the corrected value of 339,450 kcal which he obtained for the fermentation of 2559 kg of sucrose is lower than that which would be expected on the basis of the Gay-Lussac equation by about 38%. There are several reasons why he might have obtained a lower value, but the main reason is probably that the inoculated yeast cells *grew* in the medium he used, using as their carbon source some of the sucrose which disappeared.

If the quantity of sucrose utilized is converted into equivalents of glucose, the 14,964-mole equivalents of fermented glucose divided into the $-339{,}450$ kcal of heat that he measured gives a value of -22.7 kcal $mole^{-1}$ of glucose fermented as the heat of fermentation.

Because of the deficit in the amount of alcohol which Dubrunfaut obtained from his experiment compared with that to be expected on the basis of the Gay-Lussac equation, he felt compelled to write, "[There is] an almost constant analogous deficit in all fermentations in the laboratory and in the workshop [perhaps equally well translated as brewery or winery]." (Dubrunfaut,[1] p. 946.) He then states in footnotes that, "It is this fact which has authorized us to affirm that one is not able to make experimentally the [Gay-Lussac] equation of fermentable sugars with alcohol and carbonic acid." (Dubrunfaut,[1] p. 946), and "The ratio of carbonic acid to alcohol also often differs strongly from that which is given by the chemical formulas [again referring to the Gay-Lussac equation]. It varies above all with the ratio of ferment to sugar." (Dubrunfaut,[1] p. 946.) Dubrunfaut is making here a strong statement to the effect that an alcoholic fermentation according to the Gay-Lussac equation is, in fact, seldom observed. The habit with subsequent experimenters of using large quantities of yeast to initiate their fermentations continued to confound their results for many decades.

3.2. BOUFFARD'S CALORIMETRIC DETERMINATION OF THE HEAT OF FERMENTATION

The next attempt at measuring the heat of fermentation was not carried out until 1895, when A. Bouffard[2] performed what must be considered, even by today's standards, a remarkably accurate series of determinations. Finding that an "artificial" (defined?) medium gave too slow a fermentation, he decided to use grape juice containing about 130 g of sugar and supplemented with 1 g of ammonium phosphate per liter. One liter of medium was then inoculated with a pure culture of yeast and placed in a 1200-ml cylindrical tin bottle. A calorimetric thermometer, to which a perforated, plate-shaped agitator made of tin was horizontally attached, was also placed in the bottle. By this means if the thermometer were gently moved up and down, the medium could be mixed and the temperature recorded with a minimum of external thermal disturbance to the system. After the fermentation had become active the bottle with its contents was weighed and placed in a double-walled container which retarded heat loss to the environment. A double-walled cover having in it two holes was then placed over the container. One hole permitted the thermometer to pass through. A tube through the other hole permitted the collection of carbon dioxide gas. Temperature measurements were made every 5 min. After an experiment was terminated the specific heat of the system was determined and the bottle with its contents again weighed, so that the amount of water lost by evaporation could be determined by difference. Corrections were made for the heat of vaporization of some of the alcohol formed during the fermentation, for loss of heat to the environment, for the evaporation of water from the calorimeter, and for carbon dioxide gas passing into the atmosphere. A potential source of error may have been introduced in that the weight of sugar fermented was calculated on the basis of the weight of carbon dioxide produced according to a ratio established by Pasteur[3] which took into account the formation of small amounts of glycerol and succinic acid in addition to carbon dioxide and ethanol. Nevertheless, four experiments resulted in values of -23.7, -23.5, -23.6, and -23.4 kcal mole^{-1} of glucose fermented, with an average of -23.5 kcal, in good agreement with Dubrunfaut's value of -22.7 kcal mole^{-1}.

By the time Bouffard carried out his biological determination of the heat of fermentation work had also been done by two chemists toward determining what the heat of fermentation could be expected to be on the basis of straight chemistry, using the Gay-Lussac equation. Berthelot[4] was the first to do this, finding that whereas 1 mole of glucose yielded −713 kcal on combustion, the 2 moles of alcohol produced from it during fermentation yielded only −642 kcal, the difference of −71 kcal being attributed to the heat of fermentation. Von Rechenberg[5] followed a different approach. Using known values for the heats of combustion of the elements concerned and for glucose, he arrived at a value of −269 kcal $mole^{-1}$ for the heat of formation of glucose. A similar method was used to calculate the heat of formation of ethanol as −74 kcal $mole^{-1}$. The heat of formation of carbon dioxide was measured directly as −94 kcal $mole^{-1}$. He used these values to calculate that the heat of fermentation according to the Gay-Lussac equation was −67 kcal $mole^{-1}$. This was closer to the real value but still in error due to technical difficulties with the bomb calorimeters then in use.

In 1895 Berthelot[6] made what he believed to be more accurate determinations of the heats of combustion of glucose and ethanol, and on the basis of these calculated the heat of fermentation to be −32.07 kcal $mole^{-1}$ of glucose fermented. Curiously, in spite of the skill and care with which Bouffard conducted his measurements and the remarkable precision of his results, he seems to have been hesitant to accept his own results in the light of Berthelot's calculations. Instead he suggested that the true heat of fermentation probably lies somewhere between the value he obtained and that obtained by Berthelot. There was no reason to mistrust his results simply because they were not in agreement with those of the great chemist. Bouffard was apparently unaware of Dubrunfaut's experiment. Had he not been so he might have been more trusting of his own results. As we now know Bouffard's results could never have agreed with those of Berthelot because Bouffard was working with an aqueous biological system, and Berthelot was working with a system involving substances in their standard states. If nothing but alcoholic fermentation were taking place, the ΔH° for the system Berthelot studied is now known to be −16.24 kcal $mole^{-1}$ of glucose fermented, whereas for the system where all substances are in their aqueous standard states the $\Delta H^{\circ\prime}$ is −33.12 kcal $mole^{-1}$. If the yeasts are actually growing on the glucose they ferment, the $\Delta H^{\circ\prime}$ drops to about −23 kcal $mole^{-1}$, which is what Dubrunfaut and Bouffard observed. Quite certainly the composition of the medium in Bouffard's calorimeter would have permitted the growth of yeasts. However, the extent to which these yeasts might have used the sugar in the medium as a carbon source or other, non-fermentable substances introduced with the grape juice, is not known. The heat of fermentation would be expected to be considerably higher in the latter than in the former case.

3.3. THE MEASUREMENT BY A. J. BROWN OF THE HEAT OF FERMENTATION

One of the most serious difficulties associated with the early experiments in biological calorimetry was heat loss to the environment, since such experiments were usually done at higher than ambient temperatures. For example, in Bouffard's determinations this correction alone amounted to about 25% of the observed heat of fermentation. It occurred to A. J. Brown[7] that by using a brewing vat this difficulty could be minimized. In an article published in 1903 he pointed out that heat exchange can take place only through a surface, and that by using a large fermentation vessel the surface-to-volume ratio could be considerably decreased. The vessel he used measured 9 ft long × 4 ft wide × 6 ft high. It thus contained only about one-fourth of the fluid volume of Dubrunfaut's vat. There was no cover on Brown's vat because he felt that the head of foam which develops when beer is actively fermenting would provide a good insulation. Otherwise, the vat was made of thin copper metal encased in wood about 1 in. thick. Beer wort largely containing maltose as the fermentable sugar rather than glucose, which is found in grape juice, was fermented. Thus, the hydrolysis of maltose would be expected to result in the production of a small amount of heat which should be applied as a correction to make the fermentation of beer wort and grape juice comparable. Brown knew this, but could not make the correction because he did not know the heat of hydrolysis of maltose. The amount of carbon dioxide produced was measured as a function of the loss in specific gravity of the wort, since carbon dioxide was the only substance to be separated from the solution during fermentation. Corrections were made for evaporation of water from the vat and for heat lost by cooling. To be able to compare his result with Bouffard's, Brown made his calculations on the basis of moles of glucose fermented, rather than maltose, thus obtaining a value of -21.4 kcal mole^{-1} equivalent of glucose fermented.

3.4. THE EXPERIMENTS OF NAWIASKY ON THE TRANSFORMATION OF AMINO ACIDS BY *PROTEUS VULGARIS*

After Rubner[8] finished his calorimetric studies on the lactic fermentation of milk the only person ever again to use a Rubner calorimeter was P. Nawiasky, one of Rubner's assistants. In his 1908 studies of the decomposition of amino acids by *Proteus vulgaris,* Nawiasky[9] pointed out that all of the previous studies on the energy metabolism of bacteria had been carried out with highly complex materials such as peptone and albumin. These substances were first broken down to simpler substances which were then utilized by the bacteria, but it was not known specifically which these were. For the most part Nawiasky's studies did not involve growth, but rather the incubation of large quantities of *Proteus* with solutions of various individual amino acids. During the incubation periods the bacterium would decompose the amino acids to produce various substances depending on the nature of the amino acid. The products were determined when the incubation period was over. In some cases the amino acid was completely oxidized to carbon dioxide, water, and ammonia. In other cases additional substances were also formed. In one experiment three calorimeters were filled with 250 ml of 5% asparagine solution and with 2, 4, and 8 g (dry weight?), respectively, of the test organism. The maximum temperature rise within these calorimeters occurred at about 15, 7, and 3 hrs, respectively, which is roughly what one would expect if the rate of reaction were proportional to the quantity of bacteria carrying out the reaction. Nawiasky also found that carbon dioxide, ammonia, acetic acid, and succinic acid were produced during the incubation. The experiments demonstrated conclusively that bacteria could produce significant amounts of heat as a result of the metabolism of a single amino acid. Two experiments with growing cultures were carried out, but not calorimetrically. In one experiment a solution consisting of 0.5% asparagine, 0.5% NaCl, 0.2% bipotassium phosphate, and 0.05% magnesium sulfate was inoculated with a loopful of a pure culture of *P. vulgaris,* the result being no growth within four weeks. In a second experiment with 2.5% asparagine in the same mineral solution and with a larger inoculum, the result was no better. Thus, although this bacterium could well metabolize asparagine, it could not use it as a sole source of carbon and energy in the process of growth.

Nawiasky's experiments are important in that they represent a recognition of the need for a more simple system with which to study growth and metabolism in bacteria. He was also the first to combine chemical studies on the metabolism of a single substance in a system in which growth was not involved, with a measurement of the concomitant heat production.

3.5. THE MEASUREMENTS BY H. T. BROWN OF THE HEAT OF FERMENTATION

In a series of studies on yeasts published in 1914 H. T. Brown[10] repeated A. J. Brown's[7] experiments and determined the heat of fermentation of maltose using, however, even larger vessels varying in volume from 1400 to 14,000 gallons, and estimating the amount of sugar fermented by estimating the amout of alcohol formed. The result of 12 experiments was that the heat of fermentation of 1 g of maltose, including its heat of hydrolysis, was about -125 cal g^{-1}. This is equivalent to -22.5 kcal $mole^{-1}$ of glucose fermented, almost duplicating Dubrunfaut's value obtained 58 years earlier. Brown went on to measure calorimetrically the specific rate of yeast fermentation as a function of temperature and commented succinctly on the large liberation of energy which accompanies the metabolic processes of cells. He concluded that the specific rate of heat production by a suspension of fermenting yeast cells at 37°C was about 70 times that of a man at rest. Brown then asked what the physiological significance was of the extremely high metabolism, made evident by this enormous liberation of heat energy which the yeast cells exhibited while pursuing their life functions, seemingly so disproportionate to their needs for nutrition and reproduction. Could this phenomenon be regarded as being accidental "due to the fact that a living yeast cell requires some small but essential part of the sugar molecule, the removal of which brings about a rearrangement of the residual portion of the molecule with the consequent transformation of potential into kinetic energy?" (Brown,[10] p. 223.) Considering that the chemical pathway of glycolysis was only then beginning to be studied and was far from completely known, Brown's thoughts were almost prophetic.

Brown went on to state in an addendum, ". . . if we could devise means for measuring with exactness the amount of heat evolved during the disappearance of 1 grm. of sugar under the influence of yeast, whilst the cells are in the process of active reproduction, this amount of heat ought to fall short of that produced by the fermentation of the same amount of sugar when there has been no yeast reproduction, the difference being due to the energy required for the observed yeast production." (Brown,[10] p. 226.) He was thus the first to consider the relationship between the energetics of a fermentation reaction during which growth does not occur, and the same reaction during which growth does occur.

It is worth noting, however, that Brown's statement introduced a problem in semantics with respect to the nature of energy and what it does. When we speak of energy being required to accomplish something we usually mean that energy becomes used up, or expended. The nature of the expenditure depends on the system. In Brown's system, during the process of fermentation chemical energy became expended in that it was converted into heat and (as we now know) entropy. Thus, although Brown used the words ". . . energy required for the observed yeast production," this cannot be what he actually meant; if it were, the required energy would have been converted into heat and entropy and there would have been no observable difference between the heat production with growing and with nongrowing cells. He probably meant that because a small portion of the sugar had to serve as a source of carbon for the yeast cells which were produced during growth, an amount of heat equivalent to what would be observed were this sugar to be fermented would constitute the difference in heat production by nongrowing and growing cells utilizing the same initial quantity of sugar. The difference, as Battley[11] demonstrated 46 years later, represents the energy *conserved* within the substance of the yeast cells, not the energy *required* to accomplish this conservation.

3.6. THE STUDIES OF MEYERHOF ON THE HEAT OF GROWTH OF NITRIFYING BACTERIA

In 1911 O. Meyerhof published three articles on heat production by the semen, eggs, and larvae of *Stronglycentrotus lividus* (a sea urchin) and by the eggs and larvae of *Aplysia limacina* (a sea hare) all of these being either single animal cells or small, developing, multicellular animal organisms (larvae) that could be easily obtained and used in a heavy enough suspension to produce a measurable quantity of heat. First, Meyerhof[12] defined a "caloric quotient" as the ratio of calories produced per milligram of oxygen consumed. This value could be used to make comparisons of the energy output of various experimental preparations. He also described the calorimetric procedure for the research presented in the other two publications. Meyerhof's calorimeter vessel consisted of a spherical Dewar flask. This was closed by a rubber stopper through holes in which were passed a gas-escape tube, a sensitive Beckman thermometer, and a third tube for the delivery of oxygen gas to a suspension of cells contained in the flask. When in use the calorimeter vessel was immersed in an accurately controlled thermostat. Also immersed in the thermostat were a gas-washing bottle and a coil of glass tubing. Oxygen gas was first passed through the sea water and then through the coiled glass tube. It would thus contain water vapor at the same partial pressure as that of the gas phase above the cell suspension in the flask, and would be at the same temperature. This eliminated heat changes within the calorimeter vessel due to evaporation. Meyerhof found that the sea urchin eggs tended to settle out. To avoid this he devised an ingenious stirring mechanism whereby a rubber bulb was connected to the oxygen delivery tube. By clamping off the rubber tube from the oxygen source and subsequently squeezing the rubber bulb, the fluid rising into the oxygen delivery tube was made to move up and down within the tube, causing fluid also to flow in and out of its lower orifice and creating convection currents which swirled the eggs up into suspension once again. This procedure was repeated when necessary. The use of this method of stirring resulted in an increase of the cooling coefficient for the calorimeter from 3.0 to 3.4%, because of the movement of the column of fluid up and down within the oxygen delivery tube. A diagram of Meyerhof's calorimeter is shown in Fig. 3.1. It could be calibrated electrically by passing a current through a wire of known resistance which could be immersed in the fluid within the calorimeter vessel.

The eggs of marine invertebrates are in general available only in season, and it is not surprising that Meyerhof turned his attention toward studying heat production by other types of cells which could be obtained in liquid suspension and which were available throughout the year. His next experiments were therefore on heat production by red blood cells, the results of which were published in 1912 (Meyerhof[13]). In this same year he published measurements of heat production by a suspension of *Vibro metchnikoff* into which he had ingeniously mixed goose erthrocytes to serve as a source of oxygen for the bacteria (Meyerhof[14]). The medium consisted of a solution of peptone which would well have supported the growth of the bacteria, and one of his graphs of temperature rise plotted against time clearly demonstrated that exponential growth was taking place. Just as with previous experiments by other investigators who had used undefined media, only limited interpretations could be made as to the energetics of microbial growth. Measurements of oxygen consumption by identical suspensions of bacteria and erthrocytes were made while these were incubated in flasks immersed in the calorimeter thermostat at the same time that the heat production of the same suspension was being measured. The quantity of oxygen in the suspensions was determined by the Winkler

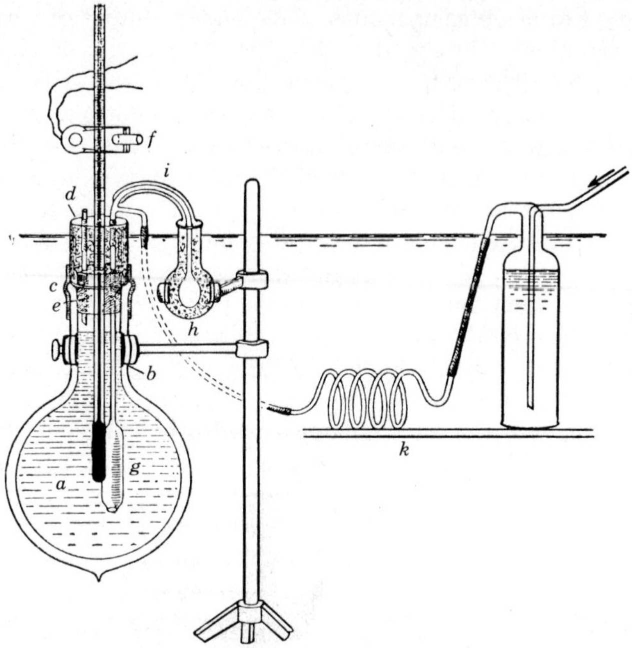

FIGURE 3.1. Diagram of Meyerhof's microcalorimeter. *a,* Dewar flask; *b,* support clamp; *c,* thick rubber collar supporting glass tube *d* packed with cotton; *e,* rubber stopper through which pass a Beckman thermometer read with telescope *f,* a short glass tube, and pipet tube *g* which is connected to rubber ball *h* through double-walled tube *i; k,* spiral glass tube which warms to thermostat temperature gas passed through the bubbler. The rubber ball is itself contained in a cotton-packed rubber enclosure which, when squeezed, causes fluid to move in and out of the pipet. The whole apparatus is immersed in a thermostated water bath. Reprinted from Reference 12 with permission.

method. From the values obtained Meyerhof concluded that the caloric quotient for bacteria was about 4.1, considerably higher than the values of 3.25 for goose erthrocytes, 3.15 for sea urchin sperm, 3.0 for sea hare eggs, and 2.7 for sea urchin eggs.

In 1916 Meyerhof[15] published the results of an extensive investigation of the respiratory processes of nitrifying bacteria. This was the first major study of these bacteria since Winogradsky[16] had summarized the then-existing knowledge of them in 1907. Meyerhof confirmed Winogradsky's[17] original observation that these bacteria were capable of development in culture solutions from which all traces of organic substances had been rigorously excluded, and that they were therefore autotrophic, the energy for their growth being derived exclusively from the oxidation of ammonium or nitrite. Using somewhat more favorable culture conditions than did Winogradsky, Meyerhof found that the ratio of the weight of nitrite nitrogen oxidized to the weight of carbon dioxide carbon utilized in synthesis (the N/C ratio) by *Nitrobacter* was 101.

Meyerhof's study included a measurement of the heat produced during the oxidation of nitrite by *Nitrobacter* cells suspended in culture medium. For this he used the calorimeter he had previously designed (Meyerhof[12]). To the extent that these bacteria actually grew within the calorimeter vessel these experiments could be considered as measurements of their heat of growth. Young cultures were used, and a slow stream of oxygen gas was constantly bubbled through the suspension in the calorimeter and through an identical suspension in a flask immersed in the calorimeter thermostat. One of the difficulties in working with nitrifying bacteria is their extremely slow growth rate. Thus, it was impossible to actually grow *Nitrobacter* from the start in Meyerhof's calorimeter since the experiments would have taken far too long. His experimental procedure was to grow up a large batch of culture, centrifuge it, and resuspend the culture in fresh medium. Part of this suspension was used for a calorimetric measurement, another part being placed in the control flask and used for measurements of nitrite consumption. Other aliquots were placed in Warburg vessels and used for the manometric determination of oxygen uptake. No determinations were made during these experiments of the amount of carbon dioxide fixed by the cells, a N/C ratio having been determined during previous experiments.

The heat of combustion of *Nitrobacter* cells had not been measured when Meyerhof carried out his research. Because of this he assumed for purposes of calculation that these cells had the same heat of combustion as glucose, which was 113,000 cal g atom of carbon.$^{-1}$ With a N/C ratio of 101 in the fixation of $14/12 \times 101$ g atoms of carbon, the heat of combustion would be expected to be

$$\frac{-113{,}000}{\frac{14}{12} \times 101} = -959 \text{ cal}$$

The equation representing the oxidation of nitrite to nitrate is as follows:

$$HNO_2 + 0.50O_2 \rightarrow HNO_3$$

Meyerhof calculated that this reaction would result in the liberation of 21,600 cal mole^{-1} of heat energy. Using this value he calculated the efficiency of growth of *Nitrobacter* to be

$$\frac{-959 \times 100}{-21{,}600} = 4.44\%$$

The results of Meyerhof's calorimetric determinations are shown in Table 3.1. It is apparent that the quantity of heat produced is always less than that expected from the complete oxidation of nitrite to the quantity of nitrate found after the experiment was over, the average value being a difference of 5%.

TABLE 3.1. The results of Meyerhof's calorimetric measurements of the heat produced during nitrite oxidation by "nitrate-formers" compared with the heat production to be expected on the basis of the complete oxidation of the nitrite utilized

Exp. #	(1) Calorimetric Measurements (cal)	Nitrite Determinations (mg NO_2^-–N)	(2) Heat Equivalent of Nitrite Oxidized (cal)	Difference (2) − (1) (%)
1	31.6	0.965	32.4	−2.5
2	30.0	0.855	31.8	−6
3	29.1	0.810	29.8	−2.5
4	23.5	0.680	25.5	−9
			Average	−5

Source: Reference 15.

Thus, considering the assumptions involved, the corrections that had to be applied to the calorimetry, the difficulties of working with the microorganisms concerned, and the considerable lack of precision of the experimental determinations, the agreement between the theoretically expected value and that which was experimentally found was surprisingly good.

3.7. THE COMMENTS BY GIAJA ON NONESSENTIAL HEAT PRODUCTION BY YEAST

In 1920 J. Giaja[18] published some comments on the nature of yeast energetics. He began by citing Rubner's observation that, per unit quantity of nitrogen, yeast cells placed in a solution of pure sugar released 58 times more energy than a man and 157 times more energy than a horse. According to the ideas prevalent at that time, this enormous amount of energy served to satisfy the energetic needs of this unicellular being. But what are the needs of an organism under these conditions, when effectively no mechanical work is being done by the yeast and in which no growth is taking place? Does this energy expenditure satisfy the need of the yeast cell to maintain itself?

Giaja divided some well-washed yeast cells into two parts, placing one part in pure water and the other part in a sugar solution. The latter suspension soon began to ferment the sugar, releasing quantities of energy, whereas the former only produced a minimum quantity of carbon dioxide gas corresponding to a very much lower quantity of energy. However, this very small quantity of energy was apparently sufficient to maintain intact the vitality of the yeast cells because when both sets of cells were thoroughly washed and resuspended in a sugar solution, they both exhibited the same fermentative power. Giaja concluded from this that the energy released by yeast when it is suspended in a sugar solution ". . . is not the expression of the energetic needs of this organism, but the result of its catalytic power which is not regulated according to the physiological needs of the yeast, . . ." (Giaja,[18] p. 1480.) He thus was the first to recognize that much of the heat production by yeasts may, depending on the environmental conditions, be due to an uncoupled metabolism wherein the energy made available by catabolic (obviously also catalytic) activity is simply wasted. This interpretation was totally different from any which had been made previously as to the cause of heat production in yeasts, and pointed out that an observed heat production need not have anything to do with maintenance or growth.

3.8. THE EARLY OBSERVATIONS OF PASTEUR ON THE GROWTH OF MICROORGANISMS IN A MEDIUM CONTAINING A SINGLE CARBON SOURCE

As early as 1858 L. Pasteur[19] observed for the first time that yeast cells were capable of fermentation and growth on what is effectively a nutrient medium containing a single carbon source, as follows.

> In a solution of pure sugar I place on the one hand an ammonium salt, for example ammonium tartrate [the organic part of this molecule could not be used anaerobically], and on the other hand the mineral which enters into the composition of beer yeast [yeast ash], and then a quantity of fresh yeast cells too small to be weighed. Remarkably, the cells inoculated in these conditions develop, multiply, and ferment the sugar, while the mineral dissolves little by little and the ammonium disappears. In other words, the ammonium becomes transformed into albuminoid [proteinaceous] material which enters into the composition of the yeast, and at the same time the phosphates [see below] give to the cells their principal minerals. As for the carbon [in the yeast cells], it is evidently furnished by the sugar. [Pasteur,[19] p. 1012.]

Pasteur then went on to state,

> If one leaves out from the composition of the nutrient medium the mineral matter, the ammonium salt, or both at the same time, the inoculated cells do not multiply at all, and there is absolutely no fermentation. One is able to use either mineral or organic salts of ammonia [for example, ammonium chloride or ammonium tartrate]. The phosphates can be obtained from the ashes of ordinary yeast or from precipitates of purely mineral origin. The double phosphate of magnesium and ammonium is able to serve as a common source of mineral material for the yeast and as a source of [nitrogen for] albuminoid material. However, one observes very significant differences in the energy of the yeast which is grown, according to whether it is given a food [nutrition] more or less appropriate to its true nature. [Pasteur,[19] p. 1013.]

Pasteur's recognition at this early stage in his research that microorganisms need minerals, and his use of yeast ashes to provide the minerals that yeast cells need, are examples of his characteristic brilliance. That he seems to identify yeast ashes with "phosphates" probably stems from his success in using mineral phosphates as a sole source of minerals. This is explicable if one realizes that even "pure" substances were, in 1858, undoubtedly sufficiently impure as to contain a great many other minerals in sufficient quantity to provide all that most microorganisms might need.

By nearly identical experiments published in 1859 Pasteur's observations were extended (Pasteur[20]) to include the "lactic yeast" (a then unidentified microorganism which fermented sugar to lactic acid) in which he cleverly included calcium carbonate to prevent the nutrient medium from becoming too acidic, and in 1860 (Pasteur[21]) to include *Penicillium glaucum,* which he recognized as having an aerobic rather than an anaerobic habit of growth. The actual use of a microorganism such as *P. glaucum* as a tool for the study of energy exchanges accompanying growth *because* it could be grown on a single carbon source did not take place until 1921, when Terroine and Wurmser commenced their research.

3.9. THE EXPERIMENTS OF TERROINE AND WURMSER, AND OF TERROINE, TRAUTMAN, BONNET, AND JACQUOT ON THE ENERGETICS OF *ASPERGILLUS NIGER* GROWTH ON A DEFINED MEDIUM

It was not until the early 1920s that experiments in microbial energetics were performed which made use of aerobic microorganisms grown in nutrient media containing a single organic substance as the source of carbon and energy. Prior to this time the determination of energy changes accompanying microbial growth had been made with fermentations (alcoholic or lactic) that were carried out in complex media such as grape must or beer wort in which the components were for the most part unidentified. These latter did contain, however, fermentable sugars and the vitamins, minerals, and trace elements necessary for the anaerobic growth of the microorganisms carrying out the fermentations. They also contained a great many other organic substances which could not be fermented. This made little difference when fermentations were being studied since only the sugars could be fermented. Moreover, because the products of the alcoholic and lactic fermentations were well known and easy to determine, material and energy balances could be constructed. If, however, similar complex media were to be used for studies of aerobic microbial growth, all organic substances in the media, not just the sugars, would be susceptible to oxidation and therefore it could not be known exactly on what substance a given microorganism was growing. For this reason it became necessary to resort to defined media using single carbon and energy sources. There were also other advantages. The analyses of fermentation products no longer became necessary since organisms could be chosen that were capable of oxidizing a given substrate completely to carbon dioxide and water. The only organic product of growth was thus the cells. In addition, microbial oxidations resulted in a great deal more heat production per unit quantity of substrate than did fermentations, and this was of considerable advantage when carrying out calorimetric studies.

In 1921 E. F. Terroine and R. Wurmser[22] performed a series of experiments in which they grew *A. niger* in an aqueous solution of minerals containing 3% glucose as the sole source of carbon and of energy. They found that the ratio of the dry weight of *Aspergillus* formed during growth to the weight of sugar disappearing from the culture medium, which they called the "utilization ratio," was constant at about 0.44 even if the rate of growth were varied by modifying the temperature between 22 and 38°C. This ratio was not constant if the rate of growth were varied by growing the organism outside these temperature limits.

In 1922 they described extensive studies on the growth and energetic efficiency of *A. niger* (Terroine and Wurmser[23]). Using the same medium previously used they found that the yield of dried mycelium per unit weight of sugar consumed, now called the "coefficient of utilization," was found to vary only slightly within the ranges of 22–38°C (confirming their previous experiments) and pH 1.7–7.7 and with respect to several different pure substances each serving as a sole source of carbon and energy. The average values fell between 0.40 and 0.44 when ammonium was used as the nitrogen source. However, when nitrate was used as the nitrogen source the coefficient of utilization fell to 0.34, indicating that some of the energy that could otherwise be used for growth had to be used to reduce the nitrate to ammonia since it was this latter form of nitrogen that was used for cellular synthesis. The constancy of the yield of *Aspergillus* when grown within such wide ranges of environmental conditions made it an ideal tool for the study of the material and energetic efficiency of growth.

In connection with their studies on the material efficiency of growth, Terroine and Wurmser found that, except for the first hours of development during which the culture was getting "under way" (i.e., the acceleration phase), the amount of growth was proportional to the age of the culture. Thus,

$$p = Kt$$

where p represents the dry weight of the mycelium formed from glucose as the substrate during time t, and K is a rate constant. Having established this, they then reasoned that the weight of glucose consumed is the sum of two terms. One term is proportional to the rate, K, of growth and represents the quantity of glucose that becomes used for the construction of the mycelium. The other term is proportional to the weight of the mycelium already formed at a given moment, and represents the quantity of glucose utilized for cell maintenance. The consumption c of glucose at any given time t is therefore

$$\frac{dc}{dt} = K(a + bt)$$

where the constants a and b represent the trophic coefficient and the coefficient of maintenance, respectively. Integrating between time 0 when the period of growth begins and time t, the total glucose consumed during the growth period, C, becomes

$$C = p(a + \tfrac{1}{2}bt) \tag{3.1}$$

In order to evaluate the constants it is necessary to vary the rate of growth. Letting C_1 represent the consumption of glucose necessary to obtain dry weight p of mycelium after time t_1, and C_2 the consumption of glucose necessary to obtain the same weight of mycelium after time t_2, it is found that

$$b = 2\frac{C_2 - C_1}{p(t_2 - t_1)} \tag{3.2}$$

from which a can be evaluated by substitution of b into Eq. 3.1. By comparing the time taken to grow the same dry weight of mycelium at pH 5.0 and at pH 1.2 and by measuring the quantity of glucose consumed, they found that b had a value of 0.011 g glucose g mycelium^{-1} hr^{-1} and that a had a value of 2.2 g. The curious aspect of their data is that whereas they found experimentally that p increased arithmetically with time, in retrospect what they *ought* to have found was that p increased logarithmically, something which had been known with respect to *Aspergillus* since 1898 when Duclaux[24] first distinguished between the quantity of sugar used for construction and that used for maintenance. This publication was not cited in their bibliography and they may have not been aware of it, which is unfortunate since their data (though not necessarily their method of calculating b) are of little use.

The energetic aspects of the growth of *A. niger* followed as a result of the research that Terroine and Wurmser[25] carried out in 1922 on the material balance accompanying the growth of this organism. To make an approximation of the energetic balance they determined the heat of combustion of dried *Aspergillus* mycelium and of glucose and found them to be 4.8 and 3.76 kcal g^{-1}, respectively. Knowing from their previous calculations that the trophic coefficient was 2.2 g, the energetic balance per gram of dried mycelium could be calculated as

$$\frac{\text{heat content (kcal g}^{-1}\text{) dried mycelium}}{\begin{array}{c}\text{heat content (kcal of glucose)}\\ \text{utilized g dried mycelium}^{-1}\end{array}} \times 100 = \%\ \text{utilization}$$

$$= \frac{4.8}{2.2 \times 3.76} \times 100$$

$$= 58\% \tag{3.3}$$

that is, 58% of the energy content of the substrate appeared in the mycelium. However, they then discovered that there were organic substances other than glucose still remaining in the medium. This resulted in a determination of the caloric content of the medium before, U_1, and after, U_2, growth, the difference $(U_1 - U_2)$ being the amount of energy, U, which was totally utilized. In addition, they measured the amount of heat, U', resulting from the combustion in a bomb calorimeter of the dried mycelium formed during growth. The amount of heat resulting from the respiration of the growing mycelium was determined by measuring the amount of carbon dioxide that was formed during growth, and from this the amount of glucose that had been completely respired (biologically combusted) to carbon dioxide and water, together with the energy equivalent, U'', of this combustion. The energy balance for the process of growth should therefore be

$$U = U' + U'' \tag{3.4}$$

In two experiments the average value for $(U' + U'')$ was within 6.3% of the average value for U, which was not a bad balance considering the state of biological calorimetry at that time. The efficiency with which energy became incorporated into the dried mycelium could then be calculated with the following equations:

$$\text{eff.} = \frac{U'}{(U - U'')}$$

or,

$$\text{eff.} = \frac{U'}{(U - R)}$$

where R represents the energy lost as a result of respiration. Using either of the above equations the energetic efficiency of growth was determined in two experiments to be 59.6 and 59.2% rather than the 58% as originally calculated using Eq. 3.3.

In retrospect there seems to be nothing wrong with the use of the term U'' and R to represent the quantity of energy resulting from the formation of CO_2 during respiration from whatever source. Nonetheless, Terroine and Wurmser felt that efficiencies of growth so calculated gave values that were more apparent than real. In order to approach the real efficiency of growth it was necessary to consider the following:

1. The variation of energy in the substrate: $U_1 - U_2$.
2. The energy contained in the dried mycelium formed as a result of growth: U'.
3. The energy remaining in the culture medium after growth has taken place: U_R (this would take into account any products resulting from growth other than the cells).

4. The energy represented by the quantity of CO_2 evolved during growth on a given substrate: U''.
5. The energy of maintenance, if possible directly measured: U_E.

Using the above notations the balance of energy before and after growth could therefore be expressed by the equation

$$U_1 - U_2 = U' + U_R + U'' + \text{work}$$

and the energetic efficiency of growth by

$$\text{eff.} = \frac{U'}{(U_1 - U_2) - (U_R + U_E)} \tag{3.5}$$

In Eq. 3.5 Terroine and Wurmser appear to have equated U_E with (U'' + work). Values for all of the terms in Eq. 3.5 could be obtained easily except for U_E. For some reason they preferred to obtain some form of direct measurement of U_E as opposed to calculating this value by means of Eq. 3.2. They first attempted to determine the rate of carbon dioxide production per gram of dried mycelium by incubating a well-washed living mycelium in a culture medium containing glucose but without a nitrogen source. Their reason for doing this was to place the mycelium in an environment in which no growth could take place, any carbon dioxide production then being hypothesized to be due to the utilization of glucose for maintenance energy. They were astonished to discover that even in the absence of a nitrogen source the mycelium gained considerable weight over a period of 45 hr. The change in the percentage of nitrogen in the mycelium led them to conclude that this augmentation of weight was not due to growth, but to the accumulation of organic substances within the cells. Abandoning this method they then placed a well-developed mycelium in a mineral nutrient solution lacking either a carbon or a nitrogen source. The carbon dioxide they observed the mycelium to produce was due to what later came to be called "endogenous respiration." The major difficulty with this technique was that the carbon dioxide production per gram hour was not constant, and Terroine and Wurmser came to the conclusion tht it was preferable not to use even average values. Finding direct measurements of U_E impossible, they then went back to using Eq. 3.2. The results of two experiments in which U_1, U_2, U'', U_R, and U_E were determined gave energy balances agreeing within about 6%, which is about the same as when the much more simple approach using Eq. 3.4 was used.

Terroine and Wurmser's research dealt more extensively with the subject of the energy of growth than had been done previously. They were the first to study the growth characteristics of a microorganism that could completely convert a single source of carbon and energy into cells, carbon dioxide, and water, and to establish environmental conditions for growth such that it was only the quantity of substrate which determined the quantity of microorganism produced. Consequently they could define a "trophic coefficient," or "coefficient of utilization" as the ratio of the quantity in grams of dried microorganism formed during growth to the quantity in grams of the carbon source utilized. They also recognized that the quantity of substrate consumed was equal to the quantity of substrate transformed into microorganism plus the quantity of substrate that is degraded to provide energy for the synthesis and maintenance of the cells. This could not be analyzed quantitatively when microorganisms were grown aerobically on complex media because of the difficulty in determining exactly which organic substances the cells were using for synthesis and for energy. Quantitative studies could only be carried out when single organic substances were used in this manner.

In 1925 Terroine, Trautman, Bonnet, and Jacquot[26] continued on to compare the energetic balance with respect to the development of *Sterigmatocystis nigra* (now *A. niger*) and *Aspergillus oryzae* on several different substrates, each serving as a sole source of carbon and of energy. One of the things they realized was that the composition of the microorganisms did not vary appreciably no matter what the substrate on which they were grown, as demonstrated by the constancy of the percent dry weight of nitrogen in the mycelia. They also concluded that in carrying out research of this kind it was advisable to use only *young* cultures. For their energy balances they used the equation:

$$\text{eff.} = \frac{\text{energy present in the mycelium after harvest}}{\begin{matrix}\text{initial energy of} \\ \text{the culture medium}\end{matrix} \quad \begin{matrix}\text{residual energy in the medium} \\ \text{after harvesting the mycelium}\end{matrix}} \times 100 \qquad (3.6)$$

all data being obtained by bomb calorimetry. The values they obtained for the initial energy of the medium agreed with the sum of the values for the energy present in the mycelium after growth had taken place plus that remaining in the medium within 5%. Whereas the efficiency as expressed by means of Eq. 3.6 was 58% when the molds were grown on glucose, it was only 39% when the molds were grown on gelatine, peptone from silk, or individual amino acids. The heat of combustion per gram dry weight of mycelium averaged 4.75 kcal (± 0.10, $n = 11$) for *S. nigra* and 4.84 kcal (± 0.08, $n = 11$), for *A. oryzae*, or 4.79 kcal if the values for both microorganisms are averaged.

The work of Terroine *et al.* represents success where Nawiasky[9] had failed in that they were able to investigate the growth of a microorganism in pure culture on a strictly defined medium containing individual amino acids as the sole source of carbon and energy. Their finding that the efficiency of growth with glucose was considerably higher than with amino acids was not explainable on the basis of what they knew about the energetics of growth. Nevertheless, the importance of their work and that of Terroine and Wurmser[25] is that their experiments were precursors to the work by Tamiya and his collaborators in the 1930s which led to accurately balanced equations representing growth and the eventual possibility of thermochemically defining the initial and final states of a growth-reaction system.

3.10. THE EXPERIMENTS OF MOLLIARD ON THE UTILIZATION OF RESPIRATORY ENERGY DURING THE GROWTH OF *STERIGMATOCYSTIS NIGRA* (*ASPERGILLUS NIGER*)

Because of the appearance of Terroine and Wurmser's[25] publication on the energetic efficiency of growth, M. Molliard[27] in 1922 was induced to publish prematurely the results of his own studies on the energy liberated during the respiration accompanying the growth of *S. nigra*. The methods he employed were identical to those of Terroine and Wurmser. Molliard determined the difference in the caloric content of the medium before (Q) and after (q) growth had taken place and found the value of $(Q - q)$ to be 25.16 kcal. The combustion of the dried mycelium formed during growth gave a value (M) of 13.90 kcal, and a measurement of the carbon dioxide gas evolved during growth showed that an amount of glucose had been oxidized during respiration (R) equivalent to 11.71 kcal. The energetic efficiency of growth was therefore

$$\text{eff.} = \frac{M}{(Q - q)} \times 100$$
$$= 55.2\%$$

a value very close to that obtained by Terroine and Wurmser for *A. niger*. Observing that $(Q - q)$ was, within experimental error, almost exactly equal to $(M + R)$, he then concluded that "... respiratory energy is uniquely utilized towards the production of heat, electricity, and mechanical work corresponding to the growth and maintenance of the mycelium, without being placed, even to a small extent, in reserve in the mycelium in the form of chemical energy." (Molliard,[27] p. 221.) Molliard thus made the first statement after Berthelot[28] to the effect that the energy contained in that part of the substrate utilized for the construction of a microorganism, and the energy utilized in respiration are two separate entities associated with two completely different chemical reaction systems, one being growth and the other respiration.

3.11. ALGERA'S DETERMINATION BY DIRECT CALORIMETRY OF THE HEAT OF GROWTH OF *ASPERGILLUS NIGER*

In 1932 there appeared a lengthy publication by L. Algera[29] on the subject of heat production by growing cultures of *A. niger*, included in which was an extensive review of the previous literature on heat production in living organisms and methods of measuring it. The main body of the article was, however, devoted to the description of an elaborate microcompensating calorimeter that Algera had designed for this research. It was without doubt the most accurate and sophisticated microcalorimeter yet constructed. Algera had come to the conclusion that direct calorimetry of thermogenesis was preferable to indirect calorimetry as employed by Terroine and Wurmser,[22,23,25] Terroine et al.,[26] and by Molliard[27] for two reasons. One was that any heat output in addition to that of respiration could be detected, the other being that any energy change that occurred when the mycelium was dried prior to bomb calorimetry could be eliminated, since in determining the heat of growth by direct calorimetry the mycelium would not have to be dried. Otherwise, the same sorts of determinations could be made that were made by Terroine and Wurmser.

Algera's calorimeter was designed to be isothermal and to operate on the same principle as that used by v. Korosy[30] in that the heat generated by growing and/or metabolizing microorganisms was removed from the calorimeter vessel as the latent heat of vaporization of a volatile liquid. The only temperature at which v. Körösy's calorimeter could be operated was that of the boiling point at 760 mm Hg (34.6°C) of the ethyl ether used to provide evaporative cooling. Instead of ether Algebra used water, a nontoxic substance having a high latent heat of vaporization. Although he obviously could not use his calorimeter at the boiling point of water he could, in theory, use it at any temperature at which water was liquid, provided that he used some method other than boiling to evaporate the water. The method he adopted was to bubble dry air through water inside the calorimeter vessel so that it was saturated by the time it reached the surface of the water. In this manner, at any given temperature, the amount of cooling was proportional to the amount of air passed through the water. As the growing microorganisms generated heat a very small temperature rise inside the calorimeter vessel above that of the water bath in which the calorimeter vessel was immersed was detected by a 20-junction constantan–manganin thermopile connected to a sensitive mirror galvanometer. A light beam reflected off the mirror was perceived by a photoelectric cell connected to a low-frequency amplifier so that the output of the circuit was proportional to the amount of light falling on the cell. At a certain current strength the output of the amplifier activated a relay which set an air pump into motion. This forced air at the temperature of the water bath through the water in the calorimeter vessel, causing some of the water to evaporate and by evaporative cooling to cool the contents of the vessel. When this happened the light beam drifted off the photoelectric cell, the relay was inactivated, and the pump stopped. The number of calories taken up as the latent heat of vaporization of the water was thus proportional to the number of strokes of the air pump, which were registered by means of a counter. Diagrams of the calorimeter vessel assembly and of the whole apparatus are given in Figs. 3.2 and 3.3, respectively. It was important that the temperature of the water bath in which the calorimeter vessel was immersed be very carfully controlled, since the thermopile constantly compared the temperature of the inside of the calorimeter vessel with that of the water bath. Algera's water bath was set at 22°C and fluctuated not more than 0.001°C. The calorimetric system could be calibrated electrically and Algera claimed that the maximum error ranged from 2.7% at a heat input of 5 cal hr^{-1} down to 1.5% at 40 cal hr^{-1}. This can be compared with a maximum error of 3.2% at a heat input of

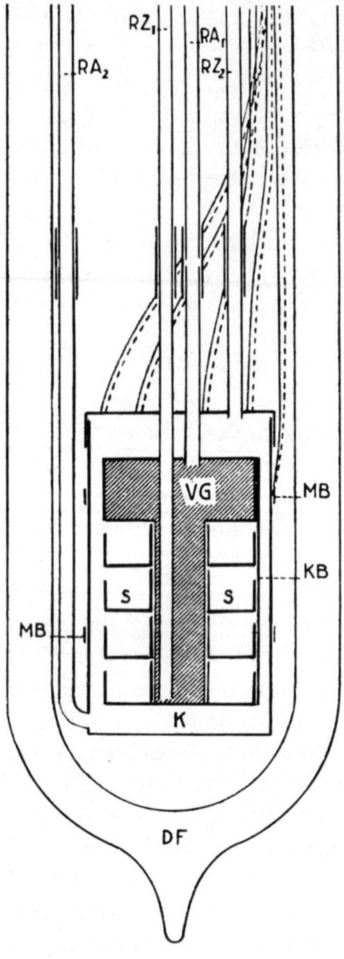

FIGURE 3.2. Diagram of the culture flask in Algera's microcalorimeter. DF, Dewar flask; K, culture flask; KB, copper strips to facilitate heat exchange between culture containers; MB, thermocouples; S, circular culture containers; VG, evaporator. Reprinted from Reference 30 with permission.

21–112 cal hr^{-1} for Rubner's calorimeter and of about 3% at a heat input of 23 cal hr^{-1} for Bayne-Jones's "Hill-type" calorimeter. Algera had thus constructed a remarkably sensitive and accurate instrument that was all the more notable because of the ingenuity of its design. It seems unfortunate that after spending all this effort at instrumentation Algera only carried out one biological measurement.

In the one experiment that Algera published *A. niger* was grown in the calorimeter vessel in 40 ml of an aqueous solution of minerals plus 15% glucose as the sole source of carbon and energy. The medium was inoculated with

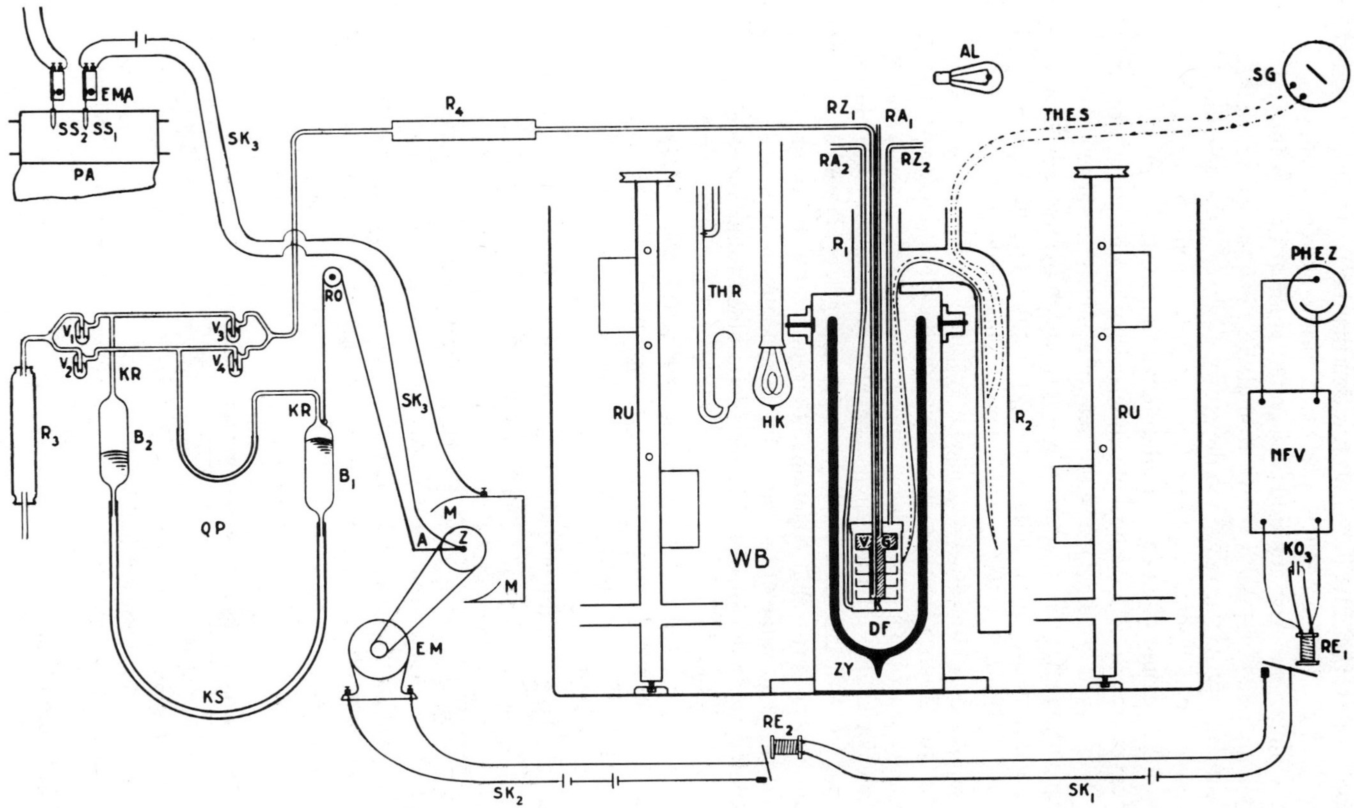

FIGURE 3.3. Diagram of Algera's microcalorimeter. A, arm; AL, automobile lamp; B_1, B_2, reservoirs; DF, Dewar flask; EM, electric motor; EMA, electromagnet; HK, heat source for thermostat; K, culture flask; KO_3, condenser; KR, capillary tube; KS, rubber connecting tube; M, electrical contacts; NFV, low frequency amplifier; PA, paper roll; PHEZ, photoelectric cell; QP, mercury pump; R_1, tube; R_2, tube containing parafine oil; R_3, R_4, tubes containing P_2O_5; RE_1, RE_2, relays; RO, roller; RU, stirrer; RZ_1, RZ_2, RA_1, RA_2, inlet and outlet tubes; SG, mirror galvanometer; SK_1, SK_2, SK_3, SK_4, electrical circuits; SS_1, SS_2 event-marking pens; THES, thermoelectric circuit; THR, thermoregulator; V_1, V_2, V_3, V_4, valves; VG, evaporator; WB, waterbath; Z, disc; and ZY, copper cylinder. Reprinted from Reference 30 with permission.

A. niger spores in a sufficiently low concentration that no heat output could be observed as a result of growth before the calorimeter vessel came into temperature equilibrium. Air saturated with water vapor at the temperature of the water bath could be passed slowly around the small dishes containing the medium, thereby providing oxygen to the growing culture. Determinations were made as shown in Table 3.2. The difference between the heat of combustion (Q) of the amount of glucose in solution before growth began and the sum of the heat of growth (W), the heat of combustion of the mycelium formed during growth and then dried (M), and the heat of combustion of the glucose remaining in solution after the experiment was terminated (q) was only 95 cal. The energy balance between the initial state (Q) and the final state ($W + M + q$) was thus $(95/19,560) \times 100 = 0.5\%$, a most remarkable result. A value for the heat equivalent of the glucose utilized in respiration (R) was calculated by measuring the amount of carbon dioxide gas produced during growth, as was also done by Terroine and Wurmser[23] and by Molliard.[27] This was found to be $-3,518$ calories. $R - W$ thus equals 219 cal, which value is 6.0% of the value or R and 6.6% of the value of W, both percentages being well outside the maximum limits of error that Algera had determined for the experiment. What, then, was the meaning of the inequality between R and W?

Algera pointed out that Molliard[27] had also determined values for R, Q, q, and M and had found that $[R - (Q - q - M)]/Q = 1.6\%$. This value was sufficiently close to zero for him to conclude that $R = (Q - q - M)$. If this were so, then absolutely none of the energy of respiration would be found as chemical energy in the mycelium formed during growth. Algera felt that $(Q - q - M)$ would better be equated with W than with R because W required three measurements, each subject to experimental error. He also felt that the difference of 219 cal for $(R - W)$ which he observed in his experiment was real. In consideration of this latter he objected to the use of R on the basis of its method of calculation, in which all the respired glucose is assumed to be oxidized to carbon dioxide and water. According to Algera this need not be the case if a part of the glucose becomes converted into cellular material. This latter is undoubtedly true. Algera came very close to the right interpretation when he stated ". . . the calculation of respiratory energy on the basis

TABLE 3.2. The results of Algera's calorimetric measurements

Heat of combustion of the mycelium		$= M =$	5,606 cal
Observed heat of growth		$= W =$	3,299 cal
Heat of combustion of the substances in the nutrient solution remaining after growth		$= q =$	10,750 cal
	Total	$=$	19,655 cal
Heat of combustion of the substances in the nutrient solution before growth		$= Q =$	19,560 cal

Source: Reference 30.

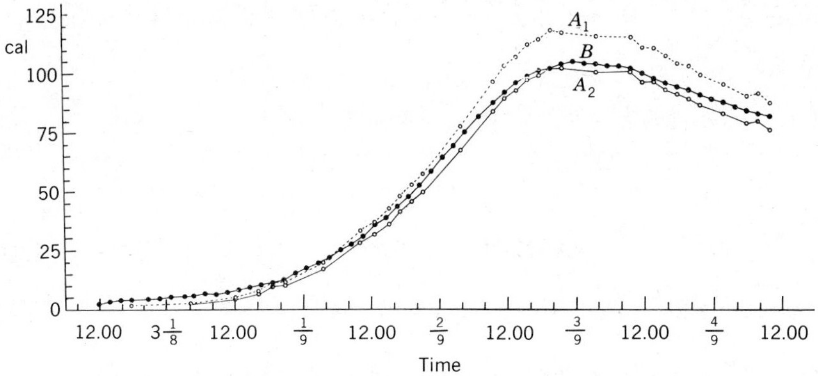

FIGURE 3.4. Algera's thermograms of heat production in calories plotted against time in days for the growth of *Aspergillus niger* in a defined medium containing glucose as the sole source of carbon and energy. A_1, calculated heat of respiration; A_2, corrected heat of respiration; B, heat production measured microcalorimetrically. Reprinted from Reference 30 with permission.

of carbon dioxide production leads to a too high value if a part of the glucose becomes incompletely decomposed. It is only permissible if the combustion is complete or if in the case of an incomplete combustion the nature and the quantity of the decomposition products are known." (Algera,[29] p. 149.) If the acceptance of a complete glucose decomposition (oxidation to carbon dioxide and water) is incorrect, the R must be too high and this could explain why Molliard had experimentally found R slightly larger than $(Q - q - M)$. Algera went to considerable lengths to rationalize the difference between R and W and developed a correction factor which, when applied to his data, made the thermogram of R very nearly identical to that of W. This is illustrated in Fig. 3.4. From this he concluded that the interpretation of Molliard was correct, and that absolutely no respiration energy becomes incorporated as chemical energy in the mycelium of an *Aspergillus* culture which is using glucose as the only source of carbon and energy.

Algera's work is important not so much because of the remarkably fine calorimeter he constructed or because of the data he obtained (he published only one experiment), but because it represents the first attempt at correlating the heat exchange accompanying the growth of a microorganism on a known quantity of a known pure substance with a chemical reaction in which one of the products is cells. From the nitrogen content given by Terroine, Trautman, Bonnet, and Jacquot[26] it was calculated that the mycelium of *Sterigmatocystis nigra* (*A. niger*) grown in an aqueous solution of minerals and glucose had a composition of about 40% protein and about 60% cellulose. Algera regarded the cellulose as being formed from the condensation of glucose molecules approximately according to the following equation:

$$C_6H_{12}O_6 \longrightarrow C_6H_{10}O_5 + H_2O - 1840 \text{ cal} \qquad (3.7)$$

Protein could be regarded as being synthesized according to the following scheme:

$$C_6H_{12}O_6 \longrightarrow 2CH_3{\cdot}CO{\cdot}COOH + 4H - 22{,}900 \text{ cal} \quad (3.8)$$

$$\left.\begin{array}{l} 2CH_3 \cdot CO{\cdot}COOH + 2NH_3 \\ + 4H \end{array}\right\} \longrightarrow \begin{array}{l} 2CH_3 \cdot CH(NH_2) \cdot COOH \\ + 2H_2O + 101{,}800 \text{ cal} \end{array} \quad (3.9)$$

$$2CH_3 \cdot CH(NH_2) \cdot COOH \longrightarrow \text{protein} + 2H_2O - 17{,}200 \text{ cal} \quad (3.10)$$

from which adding Eqs. 3.8 through 3.10 gives

$$C_6H_{12}O_6 + 2NH_3 \longrightarrow \text{protein} + 4H_2O + 61{,}700 \text{ cal} \quad (3.11)$$

Using percentages from Terroine et al.,[26] it can be calculated that the mycelium formed during Algera's experiment consisted of 0.73 g cellulose and 0.49 g protein. The molecular weight of a $C_6H_{10}O_5$ cellulose unit is 162 daltons. According to Eq. 3.7 the amount of heat released in the formation of 0.73 g of cellulose would be expected to be $(0.73/162) \times -1840 = \sim -8$ cal. Similarly, the amount of protein formed in the reaction represented by Eq. 3.11 was determined by difference to be 142 g and the amount of protein present in the mycelium would therefore be expected to be $(0.49/142) \times 61{,}700 = \sim 213$ cal. Adding these two values together gives 205 cal as the heat exchange accompanying mycelial growth. Although Algera did not do so himself, if this value is subtracted from the value of -3518 cal he calculated for R according to the assumption that all the carbon dioxide formed during growth was the result of the complete oxidation of glucose to carbon dioxide and water (which Algera did not believe), the value of -3313 cal is obtained. This is extremely close to the value of -3299 cal actually determined for W, with no corrections being necessary. On the other hand, this may be purely fortuitous. We do not know enough about the sources of Algera's thermal data in Eqs. 3.7 through 3.10 to be able to draw any firm conclusions. It is worth noting that in Algera's interpretation of the chemical events that take place during mycelial synthesis as represented by Eqs. 3.8 through 3.10 there is no evidence of the participation of carbon dioxide either as a reactant or a product. If this is to be what really happens, then the only reasonable method of calculating R is in fact on the basis of the assumption that the glucose not used for synthesis is completely respired to carbon dioxide and water. On the other hand, although Algera believed that the calculation of R on this basis gave values which were too high and in this respect seems to have misinterpreted his data, he did recognize that carbon dioxide production is proportional to the weight of mycelium produced during growth.

In considering Algera's contributions to the study of the energetics of microbial growth it becomes apparent that he made a number of correct observations, but that these were not correctly correlated. He possessed the data to construct a single equation representing the growth process but did not do so. Had he done this he might have recognized, as Tamiya did a few years later (see Chapter 6), that carbon dioxide can be a product in the reaction whereby cells are synthesized from a pure substrate.

3.12. SUMMARY

By 1932 the techniques of measuring microbial thermogenesis resulting from the utilization of specific substrates can be considered to have advanced to the point where it was no longer a question of the accuracy of the measurements, but of how to interpret them. What was measured was the sum of all the exothermic and endothermic reactions that take place during the process of microbial metabolism and growth. There are at least three individual processes (as opposed to reactions) which could be considered to contribute to microbial thermogenesis. These are the heat exchanges which have to do with microbial growth, respiration, and maintenance, respectively. The difficulty lies in knowing the magnitude of each. On the other hand, given different experimental conditions one or the other of these processes may not exist. Do exponentially growing cells need to be maintained? Perhaps respiration is a part of the process of growth, providing energy for growth to take place. About the most that can be said is that whatever considerations are made as to what processes are taking place, if these former are correct the sum total of the heat exchanges accompanying them must equal the measured thermogenesis after corrections have been made, as in the case of direct calorimetry, for any extraneous side reactions.

REFERENCES

1. M. Dubrunfaut, Note sur la chaleur et le travail méchanique produit par la fermentation vineuse, *C.R. Hebd. Séances Acad. Sci.* **42**, 945–948 (1856).
2. A. Bouffard, Détermination de la chaleur dégagée dans la fermentation alcoolique, *C.R. Hebd. Séances Acad. Sci.* **121,** 357–360 (1895).
3. L. Pasteur, Mémoire sur la fermentation alcoolique, *Ann. Chem. Phys.* **58,** 323–426 (1860).
4. M. Berthelot, Sur les quantités de chaleur dégagées dans la fermentation des composés organiques, *Ann. Chem. Phys.* **6,** 329–442 (1865).
5. C. von Rechenberg, Ueber die Verbrennungswärme organischen Verbindung, *J. prakt. Chem. N.F.* **22,** 1–45 (1880).
6. M. Berthelot, *Chaleur Animale*, Vol. I, 1895, p. 67.
7. A. J. Brown, The heat of fermentation, *Inst. Brew. London, J.* **7,** 93–103 (1903).
8. M. Rubner, Ueber spontan Wärmebildung in Kuhmilch und die Milksäuregärung, *Arch. Hyg.* **57,** 244–268 (1906).
9. P. Nawiasky, Über die Umsetzung von Aminosäuren durch *Bac. proteus vulgaris*, *Arch. Hyg.* **56,** 209–243 (1908).
10. H. T. Brown, Some studies on yeast, *Ann. Bot.* **28,** 197–226 (1914).

11. E. H. Battley, Enthalpy changes accompanying the growth of *Saccharomyces cerevisiae* (Hansen), *Physiol. Plant.* **13,** 628–640 (1960).
12. O. Meyerhof, Untersuchungen über die Wärmetönung der vitalen Oxydationsvorgange in Eiren, I, *Biochem. Z.* **35,** 246–279 (1911).
13. O. Meyerhof, Über Wärmetönungen chemischer Prozesse in lebenden Zellen. Versuche an Blutzellen, *Pfleuger's Archiv gesamte Physiol. Menschen Tiere* **146,** 159–185 (1912).
14. O. Meyerhof, Über den Energiewechsel von Bakterien, Heidelberger Akademie der Wissenschaften, Mathematischnaturwissenschaftlichen Klasse, *Sitzungsberichte. Abt. B, Abh. 1*, pp. 1–18 (1912).
15. O. Meyerhof, Untersuchungen über den Atmungsvorgang nitrifizierenden Bakterien. I. Die Atmung des Nitratbildners, *Pfleuger's Archiv gesamte Physiol. Menschen Tiere* **164,** 353–427 (1916).
16. S. Winogradsky, Die Nitrifikation. *Handbuch der Technische Mycologie*, herausgegeben von Lafar., Bd. III (1904).
17. S. Winogradsky, Recherches sur les organismes de la nitrification, 2è Mémoire, *Ann. Inst. Pasteur* **4,** 257–275 (1890).
18. J. Giaja, Sur l'énergétique de la levure, *C.R. Séances Soc. Biol. Fil.* **83,** 1479–1480 (1920).
19. L. Pasteur, Nouveaux faits concernant l'histoire de la fermentation alcoolique, *C.R. Hebd. Séances Acad. Sci.* **47,** 1011–1013 (1858).
20. L. Pasteur, Nouveaux faits pour servir à l'histoire de la levure lactique, *C. R. Hebd. Séances Acad. Sci.* **48,** 337–338 (1859).
21. L. Pasteur, Note relative au *Penicillium glaucum* et de la dissymétrie moléculaire des produits organiques naturels, *C. R. Hebd. Séances Acad. Sci.* **51,** 298–299 (1860).
22. E. F. Terroine and R. Wurmser, Influence de la température sur l'utilisation du glucose dans le développement de l'*Aspergillus niger*, *C. R. Hebd. Séances Acad. Sci.* **173,** 482–483 (1921).
23. E. F. Terroine and R. Wurmser, La rendement énergétique dans la croissance de l'*Aspergillus niger*, *C. R. Hebd. Séances Acad. Sci.* **174,** 1435–1437 (1922).
24. E. Duclaux, *Traité de Microbiologie*, Vol. I, Masson et Cie., Editeurs, Paris, 1898.
25. E. F. Terroine and R. Wurmser, L'énergie de croissance. I. Le développement de l'*Aspergillus niger*, *Bull. Soc. Chim. Biol.* **4,** 519–567 (1922).
26. E. F. Terroine, S. Trautman, R. Bonnet, and R. Jacquot, Rendéments énergétiques comparés dans le développement de moisissures sur divérs aliments organiques et méchanism de l'action dynamique specifique, *Bull. Soc. Chim. Biol.* **7,** 351–379 (1925).
27. M. Molliard, Recherches calorimétriques sur l'utilization de l'énergie au cours du développement d'une culture de *Sterigmatosystis nigra*, *C.R. Séances Soc. Biol. Fil.* **87,** 219–221 (1922).
28. M. Berthelot, Remarques sur la troisième réponse de *M.* Pasteur, *C.R. Hebd. Séances Acad. Sci.* **88,** 197–201 (1879).
29. L. Algera, Energiemessungen bei *Aspergillus niger* mit Hilfe eines automatischen mikro-kompensations-Calorimeters, *Rec. Trav. Bot. Neerl.* **29,** 47–163 (1932).
30. K. von Körösy, Mikrokalorimeter zur Bestimmung der Wärmeproduktion von Bakterien, *Z. physiol. Chem.* **86,** 383–400 (1913).

4

EARLY STUDIES ON FREE-ENERGY CHANGES ACCOMPANYING BIOLOGICAL PROCESSES

In Chapters 2 and 3 much of the early literature on heat exchanges accompanying the growth of microorganisms has been reviewed. It is apparent from this review that the concept of free energy is not emphasized. This should not be surprising since this concept as developed by Gibbs and by Helmholtz in the late 1870s and the 1880s was not systemically applied with respect to chemical calculations until 1913 by Lewis[1] and later in 1917 by Lewis and Gibson.[2] It was still another step for biologists to adopt the teachings of chemists, and to accept the idea that it was *not* the heat of a reaction but its free-energy change (ΔF, now ΔG) at constant temperature and pressure which was the true measure of the energy change of a biological reaction. Thus, there was some tendency for biologists to continue with the old notion that the energy changes accompanying the growth of microorganisms were reflected in the amount of heat which was produced during this process. Not until the publication in 1923 of the classic text by Lewis and Randall[3] did biologists and more especially biochemists begin to take advantage of the principles of chemical thermodynamics. Quite predictably, these principles were first applied to enzyme systems, since these most closely resembled the equilibrium reaction systems with which chemists were familiar. The application of chemical thermodynamics to microbial systems involving growth, which are irreversible, did not take place until much later.

Although the Second Law of Thermodynamics and the concept of free energy were well known in the early 1900s, there was considerable discussion as to their application to living systems. There was no doubt by 1902 when Rubner[4] summarized the results of his brilliant research on animal energetics that the First Law of Thermodynamics, the Law of the Conservation of Energy, was applicable in biology as well as in physics. Attention then became focused on the cell as a self-reproducing catalyst capable of transforming chemical energy into heat and work and on how the cell could be regarded thermodynamically. The typical cell appeared to be so extremely complex that surely there must be a considerable decrease in entropy associated with its formation. As a consequence it had to be determined that the Second Law, which states that as the entropy of an isolated system increases the free energy decreases, was also applicable to the cell. This was not particularly easy to demonstrate, and led to many speculations as to whether cells acted in accord with the laws of classical thermodynamics. Some writers even went so far as to propose that cells might act as "Maxwell demons" by an ability to transfer energy to a higher potential without the necessity of work being supplied from a source outside the system. These discussions have been reviewed by writers such as Wachtel (1918),[5] Höber (1924),[6] Oppenheimer (1925),[7] and Guye (1925).[8] They are now of only historical interest.

4.1. THE ATTEMPT BY SIMONSON TO CALCULATE THE FREE-ENERGY EFFICIENCY OF BACTERIAL METABOLISM

Since all of the work on thermodynamics prior to 1900 had been done by physicists or physical chemists, it should not be surprising that the first biologist to have attempted a calculation of true thermodynamic efficiency (as opposed to heat efficiency) adopted their methods. In 1824 S. Carnot calculated the maximum amount of heat that any engine could convert into mechanical work. According to Carnot,

$$W = Q\frac{T_1 - T_2}{T_1}$$

Where T_1 and T_2 represent the absolute temperatures of the heat source and the heat sink, respectively, Q represents the quantity of heat taken up by the heat source at T_1, and W represents the work performed by the engine. The ratio $(T_1 - T_2)/T_1$ thus represents the efficiency with which a heat engine can function, that is, the maximum amount of heat energy that can be converted into work. In 1911 Simonson[9] attempted to apply this relationship to the metabolism of bacteria. He claimed that the efficiency (E) with which bacteria live depends upon the temperature difference between themselves (T_1) and the temperature of the environment in which they live (T_2) according to the equation

$$E = \frac{T_1 - T_2}{T_1}$$

Simonson explained the rise in body temperature which accompanies certain bacterially caused diseases as being due to an attempt by the body to raise the temperature (T_2) of the environment of the bacteria, thus decreasing the value of $(T_1 - T_2)/T_1$. To do this he also had to postulate that above a certain body temperature (presumably when the bacterial efficiency gets too low), they cease to live.

4.2. THE OBSERVATIONS OF A. V. HILL ON BIOLOGICAL FREE-ENERGY CHANGES

One of the classic writings on muscle physiology is that published by A. V. Hill[10] entitled, "The Heat-Production of Surviving Amphibian Muscles During Rest, Activity, and Rigor." At the beginning of this article he stated emphatically that, "The muscle is undoubtedly a chemical machine: by no stretch of the imagination can it be supposed to be a heat engine." (Hill,[10] p. 466.) This, of course, was completely contrary to what Simonson had attempted to demonstrate. Hill made it perfectly clear that although the measurement of heat production in muscles was a necessary investigation, it was actually the free energy of whatever reactions were taking place in the muscles which was a measure of the maximum amount of mechanical work that could be expected of them. Moreover, one senses his frustration that so little was then known about the free energy of formation of substances likely to be implicated in muscle tissue chemistry. In an appendix to this publication Hill went to some length to explain the concept of free energy as applied to biological processes in terms a biologist could understand. First he emphasized that there was little hope for understanding the mechanism of muscular activity until more was known about the fundamental physics and chemistry of muscles. However, there was one avenue along which muscle physiology could progress as an exact science and this was by way of energy and thermodynamics. Unfortunately, there had been ". . . many and grievous misconceptions as to the application of the laws of thermodynamics." (Hill,[10] p. 507.) One of these was the study of Simonson, which Hill proceeded to demolish absolutely. He objected that, ". . . the word "work" has been used to confuse the issue as between chemical and mechanical processes by its employment in the meaningless expression 'chemical work,' the 'chemical work' which a process can do being supposed given by the efficiency." (Hill,[10] p. 507.) He pointed out that $(T_1 - T_2)$ *must* be inconceivably small, since with respect to bacteria these temperatures would occur at two points that at most could not be more than 10^{-3} mm apart. He stated further, "It is to be regretted that such misconceptions as these should have given the impression either (a) that the laws of thermodynamics do not apply to the animal body, or (b) that they apply only to a heat engine which the living organism certainly is not." (Quoted by permission from Reference 10, p. 507.)

One further quotation from Hill illustrates clearly his general attitude toward the concept of free energy.

> The greatest successes achieved by the application of the laws of thermodynamics are in Chemistry: and these successes centre around the conception of the "free-energy" of a chemical reaction. By means of changes in osmotic pressure, surface tension, volume, or vapour pressure, or by means of electrical effects set up, every chemical reaction in the universe which goes on spontaneously can be made to do external work. The maximum mechanical work which any given reaction can do is named the "free-energy" of that reaction: and the Second Law of Thermodynamics tells us that *whatever be the mechanism of the work production* the maximum work that can be done is the same. This is of fundamental interest to the physiologist; at present we are unable to say by what mechanism the organism can perform mechanical work; but provided that we know what chemical processes the organism carries out we can calculate in general the maximum mechanical work which these processes can yield, independent of the mechanism by which that work is performed. For most chemical reactions this calculation is at present difficult, but there is no reason to doubt that with the advance of physical chemistry it will be rendered possible for *all* chemical bodies. [Quoted by permission from Reference 10, p. 507.]

The problem Hill faced with muscle physiology, that the chemical mechanisms whereby muscles contract were then largely still unknown, is similar in principle to that faced by microbial physiologists in that we still do not know a great deal about the processes by which growth takes place. No matter, if we are interested in the thermodynamics of the growth process the calculation of the maximum work that can be obtained for the purpose of growth will be independent of the mechanism by which it takes place. At least, according to classical thermodynamics this should be so.

Hill did not make any biological free-energy calculations. The importance of his work lies not only in the brilliance of his experimental studies on heat production in muscle but in his recognition that a knowledge of free-energy changes is absolutely essential to an understanding of biological energy exchanges. His statement that, "The muscle is undoubtedly a chemical machine working at constant temperature" (Hill,[10] p. 507) can be extended to include microbial cells. Exactly what the concept of "work" means, as applied to microbial growth, is something which is still a matter of discussion at the present time. Hill[11] did not mention "free-energy" or "work" in his article published less than one year earlier on the heat production accompanying milk fermentation.

4.3. THE FREE-ENERGY CALCULATIONS OF BARON AND POLANYI

In 1913 J. M. Baron and Polanyi[12] published the first attempt to compare the heats of reaction and the free-energy changes accompanying biological processes. To accomplish this they used the heat theorem formulated by Nernst[13] and calculated the free-energy changes that accompany the combustion of protein, fat, or carbohydrate. These values were then compared with the measured heats of combustion of these substances, with the result that for any given combustion the values for the free-energy change and the heat of combustion were found to be nearly equal. This has been found subsequently to be true for many organic substances, and may have given rise to the still commonly held opinion that since heats of reaction are far easier to obtain than free-energy changes, a measurement of the former will closely reflect changes in the latter. This of course is manifestly untrue for many chemical reactions.

4.4. THE FREE-ENERGY CALCULATIONS OF LINHART

In 1920 G. A. Linhart[14] published the first free-energy calculations of biological processes using the methods of treating thermodynamics developed by Lewis[1] and by Lewis and Gibson.[2] Linhart pointed out that any natural process is accompanied by a transfer of energy, part of which can be used for accomplishing useful work and part of which becomes dissipated as heat or becomes latent due to an entropy change. With respect to the utilization of energy, some processes, such as the burning of coal in a steam engine, are quite inefficient with only a small percentage of the available energy being used, the rest being lost as heat and entropy. The research he described was an attempt to learn something about the efficiency with which energy could be utilized in life processes. The subject of his study was the fixation of nitrogen gas by *Azotobacter*.

According to Linhart there are three general questions that can be asked with respect to any biological process. These are:

1. What is the transfer of energy? This depends solely on the initial and final states of the system being considered.
2. What are the mechanisms of the reactions involved in the process?
3. What are the rates of these reactions?

It was Linhart's opinion that most studies of life processes prior to 1920 took into consideration only their mechanisms and rates of reaction, possibly because most biologists did not possess a knowledge of thermodynamics. It should have been possible for biologists to approach physical chemists with their problems related to biological energetics. Unfortunately, whenever this happened the chemists usually answered that, "Biological processes are so highly complex that they are not susceptible to thermodynamic treatment." (Linhart,[14] p. 248.)

Even in the 1980s it is certainly true that some biological systems do indeed seem complex. However, it is not true that because of this they are necessarily not susceptible to thermodynamic treatment. Every mass has thermodynamic properties including living cells, and the fact that these latter are not pure, crystalline solids ought to have posed a considerable challenge to physical chemists. As it has happened this challenge has been generally ignored.

Linhart expressed clearly what was needed.

> In order to find the maximum or free-energy of any chemical process, such as the transformation of mannite into carbon dioxide and water, which certain bacteria utilize for the work required in decomposing sulfates, phosphates, nitrates, or in the "fixing" of nitrogen, we must know not only the change in the heat content attending the process, but also the change in entropy from the absolute zero up to the given temperature, T, at which the process takes place. We can then calculate the free-energy of the process with the aid of the thermodynamic equation
>
> $$\Delta F = \Delta H - T\,\Delta S$$
>
> where ΔF [now ΔG] denotes the free-energy change, ΔH the change in heat content, and ΔS the change in entropy of the isothermal process in question. [Linhart,[14] p. 248.]

One of the difficulties that beset Linhart in his research was the lack of good thermochemical data. Although reasonably accurate values for the heats of combustion of many organic compounds were available and from these the heats of formation of these same compounds, the measurement of entropy values was only beginning in 1920 and most of the compounds for which entropies were being determined were not of any importance biologically. What he wanted to calculate was the free-energy efficiency with which *Azotobacter* sp. fixed nitrogen gas using mannitol as a source of energy. What he needed were data on the free energy of formation, the heat of formation, and the entropy of mannitol. Data for the other substances entering into his calculations were already known.

The heat of combustion of mannitol at 25°C had previously been determined to be -728.40 kcal mole^{-1}, this combustion being represented by the equation

$$C_6H_{14}O_6(s) + 6.5O_2(g) \longrightarrow 6CO_2(g) + 7H_2O(l) \qquad (4.1)$$

The standard heats of formation of oxygen, carbon dioxide, and water were also known to be 0.000 (by convention), -94.251, and -68.280 kcal mole^{-1}, respectively. The standard heat of formation of mannitol could therefore be calculated in the following manner with respect to Eq. 4.1:

$$\Delta H^\circ = 6\,\Delta H_f^\circ CO_2 + 7H_f^\circ H_2O - \Delta H_f^\circ \text{mannitol} - H_f^\circ O_2$$

$$\Delta H_f^\circ \text{mannitol} = 6(-94.250) + 7(-68.280) - 6.5(0.000) - (-728.40)$$

$$= -315.06 \text{ kcal mole}^{-1} \qquad (4.2)$$

There were no data on the entropy of mannitol and thus an approximation had to be used. The entropy of any substance is related to its average atomic heat capacity at constant volume (C_v) by the following equation:

$$S = \int_{T_0}^{T} C_v d \ln T = 2.3 \int_{T_0}^{T} C_v d \log T \tag{4.3}$$

The average atomic heat capacity for any substance can be obtained by multiplying its specific heat capacity at temperature T (usually given in the literature at 25°C and often at 0°C) by its molecular weight and dividing this product by the number of atoms in the molecule. This value was known at 25°C for mannitol. By comparing this with those of other substances with similar average atomic heat capacities at 25°C the entropies of which were known, an approximate value could be found for the entropy of mannitol. The average atomic heat capacity of mannitol fell between those of solid crystalline benzene and of graphite, and since for these the plots of C_v against log T between 0 and 298°K did not intersect, the average atomic heat capacity curve for mannitol could be located between these two curves. The area under this curve was approximately 0.95 entropy units (e.u., cal deg^{-1}) and if this value were multiplied by the number of atoms in mannitol (26) and by 2.3 (see Eq. 4.3) the entropy of mannitol was found to be 56.8 e.u. mole^{-1}.

One definite uncertainty about Linhart's calculations is that benzene is not a solid crystalline substance at 25°C and undergoes a phase change from solid to liquid at 5.5°C. What he did about this is not stated. He may have extrapolated the mannitol curve so that it paralleled that of graphite, but we do not actually know.

The standard entropy of formation, ΔS_f°, of mannitol was found with the use of the following equation:

$$6C(s, \text{graphite}) + 14H(g) + 6O(g) \longrightarrow C_6H_{14}O_6(s) \tag{4.4}$$

from which

$$\Delta S_f^\circ = S^\circ_{C_6H_{14}O_6} - 6S^\circ_C - 14S^\circ_H - 6S^\circ_O \text{ (all at 25°C)} \tag{4.5}$$

where S°_C, S°_H, and S°_O are the standard entropies of carbon (graphite), hydrogen, and oxygen having the values of 1.3, 15.9, and 24.1 e.u., respectively, at 25°C. Insertion of these values into Eq. 4.5 gives

$$\begin{aligned} S_f^\circ &= 56.8 - 7.8 - 222.6 - 144.6 \\ &= -318.2 \text{ cal deg}^{-1} \text{ mole}^{-1} \end{aligned} \tag{4.6}$$

The standard free energy of formation, ΔF_f°, of 1 mole of solid mannitol could then be calculated by means of the following equation and the results of the calculations using Eqs. 4.2 and 4.6. Thus,

$$\begin{aligned}\Delta F_f^\circ &= \Delta H_f^\circ - T\,\Delta S_f^\circ \\ &= -315.06 \text{ kcal mole}^{-1} - (298.2 \text{ deg} \times -318.2 \text{ cal deg}^{-1} \text{ mole}^{-1}) \\ &= -220.18 \text{ kcal mole}^{-1} \qquad (4.7)\end{aligned}$$

This value is thus the free-energy change accompanying the formation of solid mannitol from the elements at 298.15°K and 1 atm, according to Eq. 4.4. However, it is not solid, crystalline mannitol that would be utilized by *Azotobacter* sp. under natural conditions, but mannitol that had been dissolved in aqueous solution at a low concentration, for example, 0.1 *m* (molal concentrations are utilized in physical chemistry; however, in most cases the difference between the concentrations of molal and molar solutions can be ignored for purposes of biological applications). Any substance in saturated aqueous solution has a free energy of formation equal to that of the same substance in its standard state. If this substance becomes diluted from the saturation concentration to a lower concentration energy will be expended in the process. This process of dilution can be represented as follows.

$$C_6H_{14}O_6(\text{sat'd sol'n, } 1.49\ m) \longrightarrow C_6H_{14}O_6(0.1\ m)$$

This reaction, being spontaneous, will be accompanied by a negative free-energy change that can be calculated in the following manner.

$$\Delta F' = -RT \ln \frac{\text{conc. at saturation} \times \text{activity coefficient}}{\text{desired conc.} \times \text{activity coefficient}} \qquad (4.8)$$

where $\Delta F'$ is the free-energy change accompanying the particular dilution process, R is the gas constant, and T is the absolute temperature at which the process of dilution takes place. Linhart made the assumption that for mannitol concentrations may be assumed to equal activities. Thus, for Eq. 4.8,

$$\begin{aligned}\Delta F' &= -1.987 \text{ cal deg}^{-1} \text{ mole}^{-1} \times 298.2 \text{ deg} \times \ln \frac{1.49\ m \times 1}{0.10\ m \times 1} \\ &= -1600 \text{ cal mole}^{-1} \qquad (4.9)\end{aligned}$$

Adding the results of the calculations using Eqs. 4.7 and 4.9 gave -221.78 kcal mole^{-1} as the free energy of formation of 1 mole of mannitol at a concentration in aqueous solution of 0.10 *m*. The new representation of the combustion of mannitol then becomes

$$C_6H_{14}O_6(0.1\ m) + 6.5O_2(g) \longrightarrow 6CO_2(g) + 7H_2O(l) \qquad (4.10)$$

Since the free energies of formation of oxygen, carbon dioxide, and water were known to be 0.000 (by convention), −94.360, and −56.620 kcal mole^{-1}, respectively, the free-energy change accompanying the reaction represented by Eq. 4.10 was calculated as follows.

$$\Delta F'' = 6(-94.360) + 7(-56.620) - (-221.78) - 6.5(0.000)$$
$$= -740.6 \text{ kcal mole}^{-1}$$

This quantity represents the maximum amount of energy (free energy) that Linhart felt would be available to *Azotobacter* cells from 10,182.1 g of mannitol solution, 182.1 g of which were mannitol, if all this solute were to be oxidized to carbon dioxide and water. Although this is what he stated, what he probably meant to state was that this amount of energy represents the maximum amount available from 1 mole of mannitol dissolved in a sufficient volume of water to give a 0.1 *m* solution. With these data he made a calculation which showed that this quantity of energy would be sufficient to form about 10 moles of ammonia from water and nitrogen gas, provided that the bacteria merely acted as catalytic agents, but the equations he used in making this calculation were not shown. Greaves[15] had previously shown that *Azotobacter* was capable of fixing nitrogen only to the extent of 1% of the weight of the mannitol utilized. If this were true, then instead of 10 moles only about 0.1 mole of nitrogen gas would actually be fixed.

One criticism can be leveled at Linhart's study. If he had wished to use values for the thermodynamic properties of substance in aqueous solution he should have calculated these for all the reactants and products in Eq. 4.10 rather than just for the free energy of formation of mannitol in 0.10 *m* aqueous solution. He probably did not do this because not enough physico-chemical data were available. Nevertheless, this study is extremely important. Linhart emphasized that it is the free-energy change that is the true measure of the energy change of a biological process and demonstrated in a manner that was comprehensible to biologists how free-energy changes could be calculated. In addition, he realized that the standard states of the substances generally chosen for study by physical chemists, namely pure gases and liquids and pure crystalline solids, were not appropriate for the study of biological processes. The appropriate state was that of a substance in aqueous solution at low concentration. The appearance of an accepted "aqueous standard state" represented by a 1 mole quantity of substance (compound) dissolved in water at unit activity was yet to come.

4.5. THE REVIEW BY BAAS-BECKING AND PARKS

Because of an increasing interest by biologists in biological energetics there appeared a number of publications which, like Linhart's,[14] attempted to describe energy relations in microbial metabolism. The autotrophic bacteria constitute an interesting physiological group of microorganisms which oxidize inorganic compounds as sources of energy and synthesize their cellular substance from carbon dioxide. By the early 1920s the chemical thermodynamic properties of most of these inorganic energy sources were known, and so it became possible to calculate the free energy that was available to these bacteria for the purpose of growth. In 1927 L. G. M. Baas-Becking and G. S. Parks[16] wrote an article reviewing much of what was then known about the metabolism of these organisms. They separated the autotrophic bacteria into five groups according to the inorganic compounds that were capable of serving as sources of energy, these being hydrogen gas, carbon compounds, nitrogen and nitrogen compounds, sulfur and sulfur compounds, and ferrous and manganous salts. Further considerations were then given to the thermodynamics of each. First, however, they described in considerably greater detail than had Linhart the basic elements of biological thermodynamics. One of the points they emphasized was the magnitude of the difference that may exist between ΔF° and ΔH°. In some cases this difference could be extremely small. As an example, for the combustion of graphite carbon represented by the following equation,

$$\mathrm{C(graphite)} + \mathrm{O_2(g, 1\ atm)} = \mathrm{CO_2(g, 1\ atm)}$$

ΔF° and ΔH° have virtually the same values, being -94.260 and -94.250 kcal mole^{-1}, respectively, at 25°C. This is not true for the combustion of methane at the same temperature, represented as follows.

$$\mathrm{CH_4(g, 1\ atm)} + \mathrm{2O_2(g, 1\ atm)} = \mathrm{CO_2(g, 1\ atm)} + \mathrm{2H_2O(l)}$$

Here the ΔF° and ΔH° values are -194.600 and -212.600 kcal mole^{-1}, respectively, a difference of 9.2% (compared to the ΔF° value). For other substances the difference is less extreme. The following equation represents the combustion of glucose with the reactants and products all being in their standard states at 25°C.

$$\mathrm{C_6H_{12}O_6(s)} + \mathrm{6O_2(g, 1\ atm)} = \mathrm{6CO_2(g, 1\ atm)} + \mathrm{6H_2O(l)} \qquad (4.11)$$

For this reaction the ΔF° is -685.800 kcal mole^{-1} and ΔH° is -674.000 kcal mole^{-1}, the difference being 1.7% (compared to the ΔF° value).

Another important aspect of biological thermodynamics which they considered was the appropriate values to adopt with respect to certain substances as they occur in nature. This had been previously considered by Linhart when he calculated the ΔF_f° of mannitol at a concentration of 0.1 m. Baas-Becking and Parks extended their calculations to include O_2, N_2, and CO_2 under the conditions in which they are found in the natural environment. Thus, naturally occurring O_2(g) has a partial pressure of 0.2 atm, N_2(g) of 0.8 atm, and CO_2(g) of 0.0003 atm. An example will make this clear.

Dalton's Law of Partial Pressure states that each component of a gas mixture behaves in a manner completely independent of the other components of the mixture. Air is such a mixture, being composed (for the sake of this calculation) of about 20% oxygen and 80% nitrogen. If air is considered at sea level to have a total pressure of 1 atm, then oxygen and nitrogen gas will have partial pressures of 0.2 and 0.8 atm, respectively. Baas-Becking and Parks used Dalton's Law as a basis for calculating the ΔF_f° of oxygen gas in the atmosphere at a partial pressure of 0.2 atm. To understand this, it must be imagined that 1 mole of pure oxygen at atmospheric pressure and 298.2°K is allowed to expand into a vacuum of sufficient volume such that when the temperature of the gas has again returned to 298.2°K by the absorption of heat from the environment its pressure will be 0.2 atm. This reaction system can be represented by the following equation.

$$O_2(\text{g, 1 atm, 298.2°K}) \longrightarrow O_2(\text{g, 0.2 atm, 298.2°K})$$

Since this is a spontaneous reaction it will be accomplished by a negative free-energy change which can be calculated by means of the following equation:

$$\begin{aligned} \Delta F &= RT \ln \frac{P_2}{P_1} \\ &= 1.987 \text{ cal deg}^{-1} \text{ mole}^{-1} \times 298.2 \text{ deg} \times \ln \frac{0.2 \text{ atm}}{1.0 \text{ atm}} \\ &= -955 \text{ cal} \end{aligned}$$

It is by this means that Baas-Becking and Parks arrived at the values in Table 4.1, which they adopted as the standard state values for the substances

TABLE 4.1. Free energy of formation values adopted by Baas-Becking and Parks as representing the "natural states" of the following substances at 298.15°K, as opposed to their "standard states"

Substance	ΔF_f
O_2, gas at 0.2-atm pressure	−955 cal
N_2, gas at 0.8-atm pressure	−132 cal
CO_2, gas at 0.0003-atm pressure	−99,060 cal
H_2CO_3, dissolved in water in equilibrium with a CO_2 pressure of 0.0003 atm	−155,620 cal
H_2S, dissolved in water at 10^{-6} mole liter^{-1}	−14,700 cal
Fe^{2+}, at a concentration of 10^{-5} mole liter^{-1}	−27,700 cal
Fe^{3+}, at a concentration of 10^{-2} mole liter^{-1}	−19,500 cal

Source: Reference 16.

listed. Using these values rather than the usual values for the standard states and also the values of -219.100 kcal mole^{-1} (Parks and Kelly[17]) and -56.560 kcal mole^{-1} (Lewis and Randall[3]) for glucose ($C_6H_{12}O_6$)(s) and H_2O(l), respectively, they calculated the free-energy change accompanying the synthesis of glucose autotrophically at 25°C by means of the following equation.

$$6CO_2(0.0003\ \text{atm}) + 6H_2O(l) = 6O_2(0.2\ \text{atm}) + C_6H_{12}O_6(s) \qquad (4.12)$$

The result was a valuc of $+708.900$ kcal mole^{-1} of glucose synthesized. This is a quite different value than that of $+658.800$ kcal mole^{-1} which would be obtained were the reaction represented by Eq. 4.11, in which O_2 and CO_2 are present at a pressure of 1 atm, to be reversed.

In carrying out their study Baas-Becking and Parks extended the considerations of Linhart by trying to understand the thermochemical nature of substances in whatever form the environment presents these to microbial cells. However, it is interesting that they did not extend their ideas further. All cells live in an aqueous environment, and thus it is not CO_2(g) or O_2(g) with which they interact, but CO_2(aq) and O_2(aq). Similarly, if it is actually glucose which is being represented in Eq. 4.12 it should not be represented as $C_6H_{12}O_6$(s), but rather as $C_6H_{12}O_6$(aq). The use of free energies of formation of substances in solution at given concentrations would alter the value for the free-energy change even further.

Where data were available Baas-Becking and Parks attempted to calculate the free-energy efficiencies for the growth of lithoautotrophic bacteria. As an example, Ruhland[18] had found that for every ml of CO_2 gas assimilated by hydrogen bacteria 8 ml of hydrogen gas needed to be oxidized. Baas-Becking and Parks used these data to calculate a free-energy efficiency as follows:

$$\begin{aligned}\text{eff.} &= \frac{\Delta F^\circ\ \text{glucose oxidation}/6}{8 \times \Delta F^\circ\ \text{hydrogen oxidation}} \times 100 \\ &= \frac{-118.100}{8 \times -56.000} \times 100 \\ &= 26.4\%\end{aligned}$$

where the value for the ΔF° of glucose oxidation was taken as that for the reverse of the reaction represented by Eq. 4.12. Inherent in this was the assumption that the product of CO_2 fixation in the hydrogen bacteria was the same as that for green plant photosynthesis. For whatever reason the ΔF° for hydrogen oxidation was calculated using a gas pressure of 1 atm. Similar calculations were made for other groups of lithoautotrophic bacteria. These data are summarized in Table 4.2. This was the first compilation of data of this type, and the individual values should be accepted as approximations.

TABLE 4.2. Free-energy efficiencies for the growth of autotrophic bacteria which have been grown using the following reactions as sources of energy

Reaction	Author	Percent Free-Energy Efficiency
$H_2 + 0.5O_2 = H_2O$	Ruhland 1924	26.4
$CH_4 + 2O_2 = CO_2 + 2H_2O$	Söhngen 1906	0.6–29.6 (limits)
$NH_4^+ + 1.5O_2 = NO_2^- + H_2O + 2H^+$	Meyerhof 1916	7.9
$NO_2^- + 1.5O_2 = NO_3^-$	Meyerhof 1916	5.9
$S + 1.5O_2 + H_2O = H_2SO_4$	Winogradsky 1881	8.3
$6KNO_3 + 5S + 2CaCO_3 = 3K_2SO_4$	Beyerinck 1920	5.0
$+ 2\,CaSO_4 + 2CO_2 + 2N_2$	Starkey 1925	
$5Na_2S_2O_3 + 8KNO_3 + 2NaHCO_3 =$ $6Na_2SO_4$	Lieske 1912	9.0
$+ 4K_2SO_4 + 4N_2 + 2CO_2 + H_2O$		

Source: Reference 16.

With the exception of the hydrogen and possibly the methane bacteria all the lithoautotrophic bacteria studied seemed to exhibit free-energy efficiencies of growth which are of the same order of magnitude and are under 10%.

4.6. THE STUDIES BY BURK OF THE FREE-ENERGY EFFICIENCY OF NITROGEN FIXATION

One of the most active investigators to consider the concept of free-energy changes was D. Burk, who published a number of articles dealing with nitrogen fixation by living organisms. Nitrogen-fixing organisms are generally considered to be heterotrophic, using organic substances as sources of energy and carbon but being capable of fixing gaseous nitrogen as a nitrogen source if no fixed nitrogen is present in the environment. It was also generally thought that some of the energy from the oxidation of the carbon source was utilized for the reduction of nitrogen gas to ammonium, in which form it could be used by the organism, if necessary. In 1927 Burk[19] published a study in which he suggested that, to the contrary, considerable energy might be released during nitrogen fixation, and that this could theoretically be used in the same manner that energy from the oxidation of organic substances could be used. At the time Burk wrote this article the biochemical mechanism of nitrogen fixation had not been worked out, and the reactions that were used to illustrate his points, while thermochemically possible, had not been demonstrated to take place. Nevertheless, the points he made could not be ignored.

As an example, Burk stated that theoretically nitrogen can be fixed according to the following reaction equation:

$$0.5N_2(g) + 1.25O_2(g) + 0.5H_2O(l) = HNO_3(aq)$$

for which the $\Delta F°$ is 1.780 kcal mole^{-1}. Here the reactants and products are present in their usual standard states and a small amount of free energy will have to be put into the reaction in order for it to proceed. Under the more environmental conditions of N_2 at a partial pressure of 0.8 atm, O_2 at a partial pressure of 0.2 atm, and assuming the activity of the water in the plant protoplasm to be that of pure water and the activity of dilute nitric acid to be unaffected by other substances in the protoplasm, then

$$\Delta F° = -RT \ln K = -1364.9 \log K$$

but,

$$\Delta F° = 1780 \text{ cal mole}^{-1}$$

Therefore,

$$\log K = \frac{1780}{-1364.9} = -1.304,$$

from which $K = 0.0497$, and

$$\text{the activity of } HNO_3 = K \times (0.8N_2)^{0.5} \times (0.2O_2)^{1.2}$$
$$= 0.0067$$

This activity corresponds to a concentration of nitric acid in aqueous solution of just 0.1 *M*. Thus, provided that the plant can catalyze the reaction, nitric acid (nitrate) will be formed in the plant with a negative free-energy change (spontaneous reaction) so long as its concentration remains below this value, which is equivalent to 6200 ppm by weight. At this point ΔF will have a value of zero, since the reaction will be in a condition of equilibrium. The smaller the concentration of nitrate, the greater the theoretical amount of work which will be available from this reaction. If the nitrate is removed as soon as it is formed such that its concentration would never rise above, say, 0.0001 *M* (6 ppm) and if the activity coefficient of the nitrate at this concentration is unity, then, according to Burk,

$$K = \frac{(0.0001)^2}{(0.8)^{0.5}(0.2)^{1.2}} = 8.6 \times 10^{-8}$$

and

$$\Delta F = \Delta F^\circ + RT \ln K = 1780 + 1364.9(-7.07)$$
$$= -7.870 \text{ kcal mole}^{-1}$$

Assuming that the heat of reaction does not significantly change during the dilution process from its standard value of -14.940 kcal mole^{-1}, were the nitrate concentration to be lowered in the plant protoplasm to a steady-state value of 7×10^{-6} M (0.4 ppm), the free-energy change would equal the heat of reaction and, according to Burk, the process would then become 100% efficient. Burk's comparison of the ΔF of -7.870 kcal mole^{-1} at a nitrate concentration of 6 ppm with that of -14.940 kcal mole^{-1} at a nitrate concentration of 0.4 ppm illustrates clearly how rapidly the free-energy change increases as the nitrate concentration decreases. A quotation was also cited from Lewis and Randall[3] (p. 568): "It is to be hoped that nature will not discover a catalyst for this reaction, which would permit all of the oxygen and part of the nitrogen of the air to turn the oceans into dilute nitric acid."

Burk also presented several similar reactions involving nitrogen fixation which could theoretically take place with a negative free-energy change if appropriate concentrations could be employed. It seems somewhat unfortunate that his imaginative approach to the subject of nitrogen fixation energetics received little more attention than the acknowledgment that he made it. He was perhaps ahead of his time in his considerations. It is significant that in this publication the states of the reactants and products [i.e., H_2(g), HNO_3(aq)] appear consistently with respect to each equation and that the suffix (aq) appears with greater frequency. It is apparent, however, that Burk still considered gases such as N_2, H_2, and O_2 to react in biological processes in the gaseous state rather than in the aqueous state. It should be pointed out that uncertainties about this still exist at the present time. The question may well be asked whether the free energy of solution has anything to do with the energetics of microbial growth. Does the free energy of solution component of the free energy of formation of a substance in solution actually enter into the driving force of a chemical reaction?

Another publication by Burk[20] in 1931 is of interest in that he discussed the reversibility of coupled reactions in biological systems and the Second Law of Thermodynamics. It may seem that by this time such an article would be redundant. However, Burk's reason for writing this article was that in fact scientists were still thinking about whether the Second Law applied to biological systems of small dimensions. As evidence of this he offered two quotations. According to Lewis and Randall, "The second law of thermodynamics is a principle that has never failed to satisfy the severest test of experiment." (Lewis and Randall,[3] p. 120.) According to Donnan, who came to his conclusions on the basis of probability calculations, "It seems, therefore, very probable that there exist biological systems of such minute dimensions that the laws of classical thermodynamics are no longer applicable to them." (Donnan,[21] p. 688.) In this article Burk attempted to offer support for the position of Lewis and Randall. Much of what he wrote is not directly applicable to the thermodynamics of individual reaction systems. However, some of it was exceedingly pertinent. According to Burk:

> There is only one general class of biological reactions where, so far as experiment has been capable of deciding, it is certain that the second law is operating. This class may be designated as "the entire life processes of an organism." It has been universal experience that the thermodynamic work done by any organism has always been considerable, i.e., in net effect heat has always been given to, and not taken from, the environment in amount greater than that minimum required by the second law. No perfectly reversible "whole life process" of an organism has ever been observed; the chemical energy of the oxidizable organic matter supplied autotrophic organisms has always been gradually, continuously, and often completely dissipated as heat. [Burk,[20] p. 432.]

Burk then went on to define two types of "free-energy efficiency." One is the *machine efficiency*, defined as ". . . the work done in any specific reaction occurring in a biological system divided by the free-energy consumed by the system as a whole." (Burk,[20] p. 433.) He cited the free-energy efficiencies calculated in the manner of Baas-Becking and Parks[16] as being examples of machine efficiencies. The other was the *second law efficiency* defined as,

> . . . the efficiency obtained upon correcting the machine free-energy efficiency for the irrelevant losses involved in simultaneous extraneous metabolism (i.e., the work done in any specific reaction is divided by the free-energy dissipated in that reaction only); specifically, it measures, in per cent, the real reversibility of an isolated reaction. Where reversibility is perfect and the second law operates, the efficiency attains one hundred per cent, but values higher than this may result in event that the second law does not obtain. [Burk,[20] p. 433.]

Burk considered that, "The main thermodynamic work which microorganisms perform is that of growth." (Burk,[20] p. 451.) This was based on the publications of v. Angerer[22] and Ludwig[23] who showed that in motile microorganisms only an inconsequential fraction of their metabolic energy (about 0.1%) can be accounted for by their motility. In developing his argument he used the same experimental data of Ruhland[18] as had Baas-Becking and Parks,[16] but in a more exact way. This was because Ruhland had not only measured the hydrogen and carbon dioxide changes accompanying the growth of *Bacillus pycnoticus* autotrophically, but also the oxygen changes. Thus, the oxygen consumption provided a measure of the amount of hydrogen oxidized for purposes of respiration, and the carbon dioxide consumption a measure of the energy required for the reduction of this gas to organic material in the cells.

The overall reaction in which *B. pycnoticus* was grown autotrophically on hydrogen gas was represented by the following equation:

$$2H_2(g) + CO_2(g) = \tfrac{1}{6}C_6H_{12}O_6(s) + H_2O(l) \qquad \Delta F^\circ = 1.184 \text{ kcal} \tag{4.13}$$

Here all the reactants and products are in their standard states, and the formula for glucose represents the organic material synthesized as a result of carbon dioxide reduction. By now (1931) it seemed to be a generally accepted practice to represent synthesized cellular material as being approximately at the level of reduction of carbohydrate, and it eventually came to be represented by the symbol CH_2O. Burk used the equivalent representation of $\tfrac{1}{6}C_6H_{12}O_6(s)$. For whatever reason he did not use a value representing ΔF_f° for $\tfrac{1}{6}C_6H_{12}O_6(aq)$.

The reaction represented by Eq. 4.13 can be considered theoretically as a coupled reaction having two components, one being a reaction in which hydrogen gas is oxidized to provide energy for the second component, which is a reaction in which cellular material becomes synthesized from carbon dioxide gas utilizing the energy provided by the first reaction. The first reaction is thus exergonic and the second endergonic. Burk felt that if the two energy values for these coupled reactions were equal and opposite, the system exhibited a perfect reversibility. The two component reactions were represented using the states of the reactants and products as they actually existed during Ruhland's experiments.

The oxidation of hydrogen using standard states at 298.2 °*K and* 1 *atm:*

$$2H_2(g) + O_2(g) = 2H_2O(l) \qquad \Delta F^\circ = -113.120 \text{ kcal} = -56.560 \text{ kcal mole}^{-1} \tag{4.14}$$

The oxidation of hydrogen using conditions pertaining to Ruhland's experimental procedure:

$$2H_2(0.35 \text{ atm}) + O_2(0.06 \text{ atm}) = 2H_2O(l) \tag{4.15}$$

$$\begin{aligned} \Delta F &= \Delta F^\circ - RT\ln(0.35)^2 - RT\ln(0.06) - RT\ln\Delta N \\ &= (2 \times -56.560) + 1250 + 1630 + 1780 \\ &= -108.460 \text{ kcal} \end{aligned} \tag{4.16}$$

where $\Delta F^\circ = -113.120$ kcal (from Eq. 4.14); $RT\,\Delta N$ is a factor that corrects for the fact that the experimental reaction took place at constant volume (i.e., without external pressure–volume work), N being the mole volume change $=$ -3 in this case; and the gas pressures are the averages between the initial and final pressures during the experimentation.

The reduction of carbon dioxide using standard states at 298.2 °*K and* 1 *atm:*

$$CO_2(g) + H_2O(l) = \tfrac{1}{6}C_6H_{12}O_6(s) + O_2(g) \qquad (4.17)$$
$$\Delta F^\circ = 114.300 \text{ kcal mole}^{-1}$$

The reduction of carbon dioxide using conditions pertaining to Ruhland's experimental procedure:

$$CO_2(0.06 \text{ atm}) + H_2O(l) = \tfrac{1}{6}C_6H_{12}O_6(10^{-6}\,M) + O_2(0.06 \text{ atm}) \qquad (4.18)$$
$$\Delta F = \Delta F^\circ - RT\ln(0.06) + RT\ln(0.06) + RT\ln(10^{-6})^{1/6}$$
$$= 114.300 + 1.630 - 1.630 - 1.365 \qquad (4.19)$$
$$= 112.935 \text{ kcal}$$

where $\Delta F^\circ = 114.300$ (from Eq. 4.17), and the activity of glucose is considered to be about 10^{-6} M per $C_6H_{12}O_6$ (see the original article for the more complicated details). The addition of Eqs. 4.15 and 4.18 gives Eq. 4.13, but the addition of the ΔF values for Eqs. 4.16 and 4.19 does not give the ΔF° value for Eq. 4.13, being 3.291 kcal more positive. Looking deeper into the experimental values Burk then found that the formula $C_6H_{12}O_6$ did not truly represent the organic material formed during carbon dioxide reduction which, based on a hydrogen balance, would be better represented by $C_6H_{0.931\times12}O_6$, or $C_6H_{11.172}O_6$ (again, see the original article for more specific details). Thus, Eq. 4.17 would be better written

$$CO_2(0.06 \text{ atm}) + 0.931H_2O(l) = \tfrac{1}{6}C_6H_{11.172}O_6 + 1.931O_2(0.06 \text{ atm}) \qquad (4.20)$$

Equation 4.20 plus Eq. 4.15 no longer equals Eq. 4.13 but gives a different equation which Burk felt more closely represented what actually happens.

Burk reasoned that all the hydrogen consumed during the growth of *B. pycnoticus* was used to reduce carbon dioxide whereas all the oxygen consumed was used in respiration (metabolism) and was a measure of metabolic energy consumption. As he described it;

> Oxygen consumption measures that fraction of the H_2 used for reducing the CO_2 which is later reoxidized back to CO_2 in the metabolic processes of the organism. The remaining H_2 is a measure of the H_2 required to reduce the CO_2 to organic C found in the cells at the end of the experiment. [Burk,[20] p. 436.]

From a table of Ruhland's data Burk calculated that this latter ratio $(H_2 - 2O_2)/2CO_2$ had a value of 0.966 ± 0.012. Using the values he obtained for the free-energy changes accompanying the reactions represented by Eqs. 2.49 and 2.45 he found a ratio of $105.140/108.460 = 0.970 \pm 0.000$ (assigning a probable error, in distinction to any other type of error, of ± 0.000 to the thermodynamic data). Apparently slightly less hydrogen was oxidized than was necessary to produce the energy required for the reduction of carbon dioxide, the efficiency of the process being

$$\text{eff.} = \frac{0.970}{0.966} = 100.4 \pm 1.2\%$$

Burk then concluded that (1) the reduction of carbon dioxide by hydrogen is to all intents and purposes perfectly reversible, and (2) that the second law is not transgressed, the free-energy efficiency being not significantly greater than 100%. He then went on to state in a footnote;

> By perfect reversibility is here meant the perfect reversibility of the isolated process of stored carbon dioxide reduction, and that *all* of the free-energy of hydrogen oxidation concerned has been stored in the free-energy of the stored reduced CO_2, rather than being irreversibly converted into a certain amount of heat. A certain amount of heat is of course given off in the isolated process since ΔF of Equation 3 [in this present text, Eq. 4.16] is considerably smaller than ΔH. [Burk,[20] p. 439.]

It is apparent, Burk felt, that although "Oxygen consumption measures that fraction of the H_2 used for reducing the CO_2 which is later oxidized back to CO_2 . . ." (Burk,[20] p. 436), there was no direct oxidation of H_2 by O_2. Rather, CO_2 was reduced to organic material some of which was oxidized back to CO_2 by O_2 during the course of respiration and metabolism. However, since classical thermodynamics of the kind Burk was using deals only with initial and final states, it was impossible to know whether the mechanism Burk imagined operated or not.

Burk also pointed out that since the machine efficiency of growth for *B. pycnoticus* is 29%, 29 + 71 ml of CO_2 are required to produce 29 ml equivalents of stored reduced CO_2, or 29 + 71 g of sugar are required to produce approximately 29 g of hydrogen bacteria, giving a ratio of 3.4:1. He suggested that this agreed well with the best values from Ruhland's data where $(O_2{:}CO_2) + CO_2 = 2.4 + 1 = 3.4$, and showed that, "the inefficiency of the bacteria considered as a machine for storing CO_2 is entirely taken care of by its normal metabolic needs as judged by the needs of comparable organisms, and hence the reversibility of the hydrogen–carbon dioxide reaction must be fairly high." (Burk,[20] p. 441.)

This latter investigation of Burk's is important in that for the first time an attempt was made to separate the energy-yielding part of the growth metabolism of a microorganism from the energy-consuming part, each being represented as a single reaction equation. His representation was correct even though his interpretation was not (he felt that hydrogen and oxygen did not directly interact, the amount of oxygen being consumed being only a measure of the amount of hydrogen used to reduce carbon dioxide to organic material).

This investigation is also important in that, again for the first time, some attention was addressed toward achieving a satisfactory representation of what was synthesized during microbial growth. Possibly by analogy with green plant photosynthesis in which carbon dioxide is also fixed, Burk thought that the organic product of CO_2 fixation in the hydrogen bacteria could be considered a carbohydrate and he represented this as glucose. However, on the basis of Ruhland's experimental data he subsequently realized that the product of CO_2 fixation would be better represented by the formula $C_6H_{11.172}O_6$ than by the formula $C_6H_{12}O_6$ (glucose). He carried this development no further, but it is here for the first time that a representation was made of the composition of cells which was not in terms of small, whole numbers (as in $C_6H_{12}O_6$); and although Burk did not mention it, this can be considered in retrospect as the first step in considering the "average degree of reduction" of the cellular material formed during microbial growth. Burk was apparently still thinking of the reduction of CO_2 as forming a single organic substance that was subsequently oxidized by gaseous oxygen (just as in heterotrophic growth) to provide energy for the synthesis of cellular material.

4.7. THE TREATISE OF BUCHANAN AND FULMER

In 1928 Vol. I of a treatise entitled *Physiology and Biochemistry of Bacteria* by R. E. Buchanan and E. I. Fulmer[24] appeared, in which about 20% of the text is devoted to energy relationships involved in the growth and movement of microorganisms and to energetics. Included in it are several tables of data of thermochemical interest, for example, heats of formation, heats of solution, and free energies of formation, and it is apparent that in the eight years since Linhart[14] published his study much more thermochemical information had been accumulated. Nevertheless, these authors considered it unfortunate that so little information was then available with respect to the ΔF_f° of organic compounds of biological interest. They described many sources of chemical energy that could be utilized by microorganisms in the process of growth or metabolism. These included fermentations as well as oxidations of both organic and inorganic substances, together with calculations of their respective heats of reaction and free-energy changes. One gains the impression that such calculations were no longer considered to be an unusual thing, and that the importance of the free-energy change as opposed to the heat of reaction as the true measure of the energy exchange of a biological process was well recognized in 1928.

Buchanan and Fulmer pointed out that using Linhart's[14] calculations for the oxidation of mannitol and a more accurate value for its free energy of formation, the difference between ΔF° and ΔH° for the oxidation of mannitol may well be within the limits of error for the determination of ΔH°. This similarity of the values for ΔF° and ΔH° was noticed with respect to the oxidation of many organic compounds, in confirmation of the initial observation of this fact by Baron and Polanyi.[12] Thus, although Buchanan and Fulmer state that, "It seems evident that for the accurate determination of the maximum work that may be secured as the result of the energy released by a chemical reaction it is necessary to use the free-energy decrease." (Buchanan and Fulmer,[24] p. 408) they also state, "It should be noted, however, that in the general group of compounds most commonly used by bacteria, the error [i.e., the difference between ΔF and ΔH] is probably not so large as to invalidate general conclusions drawn from the use of the heats of reaction." (Buchanan and Fulmer,[24] p. 409.) While this may be true, statements such as this latter probably also served to draw the attention of subsequent investigators away from the importance of the free-energy change in a chemical reaction. The heat production accompanying the growth of a microorganism could be relatively easily and accurately measured. The free-energy change accompanying the same reaction could not, since the thermodynamic properties of the mass of cellular material produced during growth were not known and could not be calculated until relatively recently. It was all too easy to suggest that the free-energy change was probably very close in value to the heat of reaction, and therefore the fact that it could not be determined need not be of concern. On the other hand, Buchanan and Fulmer *did* point out that in fermentation reactions, such as alcoholic fermentation, the ΔF and the ΔH were very much different, being -55.4 and -31.2 kcal mole^{-1}, respectively, using standard values for their calculations.

4.8. THE CALCULATIONS OF FULMER AND LIEFSON OF THE FREE-ENERGY CHANGE ACCOMPANYING ALCOHOLIC FERMENTATION

Almost simultaneously with the publication of the treatise by Buchanan and Fulmer,[24] E. I. Fulmer and E. Liefson[25] summarized the methodology which is used for the calculation of free-energy changes and then proceeded to demonstrate this by calculating the free-energy change accompanying the fermentation of sucrose, represented by the following equation.

$$C_{12}H_{22}O_{11} + H_2O = 4C_2H_5OH + 4CO_2$$

As they emphasized:

> During the course of the reaction the concentration of the reactant is decreasing while the concentrations of the resultants [products] are increasing. These changes all lead to a decrease in the available free energy for the reaction. It is obvious then that not only cannot standard states be used for calculating the free energy decrease for the reaction, but that variation in concentrations must be taken into consideration. Other factors further complicate the situation; for example, the effect of the salts on the medium, of metabolic products, change in pH and so on, upon the activities of the reacting materials. [Fulmer and Liefson,[25] p. 169.]

Fulmer and Liefson did not attempt to take into account all the factors they could imagine which might affect the activities of the reactants and products. They did, however, consider the effect of the alcohol produced during the fermentation upon the solubility of any sucrose present, and hence upon its activity. One of the difficulties with respect to calculations of this sort became evident upon consideration of the appropriate concentrations to choose for making calculations. For example, at the start of an alcoholic fermentation when the yeast is inoculated into a solution of sucrose and minerals the concentration of alcohol in this solution is zero as is that of carbon dioxide, the two products of the fermentation reaction. Theoretically, as the concentration of a solute approaches zero its free energy of formation approaches infinity. Obviously, such values cannot be used, and Fulmer and Liefson made the simplifying assumption that when the medium is inoculated the inoculum carries with it a given small concentration of alcohol. It was also assumed that none of the alcohol leaves the solution, so that its concentration remains at a finite value. They assumed further that the nutrient solution is already saturated with carbon dioxide at its partial pressure in the atmosphere, and that it remains constant at this value (i.e., the CO_2 diffuses into the atmosphere even though the alcohol is assumed not to do so). As the fermentation becomes complete the concentration of sucrose approaches zero and theoretically its free energy of formation approaches infinity. The only way to avoid this is to stop the fermentation when the sucrose is still at some finite value. Taking all these things into consideration Fulmer and Liefson came to the conclusion that:

> . . . under the conditions specified the maximum free energy decrease for the conditions specified will be about 23% higher than that calculated for the standard state. Considering the probable inaccuracies in the values in free energy decrease of formation of the compounds . . . and the fact that the equation assumed for the fermentation does not represent accurately the products formed by biological action, the corrections from the standard state in this case are not of relatively great importance. However, it is worth while, in each instance, in the use of free energy changes of biological processes to analyze the situation along the lines presented above. [Fulmer and Liefson,[25] p. 175.]

It is apparent that this last advice is as pertinent now as it was when Fulmer and Liefson wrote it.

4.9. THE STUDY BY JOHNSON, PETERSON, AND FRED ON THE ENERGY CHANGES ACCOMPANYING THE ACETONE–BUTYL ALCOHOL FERMENTATION

Yeast and bacterial fermentations have always been of great interest to the brewing and solvent industries because of the possibilities of commercial application. Possibly because of this a considerable amount of work was done from the 1850s through 1931 (and, of course, later) on the analysis of fermentation products and on chemical representations of the various fermentations. At first the fermentations studied were relatively simple. Examples of these are the alcoholic fermentation of sugars by yeasts and the lactic fermentation of milk by lactic acid bacteria. Analyses of these fermentations showed that they could be represented chemically by equations in which the coefficients were small, whole numbers. Thus, for alcoholic fermentation the appropriate representation was the Gay-Lussac equation.

$$\underset{\text{Glucose}}{C_6H_{12}O_6} \longrightarrow 2CO_2 + \underset{\text{Ethanol}}{2CH_3CH_2OH} \tag{4.21}$$

The homolactic fermentation (*Streptococcus* sp. and some *Lactobacillus* sp.) could be represented by the equation

$$\underset{\text{Glucose}}{C_6H_{12}O_6} \longrightarrow \underset{\text{Lactic acid}}{2CH_3CHOHCOOH} \tag{4.22}$$

and for one of the heterolactic fermentations (*Leuconostoc* sp. and some *Lactobacillus* sp.) an appropriate equation was

$$\underset{\text{Glucose}}{C_6H_{12}O_6} \longrightarrow \underset{\text{Lactic acid}}{CH_3CHOHCOOH} + \underset{\text{Ethanol}}{CH_3CH_2OH} + CO_2 \tag{4.23}$$

An obvious feature about the above equations is that there is no representation of the microorganisms that bring about the fermentations, perhaps for two reasons. First, it was simpler to represent chemical reactions by equations in which the coefficients are small, whole numbers. This was done in general for equations representing both inorganic and organic reactions, and there was no reason not to do the same with respect to equations representing fermentation reactions. Second, chemical analyses of the fermentations often yielded fermentation balances that closely approached 100% in terms of the reactants and products as represented by, for example, Eqs. 4.21, 4.22, and 4.23. Everyone knew, of course, that microorganisms were needed for the fermentations. But what the fermentation reaction equations represented was what the microorganisms *did*, and this was to chemists more important than the microorganisms themselves. Thus, the idea that microorganisms are also products of these fermentations never seemed to be very seriously considered.

The fermentation industries gained great importance in Europe during World War I for the production of glycerol in Germany and acetone in England for use in the manufacture of explosives. The latter fermentation was carried out by strains of *Clostridium acetobutylicum* using carbohydrates from various sources as substrates. The particular fermentation caused by this microorganism is called an acetone–butyl (sometimes also a butyric–butylic) fermentation. It is far more complex than alcoholic or homolactic fermentation, for example, and produces a variety of fermentation products including carbon dioxide, hydrogen, butyl alcohol, acetone, ethyl alcohol, acetic acid, and butyric acid in varying proportions. This fermentation was studied in considerable detail by many investigators, one of the best studies being that of M. J. Johnson, W. H. Peterson, and E. B. Fred[26] in 1931. The microorganism was grown anaerobically in a medium consisting of minerals, 0.5% peptone, and varying quantities of carbohydrate. The results of the fermentation analysis were summarized by Wilson and Peterson[27] in the same year and are shown in Table 4.3. This is a typical example of the kind of highly accurate fermentation chemistry that was done during the 1920s with far more cumbersome analytical procedures than are presently employed. Even with such a complex fermentation the carbon and hydrogen balances in Table 4.3 are 100%, and this was not unusual. Equally good fermentation balances were recorded by Harden and Norris[28] and Braak[29] for the fermentation of glycerol by *Bacterium coli*; Grey[30] and Scheffer[31] for the fermentation of glucose by *Bacterium coli*; Gayon and Dubourd,[32] Van Steenberge,[33] and Stiles et al.[34] for the fermentation of various sugars by *Lactobacillus pentoaceticus*; Van Niel[35] and Foote et al.[36] for the fermentation of various substances by *Propionibacterium* sp.; and Smit[37] for the fermentation of glucose by *Sarcina ventriculi*. Many other fermentation balances found in the literature closely approach 100% but need not be cited

TABLE 4.3. Product and energy balance for *Clostridium acetobutylycum* fermenting carbohydrates

Carbohydrates	Glucose (millimoles)	Mannitol (millimoles)	Arabinose (millimoles)
Total fermented	1000	1000	1000
Products formed:			
Acetic acid	100	275	220
Butyric acid	12	153	3
Acetone	313	78	394
Ethyl alcohol	288	60	247
Butyl alcohol	489	604	1702
Carbon dioxide	2282	2070	946
Hydrogen	1382	2140	946
Water	122	263	216
	kcal	kcal	kcal
$-\Delta H$ of substrate fermented	302.5	317.9	253.8
$-\Delta H$ products × mole fraction of each	314.5	322.8	265.5
$-\Delta H$ of fermentation per mole of substrate	12.0	4.9	11.7
$-\Delta F°$ of substrate fermented	219.0	226.1	184.4
$-\Delta F°$ products × mole fraction of each	277.8	282.5	231.7
$-\Delta F°$ of fermentation per mole of substrate	58.8	56.4	47.3

Source: Reprinted with permission from *Chem. Rev.* **8**, 427–480. Copyright 1931 American Chemical Society.

here. It is apparent that the chemical analysts working during this period took pride in their work. Anyone not coming up with a 100% fermentation balance had obviously missed some part of the analysis.

Johnson et al.[26] attempted to represent the process of fermentation by the summation of a set of equations rather than by a single equation such as could be used to represent alcoholic fermentation or the lactic fermentation, in the following manner.

$$(1)\ C_6H_{12}O_6 = 2C_2H_5OH + CO_2$$

$$(2)\ C_6H_{12}O_6 = C_4H_9OH + 2CO_2 + H_2O$$

$$(3)\ C_6H_{12}O_6 + H_2O = CH_3COCH_3 + 3CO_2 + 4H_2$$

$$(4)\ C_6H_{12}O_6 + 2H_2O = 2CH_3COOH + 2CO_2 + 4H_2$$

$$(5)\ C_6H_{12}O_6 = C_3H_7COOH + 2CO_2 + 2H_2$$

According to these authors:

> It should be noted that these equations are not intended to mean that certain definite molecules of glucose are converted exclusively to one end product. They are, rather, a statement that the stoichiometric relations involved in the conversion of the common intermediates to the various end products are such that the quantitative relationships connecting glucose destroyed, gases evolved, and solvents produced are the same as those which would obtain if the equations were a literal statement of fact. [Johnson et al.,[26] p. 576.]

From the above quotation it is evident that Johnson et al. still did not like to think of fermentation reactions in which there were fractional coefficients rather than coefficients that were small, whole numbers.

4.10. THE REVIEW BY WILSON AND PETERSON ON THE ENERGETICS OF HETEROTROPHIC BACTERIA

In a review article published in 1931 P. W. Wilson and W. H. Peterson[27] attempted to do with respect to heterotrophic bacteria what Baas-Becking and Parks had done with respect to autotrophic bacteria. One of the most interesting aspects of this review is their originality in calculating the free-energy changes and the heats of reaction accompanying several complex fermentations. The most accurate of these makes use of the data of Johnson et al.[26] to calculate the energy changes accompanying the acetone–butyl acetone fermentation of glucose. These data showed that from the tenth to the thirty-second hour of an experimental fermentation the concentrations [ratio?] of acetic and butyric acids varied so little that they [it?] could be considered as constant during this period. During this same period 40% of the glucose disappeared, and the accompanying free-energy change could be taken as being representative of the entire fermentation. Calculations of the free-energy changes for the production of the acids showed that they were not markedly affected by the change in glucose concentration. Because of this the free-energy changes accompanying the formation of the solvents during the fermentation were calculated using the average value of the glucose concentration. As with the representation of Johnson et al., Wilson and Peterson used one equation to represent the formation of one type of fermentation product from one molecule of glucose, with the understanding that, "In the actual fermentation these products undoubtedly arise from a common intermediate compound rather than from a discrete glucose molecule." (Wilson and Peterson,[27] p. 457.) The equations of Wilson and Peterson representing the subreactions in the acetone–butyl alcohol fermentation are shown in Table 4.4. Using these data they then

TABLE 4.4. Equations representing the acetone–butyl alcohol fermentation of glucose, together with their accompanying free-energy and enthalpy changes

Acetic Acid Formation

$$C_6H_{12}O_6\left\{\begin{matrix}0.115\\0.060\end{matrix}\ M\right\} + 2H_2O(l) \rightarrow 2CH_3COOH(0.035\ M) + 2CO_2(0.65\ \text{atm}) + 4H_2(0.35\ \text{atm})$$

$$-\Delta F = \left\{\begin{matrix}54.3\\53.9\end{matrix}\right.\ \text{kcal}, \qquad -\Delta H = -17.5\ \text{kcal}$$

Butyric Acid Formation

$$C_6H_{12}O_6\left\{\begin{matrix}0.115\\0.060\end{matrix}\ M\right\} \rightarrow C_3H_7COOH(0.015\ M) + 2CO_2(0.65\ \text{atm}) + 2H_2(0.35\ \text{atm})$$

$$-\Delta F = \left\{\begin{matrix}69.0\\68.6\end{matrix}\right.\ \text{kcal}, \qquad -\ \Delta H = 17.2\ \text{kcal}$$

Acetone Formation

$$C_6H_{12}O_6(0.087\ M) + H_2O(l) \rightarrow CH_3COCH_3\left\{\begin{matrix}0.003\\0.020\end{matrix}\ M\right\} + 3CO_2(0.65\ \text{atm}) + 4H_2(0.35\ \text{atm})$$

$$-\Delta F = \left\{\begin{matrix}49.5\\48.4\end{matrix}\right.\ \text{kcal}, \qquad -\Delta H = -26.4\ \text{kcal}$$

Butyl Alcohol Formation

$$C_6H_{12}O_6(0.087\ M) \rightarrow C_4H_9OH\left\{\begin{matrix}0.004\\0.034\end{matrix}\ M\right\} + 2CO_2(0.65\ M) + H_2O(l)$$

$$-\Delta F = \left\{\begin{matrix}70.7\\69.4\end{matrix}\right.\ \text{kcal}, \qquad -\Delta H = 35.0\ \text{kcal}$$

Ethyl Alcohol Formation

$$C_6H_{12}O_6(0.087\ M) \rightarrow 2C_2H_5OH\left\{\begin{matrix}0.003\\0.017\end{matrix}\ M\right\} + 2CO_2(0.65\ \text{atm})$$

$$-\Delta F = \left\{\begin{matrix}58.7\\57.6\end{matrix}\right.\ \text{kcal}, \qquad -\Delta H = 18.1\ \text{kcal}$$

Source: Reprinted with permission from *Chem. Rev.* **8**, 427–480. Copyright 1931 American Chemical Society.

calculated that 0.055 mole of glucose was utilized from the tenth to the thirty-second hour of the experiment with an accompanying free-energy change of −3.34 kcal, from which $\Delta F = -60.8$ kcal $mole^{-1}$ of glucose fermented. This is in close agreement with the value of −58.8 kcal for the same fermentation found in Table 4.3. They also pointed out that this value is less than one-tenth of the −685.8 kcal $mole^{-1}$ of free energy "potentially" available from the complete oxidation of glucose were it in the standard state.

The calculations of Wilson and Peterson shown in Table 4.4 represent a considerable advance toward understanding the energetics of biological processes in which microbial growth undoubtedly took place but was never represented.

Other calculations were made of the free-energy changes and the heats of reaction accompanying the fermentation of glycerol and of glucose by coliform bacteria, of glucose by the propionibacteria, and of several sugars by mannitol-forming bacteria. The free energies and heats of formation of the reactants and products were multiplied by their respective mole fractions and these values used in turn to calculate the free-energy change and the heat of reaction for the individual fermentations. The results are astonishing! For example, using data from two strains of *Propionibacterium* sp. studied by Van Neil[34] and from three strains studied by Foote et al.,[35] Wilson and Peterson calculated the ΔF and ΔH values for the propionic acid fermentation of glucose to be -70.98 ± 0.85 ($n = 5$) and -46.48 ± 1.10 ($n = 5$) kcal $mole^{-1}$, respectively. What is astonishing is that with five different strains of the bacterium and two different investigations, and with each strain exhibiting slightly different ratios of fermentation products, the calculated ΔF values agree within about 1% and the ΔH values within about 2%. If these were to be calculated using presently accepted values for the free energies and heats of formation the agreement might be even better. Since we know that bacterial growth was involved in these reactions, might not some thermochemical representation of the process of microbial growth give values that are equally precise?

4.11. SUMMARY

From the presentation of the foregoing material it is evident that the consideration of free-energy changes in biological processes has had a far shorter history than that of their heats of reaction. Nevertheless, by 1931 investigations of microbial energetics had progressed to the point where, at least as far as the methods of calculating free-energy changes and heats of reaction were concerned, there was little more to do. Further advances had to keep step with the acquisition of data in the form of chemical thermodynamic properties of a much larger number of organic substances of biological interest. This was slow in coming.

REFERENCES

1. G. N. Lewis, The free-energy of chemical substances, *J. Am. Chem. Soc.* **30,** 1–30 (1913).
2. G. N. Lewis and G. E. Gibson, The entropy of the elements and the third law of thermodynamics, *J. Am. Chem. Soc.* **39,** 2554–2581 (1917).
3. G. N. Lewis and M. Randall, *Thermodynamics and the Free Energy of Chemical Substances*, McGraw-Hill, New York, 1923.
4. M. Rubner, *Die Gesetze des Energieverbrauchs bei der Ernährung*, Franz Deuticke, Leipzig and Wein, 1902.
5. C. Wachtel, Über die Anwendbarkeit des Zweiten Hauptsatzes der Thermodynamik auf Vorgänge im tierschen Organismus, *Pfluger's Arch.* **171,** 66–85 (1918).
6. R. Höber, *Physikalische Chemie der Zelle und der Gewebe*, 5th ed., Wilhelm Engelmann, Leipzig, 1924.
7. C. Oppenheimer, *Handbuch der Chemie*, 2nd ed., Vol. II, Gustav Fischer, Jena, 1925.
8. C. E. Guye, *Physico-chemical Evolutions*, trans. by J. R. Clarke, E. P. Dutton and Co., London, 1925.
9. E. Simonson, Die Bedeutung des zweiten Hauptsatzes der Thermodynamik für die Fiebers, *Zentralbl. Physiol.* **24,** 702–705 (1911).
10. A. V. Hill, The heat-production of surviving amphibian muscles during rest, activity, and rigor, *J. Physiol.* **44,** 466–513 (1912).
11. A. V. Hill, A new form of differential calorimeter, for the estimation of heat production in physiological, bacteriological, or ferment actions, *J. Physiol.* **43,** 261–285 (1911–1912).
12. J. Baron and M. Polanyi, Uber die Anwendung des zweiten Hauptsatzes der Thermodynamik auf Vorgänge im tierschen Organsimus, *Biochem. Z.* **53,** 1–20 (1913).
13. W. Nernst., Nachr. kgl. Ges. Wiss. Gottingen, Math.-physik. Klass. pp. 1–40, 1906; later summarized by Nernst in *Die theroetischen und experimentellen Grundlagen des neues Wärmesatzes*, W. K. Napp, Halle, 1918; trans. by G. Barr as *The New Heat Theorem*, E. P. Dutton and Co., New York, 1926.
14. G. A. Linhart, The free energy of biological processes, *J. Gen. Physiol.* **2,** 247–261 (1920).
15. J. E. Greaves, Azofication, *Soil Sci.* **6,** 163–217 (1918).
16. L. G. M. Baas-Becking and G. S. Parks, Energy relations in the metabolism of autotrophic bacteria, *Physiol. Rev.* **7,** 85–106 (1927).
17. G. S. Parks and K. K. Kelly, Thermal data on organic compounds. II. The heat capacities of five organic compounds. The entropies and free energies of some homologous series of aliphatic compounds, *J. Am. Chem. Soc.* **47,** 2089–2097 (1925).
18. W. Ruhland, Beiträge zur Physiologie der Knallgasbakterien, *Jahrb. Wiss. Bot.* **63,** 321–398 (1924).
19. D. Burk, The free-energy of nitrogen fixation by microorganisms, *J. Gen. Physiol.* **10,** 559–573 (1927).

20. D. Burk, The reversibility of coupled reactions in biological systems and the second law of thermodynamics, *J. Phys. Chem.* **35,** 432–455 (1931).
21. F. G. Donnan, Concerning the applicability of thermodynamics to the phenomenon of life, *J. Gen. Physiol.* **8,** 685–688 (1927).
22. K. v. Angerer, Über die Arbeitsleistung eigenbeweglicher Bakterien, Arch. Hyg. Bakt. **88,** 139–145 (1919).
23. W. Ludwig, Der Betriebstoffwechsel von *Paramecium caudatum* Ehrbg. Zugleich ein Beitrag zur Frage nach der Funktion der kontraktilen Vakuolen, *Arch. Protistenkd.* **62,** 12–40 (1928).
24. R. E. Buchanan and E. I. Fulmer, *Physiology and Biochemistry of Bacteria*, Vol. I, The Williams and Wilkins Co., Baltimore, 1928.
25. E. I. Fulmer and E. Leifson, The free energy decrease of alcoholic fermentation, *Iowa St. J. Sci.* **2,** 159–177 (1928).
26. M. J. Johnson, W. H. Peterson, and E. B. Fred, Oxidation and reduction relations between and substrate and products in the acetone-butyl alcohol fermentation, *J. Biol. Chem.* **91,** 569–591 (1931).
27. P. W. Wilson and W. H. Peterson, The energetics of heterotrophic bacteria, *Chem. Rev.* **8,** 427–480 (1931).
28. A. Harden and D. Norris, The bacterial production of acetylmethylcarbinol and 2,3 butylene glycol from various substances, II, *Proc. R. Soc. Lon.* **85B,** 73–78 (1912).
29. H. R. Braak, *Onderzoekingen over Vergisting van Glycerine*, Doctoral thesis, Technische Hoogeschool, Delft, 1928.
30. E. C. Grey, The enzymes which are concerned in the decomposition of glucose and mannitol by *Bacillus coli communis*, *Proc. R. Soc. Lon.* **87B,** 472–484 (1914).
31. M. A. Scheffer, *De suikervergisting door bacterien der coligroep*, Doctoral thesis, Technische Hoogeschool, Delft, 1928.
32. U. Gayon and E. Dubourd, Nouvelles recherches sur le ferment mannitique, *Ann. Inst. Pastuer* **15,** 527–569 (1901).
33. P. Van Steenberge, Les propriétés des microbes lactique; leur classification, *Ann. Inst. Pasteur* **34,** 803–870 (1920).
34. H. R. Stiles, W. H. Peterson, and E. B. Fred, Fermentation products of certain mannitol-forming bacteria, *J. Biol. Chem.* **54,** 643–653 (1925).
35. C. B. Van Niel, *The Propionic Acid Bacteria*, Doctoral thesis, Technische Hoogeschool, Delft, 1928.
36. M. Foote, W. H. Peterson, and E. B. Fred, The fermentation of pentoses by certain propionic acid bacteria, *Zentralbl. Bakteriol. Parasitenkd. Inkektionskr. Hyg, Sec. II* **58,** 379–389 (1930).
37. J. Smit, *Die Gürungssarcinen*, Gustav Fischer, Jena, 1930.

5

THE DEVELOPMENT OF REPRESENTATIONS OF GROWTH, 1932–1954

As described in Sections 4.9 and 4.10, in nearly all cases in which a given microorganism was grown *anaerobically* on a single substrate, what can be regarded as a 100% material balance was obtained with respect to the fermentation. All of the carbon, hydrogen, and oxygen in the substrate could be accounted for in fermentation products other than the cells. This is remarkable because of the fine analytical chemistry that must have been entailed in the analyses of these complex reactions. It is even more remarkable in that we must therefore conclude that none of the substance of the substrate could have become incorporated into cellular material, in spite of the fact that growth must have taken place. Although it never seems to have been emphasized anywhere in the literature of that time, the only reasonable conclusion is that the substance of the cells must have come from somewhere else. A search through the literature shows that in all cases in which a 100% material balance was obtained for a fermentation there were other carbon sources present in the nutrient media in which the microorganisms were grown. These were for the most part in the form of peptone, yeast extract, or both, that were added to the media to serve as a source of nitrogen, minerals, vitamins, or growth factors needed by the cells. The microorganisms would not grow anaerobically on peptone or yeast extract alone, and thus whatever was in these complex additives could not serve as a source of energy for growth. Good growth did occur, however, in the presence of these additives plus a fermentable substrate, and it seems likely that a general opinion prevailed in the 1920s and 1930s that peptone and yeast extract contributed nitrogen and small amounts of other substances for the cells. Since attention was always focused on the fermentation and not on the cells, even though these were necessary to carry out the fermentation, a material balance including the cells was never made. The emphasis was on the chemistry of the fermentation, not on the process of cellular growth. For this reason it apparently never occurred to anyone that cells could be regarded as products of a fermentation, and equations representing fermentations never included a term representing cells.

It is of historical interest how two different approaches to the study of fermentations led to the same interpretation. For example, Pasteur[1] in 1860 attempted to account for all the products of an alcoholic fermentation *including* any cellular material that might be formed. He did this by incubating 9.998 g of sucrose with 2.026 g of yeast paste and 100 g of water. After 10 days all the sugar had disappeared and an analysis was begun. He found that from the initial quantity of sugar there were formed 5.100 g of alcohol, 4.911 g of carbon dioxide, 0.34 g of glycerine, 0.065 g of succinic acid, and 0.13 g of cellulose and indeterminate material adding up to 10.546 g total. This represented a good material balance, and since growth seemingly did not occur under these experimental conditions the data represented accurately what happened during fermentation. To be sure, small quantities of glycerine and succinic acid were formed, but most scientists tended to ignore this fact. The Gay-Lussac equation, which until this time had been considered as much a theoretical representation of alcoholic fermentation as a factual one, could still be written

$$C_6H_{12}O_6 \longrightarrow 2CH_3CH_2OH + 2CO_2 \tag{5.1}$$

Thus, although Pasteur was the first to demonstrate products of an alcoholic fermentation other than carbon dioxide and alcohol, it was little noticed. Nevertheless, the question remained in Pasteur's mind, for he wrote,

> What matters here is the equation between the sugar and the sum of the principal products. With regard to a chemical equation formulated in equivalents, is it at all possible to establish this for such a complex act? At the moment when alcohol is formed there is simultaneously formed carbonic acid, succinic acid, glycerine, cellulose, fatty materials, and without doubt many other substances perhaps as essential as these last two to the life of the cells and in consequence to the phenomenon of the chemical transformation of the sugar, even though their total weight ought to be extremely small. [Pasteur,[1] p. 352.]

He then stated, "Science is too little advanced to hope to put into a rigorous equation an act correlated with a vital phenomenon." (Pasteur,[1] p. 352.)

Pasteur's use of a nongrowing suspension of yeast to carry out a fermentation was quite different from that used by the fermentation analysts of the 1920s. These latter almost universally worked with growing cell suspensions, since all the necessary substances for growth except an energy source were present in the peptone or yeast extract that was also added to the nutrient medium. However, because the cells apparently used substances in the peptone and yeast extract as sources of carbon (this could not have been tested during this period because of the nonavailability of radio active tracers), all of the substrate could be fermented with a resulting 100% material balance. Once again the interpretation did not permit the inclusion of a term representing cells.

While the above fermentation chemistry was going on other investigators were studying the aerobic metabolism of substrates by microorganisms. In most cases small molecular weight substances such as glucose, acetic acid, alcohol, amino acids, and so on, were found to be oxidized completely to carbon dioxide and water. In some cases the oxidation was incomplete, as in the oxidation of alcohol to acetic acid, isopropyl alcohol to acetone, glycerol to dihydroxyacetone, sorbitol to sorbose, and so on, which can be carried out by *Acetobacter suboxydans*. In all cases equations representing these oxidations could be written conveniently without the inclusion of a term representing cells. Two examples are as follows:

The complete oxidation of glucose:

$$C_6H_{12}O_6 + 6O_2 \longrightarrow 6CO_2 + 6H_2O$$

The partial oxidation of ethyl alcohol:

$$CH_3CH_2OH + O_2 \longrightarrow CH_3COOH + H_2O$$

It was not until the 1930s that representations of cells or of cellular materials began to appear in equations representing the metabolism of cells. These equations were of two kinds, one representing the phenomenon of assimilation and the other representing the phenomenon of growth. They are both equally important with respect to a consideration of the energetics of microbial growth since one must be differentiated from the other.

5.1. THE HISTORICAL DIFFERENCE BETWEEN THE CONCEPT OF GROWTH AND THAT OF ASSIMILATION

In order to study growth the use of the word should first be defined, not an easy thing to do since the term is a conceptual one that may have more than one specific meaning. In general, the concept of growth denotes an increase, or an act of increasing. Physical growth involves an increase in mass, but in the biological sense an increase in mass may be brought about by different phenomena. For the purpose of studying the energetics of microbial growth we will define growth here as an increase in mass of a population of microorganisms brought about by an increase in the number of individuals, *exclusive* of the mass of intracellular storage products comprising the primary products of fermentative or oxidative assimilation. What contributes to the former increase in mass is thus only cellular structural material, perhaps better referred to as the "fabric" of the cells by analogy with the same term as applied to buildings. The fabric of a building does not include things put into it, such as furnishings, stored material, or other contents unrelated to the support or form of the structure. We shall have to stretch the analogy somewhat to include as part of the fabric such things as soluble or particulate intracellular enzyme systems, organelles, the nucleus and its contents, and nucleic acid moieties present in the cytoplasm. The fabric of cells thus includes (by definition) cell walls where present, the plasma membrane, the cytoplasm, and anything extended from it or suspended in it. It does not include the products of primary assimilation such as glycogen, starch, poly-betahydroxybutyric acid, lipids, and so on.

Primary assimilation is different from growth. It involves an increase in mass of a population of microorganisms which is not brought about by an increase in the number of individuals, but by an increase in the mass of each individual as the result of the accumulation of storage products.

It is apparent from the difference between the two types of mass increase involved in growth and in primary assimilation that simply measuring the increase in mass of a population of cells before and after an experiment will not distinguish between them. Other methods must be used to ensure that it is only growth and not growth plus assimilation, or just assimilation, which is being studied.

5.2. THE CONCEPT OF MICROBIAL ASSIMILATION

Assimilation usually means the permeation, absorption, or taking-in by cells of nutritive substances from their environments, plus something else. Beyond this the concept becomes unclear. At least as evident from common usage of the word, assimilation may or may not include an actual incorporation of assimilated substances or their derivatives into the structural or catalytic materials of which cells are composed. Assimilated substances, by definition, must pass through the limiting boundary of a cell to its interior. The concept of assimilation does not seem to include the surface adsorption of substances to the outside of the plasma membrane. Except in cases of assimilation by means of pinocytosis or phagocytosis, assimilated substances are usually of low molecular weight. If these latter cannot be removed from solution within the cell as rapidly as they are assimilated, assimilation theoretically can cause serious osmotic problems for those cells that do not have cell walls or some means of preventing an increase in the volume of the protoplast sufficient to cause them to burst. Even with those cells that do have cell walls, an excessive increase in turgor pressure due to osmosis may create metabolic problems. One way to prevent an osmotic imbalance is for assimilated substances to be metabolized to end products as rapidly as they are assimilated. To accomplish this the overall rate of metabolism would have to equal the rate of assimilation. Another way of preventing an osmotic imbalance is for the appropriate catalytic mechanisms within a cell to bring about the polymerization of the assimilated substances or their derivatives to form insoluble products. Thus, the transient formation of insoluble intracellular substances such as starch, glycogen, poly-betahydroxybutyric acid, and oil as storage materials which can be metabolized at a later time, but are not generally considered to serve a structural or catalytic function, is usually included within the concept of assimilation.

In principle, the formation of storage materials within a cell is not greatly different from the formation of structural or catalytic materials constituting integral parts of a cell. The formation of these latter, however, is also considered to be associated with the phenomenon of growth. The concept of assimilation is thus only indistinctly differentiated from that of growth, since for cells to grow they must also assimilate and incorporate the substances assimilated. Not included in the concept of assimilation is anything related to an increase in cell number. The concept of growth, unless the growth occurs solely as the result of the metabolism of previously assimilated and stored materials, may include that of assimilation, but the concept of assimilation need not include that of growth if only polymerized or reduced storage materials are formed within the cell. The terms "oxidative assimilation" and "fermentative assimilation" as found in the literature usually refer *only* to the assimilation of small molecular weight substances under aerobic or anerobic conditions plus the formation of reduced or polymerized storage materials. Unfortunately, there has also been a tendency to equate the substance of these storage materials with the fabric of the cells, since the formula "(CH_2O)" was thought to represent carbon in a state of reduction similar to that found in cells.

Both organic and inorganic substances can be assimilated, and the process need not involve the expenditure of free energy on the part of the cell. Some substances enter a cell by simple or facilitated diffusion and as such might be said to assimilate themselves rather than to be assimilated. Most substances are probably assimilated only through the expenditure of free energy made available through cellular catabolism. This can involve anything from the activation of a substance so that it can be transported into a cell by permeases, to the enclosure of an ionic substance by an ionophore, to the loss in free energy that occurs during the process of polymerization or reduction to form storage products. Historically, studies on assimilation have for the most part dealt with the absorption of organic compounds that can be used both as a source of carbon and of free energy, the notable exception being the assimilation of carbon dioxide in autotrophic photosynthesis or chemosynthesis.

Because of the somewhat indistinct difference between the concepts of assimilation and growth there is some reason to clarify the definition of assimilation, if only for the purposes of this book. Assimilation will therefore be redefined as the permeation, absorption, or taking-in of nutritive substances by cells and the storage of these substances either in solution within the cell, in which case the substances will be "pooled" and possibly not very much will be stored in this manner, or in the form of products of the polymerization of these substances or their derivatives, or in the form of reduced substances such as fats and oils. Such a definition avoids any confusion with the phenomenon of growth, during which permanent structural components of the cell are formed rather than transient storge products.

5.2.1. Early Experiments on Assimilation

The phenomenon of assimilation, as defined above, was first clearly observed by L. Pasteur. In 1859 Pasteur[2] noted that considerably different results were obtained that depended on the dry-weight ratio of yeast to sugar when yeast was incubated anaerobically in a sugar solution. Previous experiments had shown (Pasteur[1]) that when this ratio was small it required a long time for all the sugar to be fermented, but after the fermentation was complete the quantities of carbon dioxide and alcohol produced closely approximated those predicted on the basis of the Gay-Lussac equation. On the other hand, if this ratio were large it required only a short time for all the sugar to be fermented, and far more carbon dioxide and alcohol were produced than would have been predicted on the basis of the Gay-Lussac equation. Pasteur felt that the excess gas probably came from glycogen that was present in the yeast cells which, when hydrolyzed, was changed into a fermentable sugar. This led to another type of experiment (Pasteur[3]) also published in 1859 in which he determined that 2.626 g of yeast contained 0.532 g of an extractable hydrocarbon that could be transformed into a fermentable sugar. After the same quantity of the same yeast had fermented 100 g of sugar it was found to contain 0.918 g of a similar extractable hydrocarbon. This demonstrated that during alcoholic fermentation by nongrowing suspensions of yeast cells a part of the sugar becomes fixed within the yeast in the form of a hydrocarbon. In this experiment the ratio of yeast to sugar that Pasteur used was very small, and after the fermentation was over the yeast cells were probably loaded with what we now know to be glycogen to the point where no more could be formed within the cells. It was the fermentation of stored glycogen that was responsible for the production of excess carbon dioxide and alcohol observed by Pasteur when the dry-weight ratio of yeast to sugar was high.

More definitive research on assimilation can be found in a treatise published by Duclaux[4] in 1900, wherein, among other things, he described the phenomenon of autophagie (endogenous metabolism) in yeast. Duclaux presented the experiments of Schutzenberger and Destrem (no reference given) in which two equal weights of yeast were incubated, one in the presence of two times its weight of sugar and the other in the absence of sugar, for a period of 24 hr. At the end of this time each sample was treated with boiling water so as to separate the soluble from the insoluble fractions, which were then subjected to various analyses. The results were:

1. That there was a considerable increase in the weight of the yeast cells as a consequence of the fermentation.
2. That, proportional to the increase in weight of the yeast, there was an increase in the amount of extractable hydrocarbon.
3. That there was a decrease in the proportion of extractable nitrogen in the yeast during the fermentation.

From these results we can now conclude that there had been both an assimilation of sugar to form a hydrocarbon (carbohydrate) within the yeast, and that some growth had also occurred as shown by the decrease in the extractable pool of nitrogenous substances. Thus, there can be no doubt that by 1900 the phenomenon of what Van Niel and Anderson[5] in 1941 described as "fermentative assimilation" was already well known.

In 1888–1889 Nägeli[6,7] investigated the production of fat by yeasts and molds and discovered that when these organisms were grown in a well-aerated culture medium that was rich in carbohydrate but poor in nitrogen, the fat content of these organisms could be increased by 5 to 12%. This research indicated that when microbial growth becomes limited, in this case due to the insufficiency of a nitrogen source, the process of stored food synthesis takes place independently and at a higher rate than does the process of growth. It thus represents a case of assimilation that is taking place concurrently with growth.

Summarizing the concepts of assimilation and "disassimilation" as they were understood in 1900, Duclaux[4] wrote,

> . . . the work that we analyze [of Schutzenberger and Destrem] confirms by its data some of the results and is in agreement with some of the conceptions of our previous studies on fermentation. It establishes, moreover, from some very interesting ratios, the changes which take place in yeast left to itself or placed in the presence of sugar. Of the important phenomena of assimilation and dissassimilation that the maintenance of life demands, the first prevails in the yeast in the presence of sugar and carries along the second without growth. The second phenomenon is in its turn more noticeable in yeast left to itself; and although the first does not disappear totally, for one cannot proceed without the other, as is shown, moreover, by the proliferation of cells that is sometimes visible with a microscope, its influence and its results are completely masked by those of the first [i.e., the effects of disassimilation are completely masked by those of assimilation]. But in every case, it is the same life which continues, the same cell which assimilates and disassimilates. Its protoplasmic function while there is sugar is the same as that after, and one is even able to point out that glycogen, an intracellular nutrient, replaces after some time the sugar, as extracellular nutrient, when this has disappeared. [Duclaux,[4] pp. 327–328.]

From the above quotation it is evident that our present understanding of the overall phenomenon of assimilation has not progressed far beyond that of Pasteur and Duclaux although, of course, we now have a detailed knowledge of the actual biochemical mechanisms by which this takes place.

5.2.2. Later Studies on Assimilation

From 1900 to 1922 few studies were accomplished relative to the assimilation of substrates by what we now call "resting cell suspensions." During this time, however, work was being carried out with growing microorganisms in the brewing and solvent industries and in yeast factories, in an effort to increase production. The biochemistry of alcoholic fermentation was intensively studied, one result being that during World War I glycerol could be produced by yeast from unrefined plant extracts containing fermentable sugars by trapping with bisulfite the glyceraldehyde formed during glycolysis. References made to assimilation during this period are usually with respect to the growth of microbial cultures on various substrates and in different media. Also during the war Lindner[8] applied conditions of growth to the organism *Endomyces vernalis* which were similar to those employed by Nägeli in that the culture medium was rich in carbohydrate but poor in nitrogen, so that growth became nitrogen limited. This resulted in the fat content of this yeast being raised to the point where it could be used successfully as a supplementary source of fat and protein for human diets.

In 1922 the studies of Fürth and Lieben[9,10] demonstrated that only part of the lactic or pyruvic acid incubated with an aerated yeast suspension was oxidized to carbon dioxide and water. The other part was thought to be converted into carbohydrate or cell material. In the same year Lieben,[11] using amino acids, showed that only asparagin was readily assimilated by aerated yeast suspensions, others being less so. He conjectured that assimilated amino acids were converted into yeast structural substance. In 1923 Lieben[12] observed similar results with pyrotartaric acid. That same year Lundin[13,14] carried out experiments dealing with the effect of oxygen on the assimilatory and dissimilatory activities of yeast suspensions. He incubated yeast cells in a solution of pure glucose or fructose until the sugar had completely disappeared. The cells were observed to increase in size, but not in number. Analyses showed that no protein or fat had been synthesized, nor had there been any appreciable quantity of organic acids or aldehydes released into the suspending medium. Of the total sugar consumed during the experiments, about 33% appeared in the form of carbon dioxide, the rest being accounted for as assimilated carbohydrate and as ethanol in the medium. An appreciable deviation from an ordinary fermentation was found in that the reactions produced less carbon dioxide and alcohol than that predicted on the basis of the Gay-Lussac equation. The amount of alcohol formed attained a maximum just after all the sugar had disappeared from the medium, after which it decreased. The presence of oxygen was observed to have *no* effect on the fermentation. Lundin concluded that the utilization of sugar was preceded by the process of fermentation, which was then followed by an oxidation of the ethanol produced during the fermentation reaction. During the process of oxidation a part of the ethanol was converted into an easily hydrolyzable carbohydrate, presumably glycogen, and into a structural carbohydrate that was not easily hydrolyzed. The oxidation of the ethanol began before the alcoholic fermentation was completed, and was accompanied by a loss in energy amounting to about one-third of that originally contained in the sugar. Further studies conducted in the same manner (Lundin[15]) demonstrated that alcohol could only be assimilated aerobically, with the assimilation product being a carbohydrate. No fat was formed. Lundin[16] also found that salts of citric, malic, tartaric, and succinic acids were not assimilated. Potassium acetate was assimilated from 15 to 20%, the assimilation product being again a carbohydrate.

Smedley-Maclean and Hoffert[17–20] and Hoffert,[21] in a series of studies between 1923 and 1926 on carbohydrate and fat metabolism in yeast, confirmed and amplified the experiments of Fürth and Lieben.[9,10] The incubation of yeast cells suspended in oxygenated water with various substrates resulted in the storage of part of the substrate in the form of fat or carbohydrate and in the oxidation of part to carbon dioxide and water. Marian[22,23] in 1924 observed that yeast cells in aqueous, oxygenated suspension assimilate betahydroxybutyric acid and glycerol with little or no production of carbon dioxide and with the formation of an easily hydrolyzable intracellular carbohydrate. Experiments published in 1925 by Meyerhof[24] on the effect of oxygen on alcoholic fermentation by yeast suspensions gave results contrary to those of Lundin[13,14] in that oxygen was observed to inhibit fermentation to a marked extent. However, the oxidation of alcohol by a suspension of yeast was found to result in the formation of an intracellular product having the empirical formula of $(CH_2O)_n$.

Prior to 1928 all of the work on assimilation seems to have been done with yeast as a test organism. In 1928 Cook and Stephenson[25] used washed suspensions of *Escherichia coli* in manometric experiments during which oxygen consumption was measured. They observed that formate was oxidized quantitatively, but that only two-thirds of the expected oxygen uptake occurred with glucose and with lactate as the substrates, and three-fourths with acetate. When *Alcaligenes faecalis* was used as the test organism the results were essentially the same except that glucose was not oxidized at all. However, similar experiments in 1932 by Lineweaver[26] seemed to indicate that *Azotobacter vinlandii* completely oxidized all the substrates tested. This appears to be one of the few cases in which assimilation by a cell suspension did not occur. One year later Brücke[27] found that for the same amount of carbon, two to three times as much glycogen was formed by a yeast suspension from glucose as from alcohol. The studies of McAnally and Smedley-Maclean[28] in 1934 were a much later continuation of those by Smedley-Maclean and Hoffert[17–20] and Hoffert[21] on fat and carbohydrate assimilation in yeast. Similar experiments were conducted on a different yeast species with similar results. In 1936 Guillmet[29,30] conducted a series of alcoholic fermentations with yeasts using dilute sugar solutions (1% or less), and found that for different yeasts the amount of ethanol formed varied from 0 to 45% of that predicted on the basis of the Gay-Lussac equation.

From the preceding brief description of the work on assimilation through 1936 it is clear that a number of investigators had succeeded in describing quantitatively many examples of assimilation, and in so doing had established firmly the existence of this phenomenon. Their work can thus be considered a strong confirmation of the initial discovery of assimilation by Pasteur.[2,3] The investigations from 1922 to 1936 were carried out for the most part with yeast suspensions incubated with sugar solutions both anaerobically and aerobically and with solutions of other substrates aerobically, although a few later experiments had also been done using three genera of bacteria. That so much of this type of study was done with yeast was probably because it was then the only microorganism that could be brought by the pound, and was thus a most convenient tool and source of material for microbiological and biochemical research.

In 1936 the studies of Barker[31] with the colorless alga *Prototheca zopfii* served to call attention to assimilation as comprising as identifiable, regularly occurring phenomenon. This is historically somewhat surprising since the phenomenon of microbial assimilation had been well studied qualitatively by several investigators prior to this time, as previously described in this section. What seems to have happened is that Barker's quantitative experiments demonstrated clearly that unexpectedly large quantities (50–80%) of the substrate consumed by an aerated microbial suspension could be assimilated. In addition, Barker gave the phenomenon the name of "oxidative assimilation," and this served to identify it and to call attention to it. Finally, subsequent investigators were able to represent the phenomenon of oxidative assimilation quantitatively in the form of a chemical reaction equation.

At the beginning of his publication Barker[31] made some succinct comments on the nature of the problem he was investigating, stating that:

> The oxidative formation of cell material from a single, simple organic compound and a source of nitrogen is a very complex process. It is complex not only because of the many different organic substances which make up cell material and must be synthesized from the substrate, but also because these substances will in general be formed in different amounts and at different rates. Further, the cells once formed store food. Subsequently this stored food, as well as the basic cell material, undergoes further transformations. It is therefore inconceivable that the quantity of cells and carbon dioxide produced in a culture which has been incubated for a more or less arbitrary length of time will have any simple, interpretable relation to the amount of substrate decomposed. [Quoted by permission from Reference 31, p. 232.]

In many respects the above comments summarize the general opinions on the subject of assimilation that prevailed in the minds of most microbiologists prior to this time. Barker then went on to state:

> The complexity of the problem may, however, largely be avoided, in theory at least, by distinguishing between the initial oxidation and assimilation of the organic compound and the subsequent reactions resulting in the formation of various cell constituents. The former process may be called the primary conversion of the substrate. The primary conversion in this sense corresponds with the reduction of carbon dioxide to a carbohydrate in the process of photosynthesis. As in this analogy, the primary conversion will presumably give rise to a primary product of assimilation which is the raw material for all subsequent syntheses within the cell. [Quoted by permission from Reference 31, p. 232.]

Barker's experimental method was to measure manometrically the oxygen consumption and the carbon dioxide production corresponding to the assimilation of known quantities of glycerol, ethyl alcohol, glucose, and acetic, propionic, *n*-butyric, isobutyric, and valeric acids. From the results of these experiments and from a consideration of their respective respiratory quotients (RQ, the ratio of the amount of carbon dioxide produced to the amount of oxygen consumed during a biological reaction), he interpreted the lack of a complete oxidation of these substrates by well-washed suspensions of *P. zopfii* as being due to the formation of a carbohydrate in the cells. Although Barker[31] himself did not do so, subsequent investigators adopted the representation of oxidative assimilation in the form of an equation. As an example, were isobutyrate to be completely oxidized, this reaction could be represented by the following equation.

$$C_4H_8O_2 + 5O_2 \longrightarrow 4CO_2 + 4H_2O$$

Here it is apparent that 5 moles of oxygen gas are required for the oxidation of 1 mole of isobutyrate to 4 moles of dioxide and 4 moles of water. The RQ = 4/5 = 0.80. What Barker found was that when a suspension of *P. zopfii* cells was incubated in a solution of isobutyrate only 2.90 moles of oxygen were consumed per mole of isobutyrate that disappeared from the suspending medium, and only 1.94 moles of carbon dioxide were produced. The observed RQ was thus 1.94/2.90 = 0.67. This was sufficiently close to 0.66 that the following equation could be written to represent the whole process of isobutyrate assimilation in which the product of assimilation was a carbohydrate.

$$C_4H_8O_2 + 3O_2 \longrightarrow 2(CH_2O) + 2CO_2 + 2H_2O \qquad (5.2)$$

Equations such as Eq. 5.2 are historically important in that they constitute the first attempts apart from those of Algera[32] and Tamiya (see Chapter 6) to include as a term in an equation something related to cells. In this case it is the carbohydrate storage material that accumulates within the cells.

Barker[31] summarized the results of his investigations as follows.

> On the basis of these data and certain quantitative considerations it is concluded that the process of assimilation . . . proceeds in two experimentally distinct stages. The first, which has been called the primary process of assimilation, consists of an oxidative conversion of the substrate into a carbohydrate which is stored in the cells as glycogen. The formation of this carbohydrate is rapid compared to its subsequent decomposition in the process of cell synthesis which constitutes the second stage of assimilation. Further, it is formed with great economy; from 50 per cent to more than 80 per cent of the carbon of the substrate is assimilated depending on the organic compound used [Barker,[31] p. 250.]

If we equate Barker's "first stage of assimilation" with what Duclaux[4] in 1900 called "assimilation," and Barker's "second stage of assimilation" with

Duclaux's "disassimilation," the conclusions Barker drew relative to oxidative assimilation are exactly the same as those drawn by Duclaux for what Van Niel and Anderson[5] called "fermentative assimilation."

Barker's work initiated a great deal of experimentation designed to increase our understanding of the assimilatory process and of the relation between respiration and fermentation, and assimilation. The experiments with different *Spirillum* sp. that Giesberger[33] published in 1936 supported Barker's conclusions and showed in addition that the amount of material assimilated from a given quantity of substrate was not necessarily constant, but could vary with the species of microorganism.

Following the publications of Barker[31] and of Giesberger[33] there was a considerable number of investigations published on assimilation including one by Van Niel and Anderson[5] in 1941 in which the assimilation of sugars first described by Pasteur,[2,3] further investigated by Schutzenberger and Destrem (as described by Duclaux[4]), and well studied by Lundin,[13,14] seems to have been rediscovered. Van Niel and Anderson apparently were unaware of Lundin's work of 18 years earlier, and came to almost exactly the same conclusion: that 70% of the glucose consumed anaerobically by a yeast suspension could be accounted for as carbon dioxide and ethanol (about 66% in Lundin's experiments). Whereas Barker[31] and Lundin[13,14] had inferred the product of assimilation to be a carbohydrate, with typical caution Van Niel and Anderson did not make this claim. These last investigators did, however, give the phenomenon they studied its name.

Clifton[34–36] has reviewed the literature on assimilation through 1952, therefore it need not be further considered here. Apparently only two publications on oxidative or fermentative assimilation[34,35] have appeared since then. This may have been because it had been adequately demonstrated that with non-poisoned, washed suspensions of microorganisms the majority of substrates tested could be assimilated, the product in most cases being a carbohydrate, or they were not metabolized at all. Some investigators have concluded, however, that the assimilation product was in many cases a lipid, not a carbohydrate. In at least one case the fact that a microorganism could not assimilate a given substrate was found to be due to the experimental conditions. Barker[39] found that *P. Zopfii* was unable to oxidize keto-, hydroxy-, or dibasic acids. However, subsequent studies by Anderson[40] indicated that these compounds could be utilized if the pH of the suspending medium were lowered enough to raise the concentration of the undissociated acids. On the other hand, Van Niel and Anderson[5] demonstrated that *Streptococcus fecalis* could ferment glucose or fructose completely to lactic acid, but was unable to carry out any assimilation of these substrates.

Based on all of the work on oxidative and fermentative assimilation which has been done it is impossible not to conclude that, if the mechanism for the metabolism of a given substrate exists within a given microorganism and if the appropriate environmental conditions are present (e.g., a suitable pH and temperature), assimilation *will* take place in washed cell suspensions of most microorganisms. The complete oxidation or fermentation of a substrate may be something that microoganisms "do" from the point of view of a biochemist, but it is probably something that only rarely occurs naturally. Microorganisms must be constantly prepared for the appearance in their environments of metabolizable substrates, and if for whatever reason these substrates cannot be utilized at that time in the process of growth (e.g., no nitrogen source, or a lack of one or more required minerals, trace elements, or vitamins), they are assimilated and, as Duclaux[4] pointed out, *then become internal substrates which can be utilized at a later time for various purposes.*

What has all this discussion on assimilation to do with a study of the energetics of microbial growth? The answer is, of course, that assimilation must be avoided. Assimilation and growth as defined for the purposes of this book are two different phenomena. Stored material is not structural material in the sense of being a part of the fabric of a cell. It is apparent from the foregoing discussion that assimilation takes place when cells are washed. What washing does is to remove anything from the environment of the cells which could contribute to their growth. An inhibition of growth because of the lack of a nutrient will initiate assimilation provided that an assimilable substrate is present. An inhibition of growth because of unfavorable physical conditions or of an unfavorable pH in the environment will also initiate assimilation unless these conditions are too extreme. It then follows that the appropriate environmental conditions for the study of the energetics of microbial growth are those in which growth is *unlimited* throughout an experiment. Under these conditions a maximum growth rate will be attained and maintained and there will be, at least in theory, a maximal increase in mass of a population of cells exclusive of storage materials. This can occur in a batch culture, in which nearly to the time a substrate becomes exhausted nothing need be growth limiting if suitable experimental conditions are chosen. On the other hand if, during an experiment the pH drops sufficiently low that growth is inhibited, if the oxygen supply becomes limiting, or if something in the culture medium which is necessary for growth becomes limiting, the cells will begin to assimilate. If this occurs the population of cells will still increase in mass although the rate of cell division will drop off. If an increase in mass or energy content is used as a criterion of growth it will still appear as if growth has occurred. The problem is that growth may have stopped occurring, and what is actually being measured is oxidative or fermentative assimilation.

5.2.3. Can Assimilation Be Prevented?

Following Yamamoto's[41] demonstration in 1933 that microbial synthesis could be repressed and respiration increased by the addition of potassium cyanide, sodium fluoride, carbon monoxide, or phenylurethan to the suspending medium, Clifton[42,43] in 1937 studied the influence of various poisons on oxidative assimilation. He found that the processes of synthesis and respiration appeared to be closely connected. They could be influenced by the same agents, the extent to which this happened to either of them being related to the concentrations of the poisons used. In 1946 Clifton summarized the results of his own studies with the statement:

> Low concentrations of the poisons inhibit respiration slightly and simultaneously partially prevent synthesis. Critical concentrations of the inhibitory agent completely prevent synthesis and force the direction of the reaction to carbon dioxide and water, the rate of oxygen consumption being somewhat further reduced. Higher concentrations of the poisons reduce the rate of respiration to a negligible value. [Clifton,[34] p. 291.]

Further studies by a number of investigators confirmed that with certain poisons in suitable concentrations both fermentative and oxidative assimilation could be prevented.

The important thing to point out from these studies is that normally fermentative or oxidative assimilation *will* take place in the absence of growth *unless* the cells are suitably poisoned. In studies of the energetics of microbial growth the cells cannot be poisoned because growth will then be prevented. However, *fermentative or oxidative assimilation are almost guaranteed to happen if there is any environmental limitation of growth.*

5.3. WHAT IS MEANT BY MICROBIAL GROWTH?

In Section 5.1 microbial growth was defined as an increase in mass of a population of microorganisms which is brought about by an increase in the number of individuals, *exclusive* of the mass of intracellular storage products which are the primary products of fermentative or oxidative assimilation. It follows that to realize growth according to this definition the ideal condition to have is one in which no primary assimilation at all takes place. Under these ideal conditions what is meant by growth is the permeation, absorption, or taking in by cells of nutritive substances and the direct incorporation of these substances into the structure, or fabric, of the cells. An intermediary condition is possible in which the nutritive substances could be first assimilated in the sense previously defined, after which growth could take place utilizing the internally stored products of assimilation. This would make no difference with respect to the energetics of microbial growth provided that at the end of an experiment the mass of the products of assimilation is subtracted from the mass of the structural material of the cells.

An increase in anything cannot occur by itself, but can take place only as a result of something else. Microbial growth cannot occur without the participation of all the things that contribute to it. These latter include sources of building materials for the fabric of the cells as well as some source of energy with which to accomplish its construction. The phenomenon of microbial growth cannot be represented simply as an increasing mass of microorganisms. It must be represented in terms of a system that undergoes a change such that an increase in the mass of a population of microorganisms as defined above takes place as a result of the incorporation of a substrate, nutrients, and minerals.

5.4. ON THE USE OF THE TERM (CH_2O)

In the review by Clifton[34] on microbial assimilations published in 1946 it is evident that the term (CH_2O) had become widely accepted as representing the primary product of assimilation. It is equally evident that in many cases the experimental data did not provide adequate support for this assumption. For example, Barker[31] incubated washed cell suspensions of *P. zopfii* in the presence of acetic acid during which the oxygen consumption and the carbon dioxide production were measured. If acetic acid were to be completely oxidized to carbon dioxide and water the reaction could be represented as follows:

$$CH_3COOH + 2O_2 \longrightarrow 2CO_2 + 2H_2O \qquad (5.3)$$

In contrast to what was expected on the basis of Eq. 5.3 the results of Barker's experiment showed that for every mole of acetic acid which disappeared from the reaction system only 1.03 moles of carbon dioxide were produced and 1.00 mole of oxygen was consumed. These data indicated that the product of assimilation must have an RQ (CO_2 produced/O_2 consumed) of 0.97, by difference $(2 - 1.03)/(2 - 1)$. This was close enough to the RQ of 1.00 obtained when substances having the formula $(CH_2O)_n$ are oxidized to completion that the product of assimilation could be assumed to be a carbohydrate. The equation representing the assimilation of acetic acid by *P. zopfii* could then be written with some degree of confidence as follows, even though the product of assimilation had not been chemically identified.

$$CH_3COOH + O_2 \longrightarrow (CH_2O) + CO_2 + H_2O$$

On the other hand, if glycerin were the substrate its complete oxidation could be represented as follows:

$$C_3H_8O_3 + 3.5O_2 \longrightarrow 3CO_2 + 4H_2O$$

The results of Barker's experiments showed that for every mole of glycerin which disappeared from the reaction system only 0.61 mole of carbon dioxide was produced and 1.01 moles of oxygen consumed. The RQ of the assimilation product was therefore $(3 - 0.61)/(3.5 - 1.01) = 0.959$, indicating that it was more reduced than a carbohydrate. These data *could* have been represented accurately by the equation

$$C_3H_8O_3 + 1.01O_2 \longrightarrow 2.39(CH_{2.167}O) + 0.61CO_2 + 1.41H_2O$$

in which ($CH_{2.167}O$) represents the assimilation product in its appropriate state of reduction. However, Barker was apparently convinced that the assimilation product either was or ought to be a carbohydrate so that an appropriate equation for representing the oxidative assimilation of glycerin by *P. zopfii* is

$$2C_3H_8O_3 + 2.00O_2 \longrightarrow 5(CH_2O) + 1.00CO_2 + 3H_2O \qquad (5.4)$$

The result is an error of 18.0% in the coefficient for carbon dioxide. Equation 5.4 is a good example of the tendency of most subsequent investigators of the subject of fermentative or oxidative assimilation for the next 15 years to use coefficients in their equations which were small, whole numbers. This was partly due to the universal use of the term (CH_2O) to represent the product of assimilation, usually always assumed to be a carbohydrate. When this term is incorporated it becomes difficult to write equations that are anything more than gross approximations to represent experimental results where the product of assimilation is not, in fact, at the same level of reduction as that of carbohydrate.

There is no doubt that Barker and the many investigators who subsequently studied oxidative assimilation were aware that representations such as Eq. 5.4 were only approximations. Thus, in 1939 Clifton and Logan[44] studied the oxidation of several organic compounds in bicarbonate or phosphate buffer by washed suspensions of *E. coli,* concluding that a portion of the substrate was always assimilated by the cells. Their experimental results suggested the equations in Table 5.1. Apart from the carbon dioxide and water that are the usual products of an oxidation, the most striking aspect of the equations in this table is the term (CH_2O). Because of the universal use of this term to describe the product of oxidative assimilation, experimental results notwithstanding, it is not surprising that it came to be almost automatically considered a carbohydrate, and in some cases this was even demonstrated. But what is more surprising is the tendency seen in the literature toward equating (CH_2O), a storage product, with the products of cellular synthesis.

What followed naturally from considerations of primary fermentative or oxidative assimilation were considerations of secondary assimilation. Second-

TABLE 5.1. Equations Representing the Oxidative Assimilation of Several Substrates by *Escherichia coli*

Substrate	Reaction Equation
Acetate	$2CH_3COOH + 3O_2 \rightarrow (CH_2O) + 3CO_2 + 3H_2O$
Pyruvate	$2CH_3COCOOH + 3O_2 \rightarrow 2(CH_2O) + 4CO_2 + 2H_2O$
Fumarate	$C_4H_4O_4 + 2O_2 \rightarrow (CH_2O) + 3CO_2 + H_2O$
Succinate	$C_4H_6O_4 + 2.5O_2 \rightarrow (CH_2O) + 3CO_2 + H_2O$
Glucose	$C_6H_{12}O_6 + 3O_2 \rightarrow 3(CH_2O) + 3CO_2 + 3H_2O$

Source: Data from Reference 44.

ary assimilation was what was thought to happen during the synthesis of cellular material from the products of primary assimilation. For this reason comparisons were made of the chemical activities of growing cells and of washed cell suspensions of the same microorganism on the same substrate. For example, Clifton and Logan felt as a result of their studies that ". . . the same relation between the extent of oxidation and assimilation was observed in actively proliferating cultures as in washed suspensions of *E. coli.*" (Clifton and Logan,[44] p. 539.) This can be interpreted in two ways. One is that both washed cell suspensions and actively proliferating cultures first assimilate a substrate and form a primary product of assimilation from it, after which the proliferating cultures use this primary assimilation product as an internal carbon source for the synthesis of cellular structure. The other is that primary assimilation occurs with washed cell suspensions but that growth takes place without the necessary prior formation of a primary assimilation product. Clifton and Logan favored the first interpretation. The second, however, is equally possible. A confusion eventually came to exist as to the meaning of the word "assimilation." Did it mean the formation of stored carbohydrate or the incorporation of the substrate carbon into the cellular structural material, or both?

Not all investigators were satisfied that the primary product of assimilation was necessarily a carbohydrate. Also in 1939 Fink, Krebs, and Lechner[45] used the following equation to represent assimilation by washed suspensions of *Torula utilis*.

$$C_6H_{12}O_6 \longrightarrow 2CH_3CHO + 2CO_2 + 4H \qquad (5.5)$$

where CH_3CHO represents the assimilation product. Since 4H does not seem to be an adequate representation of what happens to the excess hydrogen in Eq. 5.5, the addition of a molecule of oxygen on both sides of the equation gives something representing more closely the assimilation reaction.

$$C_6H_{12}O_6 + O_2 \longrightarrow 2CH_3CHO + 2CO_2 + 2H_2O \qquad (5.6)$$

Equation 5.6 as modified from Fink et al.[45] is the first representation of oxidative assimilation in which the assimilation product is more reduced than a carbohydrate, in this case at the level of reduction of acetaldehyde. An investigation in 1945 by Whelton and Doudoroff[46] of the assimilation of glucose and related compounds by growing cultures of *Pseudomonas saccharophila* gave similar results. These investigators made ". . . an attempt to compare the assimilation from certain compounds by *P. saccharophila* . . . during cell multiplication with the assimilation shown by resting suspensions of the same organism." (Whelton and Doudoroff,[46] p. 177.) It is evident that the term "assimilation" was already becoming ambiguous in that if used without further description one would not know whether it applied to growing cells or resting cell suspensions. Whelton and Doudoroff grew *P. saccharophila* on a mineral medium in the presence of low ($M/75$) concentrations of substrate and determined the percent assimilation of substrate carbon by growing cells and by resting cell suspensions. Their analysis of the elemental composition of cells grown on glucose was 54.5% C, 11.05% N, 7.4% H, and 3.3% ash, on a dry-weight basis, from which they arrived at the formula $CH_{2.12}O$ as approximating the product of assimilation (i.e., bacterial cells). Curiously, they did not attempt to incorporate this term into equations representing assimilation during growth on any of the organic substrates they tested, even though a method of constructing such equations had been brilliantly worked out 10 years earlier by Tamiya (see Chapter 6). However, because *P. saccharophila* is a facultative autotroph capable of using the oxidation of hydrogen as a source of energy, by measuring the uptake of hydrogen, oxygen, and carbon dioxide by cells grown in a mineral medium and using the formula they constructed to represent cells grown on glucose as being also representative of cells grown on hydrogen, they were able to represent growth on hydrogen by the following equation.

$$CO_2 + 11.68H_2 + 4.81O_2 \longrightarrow (CH_{2.12}O) + 10.62H_2O$$

This equation is interesting as the first equational representation in the literature on oxidative and fermentative assimilation in which the coefficients are not small, whole numbers.

In 1948 Swanson and Clifton[47] studied growth and assimilation by *Saccharomyces cerevisiae*. They apparently rediscovered what A. J. Brown[48,49] had discovered in 1892 and discussed at length in 1894: that growing yeasts possess as much or even more fermentative power when well supplied with free oxygen than they do under conditions of oxygen starvation (i.e., that the Pasteur effect as applied to growing yeasts was a misinterpretation on Pasteur's part). This phenomenon later came to be called the "Crabtree effect," and the "Negative Pasteur effect," although if anything it should be called the "Brown effect." In retrospect, it is probably the first known example of catabolite repression. In their summary Swanson and Clifton stated:

> Actively proliferating cultures of certain strains of *Saccharomyces cerevisiae* when cultivated aerobically carry on the most active initial stages of their metabolism, from the point of growth rate and synthesis of cellular carbon, by the process of fermentative assimilation. This stage predominates until hexose or other polyose molecules disappear from the medium. Oxidative assimilation then occurs from intermediate products of saccharide dissimilation, including ethanol, until practically all these have been utilized with the exception of the nonvolatile organic acids. [Swanson and Clifton,[47] p. 123.]

Here the process of fermentative assimilation appears to be associated definitely with growth. Swanson and Clifton[47] conducted some very interesting experiments which turned out to be remarkably inconclusive. They used a synthetic medium supplemented with vitamins plus 1% of a single sugar as a sole source of carbon and energy. In one experiment with glucose as the substrate they determined the oxygen consumed and the carbon dioxide, ethanol, and cellular material produced during growth. For whatever reasons their cell yield was low, with only 15.4% of the substrate carbon being incorporated into cellular material. Another experiment with sucrose as the substrate showed an incorporation of only 20.3%. These low values led them to conduct experiments with washed cell suspensions in the usual way to determine the extent of fermentative and oxidative assimilation. The values they observed were a carbon dioxide production of 75% of the theoretical value for a complete fermentation of the substrate and an increase in cellular carbon of 0.13–0.14 mg, whereas the postulated value for an increase in cellular carbon was 0.18 mg based on the following equation:

$$2C_6H_{12}O_6 \longrightarrow 3(CH_2O) + 3CH_3CH_2OH + 3CO_2$$

Under aerobic conditions there was a carbon dioxide production of 33% of the theoretical value for a complete oxidation of the substrate and an increase in cellular carbon of 0.36–0.38 mg, whereas the postulated value was 0.48 mg based on the following equation:

$$C_6H_{12}O_6 + 2O_2 \longrightarrow 4(CH_2O) + 2CO_2 + 2H_2O$$

The discrepancies between the postulated and the experimentally observed values for the quantity of assimilation product in fermentative and in oxidative assimilation are −25 and −23%, respectively. It is clear that the usual representation of the assimilation product as being a carbohydrate was not very satisfactory.

The problem with Swanson and Clifton's experiments may have been that they did not include a steroid in their medium. It was to be demonstrated 5 years later by Andraesen and Stier[50] that ergosterol was required for the anaerobic growth of *S. cerevisiae*. Thus, it is probable that during the initial fermentative metabolism of the yeast in Swanson and Clifton's experiments the yeast was not growing. This might account for the low values of assimilation they observed. It does not account for the low values of assimilation observed with washed cells.

The work of Swanson and Clifton[47] represents a good example of an attempt to extrapolate from fermentative and oxidative assimilation to secondary assimilation, or growth, or at least to compare the two processes. The results were most inconclusive. Siegel and Clifton[51–53] subsequently attempted a similar investigation. Their studies, published in 1950, were conducted largely with growing cells, and it is evident that their emphasis was changing away from the use of resting cell suspensions as experimental tools.

Siegel and Clifton cultivated *E. coli* (K-12) in a synthetic medium consisting of minerals, trace elements, and glucose. After 18 hr of growth the cultures were diluted 1:10 with the identical medium for growth experiments, which were conducted in Warburg vessels so that gas exchanges could be determined manometrically. Initial studies showed that oxygen might become limiting after a few hours of growth, and in subsequent experiments the time intervals for readings were selected so as to include the early lag phase, the late lag phase, the early logarithmic phase, and a representative logarithmic phase. Thus, whereas in earlier studies such as those described by Clifton[34] on oxidative assimilation with washed cells the oxidation was permitted to proceed until it had reached the level of the control cells, with consequent wide fluctuations in the RQ of the system, Siegel and Clifton stopped their experiments when glucose was still in the medium. Previous studies with washed cells had indicated that the oxidative assimilation of glucose by *E. coli* could be represented by the following equation:

$$C_6H_{12}O_6 + 3O_2 \longrightarrow 3(CH_2O) + 3CO_2 + 3H_2O$$

The results of several experiments using growing cultures of *E. coli* provided carbon balances indicating that the product (or products) of assimilation might be more reduced than carbohydrate, and that assimilation by growing *E. coli* could be represented better by the following equation

$$C_6H_{12}O_6 + 2.5O_2 \longrightarrow (CH_2O)_3H_2 + 3CO_2 + 2H_2O \qquad (5.7)$$

where $(CH_2O)_3H_2$ represents the assimilation product. Even though Siegel and Clifton tried to get away from tradition in that they used a fractional coefficient for the oxygen term in Eq. 5.7, they still could not quite bring themselves to abandon the use of the term (CH_2O) in their representation of the product of assimilation during growth, such was its universal acceptance.

Subsequently Siegel and Clifton[52] studied energy relationships with respect to carbohydrate assimilation by *E. coli.* They first conducted experiments with washed cell suspensions to obtain equations for the oxidative assimilation of arabinose, glucose, and lactose. These are as follows:

$$C_5H_{10}O_5 + 3O_2 \longrightarrow 2(CH_2O) + 3CO_2 + 3H_2O \qquad (5.8)$$

$$C_6H_{12}O_6 + 3O_2 \longrightarrow 3(CH_2O) + 3CO_2 + 3H_2O \qquad (5.9)$$

$$C_{12}H_{22}O_{11} + 6O_2 \longrightarrow 6(CH_2O) + 6CO_2 + 5H_2O \qquad (5.10)$$

They then determined the amounts of carbon stored in the cells as a result of aerobic growth on the three substrates. This gave them data with which to make thermodynamic calculations of the energy exchanges involved, using the values in Table 5.2. It is seen that they are here carrying on the same procedure for calculating the free energy of formation of carbon dioxide and of oxygen as had been used by previous investigators in that this was calculated for O_2 at 0.2 atm and for CO_2 at 0.0003 atm. The values for the free-energy change accompanying the complete oxidation of arabinose, glucose, and lactose to carbon dioxide and water were calculated to be −592,817, −709,300, and

TABLE 5.2. Free Energies and Heats of Formation at 25°C

Substance	$-\Delta F$, cal mole^{-1}	$-\Delta H$, cal mole^{-1}
O_2(g, 0.2 atm)	950	0
CO_2(g, 0.0003 atm)	99,060	94,052
H_2O(l)	56,560	68,317
$C_5H_{10}O_5$(s)	180,533	252,000
$C_6H_{12}O_6$(s)	218,720	305,730
$C_{12}H_{22}O_{11}$(s)	418,200	592,900

Source: Reference 52.

$-1{,}381{,}280$ cal mole^{-1}, respectively. As a first approximation Siegel and Clifton felt that the free-energy changes accompanying the reactions represented by Eqs. 5.8, 5.9, and 5.10 could be ascribed only to that part of the substrate which was oxidized completely to carbon dioxide and water, assuming that the free energy of formation of the hypothetical (CH_2O) product of assimilation is the same in all cases. It was admitted that this approximation was arbitrary, but might be useful for comparative purposes. The computation was easy to make for Eqs. 5.8 and 5.9, where the molar quantities of CO_2 and H_2O formed are the same. For Eq. 5.10 only $\frac{5}{6}$ as much water is formed as CO_2, and here they used only the value for CO_2 to make the calculation. Thus, the free-energy changes accompanying the reactions represented by Eqs. 5.8, 5.9, and 5.10 were calculated to be $-355{,}690$, $-354{,}650$, and $-690{,}640$ cal mole^{-1}, respectively. They then calculated the "high-energy phosphate bond" energy available from the oxidation of these substrates and found this to be about $-216{,}000$ cal for arabinose and glucose, and about $-432{,}000$ for lactose. This was a set of energies which was proportional to the free-energy changes calculated for the reactions represented by Eqs. 5.8, 5.9, and 5.10 in that about twice as much energy was released in "oxidative synthesis" (whatever this means) from lactose as from either arabinose or glucose. On the other hand, from the growth studies they concluded that "the proportional increase in stored carbon with growth is greatest with arabinose as the substrate, less with glucose, and least with lactose; results that would certainly not be anticipated from the energetics of the systems concerned." (Siegel and Clifton,[52] p. 580.) Siegel and Clifton then went on to offer evidence "that consideration of energetics alone does not provide adequate criteria for theoretical considerations of assimilation." However, quite apart from the approximation made in comparing reactions carried out by washed cell suspension with those by growing cells, comparisons of the efficiencies of assimilation should have been made on the basis of energy *conserved* in the assimilation product, not on the basis of energy expended during the assimilatory reactions, as Siegel and Clifton did.

Also in 1950 Siegel and Clifton[53] published the results of another investigation in which they studied the oxidative utilization by *E. coli* of succinate, fumerate, lactate, pyruvate, glycerol, and acetate. As with their previous studies, they compared the data obtained using washed cell suspensions with that using cultures growing in synthetic media. In addition, rather than using the amount of carbon dioxide formed, as indicated by the equations representing oxidative assimilations by washed cell suspensions, to calculate the amount of substrate oxidized to carbon dioxide and water as they had done previously, they used the amount of oxygen consumed. Interestingly, they did not use the amount of oxygen indicated in the equations they constructed but used the actual quantities of oxygen which were measured experimentally. As an example, the equation they constructed representing the oxidative assimilation of fumarate by washed cell suspensions of *E. coli* is as follows:

$$C_4H_4O_4 + 2O_2 \longrightarrow (CH_2O) + 3CO_2 + H_2O \qquad (5.11)$$

The complete oxidation of fumarate to carbon dioxide and water is accompanied by a free-energy change of $-349{,}302$ cal mole^{-1} and requires 3 moles of oxygen per mole of fumarate oxidized. The free-energy change accompanying the reaction represented by Eq. 5.11 would thus be $\frac{2}{3} \times -349{,}302 = -232{,}879$ cal mole^{-1}. On the basis of the actual oxygen consumption by washed cell suspensions which was measured experimentally, the oxygen uptake was 76% of that predicted for a complete oxidation, not 66% as predicted from Eq. 5.11 which Siegel and Clifton used to represent this reaction. Thus, they adopted the value of $-265{,}469$ cal mole^{-1} as a better indicator of the free-energy change accompanying the reaction represented by Eq. 5.11. It is apparent that this sort of manipulation and the use of equations such as 5.11 to make comparative calculations of energy exchanges with respect to growing cell cultures makes no sense. Nevertheless, using such calculations Siegel and Clifton again came to the conclusion that "The efficiency of assimilation of different substrates by the growing cell is seen to be generally independent of the free energy of oxidation of the substrate . . ." (Siegel and Clifton,[53] p. 592.)

A 1951 publication by Wiame and Doudoroff[37] on oxidative assimilation by *P. saccharophila* and one in 1954 by Wilner and Clifton[38] on oxidative assimilation by *Bacillus subtilis* were the results of later (and the last) investigations of the subjects of oxidative or fermentative assimilation. In 1954 Wilner and Clifton attempted to correct for the endogenous respiration frequently encountered with washed cell suspensions which had plagued research on oxidative assimilation ever since Barker[31] first described it. Their results showed nothing new.

The cessation in the work on fermentative and oxidative assimilation was an indication that the subject was too unwieldy to yield useful results. The difficulty seems to have been the impossibility of defining appropriate conditions or limits within which to carry out experiments. For example, Winzler[54] found that the product of the oxidative assimilation of acetate by *S. cerevisiae* was a carbohydrate. In contrast to this, with the same microorganism and with apparently the same experimental conditions Macleod and Smedley-Maclean,[55] White and Werkman,[56,57] and White, Krampitz, and Werkman[58] found the assimilation product to be a lipid which is formed directly from acetate without passing through a carbohydrate stage. Here are two quite different results from reliable workers, and one can only conclude that since there can be only one set of results obtainable from a given experimental system, the experimental conditions were somehow not identical. Such discrepancies are typical of work with washed ("resting") cell suspensions. They were already well known in Pasteur's time, when different amounts of ethanol and carbon dioxide were obtained from alcoholic fermentation depending on the ratio of the weight of cells to substrate.

If one can conceive of fermentative or oxidative assimilation as being *only* the formation of a "primary assimilation product" as described by Duclaux[4] or by Barker,[31] the experimental difficulties encountered by many investigators do not occur. All that happens when a well-washed suspension of cells is incubated in the presence of a metabolizable substrate is that a well-defined storage material appears in the cells concomitant with the fermentation or oxidation of the rest of the substrate. However, it is evident from Pasteur's early work that fermentative or oxidative assimilation as described above may not be the only processes to occur.

Pasteur[1] dissolved 100 g of sugar (a lot) in 750 ml of water and added to this 2.626 g (dry weight) of yeast. When the fermentation had ceased, he found that:

1. In the fermentation of 100 g of sugar by 2.626 g of yeast there was fixed about 0.4 g of hydrocarbon material that was transformable by dilute sulfuric acid into a fermentable sugar.
2. There was a measurable increase in those nitrogenous materials that are insoluble in sulfuric acid.

This latter result is an indication that during the fermentation there was a fixation into an insoluble state of nitrogenous materials that are contained within the yeast cells. Thus, during this experiment in which resting cells were used there not only occurred a fermentative assimilation but a synthesis of protoplasmic material as well from the acid soluble nitrogenous substances present within the cells. It is well known that such soluble nitrogenous substances are present in yeast. A hot, aqueous extract of yeast (yeast extract) can readily be used as a nitrogen source for the growth of yeast, as Pasteur also showed. This extension of Pasteur's conclusions can reasonably be expanded to include other kinds of cells, for a "pool" of amino acids or other nitrogenous compounds is thought to exist to a greater or lesser extent in most microorganisms which have been recently harvested from a growing culture, or at least in those microorganisms in which the phenomenon of adaptation can readily be demonstrated. The evidence indicates that more than the formation of a single, primary assimilation product may occur in recently harvested washed call suspensions. In addition to less complex assimilation products it is possible to have an actual synthesis of protoplasm which can be equated with growth. This phenomenon undoubtedly explains the change in the RQ which occurred with the passage of time in studies on oxidative assimilation. As one of many examples, in the research published by Clifton and Logan[44] on the oxidative assimilation of pyruvate by *E. coli,* the RQ varied from 1.55 down to 0.93. However, without knowing more about the system it was impossible to know how to interpret this variation.

One thing that became abundantly evident toward the end of the general research on oxidative assimilation was that experiments with growing cells were more reproducible than those using washed cell suspensions. Thus, Siegel and Clifton state, "This study, however, primarily deals with assimilation by young, growing cells, since preliminary trials indicated that more reproducible carbon balances were obtainable with them than with washed suspensions in buffer." (Siegel and Clifton,[52] p. 574.) This was obviously the approach to adopt. The product of assimilation, with the possible exception of that in well-washed, *starved* cells (which no one used) from which all pooled substances would have disappeared, could be critically represented as CH_2O in only a very few examples. In most cases the use of this term was a gross approximation. The amazing thing is how long people persisted in using it when there were other, better alternatives.

The use of oxidative assimilation equations constructed from washed cell suspension data as a first approximation toward studying oxidative assimilation in growing cell cultures was not sound scientifically. Just as it is incorrect to generalize from a reaction in which glucose is oxidized to completion by a resting yeast cell suspension poisoned with dinitrophenol to a reaction in which glucose is oxidatively assimilated by a nonpoisoned, washed cell suspension, so it is incorrect to generalize from the latter to a reaction in which cells are actively growing. The three types of reactions represent three completely different systems having completely different properties. To put this more in perspective, these reactions can be represented by equations using glucose as the substrate, as follows:

Complete oxidation:

$$C_6H_{12}O_6 + 6O_2 \longrightarrow 6CO_2 + 6H_2O \qquad (5.12)$$

Oxidative assimilation (*Swanson and Clifton*[47]):

$$C_6H_{12}O_6 + 2O_2 \longrightarrow 4(CH_2O) + 2CO_2 + 2H_2O \qquad (5.13)$$

Growth:

$$C_6H_{12}O_6 + aO_2 + bNH_3 \longrightarrow x(CH_hO_oN_n) + yCO_2 + zH_2O \quad (5.14)$$

In the growth equation a, b, x, y, and z represent coefficients that are seldom small, whole numbers, and where $(CH_hO_oN_n)$ represents an empirical formula for the cells calculated on the basis of a unit amount of carbon. It is evident that the amount of assimilation represented in Eq. 5.13 cannot be predicted from any information having to do with Eq. 5.12. It should be equally evident that the amount of cellular material formed during growth as represented by Eq. 5.14 cannot be predicted on the basis of any information having to do with Eq. 5.13. This may seem obvious, yet generalizations from reactions represented by Eq. 5.13 have frequently been made, with the term (CH_2O) representing "cellular material" (or cellular carbon). In one respect this is true: the product of assimilation (in this case a carbohydrate) *is* found inside the cell and can thus be considered to be cellular material. However, it may equally be a transient storage material serving as an internal substrate that is not strictly a part of the structure of the cell.

By 1954 even the meanings of the terms fermentative or oxidative assimilation were obscure, since they were used to designate phenomena that were observed both with washed cell suspensions and with growing cell cultures. It was never established whether, during growth, a primary product of assimilation such as glycogen was formed after which it became dissimilated to form the products of secondary assimilation, or whether cell structure could be formed directly from intermediates derived from the substrate without the necessary formation of a primary product. Energetically both of these phenomena could result in the same end product, the cell, as a result of equivalent energy exchanges provided that a primary product was formed and then synthesized into cell structure before an experiment is terminated. What is not permitted in the study of the energetics of microbial growth is to have stored materials present in the cells when an experiment is ended, without correcting for this. If this were to happen what might be considered *only* growth would actually be a combination of the reactions represented by Eqs. 5.13 and 5.14, the sum of which would only be partially representative of growth.

5.5. SUMMARY

All the work on oxidative and fermentative assimilation seems to have led nowhere. The concept originated in the Laboratorium voor Microbiologie, in Delft, Holland, in 1936 and after a brief period of investigation in Europe passed over to laboratories at Stanford University and the University of California where the influence of the "Delft School of Microbiology" was very strong. Quite a number of publications appeared in the 1940s, but except for occasional later publications such as those by Wilson and Danforth[59] and Danforth[60] on the oxidative assimilation of acetate by bleached *Euglena gracilis,* by 1954 interest in the phenomena of fermentative or oxidative assimilation had died out and at the present time it is not even mentioned in most textbooks on microbiology. On the other hand, it is important to know that oxidative and fermentative assimilation as previously defined can, under the appropriate conditions, lead to the accumulation of storage products which are not part of the structure of the cells and, if not taken into account, to erroneous values of growth yields.

REFERENCES

1. L. Pasteur, Memoire sur la fermentation alcoolique, *Ann. Chem. Phys.* **58,** 323–426 (1860).
2. L. Pasteur, Nouveax faits concernant la fermentation alcoolique, *C.R. Hebd. Acad. Sci.* **58,** 640–642 (1859).
3. L. Pasteur, Nouveaux faits relatifs à la fermentation alcoolique; cellulose et matières grasses de la levure constituées aux dépens du sucre, *C.R. Hebd. Acàd. Sci.* **48,** 735–737 (1859).
4. E. Duclaux, *Traité de Microbiologie,* Vol. III, Fermentation Alcoolique, Masson et Cie., Paris, 1900.
5. C. B. Van Niel and E. H. Anderson, On the occurrence of fermentative assimilation, *J. Cell. Comp. Physiol.* **17,** 49–56 (1941).
6. C. Nägeli and O. Loew, Ueber die chemische Zusammensetzung der Hefe, *Justus Leibig's Ann. Chem.* **193,** 322–348 (1879).
7. C. Nägeli, Ueber die Fettbildung bei den niederen Pilzen, *Sitzungsber. königl. bayer. Akad. Wiss. München* **9,** 287–316 (1879).
8. P. Lindner, Das Problem der biologischen Fettbildung und Fettgewinnung, *Z. Angew. Chem.* **35,** 110–114 (1922).
9. O. Fürth and F. Lieben, Über Milchsäurezerstörung durch Hefe und durch Blutzellen, *Biochem. Z.* **128,** 144–168 (1922).
10. O. Fürth and F. Lieben, Weitere Untersuchungen über Milchsäurezerstörung durch Hefe, *Biochem. Z.* **132,** 165–179 (1922).
11. F. Lieben, Über das Verhalten von einigen Aminosäuren gegunüber sauerstoffgelufteter Hefe, *Biochem. Z.* **135,** 180–187 (1922).
12. F. Lieben, Über das Verhalten von Brenztraubensäure und Acetaldehyd gegunüber mit Sauerstoff gelüfteter Hefe, *Biochem. Z.* **135,** 240–247 (1923).
13. H. Lundin, Über den Einfluss des Sauerstoffs auf die assimilatorische und dissimilatorische Tätigheit der Hefe, I. Teil, Verhalten der Dextrose, *Biochem. Z.* **141,** 310–341 (1923).
14. H. Lundin, Über den Einfluss des Sauerstoffs auf die assimilatorische und dissimilatorische Tätigheit der Hefe, II. Teil, Weitere Untersuchungen über das Verhalten von Monosacchariden, *Biochem. Z.* **141,** 342–369 (1923).
15. H. Lundin, Über den Einfluss des Sauerstoffs auf die assimilatorische und dissimilatorische Tätigheit der Hefe, III. Teil, Verhalten zugesetzten Alkohols in Hefesuspension, *Biochem. Z.* **142,** 454–462 (1923).
16. H. Lundin, Über den Einfluss des Sauerstoffs auf die assimilatorische und dissimilatorische Tätigheit der Hefe, IV. Teil, Verhalten einiger organischer Säuren, *Biochem. Z.* **142,** 463–492 (1923).
17. I. Smedley-Maclean and D. Hoffert, Carbohydrate and fat metabolism in yeast, *Biochem. J.* **17,** 720–742 (1923).

18. I. Smedley-Maclean and D. Hoffert, Carbohydrate and fat metabolism in yeast. II. The influence of phosphates on the storage of fat and carbohydrate in the cell, *Biochem. J.* **18,** 1273–1278 (1924).
19. I. Smedley-Maclean and D. Hoffert, The formation of fat by yeast, Food Investigation Board, report for the year 1924, *Dept. Sci. Ind. Res.* 66–68 (1925).
20. I. Smedley-Maclean and D. Hoffert, Carbohydrate and fat metabolism of yeasts. III. The nature of the intermediate stages, *Biochem. J.* **20,** 343–357 (1926).
21. D. Hoffert, Action of yeast on lactic acid, *Biochem. J.* **20,** 358–362 (1926).
22. J. Marian, Über das Verhalten sauerstoffgeschüttelter Hefe zu β-oxybuttersäure, *Biochem. Z.* **150,** 281–289 (1924).
23. J. Marian, Die Assimilation des Glycerins durch sauerstoffgeschüttelte Hefe, *Biochem. Z.* **150,** 290–303 (1924).
24. O. Meyerhof, Über den Einfluss des Sauerstoffs auf die alcoholische Gärung der Hefe, *Biochem. Z.* **162,** 43–87 (1925).
25. R. P. Cook and M. Stephenson, Bacterial oxidations by molecular oxygen. I. The aerobic oxidation of glucose and its fermentation products in its relation to the viability of the organism, *Biochem. J.* **22,** 1368–1386 (1928).
26. H. Lineweaver, Characteristics of oxidation by *Azotobacter, J. Biol. Chem.* **99,** 575–593 (1932).
27. F. T. Brücke, Über Glykogenbildung in Hefe, *Biochem. Z.* **264,** 157–162 (1933).
28. R. A. MacAnally and I. Smedley-Maclean, Note on the storage of carbohydrate and fat by *Saccharomyces frohberg* when incubated in sugar solution, *Biochem. J.* **28,** 495–498 (1934).
29. R. Guillmet, Sur le bilan de la fermentation des glucides par les levures, *Bull. Soc. Chim. Biol.* **18,** 941–958 (1936).
30. R. Guillmet, Sur le bilan de la fermentation alcoolique par les levures, *C.R. Séances Soc. Biol. Fil.* **121,** 465–567 (1936).
31. H. A. Barker, The oxidative metabolism of the colorless alga, *Prototheca zopfii, J. Cell. Comp. Physiol.* **8,** 231–250 (1936).
32. L. Algera, Energiemessung bei *Aspergillus niger* mit Hilfe eines automatischen mikro-kompensations-Calorimeters, *Rec. Trav. Bot. Neerlandais* **29,**47–163 (1932).
33. G. Giesberger, Beiträge zur kenntnis der Gattung *Spirillum Ehbg,* Doctoral Thesis, Delft, 1936.
34. C. E. Clifton, Microbial assimilations, *Adv. Enz.* **6,** 269–308 (1946).
35. C. E. Clifton, Assimilation by bacteria, in C. H. Werkman and P. W. Wilson, Eds., *Bacterial Physiology,* Academic Press, New York, 1951, pp. 531–547.
36. C. E. Clifton, Oxidative assimilation by microorganisms, in J. B. Sumner and K. Myrbäck, Eds., *The Enzymes,* Vol. II, Part 2, Academic Press, New York, 1952, pp. 912–928.
37. J. M. Wiame and M. Doudoroff, Oxidative assimilation by *Pseudomonas saccharophila* with C^{14}-labelled substrates, *J. Bact.* **62,** 187–193 (1951).
38. B. Wilner and C. E. Clifton, Oxidative assimilation by *Bacillus subtilis, J. Bact.* **67,** 571–575 (1954).
39. H. A. Barker, The metabolism of the colorless alga, *Prototheca zopfii* Krüger, *J. Cell. Comp. Physiol.* **7,** 73–93 (1935).

40. E. H. Anderson, Studies on the metabolism of the colorless alga *Prototheca zopfii, J. Gen. Phys.* **28,** 297–327 (1945).
41. A. Yamamoto, Über den Einfluss einiger Gifte und der Temperatur auf den Ausnutzungsgrad der Atmungsenergie beim Wachstum des Schimmelpilzes, *Acta Phytochim.* **7,** 65–92 (1933).
42. C. E. Clifton, Prevention of assimilation in respiring cells, *Nature* **140,** 318–319 (1937).
43. C. E. Clifton, On the possibility of preventing assimilation in respiring cells, *Enzymologia* **4,** 246–253 (1937).
44. C. E. Clifton and W. A. Logan, On the relation between assimilation and respiration in suspensions and in cultures of *Escherichia coli, J. Bact.* **37,** 523–540 (1939).
45. H. Fink, J. Krebs, and R. Lechner, Zur Theorie der Zellsubstanz—Synthese und der maximalen Hefeernten, *Biochem. Z.* **301,** 143–149 (1939).
46. R. Whelton and M. Doudoroff, Assimilation of glucose and related compounds by growing cultures of *Pseudomonas saccharophila, J. Bact.* **49,** 177–186 (1945).
47. W. H. Swanson and C. E. Clifton, Growth and assimilation in cultures of *Saccharomyces cerevisiae, J. Bact.* **56,** 115–124 (1948).
48. A. J. Brown, Influence of oxygen and concentration on alcoholic fermentation, *J. Chem. Soc. Lond., Trans.* **61,** 369–385 (1892).
49. A. J. Brown, The specific character of the fermentative functions of yeast cells, *J. Chem. Soc. Lond., Trans.* **65,** 911–923 (1894).
50. A. A. Andraesen and T. J. B. Stier, Anaerobic nutrition of *Saccharomyces cerevisiae.* I. Ergosterol requirement for growth on a defined medium, *J. Cell. Comp. Physiol.* **41,** 23–36 (1953).
51. B. V. Siegel and C. E. Clifton, Oxidative assimilation of glucose by *Escherichia coli, J. Bact.* **60,** 113–118 (1950).
52. B. V. Siegel and C. E. Clifton, Energy relationships in carbohydrate assimilation by *Escherichia coli, J. Bact.* **60,** 573–583 (1950).
53. B. V. Siegel and C. E. Clifton, Energetics and assimilation in the combustion of carbon compounds by *Escherichia coli, J. Bact.* **60,** 585–593 (1950).
54. J. R. Winzler, The oxidation and assimilation of acetate by Baker's yeast, *J. Cell. Comp. Physiol.* **15,** 343–354 (1940).
55. L. D. Macleod and I. Smedley-Maclean, The carbohydrate and fat metabolism of yeast. V. The synthesis of fat and acetic acid; the influence of metallic ions on carbohydrate and fat storage, *Biochem. J.* **32,** 1571–1581 (1938).
56. A. G. C. White and C. H. Werkman, Assimilation of acetate by yeast, *Arch. Biochem.* **13,** 27–32 (1947).
57. A. G. C. White and C. H. Werkman, Fat synthesis in yeast, *Arch. Biochem.* **17,** 475–482 (1948).
58. A. G. C. White, L. O. Krampitz, and C. H. Werkman, Assimilation of acetate by yeast, *J. Bact.* **53,** 377 (1947).
59. B. W. Wilson and W. F. Danforth, The extent of acetate and ethanol oxidation by *Euglena gracilis, J. Gen. Microbiol.* **18,** 535–542 (1958).
60. W. F. Danforth, Oxidative assimilation of acetate by *Euglena.* Carbon balance and effects of ethanol, *J. Protozool.* **8,** 152–158 (1961).

6

THE MATERIAL AND ENERGY BALANCE STUDIES OF H. TAMIYA AND HIS COLLABORATORS, 1927–1939

Even before Barker's[1] initial investigation in 1936 which led to the general concepts of oxidative and fermentative assimilation and to the use of the term (CH_2O) to represent primary assimilation, a small group of Japanese scientists had been occupied with an extensive study of microbial growth. Their work thus completely preceded all that on assimilation described in Chapter 3. Initially, studies on the metabolic physiology of *Aspergillus oryzae* were conducted, possibly because of the importance of this mold in the fermentation of oriental foods and in the production of *koji*. By the time these were completed enough experience had been obtained that their research evolved into a study of the energetic relationship between the substrate on which the mold was grown and the amount of growth obtained. Tamiya's research and that of his collaborators has never been accorded the recognition it deserves, and will be described in considerable detail here.

Beginning in 1927 H. Tamiya published a set of four articles on the physiology of *A. oryzae*, a mold which is capable of growth on a medium consisting only of minerals, trace elements, and a single source of carbon and energy. The first three of these publications (Tamiya[2–4]) dealt with the effects of various environmental conditions on growth, including the effects of different substrates, different nitrogen sources, pH,metal ions, O_2 concentration, poisons, and so on. These studies provided information on the optimal physical conditions for the growth of the mold, and resulted in the development of a technique for measuring its respiratory gas exchanges. In the fourth article of this series Tamiya[5] described a prodigious amount of work in which the growth and the accompanying gas exchanges of *A. oryzae* were measured with respect to 123 organic compounds. The mold was grown on a medium consisting of NH_4NO_3, 1g; $MgSO_4 \cdot 7H_2O$, 0.5g; $FeCl_3$ (0.5%), 1 drop; phosphate mixture

(*M*/2), 25 ml; and carbon source solution, 50 ml; all in distilled water brought up to a final volume of 100 ml. In general, the carbon source solutions were *M*/2. The pH was adjusted to 5.5–5.7, this being the midpoint between the pH extremes of 2.1 and 8.7 within which growth had been previously demonstrated to take place. By choosing this midpoint a variation in pH as a result of the utilization of substances in the medium or by the production of acids would have no effect on the growth. For the measurement of growth and respiration *A. oryzae* was grown inside the respirometer which is illustrated in Fig. 6.1. It was essentially a large syringe having a barrel 18.5 cm long and 3.1 cm inside diameter that was calibrated with 1-ml graduations up to 100 ml. In the end of the plunger was a cavity 2.3 cm in diameter holding 5 ml. The tubulated end of the syringe was connected to the gasometer illustrated in Fig. 6.2. The respirometer assembly could be sterilized by autoclaving. The experimental procedure was as follows.

Eight small cylindrical glass containers measuring 2.3 cm in diameter and

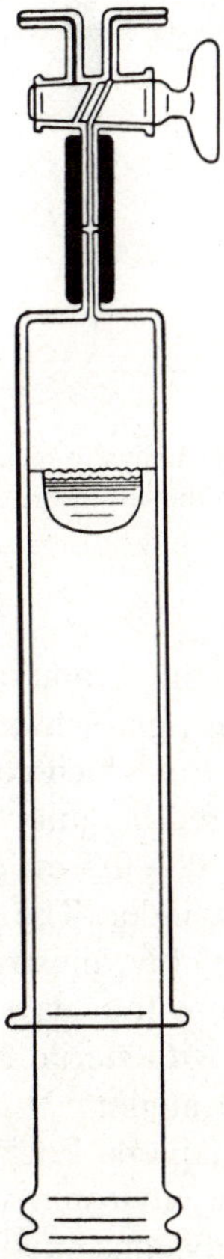

FIGURE 6.1. Diagram of Tamiya's respirometer syringe for the growth of *Aspergillus oryzae*. Reprinted from Reference 4 with permission.

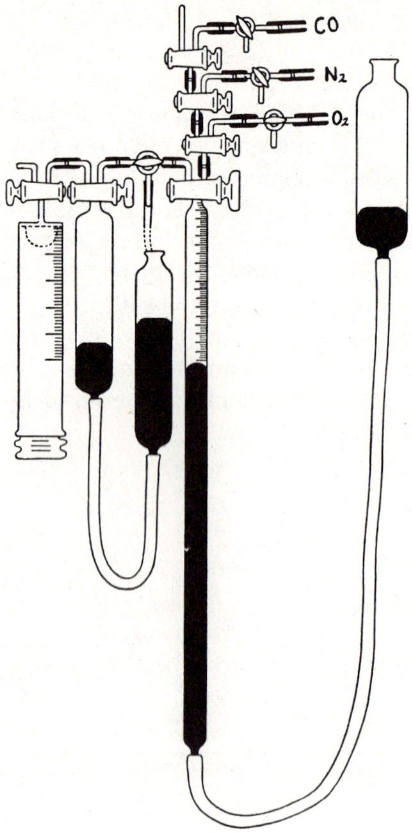

FIGURE 6.2. The gasometer used by Tamiya to measure the respiration of *Aspergillus oryzae*. The mold was grown in the syringe on the left of the diagram. Reprinted from Reference 4 with permission.

3.1 cm deep were filled with 10 ml of mineral solution containing *M*/2 sucrose at a pH of 4.2, placed on a heavy glass disc, and covered with a bell-jar which opened through a neck at the top which could be closed with a cotton plug. After this assembly was sterilized by autoclaving, and permitted to cool, the cotton plug was removed and 0.2–0.3 ml of a spore suspension was placed with a sterile pipet into each cylinder. The plug was then replaced in the top of the bell-jar. The whole assembly, illustrated in Fig. 6.3, was placed in an incubator at 30°C. After three or four days the mycelial pad that had grown in each cylinder was removed with sterile tweezers and placed separately on the surface of the same liquid medium, this time without sugar, which had been sterilized in separate containers. Each mycelial pad was then incubated overnight in this substrate-free environment to enable any reserve material within the cells to be completely consumed. Following this, some of the mycelial pads were dried to obtain an initial dry weight before growth. To measure

FIGURE 6.3. Tamiya's incubator for growing *Aspergillus oryzae* mycelial pads for use in the respirometer syringe illustrated in Fig. 6.1. Reprinted from Reference 4 with permission.

growth and respiration the plunger was withdrawn from the barrel of a sterile respirometer and the cavity at the end of the plunger filled with 5 ml of a sterile mineral solution at a pH of 5.5–5.7 containing the carbon source to be tested. A mycelial pad was then placed on the surface of the medium in the cavity, the pad already having the diameter of the cavity from having been grown in the small cylinders under the bell-jar. The plunger was then reinserted into the barrel of the respirometer and pushed completely in, after which the respirometer was connected to the gasometer. The gas burette of the gasometer (Fig. 6.2) was filled with a mixture of 85% O_2 plus 15% N_2, this mixture then being transferred to the respirometer by lowering the plunger to the 100-ml gradation. The respirometer was then disconnected from the gasometer and placed in a glass cylinder partly filled with water so that the lower end of the barrel was completely covered. This is illustrated in Fig. 6.4. The respirometer was next incubated at 30°C for 22–27 hr after which it was reconnected to the gasometer. Pushing the plunger completely in enabled the gas mixture in the respirometer to be transferred to the gasometer so that the total volume of gas, the O_2 consumed, and the CO_2 produced could be measured. The amount of substrate consumed could be determined by analyzing the residual medium for the quantity of substrate remaining and subtracting this value from the quantity originally present. Control experiments were carried out in the same manner using a mineral medium without a carbon source. After an experiment was terminated the mycelial pads were removed from each respirometer, washed

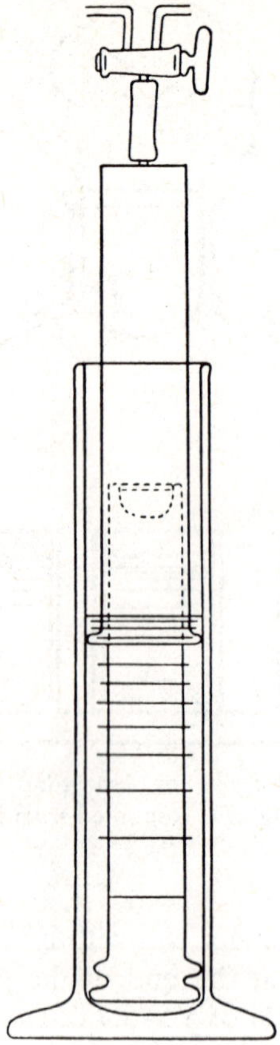

FIGURE 6.4. Diagram of Tamiya's respirometer syringe being incubated in a glass cylinder containing water. Reprinted from Reference 4 with permission.

in distilled water, dried, and weighed. The relative increase in dry weight of the mycelium was determined with the following equation.

$$w_D = \frac{M - M_0}{M_K - M_0} \times 100 \qquad (6.1)$$

where w_D represents the relative increase in weight of the mycelial pad, M_0 the initial dry weight in milligrams of a mycelial pad, M the dry weight in milligrams of the mycelial pad after growth, and M_K the dry weight of the mycelial pad in the control culture. Several control and several experimental cultures were made simultaneously, the results being averaged.

If it is assumed that there is a steady formation during growth of Δm mg dry weight of new mycelium per hour during an experiment lasting Z hr, the following relationships can be applied.

$$M_0 \times (1 + \Delta m)^Z = M$$

therefore

$$\Delta m = \sqrt[z]{\frac{M}{M_0}} - 1 \tag{6.2}$$

and

$$\log (1 + \Delta m) = \frac{1}{Z} \log\left(\frac{M}{M_0}\right)$$

Equations 6.1 and 6.2 can be combined to give

$$\begin{aligned} \Delta m &= \sqrt[z]{\frac{w_D}{100}\left(\frac{M_K}{M_0} - 1\right) + 1} - 1 \\ \text{or } \log(\Delta m + 1) &= \frac{1}{Z} \log\left[\frac{w_D}{100}\left(\frac{M_K}{M_0} - 1\right) + 1\right] \end{aligned} \tag{6.3}$$

Tamiya found that on average $M_K/M_0 = 3.5$. Inserting this value into Eq. 6.3 plus $Z = 24$ hr, the average value for Δm could be calculated if the value for w_D were known.

$$\log(\Delta m + 1) = \frac{1}{24} \log(0.025 \times w_D + 1)$$

To determine the gas exchange accompanying the growth of the mold in the respirometer the following data had to be obtained:

1. The temperature and the atmospheric pressure at the beginning and at the end of an experiment; t, p, and t', p', respectively.
2. The water vapor pressure at t and t'; Δp_W and $\Delta p'_W$, respectively.
3. The total gas volume at the end of an experiment, in milliliters; V'.
4. The volume of carbon dioxide in the gas space of the respirometer at the end of an experiment, in milliliters; v'_{CO_2}.
5. The absorption coefficient for carbon dioxide in water at t'; a.

The gas exchange could be calculated by means of the following equations.

$$I_{O_2} = 100 - \frac{(p' - \Delta p'_W)(273.1 + t)}{(p - \Delta p_W)(273.1 + t')} [V' - v'_{CO_2}] \tag{6.4}$$

$$I_{CO_2} = \frac{(p' - \Delta p'_W)(273.1 + t)}{(p - \Delta p_W)(273.1 + t')} \left[1 + \frac{5a(273.1 + t')}{273.1 \times V'}\right] v'_{CO_2} \tag{6.5}$$

in which I_{O_2} and I_{CO_2} represent the milliliters of O_2 consumed and CO_2 produced, respectively, during the course of an experiment. The volume of carbon dioxide in the gas space of the respirometer was measured by absorption in alkali after the gas had been removed from the respirometer. The coefficient before a in Eq. 6.5 is the number of milliliters of nutrient solution in the cavity of the respirometer plunger.

If the culture solution contained a volatile substance the following value, $\Delta p'$, was inserted in Eqs. 6.4 and 6.5 in place of $\Delta p'_W$.

$$\Delta p' = \frac{\Delta p'_W \times 55.5 + \Delta p'_N \times n}{n + 55.5} \tag{6.6}$$

where n represents the molarity of the volatile substance, $\Delta p'_N$ the vapor pressure of the volatile substance at t', and 55.5 the number of moles of water in 1 liter.

The rates of respiration during an experiment could be calculated in the following manner.

$$Q_{O_2} = \frac{I_{O_2} \times 1000}{[(M + M_0)/2] \times Z} \tag{6.7}$$

$$Q_{CO_2} = \frac{I_{CO_2} \times 1000}{[(M + M_0)/2] \times Z} \tag{6.8}$$

in which Q_{O_2} and Q_{CO_2} represent the quantities of oxygen or carbon dioxide, respectively, in microliters which are consumed or produced by 1 mg dry weight of mold in 1 hr, or the same quantity in milliliters which is consumed or produced by 1 g dry weight of mold hour^{-1}.

The above brief description of some of Tamiya's experimental methods has been given only to provide some idea of the extent of the research involved. The original literature must be consulted for more specific details, and also to gain some idea of the real beauty of these investigations.

Following his studies on the metabolic physiology and growth of *A. oryzae.*, Tamiya[6–7] and Tamiya and Yamagutchi[8] in 1932–1933 published a series of three articles on the physiology of respiration in molds. In the first of these (Tamiya[6]) he sought to relate what he called the "combustion quotient" (CQ) with the respiratory quotient (RQ) actually observed when *A. oryzae* was grown on a specific substrate. To do this he used data obtained from the last of his previous series of four publications (Tamiya[5]) in which I_{O_2} and I_{CO_2} were determined for the growth of this mold on many different substrates. The RQ is related either to the total amount of carbon dioxide produced and oxygen consumed, or to the specific rates of carbon dioxide production and oxygen consumption, by the following equation.

$$\mathrm{RQ} = \frac{I_{CO_2}}{I_{O_2}} = \frac{Q_{CO_2}}{Q_{O_2}} \tag{6.9}$$

The RQ was originally used in animal physiology as a means of determining something about the nature of the substances in a respiring animal which were being oxidized. For example, if carbohydrates were being burned, the RQ was 1.00 and if lipids were being oxidized the RQ was less than 1.00. If the total amount of nitrogen excreted were known, the RQ could be used to calculate the amounts of carbohydrate, protein, and lipid which were being consumed in respiration. Tamiya sought to turn this around and to relate the RQ to what was being formed during growth from a specific substrate. It was for this purpose that he defined the combustion quotient as the ratio of the amount of carbon dioxide produced to the amount of oxygen consumed when an organic substance is completely combusted to carbon dioxide and water. *The CQ is effectively the RQ for complete combustion.* Tamiya derived the calculation of the CQ in the following manner.

$$\mathrm{C}n_\mathrm{C}\mathrm{H}n_\mathrm{H}\mathrm{O}n_\mathrm{O} + a\mathrm{O}_2 \longrightarrow n_\mathrm{C}\mathrm{CO}_2 + \frac{n_\mathrm{H}}{2}\,\mathrm{H_2O} \qquad (6.10)$$

where n_C, n_H, and n_O represent the relative numbers of C, H, and O atoms, respectively, in an organic substance, and where a represents the coefficient for the oxygen consumed during complete combustion. From Eq. 6.10 it follows that

$$a = n_\mathrm{C} + \frac{n_\mathrm{H} - 2n_\mathrm{O}}{4}$$

and

$$\mathrm{CQ} = \frac{n_\mathrm{C}}{a} = \frac{4n_\mathrm{C}}{4n_\mathrm{C} + n_\mathrm{H} - 2n_\mathrm{O}} \qquad (6.11)$$

The calculation of the CQ was later formulated to include organic substances containing nitrogen. In this case the products of a complete combustion are taken to be carbon dioxide, water, and ammonia, since this latter is usually the product of nitrogen metabolism in bacteria, molds, and lower organisms. The appropriate equation for this calculation is therefore

$$\mathrm{CQ} = \frac{4n_\mathrm{C}}{4n_\mathrm{C} + n_\mathrm{H} - 2n_\mathrm{O} - 3n_\mathrm{N}} \qquad (6.12)$$

The accompanying CQ values apply to the combustions represented by the following equations.

$$\underset{\text{Glucose}}{C_6H_{12}O_6} + 6O_2 \longrightarrow 6CO_2 + 6H_2O; \qquad CQ = 1.000$$

$$\underset{\text{Formic acid}}{HCOOH} + 0.5O_2 \longrightarrow CO_2 + H_2O; \qquad CQ = 2.000$$

$$\underset{\text{Ethanol}}{CH_3CH_2OH} + 3O_2 \longrightarrow 2CO_2 + 3H_2O; \qquad CQ = 0.666$$

$$\underset{\text{Leucine}}{C_6H_{13}O_2N} + 7.5O_2 \longrightarrow 6CO_2 + 5H_2O + NH_3; \qquad CQ = 0.800$$

Tamiya then compared the CQ values for 48 substances with the RQ observed during the growth of *A. oryzae* on these same substances and discovered the following generalizations.

1. The RQ is generally greater than the CQ of the substrate if the latter is greater than 0.89.
2. The RQ is generally smaller than the CQ of the substrate if the latter is less then 0.82.
3. The RQ has about the same value as the CQ of the substrate if the latter is between 0.82 and 0.89.

There were thus certain regularities in the relationship between the RQ and the CQ that Tamiya proceeded to investigate further. He recognized that the CQ could be used to determine whether a given reaction in which an organic product was formed from an organic reactant constituted an oxidation or a reduction. Thus, if the CQ of the product is less than that of the reactant, the reaction represents a reduction.

$$\underset{\substack{\text{Glucose} \\ CQ\,=\,1.00}}{C_6H_{12}O_6} + 2H \xrightarrow{\text{reduction}} \underset{\substack{\text{Mannitol} \\ CQ\,=\,0.92}}{C_6H_{14}O_6}$$

If the CQ of the product is greater than that of the reactant, the reaction represents an oxidation.

$$\underset{\substack{\text{Lactic acid} \\ CQ\,=\,1.00}}{C_3H_6O_3} - 2H \xrightarrow{\text{oxidation}} \underset{\substack{\text{Pyruvic acid} \\ CQ\,=\,1.20}}{C_3H_4O_3}$$

If the product is nitrogenous, the same method of calculation can be used:

$$\underset{\substack{\text{Glucose} \\ CQ\,=\,1.00}}{C_6H_{12}O_6} + NH_3 - 6H \xrightarrow{\text{reduction}} \underset{\substack{\text{Leucine} \\ CQ\,=\,0.80}}{C_6H_{13}O_2N} + 4H_2O \qquad (6.13)$$

If in the conversion of one substance to another there is an exchange of water and ammonia, there is no variation in the CQ as it has been defined. Such a transformation is neither an oxidation nor a reduction, but can be called a neutral oxidation–reduction. The following equation is an example.

$$\underset{\substack{\text{Lactic acid} \\ \text{CQ} = 1.00}}{C_3H_6O_3} + NH_3 - H_2O \longrightarrow \underset{\substack{\text{Alanine} \\ \text{CQ} = 1.00}}{C_3H_7O_2N} \qquad (6.14)$$

From this point it was a logical step to consider the chemical composition of mold mycelium, this latter being an organic product that is formed from an organic substrate during a reaction in which growth takes place. Thus, cell substance could be considered in place of nitrogenous substances such as leucine or alanine in Eqs. 6.13 and 6.14. What was needed was the chemical composition of *A. oryzae* cells. As a first approach to determining this, data on the fat, protein, carbohydrate, and ash content of *Aspergillus* sp., *Penicillium* sp., and *Eurotiopsis* sp. as determined by several previous investigators were used, together with the compositions of "average" fats, proteins, and carbohydrates, to find the mycelial composition. This proved unsatisfactory and so the analyses published in 1902 by Mazé[9] on the percent dry weight composition of C, H, O, and N in *Eurotyopsis Gayoni* were selected as being the best available data. These analyses showed that this composition was nearly the same whether *E. Gayoni* were grown on sugar or ethanol. Together with the analysis of the ash composition of *A. oryzae* made by Takata[10] in 1929, the formula representing the composition of dried *A. oryzae* mycelium could be written as

$$C_{86}H_{160}O_{45}N_7P_{0.30}K_{0.25}Mg_{0.17}Fe_{0.008}S_{0.003}$$

Tamiya felt that the contribution of the ash was negligible compared to the weight of the organic components, and at this time adopted the following "ash-free" formula as representing the organic substance in *A. oryzae*.

$$C_{86}H_{160}O_{45}N_7$$

Using this formula the CQ of *A. oryzae* was calculated to be 0.875.

Now that a tentative formula had been established for *A. oryzae* mycelium, it became possible to represent the actual formation of the cellular substance of the mold from the substance of a substrate by an equation. In the case of the growth of *A. oryzae* on glucose this equation could be written

$$131C_6H_{12}O_6 + 56NH_3 = 8C_{86}H_{160}O_{45}N_7 + 98CO_2 + 230H_2O \qquad (6.15)$$

or, in the case of growth on ethanol,

$$172CH_3CH_2OH + 28NH_3 + 123O_2 = 4C_{86}H_{160}O_{45}N_7 + 238H_2O \qquad (6.16)$$

Equations 6.15 and 6.16 were the first accurate representations ever made of the conversion of a substrate into cellular material. It is apparent that this is a far better method of representation than that using (CH_2O), even if this latter method is admitted to be only an approximation.

If the CQ of a microorganism is known, it becomes possible to divide the carbon sources on which it can be grown into three groups with the CQ of the microorganism as a reference. Thus, Tamiya[7] defined a substance with a CQ greater than that of the microorganism as having a "hyperquotient," one with a CQ less than that of the microbe as having a "hypoquotient," and one with a CQ equal to that of the microorganism as having an "isoquotient." This type of classification is obviously analogous to that in which the osmolality of an environment is divided into three groups distinguishing those which are hyperosmotic (hypertonic), hypoosmotic (hypotonic), and isoosmotic (isotonic), all with reference to the osmolality of the cytoplasm of the cells. From what has been stated previously, the formation of cellular material during growth from a substrate having a hyperquotient constitutes a reduction, that from a substrate having a hypoquotient an oxidation, and that from a substrate having an isoquotient a neutral oxidation–reduction.

Tamiya[7] then proceeded to develop equations whereby the RQ, CQ, and the quantity of cellular material formed can be related. This was a marvellous piece of deductive reasoning. The derivations of these equations will not be presented here in exactly the same manner as Tamiya did for two reasons. First, his derivations have a number of Greek symbols the usage of which is less simple than that of more common notations. Second, Tamiya's equations make use of the actual volumes of gases measured in his respirometers. The reduction of these values to molar values is a part of Tamiya's derivations, making them rather complex to sort out. Therefore, to enable Tamiya's ideas to be comprehended more easily his equations have been modified from the originals.

It quickly became evident that the formula originally used to represent the cells was inaccurate. The analyses on which this formula was based had been done by Mazé[9] 30 years earlier, and techniques for elemental analysis had improved since then. For this reason S. Yamagata[11] performed analyses of dried *A. oryzae* cells which had been grown on several different substrates using NH_4Cl, KNO_2, or KNO_3 as nitrogen sources and the identical environmental conditions for growth previously used by Tamiya.[6–8] The mycelium was harvested before spore production could take place, washed, dried, and the percent by weight of carbon, hydrogen, oxygen, nitrogen, and ash determined. Carbon and hydrogen were determined by a half-micro Pregl analysis and nitrogen by a micro-Dumas method. The ash consisted of the material remaining after combustion and the percentage of oxygen was determined by difference. By multiplying the initial percentages by weight of carbon, hydrogen, oxygen and nitrogen by $100/(100 - \%\ \text{ash})$, the relative percentages of these elements in only the organic substance of the mycelium were calculated. These values are the "ash-free" dry weight percentages of these elements. The results of Yamagata's careful analyses are shown in Table 6.1.

TABLE 6.1. The elemental composition, empirical formula, empirical formula weight, and CQ of *Aspergillus oryzae* grown on four carbon sources and two nitrogen sources

Carbon Source	Nitrogen Source	Weight of C, H, O, and N 100 g^{-1} Ash-Free Dry Weight of Mycelium				Ashes 100 g^{-1} Dry Weight of Cells	Empirical Formula $Cn_cHn_HOn_ONn_N$	Empirical Formula Weight	Mold CQ (CQ_M)
		C	H	O	N				
Glucose	NH_4Cl	49.07	7.17	37.29	6.47	5.7	$C_{409}H_{717}O_{233}N_{46}$	9,997	0.935
M/2	KNO_3	49.78	7.28	37.99	4.95	8.9	$C_{415}H_{728}O_{237}N_{35}$	9,990	0.918
Mannitol	NH_4Cl	48.95	7.09	37.04	6.92	4.4	$C_{408}H_{709}O_{232}N_{49}$	10,003	0.943
M/2	KNO_3	49.00	7.27	37.69	6.04	9.0	$C_{409}H_{727}O_{235}N_{43}$	9,997	0.927
Glycerin	NH_4Cl	49.54	7.00	34.95	8.51	6.0	$C_{413}H_{700}O_{218}N_{61}$	9,998	0.953
M/2	KNO_3	49.82	7.48	36.97	5.73	5.3	$C_{415}H_{748}O_{231}N_{41}$	9,998	0.911
Ethanol	NH_4Cl	49.37	7.41	35.11	8.11	10.5	$C_{411}H_{741}O_{219}N_{58}$	9,989	0.927
M/2	KNO_3	49.45	7.52	36.35	6.68	10.0	$C_{411}H_{752}O_{227}N_{48}$	10,000	0.915

Source: Reprinted by permission from Reference 12.

Dividing each percentage by its respective atomic weight and multiplying by 100 enabled Tamiya to write an empirical formula for the cells in terms of carbon, hydrogen, oxygen, and nitrogen. From this a formula weight could be calculated as well as a value for the CQ. The reason Tamiya gave for not including the composition of the ash in these formulas was that it could be neglected as far as the balance of mass in metabolism was concerned, even though minerals were absolutely necessary for metabolism to take place. Data with respect to the empirical formulas, formula weight, and CQ of *A. oryzae* were summarized by Tamiya[12] in 1935 and are also shown in Table 6.1. The standard deviation of Yamagata's data in Table 6.1 with respect to the individual elements is worth noting. On an ash-free basis, C = 49.35 ± 0.30%, H = 7.26 ± 0.16%, O = 36.68 ± 0.99%, and N = 6.70 ± 1.06%, with n = 9. Thus, although the percentage of carbon is amazingly constant, the percentages of hydrogen and oxygen vary considerably, and that of nitrogen varies a great deal, depending on how the mold has been grown. In spite of this variation, as shown in Table 6.1 there is almost no variation in the formula weight of the mycelium, which remains almost exactly constant with an average of 9997 ± 4 daltons, n = 9. *This illustrates clearly that weights alone do not necessarily show up differences in composition.* In general, these data show that the composition of each *A. oryzae* mycelium varies slightly, depending on the kind of carbon and nitrogen source on which it is grown. This variation also shows up in the CQ values for the individual mycelia.

In developing equations relating the RQ and the CQ observed when *A. oryzae* was grown on several substrates and on different nitrogen sources, Tamiya took advantage of the observations of Terroine and Wurmser[14] and of Molliard[15] which confirmed his own findings that cells of *Aspergillus* sp. are capable of growing aerobically on a number of single sources of carbon and energy and that in so doing they oxidize the carbon sources completely to carbon dioxide and water. Thus, the cellular material is the only organic product of the overall reaction by which growth takes place.

There is only one possible equation that can be written to represent the construction of *A. oryzae* from a carbon source having a hyperquotient (where the CQ of the carbon source is greater than that of the microorganism). Such a transformation of the carbon source into the fabric of microbial cells constitutes an overall reduction. In this case the excess oxygen can be eliminated only in the form of carbon dioxide, since there can be neither the production of noncellular organic substances having a hyperquotient (the only product of growth is the cellular material) nor the fixation of organic substances having a hypoquotient (there is only one substrate in this case, and this has a hyperquotient). By the same reasoning there can be only one equation that represents the construction of this mold from a carbon source having a hypoquotient (where the CQ of the carbon source is less than that of the microorganism). This transformation constitutes an overall oxidation which, in Tamiya's opinion, could only occur if there were a fixation of oxygen gas, since there can be neither the production of noncellular organic material having a hypoquotient (again, the only product of growth is cellular material) nor the fixation of organic substances having a hyperquotient (there is only one substrate in this case, and this has a hypoquotient). This elimination of carbon dioxide in the case of microbial growth on carbon sources having a hyperquotient or the fixation of oxygen gas in the case of microbial growth on carbon sources having a hypoquotient is *in addition to* the respiratory exchange of gases which takes place in cells *apart from growth* and which has to do with the provision of energy rather than material for the growth process. Tamiya thus made a clear distinction between two events that take place during the growth process, one being the actual construction of the fabric of the cells during growth and the other being a respiratory exchange of gases which occurs at the same time.

Tamiya[12] pointed out that growth, which he regarded in general as the formation of new parts of an organism or more simply as an increase in their weight, is comparable to the construction of a building. This construction requires building materials and the work to put it together. It is possible to know quite exactly the amount of material necessary for a certain building, to arrive at a good approximation of the waste materials resulting from the construction, and to calculate the salaries of the architects, workmen, and so on, which correspond, in sum, to the "work" which must be done, since the salaries are proportional to the amount of energy that must be expended in the construction. What he hoped to determine was the type and quantity of material necessary for the formation of a given quantity of microorganism, the nature and amount of waste materials that result from this, and the amount of energy required to accomplish it. He also pointed out that the intermediary chemical processes of growth and the energy exchanges associated with it are most certainly extremely complicated in detail and that these details will probably never be completely known. But, all these partial processes can be added up and their sum constitutes ". . . a complete and independent vital phenomenon of which one can study qualitatively and quantitatively the metabolism of material and energy. One is probably not able to represent material and energetic exchanges more simply . . ." (Tamiya,[12] p. 5.)

Because, at least in the case of the growth of *A. oryzae*, the medium does not contain any trace of an extracellular product of an incomplete oxidation after growth has ceased, it can be concluded that the carbon source becomes completely oxidized. Therefore, the RQ with respect to the respiratory exchanges that take place *apart* from those directly involved in cellular synthesis *ought to be equal to the CQ of the carbon source being utilized*. The total RQ, which *includes* the respiratory gas exchanges *plus* those directly involved in cellular synthesis, will always be larger than the CQ of the substrate if this latter has a hyperquotient, and will always be smaller if this latter has a hypoquotient.

These predictions by Tamiya were accurately confirmed by Yamagata[13] in 1934 when he published the RQ values observed when he grew *A. oryzae* on a number of carbon sources having different CQ values. His results are shown in Table 6.2. In the first group the RQ_{NH_3} is always greater than the CQ of the carbon source, this latter value being 1.000 or greater. The second group includes mannitol and sorbitol as carbon sources, and here the RQ and CQ

TABLE 6.2. Values for the combustion quotients of several carbon sources and for the RQs observed when *Aspergillus oryzae* was grown on them. CQ = combustion quotient; RQ_{NH_3} = RQ when the nitrogen source was ammonia; $RQ_{NO_3^-}$ = RQ when the nitrogen source was nitrate

Carbon Source	CQ	RQ_{NH_3}	$RQ_{NO_3^-}$
Gluconic acid	1.091	1.32	1.39
Starch	1.000	1.11	1.28
Sucrose	1.000	1.22	1.46
Mannose	1.000	1.25	1.57
Glucose	1.000	1.16	1.30
Galactose	1.000	1.15	1.33
Arabinose	1.000	1.12	1.31
Dihydroxyacetone	1.000	1.09	1.19
Mannitol	0.923	0.92	1.15
Sorbitol	0.923	0.93	1.17
Adonitol	0.909	0.86	1.14
Erythritol	0.889	0.85	1.08
Glycerin	0.857	0.77	0.94
Ethanol	0.667	0.59	0.65

Source: Reprinted by permission from Reference 12.

are virtually identical. The only discrepancy to be noted here is that *if* the RQ and CQ are so close in value, then the CQ for the *Aspergillus* cells grown on these carbon sources should also be the same as that of the carbon source. From Table 6.1 it is apparent that the CQ for *A. oryzae* cells grown on mannitol is 0.943 rather than 0.923, which is the CQ for mannitol. In the third group the RQ_{NH_3} is always smaller than the CQ of the carbon source, this latter value being 0.909 or less. Thus, the agreement of Tamiya's theory with experimentally determined data is in general extremely good. It even covers those experiments in which *A. oryzae* is grown on different carbon sources with nitrate as the nitrogen source. In order for nitrate to be utilized it must first be reduced to the oxidation state of ammonia. This reduction must be coupled with an oxidation and, since the only oxidizable substance is the carbon source, this would be expected to result in an increase in the amount of carbon dioxide gas produced and would correspondingly raise the RQ. This prediction is completely verified by the data in Table 6.2, in which the values for $RQ_{NO_3^-}$ are all higher than those for RQ_{NH_3}.

Following from what has been stated previously, it is now possible to write equations based on Tamiya's[7] original derivations, representing the construction of cellular material from any carbon source on which a microorganism can be grown provided, of course, that the metabolism is aerobic, that there is only *one* carbon source, and that there is no organic product formed during growth other than the cellular material.

6.1. NOTATIONS USED IN EQUATIONS REPRESENTING THE CONSTRUCTION OF MICROBIAL CELLS FROM THE SUBSTANCE OF THE SUBSTRATE

Tamiya[6] used Greek symbols in his theoretical equations. What has been adopted here is a somewhat more simple form of notation which will be used instead. Carbon sources are represented by the empirical formulas $\mathrm{Cn_CHn_HOn_O}$ or $\mathrm{Cn_CHn_HOn_ONn_N}$, where $\mathrm{n_C}$, $\mathrm{n_H}$, $\mathrm{n_O}$, and $\mathrm{n_N}$ represent the relative numbers of carbon, hydrogen, oxygen, and nitrogen atoms, respectively, in the formula of a given carbon souce. The cellular material formed during growth is represented by the empirical formula $\mathrm{C}n_\mathrm{C}\mathrm{H}n_\mathrm{H}\mathrm{O}n_\mathrm{O}\mathrm{N}n_\mathrm{N}$, where n_C, n_H, n_O, and n_N represent the relative numbers of carbon, hydrogen, oxygen, and nitrogen atoms, respectively, in this formula. The several coefficients for the other reactants and products in the equations are represented arbitrarily as follows:

S = coefficient for a carbon source having a hyperquotient
S' = coefficient for a carbon source having a hypoquotient
N = coefficient for ammonia as a nitrogen source
N' = coefficient for nitrate as a nitrogen source
C = coefficient for carbon dioxide
W = coefficient for water when the carbon source has a hyperquotient
W' = coefficient for water when the carbon source has a hypoquotient
O = coefficient for oxygen
CQ_S = combustion quotient for a carbon source having a hyperquotient
$\mathrm{CQ}_{S'}$ = combustion quotient for a carbon source having a hypoquotient
CQ_M = combustion quotient for the cellular material (M represents mycelium)

6.2. EQUATIONS REPRESENTING THE CONSTRUCTION OF CELLULAR MATERIAL FROM A CARBON SOURCE HAVING A HYPERQUOTIENT WHEN THE NITROGEN SOURCE IS AMMONIA

With respect to carbon sources having a hyperquotient the construction of cellular material using ammonia as the nitrogen source can be represented by a general equation in the following manner:

$$S\mathrm{Cn_CHn_HOn_ONn_N} + N\mathrm{NH_3} \longrightarrow \mathrm{C}n_\mathrm{C}\mathrm{H}n_\mathrm{H}\mathrm{O}n_\mathrm{O}\mathrm{N}n_\mathrm{N} + C\mathrm{CO_2} + W\mathrm{H_2O} \qquad (6.17)$$

In working with Eq. 6.17 it is apparent that if the empirical formulas for the carbon source and for the cellular material are known, the coefficients for the noncellular reactants and products are determined and can be calculated by means of the following equations that have been modified from Tamiya's[7] derivations.

$$S = \frac{n_C + n_C\left(\dfrac{CQ_S}{CQ_M} - 1\right)}{n_C}$$

$$N = n_N$$

$$C = n_C\left(\frac{CQ_S}{CQ_M} - 1\right)$$

$$W = \frac{1}{2}\left[\frac{n_C + n_C\left(\dfrac{CQ_S}{CQ_M} - 1\right)}{n_C} n_H + 3n_N - n_H\right]$$

Taking as an example the construction of *A. oryzae* from glucose as a nitrogen source, the empirical formula for glucose is $C_6H_{12}O_6$ and that for the cellular material, taken from Table 6.1, is $C_{409}H_{717}O_{233}N_{46}$. Thus $n_C = 6$, $n_H = 12$, $n_C = 409$, $n_H = 717$, and $n_N = 46$. CQ_S can be calculated by means of Eq. 6.11 and for glucose = 1.000. CQ_M can be calculated by using Eq. 6.12, but in this case its value of 0.935 can be taken from Table 6.1. Carrying out the calculations,

$$S = \frac{1}{6}\left[409 + 409\left(\frac{1.000}{0.935} - 1\right)\right] = 72.9$$

$$N = 46.0$$

$$C = 409\left(\frac{1.000}{0.935} - 1\right) = 28.4$$

$$W = \frac{1}{2}\left\{\frac{1}{6}\left[409 + 409\left(\frac{1.000}{0.935} - 1\right)\right] \times 12 + (3 \times 46) - 717\right\} = 147.9$$

Inserting these values into Eq. 6.17 together with the empirical formulas for the carbon source and for the cellular material gives the required equation for the construction of *A. oryzae* cells from glucose and ammonia.

$$72.9C_6H_{12}O_6 + 46.0NH_3 \longrightarrow C_{409}H_{717}O_{233}N_{46} + 28.4CO_2 + 147.9H_2O \tag{6.18}$$

6.3. EQUATIONS REPRESENTING THE CONSTRUCTION OF CELLULAR MATERIAL FROM A CARBON SOURCE HAVING A HYPOQUOTIENT WHEN THE NITROGEN SOURCE IS AMMONIA

When a carbon source has a hypoquotient, the construction of cellular material using ammonia as a nitrogen source can be represented by a general equation as follows:

$$\left.\begin{array}{l} S'\mathrm{Cn_CHn_HOn_O} + N\mathrm{NH_3} \\ + O\mathrm{O_2} \end{array}\right\} \longrightarrow \mathrm{C}n_\mathrm{C}\mathrm{H}n_\mathrm{H}\mathrm{O}n_\mathrm{O}\mathrm{N}n_\mathrm{N} + W'\mathrm{H_2O} \quad (6.19)$$

As with Eq. 6.17, it is apparent that if the empirical formulas for the carbon source and for the cellular material are known, the coefficients for the non-cellular reactants and products of Eq. 6.19 are determined. The following equations have been modified from those derived by Tamiya[7] for use in calculating the coefficients in Eq. 6.19.

$$S' = \frac{n_\mathrm{C}}{\mathrm{n_C}}$$

$$N = n_\mathrm{N}$$

$$W' = \frac{1}{2}\left[\frac{n_\mathrm{C}}{\mathrm{n_C}}\mathrm{n_H} + 3n_\mathrm{N} - n_\mathrm{H}\right]$$

$$O = n_\mathrm{C}\left(\frac{1}{\mathrm{CQ}'_S} - \frac{1}{\mathrm{CQ}_M}\right)$$

Taking as an example the construction of *A. oryzae* from ethanol as a carbon source having a hypoquotient and ammonia as the nitrogen source, the empirical formula for ethanol is C_2H_6O and that for the cellular material, taken from Table 6.1, is $C_{411}H_{741}O_{219}N_{58}$. Therefore, $\mathrm{n_C} = 2$, $\mathrm{n_H} = 6$, $n_\mathrm{C} = 411$, $n_\mathrm{H} = 741$, and $n_\mathrm{N} = 58$. $\mathrm{CQ}_{S'}$ can be calculated by means of Eq. 6.11 and for ethanol = 0.667. CQ_M can be calculated using Eq. 6.12, but in this case its value of 0.927 (for cells grown on ethanol and ammonia) can be taken from Table 6.1. Carrying out the calculations,

$$S' = \frac{411}{2} = 205.5$$

$$N = 58.0$$

$$W' = \frac{1}{2}\left[\frac{411}{2} \times 6 + (3 \times 58) - 741\right] = 333.0$$

$$O = 411\left(\frac{1}{0.667} - \frac{1}{0.927}\right) = 172.8$$

Inserting these values into Eq. 6.19 together with the empirical formulas representing the carbon source and the cellular material gives the required equation representing the construction of *A. oryzae* cells from ethanol and ammonia, as follows:

$$\left.\begin{array}{l}205.5C_2H_6O + 58.0NH_3\\ + 172.8O_2\end{array}\right\} \longrightarrow C_{411}H_{741}O_{219}N_{58} + 333.0H_2O \quad (6.20)$$

6.4. EQUATIONS REPRESENTING THE CONSTRUCTION OF CELLULAR MATERIAL FROM A CARBON SOURCE HAVING A HYPERQUOTIENT, WHEN THE NITROGEN SOURCE IS NITRATE

The construction of *A. oryzae* from a given carbon source when nitrate serves as the nitrogen source can be represented in much the same manner as in Eqs. 6.17 and 6.19 except that the reduction of nitrate to ammonia must be included in the equation.

When a carbon source has a hyperquotient the construction of cellular material using nitrate as the nitrogen source can be represented by the following general equation:

$$SC\mathrm{n}_C H\mathrm{n}_N O\mathrm{n}_O + N'HNO_3 \longrightarrow \begin{array}{l}Cn_C Hn_H On_O Nn_N\\ + CCO_2 + WH_2O\end{array} \quad (6.21)$$

In this case it is Yamagata's[13] derivations which have been modified, since it was he who applied to a study of the growth of *A. oryzae* using nitrate as a nitrogen source, the ideas that Tamiya[7] had developed with respect to the growth of the same organism using ammonia as the nitrogen source.

$$S = \frac{n_C + n_C\left(\dfrac{CQ_S}{CQ_M} - 1\right) + 2CQ_S n_N}{n_C}$$

$$N' = n_N$$

$$C = n_C\left(\frac{CQ_S}{CQ_M} - 1\right) + 2CQ_S n_N$$

$$W = \frac{1}{2}\left[\frac{n_C + n_C\left(\dfrac{CQ_S}{CQ_M} - 1\right)}{\mathrm{n}_C}\mathrm{n}_H + 3n_N - n_H + \left(\frac{2CQ_S}{\mathrm{n}_C} \times n_N\right)\mathrm{n}_H - 2n_N\right]$$

Taking as a specific example the construction of *A. oryzae* from glucose as a carbon source having a hyperquotient and nitrate as the nitrogen source, the empirical formula for glucose is $C_6H_{12}O_6$ and that for the cellular material, taken from Table 6.1, is $C_{415}H_{728}O_{237}N_{35}$. Thus, $n_C = 6$, $n_H = 12$, $n_C = 415$, $n_H = 728$, and $n_N = 35$. CQ_S can be calculated by means of Eq. 6.11 and for glucose = 1.000. CQ_M can be calculated using Eq. 6.12, but in this case its value of 0.918 can be taken from Table 6.1. Carrying out the calculations,

$$S = \frac{1}{6}\left[415 + 415\left(\frac{1.000}{0.918} - 1\right) + (2 \times 1.000 \times 35)\right] = 87.0$$

$$N' = 35$$

$$C = 415\left(\frac{1.000}{0.918} - 1\right) + (2 \times 1.000 \times 35) = 107.1$$

$$W = \frac{1}{2}\left(\left\{\frac{1}{6}\left[415 + 415\left(\frac{1.000}{0.918} - 1\right)\right] \times 12\right\} + (3 \times 35) - 728\right) + \frac{1}{2}\left[\left(\frac{2 \times 1}{6} \times 35\right) \times 12 - (2 \times 35)\right] = 175.6$$

When these values are inserted into Eq. 6.21 together with the empirical formulas representing the carbon source and the cellular material, the resulting equation represents the construction of *A. oryzae* cells from glucose and nitrate.

$$87.0C_6H_{12}O_6 + 35HNO_3 \longrightarrow C_{415}H_{728}O_{237}N_{35} + 107.1CO_2 + 175.6H_2O \qquad (6.22)$$

6.5. EQUATIONS REPRESENTING THE CONSTRUCTION OF CELLULAR MATERIAL FROM A CARBON SOURCE HAVING A HYPOQUOTIENT WHEN THE NITROGEN SOURCE IS NITRATE

In the case of a carbon source having a hypoquotient the construction of cellular material using nitrate as the nitrogen source can be represented by the following general equation:

$$\left.\begin{array}{l} S'Cn_CHn_HOn_O + N'HNO_3 \\ + OO_2 \end{array}\right\} \longrightarrow Cn_CHn_HOn_ONn_N + W'H_2O \qquad (6.23)$$

The equations which have been modified from Yamagata's[13] derivations for calculating the coefficients in Eq. 6.23 are

$$S' = \frac{n_C}{\mathrm{n_C}}$$

$$N' = n_N$$

$$O = n_C\left(\frac{1}{CQ_{S'}} - \frac{1}{CQ_M}\right) - 2n_N$$

$$W' = \frac{1}{2}\left(\frac{n_C}{\mathrm{n_C}}\mathrm{n_H} + 3n_N - n_H\right) - n_N$$

Taking as an example the construction of *A. oryzae* from ethanol as a carbon source having a hypoquotient and nitrate as the nitrogen source, the empirical formula for ethanol is C_2H_6O and that for the cellular material, taken from Table 6.1, is $C_{411}H_{752}O_{227}N_{48}$. Thus, $\mathrm{n_C} = 2$, $\mathrm{n_H} = 6$, $n_C = 411$, $n_H = 752$, and $n_N = 48$. $CQ_{S'}$ can be calculated through the use of Eq. 6.11 and for ethanol $= 0.667$. CQ_M can be calculated with Eq. 6.12, but in this case its value of 0.915 (for cells grown on ethanol and nitrate) can be taken from Table 6.1. Carrying out the calculations,

$$S' = \frac{411}{2} = 205.5$$

$$N' = 48$$

$$O = 411\left(\frac{1}{0.667} - \frac{1}{0.915}\right) - (2 \times 48) = 71.0$$

$$W' = \frac{1}{2}\left[\frac{411}{2} \times 6 + (3 \times 48) - 752\right] - 48 = 264.5$$

Inserting these values into Eq. 6.23 gives the following equation representing the construction of *A. oryzae* from ethanol and nitrate.

$$\left.\begin{array}{l} 205.5C_2H_6O + 48HNO_3 \\ + 71.0O_2 \end{array}\right\} \longrightarrow C_{411}H_{752}O_{227}N_{48} + 264.5H_2O \qquad (6.24)$$

Comparing Eq. 6.18 with Eq. 6.22, it should be noted that when a carbon source with a hyperquotient is used for growth more of it is required if nitrate is the nitrogen source than if ammonia is the nitrogen source, for the same quantity of cells which are grown. This is because 19.3% more glucose is needed to reduce the required amount of nitrate to ammonia, in which form it is actually used, than would be needed were ammonia to be provided initially. However, comparing Eq. 6.20 with Eq. 6.24 it can be seen that this is not true when the carbon source has a hypoquotient. In this latter case the nitrate serves as a source of oxygen for the oxidation of the ethanol, and at the same time is incidentally reduced to ammonia.

6.6. INTRODUCING RESPIRATION

During growth some of the carbon source is used for the construction of cellular material and some is utilized in the respiration which takes place at the same time. If the empirical formulas for the carbon source and the cellular material are known, by means of Eqs. 6.17, 6.19, 6.21, and 6.23 it is possible to determine the amount of carbon source that enters into the construction of the cellular material. In order to determine the amount of carbon source which is used for respiration during growth, it is necessary also to know the total amount of oxygen consumed, the total amount of carbon dioxide produced, or the RQ. *It was a part of Tamiya's theory that the portion of the carbon source that was used in respiration would be completely respired to carbon dioxide and water.* This can be represented by a general equation in the following manner:

$$s\mathrm{Cn_CHn_HOn_O} + o\mathrm{O_2} \longrightarrow c\mathrm{CO_2} + w\mathrm{H_2O} \tag{6.25}$$

where s, o, c, and w represent the coefficients for the carbon source, oxygen, carbon dioxide, and water, respectively, and where $\mathrm{n_C}$, $\mathrm{n_H}$, and $\mathrm{n_O}$ represent the numbers of carbon, hydrogen, and oxygen atoms, respectively, in the empirical formula representing the carbon source.

6.7. CALCULATING THE AMOUNT OF OXYGEN REQUIRED FOR RESPIRATION WHEN THE CARBON SOURCE HAS A HYPERQUOTIENT AND THE NITROGEN SOURCE IS AMMONIA

When cells are grown on a carbon source having a hyperquotient and use ammonia as a nitrogen source, the amount of oxygen required for respiration can be calculated by means of the following modification of one of Tamiya's[7] derivations.

$$o = \frac{n_\mathrm{C}\left(\dfrac{\mathrm{CQ}_S}{\mathrm{CQ}_M} - 1\right)}{\mathrm{RQ}_{\mathrm{NH_3}}^{\mathrm{hyper}} - \mathrm{CQ}_S} \tag{6.26}$$

All the notations have been previously defined except $\mathrm{RQ}_{\mathrm{NH_3}}^{\mathrm{hyper}}$, which represents the respiratory quotient observed during the growth of a microorganism on a carbon source having a hyperquotient when ammonia is the nitrogen source. Taking the growth of *A. oryzae* on glucose and ammonia as an example, the empirical formula for glucose is $C_6H_{12}O_6$ and that for the cellular material is found in Table 6.1 to be $C_{409}H_{717}O_{233}N_{46}$. The value for n_C is thus 409. CQ_S can be calculated using Eq. 6.11 and for glucose equals 1.000. CQ_M can be calculated using Eq. 6.12, but in this case can be found in Table 6.1 to be 0.935. $RQ_{NH_3}^{hyper}$ (for cells grown on glucose and ammonia) from Table 6.2 is 1.16. Carrying out the calculation using Eq. 6.26,

$$o = \frac{409\left(\dfrac{1}{0.935} - 1\right)}{1.16 - 1.00} = 177.7$$

The equation representing the complete combustion of glucose is

$$C_6H_{12}O_6 + 6O_2 \longrightarrow 6CO_2 + 6H_2O \qquad (6.27)$$

If coefficient o = 177.7, then coefficient s in Eq. 6.25 will be 177.7/6 = 29.6. Inserting this value into Eq. 6.27 and completing the mass balance gives an equation that represents the respiration of glucose occurring when one empirical formula weight of *A. oryzae* cells is formed from glucose and ammonia.

$$29.6C_6H_{12}O_6 + 177.7O_2 \longrightarrow 177.7CO_2 + 177.7H_2O \qquad (6.28)$$

Although Tamiya never did this in any of his publications, it is possible to add together Eqs. 6.18 and 6.28 to give an equation representing the complete process of growth.

$$72.9C_6H_{12}O_6 + 46.0NH_3 \longrightarrow C_{409}H_{717}O_{233}N_{46} + 28.4CO_2 + 147.9H_2O \qquad (6.18)$$

$$29.6C_6H_{12}O_6 + 177.7O_2 \longrightarrow 177.7CO_2 + 177.7H_2O \qquad (6.28)$$

$$\left.\begin{matrix}102.5C_6H_{12}O_6 + 177.7O_2 \\ + 46.0NH_3\end{matrix}\right\} \longrightarrow \begin{matrix}C_{409}H_{717}O_{233}N_{46} \\ + 206.1CO_2 + 325.6H_2O\end{matrix} \qquad (6.29)$$

Equation 6.29 has an RQ of 206.1/177.7 = 1.16, which is what was measured experimentally by Yamagata.[13] In addition, Eq. 6.29 shows that 409/(102.5 × 6) × 100 = 66.5% of the carbon in the carbon source becomes incorporated into cellular material.

Possibly the reason why Tamiya never published equations such as Eq. 6.29 is because to him this did not represent growth. What Tamiya considered to be growth is represented by Eq. 6.18, whereas Eq. 6.28 represents respiration, something necessary for, but independent of, growth. This is, of course, a perfectly valid representation. A different but equally valid representation of the growth process is to include construction and respiration together in the same equation.

An inspection of Eqs. 6.18, 6.28, and 6.29 shows clearly why Tamiya felt that it was only the oxygen consumption that occurs during growth on a carbon source having a hyperquotient that is a measure of the extent to which this carbon source has been respired. The carbon dioxide production during growth on such carbon sources cannot be used because a part of this production (here, 13.8%) is a result of the construction reaction represented by Eq. 6.18. The observed carbon dioxide production is not entirely due to respiration.

6.8. CALCULATING THE AMOUNT OF OXYGEN REQUIRED FOR RESPIRATION WHEN THE CARBON SOURCE HAS A HYPOQUOTIENT AND THE NITROGEN SOURCE IS AMMONIA

For carbon sources having a hypoquotient the amount of carbon dioxide produced during the respiration that accompanies growth can be calculated by means of the following modification of one of Tamiya's[7] derivations.

$$c = \frac{\mathrm{RQ}_{\mathrm{NH_3}}^{\mathrm{hypo}} \cdot n_{\mathrm{C}} \left(\dfrac{1}{\mathrm{CQ}_{S'}} - \dfrac{1}{\mathrm{CQ}_{M}} \right) \mathrm{CQ}_{S'}}{\mathrm{CQ}_{S'} - \mathrm{RQ}_{\mathrm{NH_3}}^{\mathrm{hypo}}} \qquad (6.30)$$

All the notations have been defined previously except $\mathrm{RQ}_{\mathrm{NH_3}}^{\mathrm{hypo}}$, which represents the respiratory quotient observed during the growth of a microorganism on a carbon source having a hypoquotient when ammonia is the nitrogen source. Taking as an example the growth of *A. oryzae* on ethanol and ammonia, the empirical formula for ethanol is C_2H_6O and that for the cellular material from Table 6.1 is $C_{411}H_{741}O_{219}N_{58}$. The value for n_{C} is therefore 411. $\mathrm{CQ}_{S'}$ is calculated using Eq. 6.11 and for ethanol is 0.667. Equation 6.12 is used to calculate CQ_M, but in this case the appropriate value can be found in Table 6.1 to be 0.927. $\mathrm{RQ}_{\mathrm{NH_3}}^{\mathrm{hypo}}$ for growth on ethanol and ammonia from Table 6.2 is 0.59. Carrying out the calculations,

$$c = \frac{0.59 \times 411 \left(\dfrac{1}{0.667} - \dfrac{1}{0.927} \right) \times 0.667}{0.667 - 0.59} = 883.3$$

The equation representing the complete combustion of ethanol is

$$C_2H_6O + 3O_2 \longrightarrow 2CO_2 + 3H_2O \qquad (6.31)$$

If coefficient c is 883.3, then coefficient s in Eq. 6.25 will be 883.3/2 = 442.1. Inserting this value in Eq. 6.31 and completing the mass balance gives an equation that represents the respiration of ethanol occurring when one empirical formula weight of *A. oryzae* cells is formed from ethanol and ammonia.

$$442.1C_2H_6O + 1326.4O_2 \longrightarrow 883.3CO_2 + 1326.4H_2O \qquad (6.32)$$

Adding Eqs. 6.20 and 6.32 together results in the complete growth-reaction equation.

$$\left.\begin{matrix} 205.5C_2H_6O + 58.0NH_3 \\ + 173.2O_2 \end{matrix}\right\} \longrightarrow \begin{matrix} C_{411}H_{741}O_{219}N_{58} \\ + 333.0H_2O \end{matrix} \qquad (6.20)$$

$$442.1C_2H_6O + 1326.4O_2 \longrightarrow 883.3CO_2 + 1326.4H_2O \qquad (6.32)$$

$$\left.\begin{matrix} 647.6C_2H_6O + 58.0NH_3 \\ + 1499.6O_2 \end{matrix}\right\} \longrightarrow \begin{matrix} C_{411}H_{741}O_{219}N_{58} \\ + 883.3CO_2 + 1659.4H_2O \end{matrix} \qquad (6.33)$$

Equation 6.33 has an RQ of 883.3/1499.6 = 0.59, which is what was measured experimentally by Yamagata.[12] In addition, from Eq. 6.20 it is evident that only 411/(2 × 647.6) × 100 = 31.7% of the carbon in the carbon source becomes incorporated into cellular material. This is hardly to be expected, since the efficiency of incorporation with the same organism grown on glucose, a carbon source which is much more oxidized than ethanol, is 66.3%. If Tamiya had actually added Eqs. 6.20 and 6.32 together he might have noticed this discrepancy and wondered about it. The reason for the discrepancy will be discussed in Section 8.12.

An inspection of Eqs. 6.20, 6.32, and 6.33 shows clearly that when growth occurs on carbon sources having a hypoquotient it is the carbon dioxide production and not the oxygen consumption that is the true measure of the extent to which such carbon sources have been respired. The oxygen consumption during growth cannot be used because a part of this consumption (here, 11.5%) is the result of the construction reaction represented by Eq. 6.20. The observed oxygen consumption is not entirely due to respiration.

6.9. CALCULATING THE AMOUNT OF OXYGEN REQUIRED FOR RESPIRATION WHEN THE CARBON SOURCE HAS A HYPERQUOTIENT AND THE NITROGEN SOURCE IS NITRATE

When a microorganism is grown on a carbon source having a hyperquotient with nitrate rather than ammonia as the nitrogen source, the amount of oxygen consumed during the respiration that accompanies growth can be calculated by means of the following equation which has been adapted from one of Yamagata's[13] derivations.

$$o = \frac{CQ_S + n_C\left(\dfrac{CQ_S}{CQ_M} - 1\right) + 2n_N}{RQ_{NO_3}^{hyper} - CQ_S} \tag{6.34}$$

All the notations have been defined previously except $RQ_{NO_3}^{hyper}$, which represents the respiratory quotient observed during the growth of a microorganism on a carbon source having a hyperquotient when nitrate is the nitrogen source. Taking as an example the growth of *A. oryzae* on glucose and nitrate, the empirical formula for glucose is $C_6H_{12}O_6$ and that for the cellular material from Table 6.1 is $C_{415}H_{728}O_{237}N_{35}$. The values for n_C and n_N are thus 415 and 35, respectively. Using Eq. 6.11 $CQ_S = 1.000$. CQ_M is calculated with Eq. 6.12, but in this case can be found from Table 6.1 to be 0.918. $RQ_{NO_3}^{hyper}$ for growth on glucose and nitrate from Table 6.2 is 1.30. Carrying out the calculations using Eq. 6.34,

$$o = \frac{1.000 + 415\left(\dfrac{1.000}{0.918} - 1\right) + (2 \times 35)}{1.30 - 1.00} = 359.8$$

The complete combustion of glucose is represented by Eq. 6.27. If coefficient o is 359.8, then coefficient s in Eq. 6.25 will be $359.8/6 = 60.0$. Inserting this value in Eq. 6.27 and completing the mass balance gives an equation representing the respiration of glucose which occurs when one empirical formula weight of *A. oryzae* cells is formed from glucose and nitrate.

$$60C_6H_{12}O_6 + 359.8O_2 \longrightarrow 359.8CO_2 + 359.8H_2O \tag{6.35}$$

Adding Eq. 6.22 to 6.35 gives the complete growth-reaction equation.

$$87.0C_6H_{12}O_6 + 35HNO_3 \longrightarrow \begin{matrix} C_{415}H_{728}O_{237}N_{35} \\ + 107.1CO_2 + 175.6H_2O \end{matrix} \tag{6.22}$$

$$60.0C_6H_{12}O_6 + 359.8O_2 \longrightarrow 359.8CO_2 + 359.8H_2O \tag{6.35}$$

$$\left.\begin{matrix} 147.0C_6H_{12}O_6 + 35HNO_3 \\ + 359.8O_2 \end{matrix}\right\} \longrightarrow \begin{matrix} C_{415}H_{728}O_{237}N_{35} \\ + 466.9CO_2 + 535.4H_2O \end{matrix} \tag{6.36}$$

Equation 6.36 has an RQ of $466.9/359.8 = 1.30$, which is what Yamagata[13] found experimentally. In addition, Eq. 6.36 shows that only $415/(6 \times 147) \times 100 = 47.0\%$ of the carbon in the carbon source has become incorporated into cellular material, as compared with 66.5% when the same cells were grown on glucose with ammonia as the nitrogen source. The difference of 19.5% is due to the necessity of having to reduce the nitrate to ammonia before it can be utilized in cellular synthesis.

6.10. CALCULATING THE AMOUNT OF OXYGEN REQUIRED FOR RESPIRATION WHEN THE CARBON SOURCE HAS A HYPOQUOTIENT AND THE NITROGEN SOURCE IS NITRATE

The amount of carbon dioxide produced when a microorganism is grown on a carbon source having a hypoquotient with nitrate as the nitrogen source can be calculated through the use of the following equation.

$$c = \frac{RQ^{hypo}_{NO_3} \cdot n_C \left(\frac{1}{CQ_{S'}} - \frac{1}{CQ_M} \right) \cdot CQ_{S'} - 2n_N \cdot RQ^{hypo}_{NO_3} \cdot CQ_{S'}}{CQ_{S'} - RQ^{hypo}_{NO_3^-}} \tag{6.37}$$

All the notations have been defined previously except for $RQ^{hypo}_{NO_3^-}$, which represents the respiratory quotient observed during the growth of a microorganism on a carbon source having a hypoquotient when nitrate is the nitrogen source. Taking as an example the growth of *A. oryzae* on ethanol and nitrate, the empirical formula for ethanol is C_2H_6O and that for the *A. oryzae* mycelium from Table 6.1 is $C_{411}H_{752}O_{227}N_{48}$. The values for n_C and n_N are therefore 411 and 48, respectively. Using Eq. 6.11 $CQ_{S'}$ for ethanol = 0.667. Equation 6.12 can be used to calculate CQ_M, but in this case the appropriate value can be found in Table 6.1 to be 0.915. $RQ^{hypo}_{NO_3^-}$ with respect to growth on ethanol and nitrate is found in Table 6.2 to be 0.65. Carrying out the calculation,

$$c = \frac{0.65 \times 411 \left(\frac{1}{0.667} - \frac{1}{0.915} \right) \times 0.667 - (2 \times 48 \times 0.650 \times 0.667)}{0.667 - 0.65}$$

$$= 1811$$

The complete combustion of ethanol is represented by Eq. 6.31. If coefficient c = 1811, then coefficient s in Eq. 6.25 will be 1811/2 = 905.5. Inserting this value in Eq. 6.31 and completing the mass balance gives an equation representing the respiration of ethanol that occurs when one empirical formula weight of *A. oryzae* cells is formed from ethanol and nitrate.

$$905.5C_2H_6O + 2716.5O_2 \longrightarrow 1811CO_2 + 2716.5H_2O \tag{6.38}$$

Adding Eqs. 6.24 and 6.38 then gives the complete growth-reaction equation.

$$\left.\begin{array}{l}205.5C_2H_6O + 48.0HNO_3 \\ + 71.3O_2\end{array}\right\} \longrightarrow \begin{array}{l}C_{411}H_{752}O_{227}N_{48} \\ + 264.5H_2O\end{array} \quad (6.24)$$

$$905.5C_2H_6O + 2716.5O_2 \longrightarrow 1811CO_2 + 2716.5H_2O \quad (6.38)$$

$$\left.\begin{array}{l}1111C_2H_6O + 48.0HNO_3 \\ + 2787.8O_2\end{array}\right\} \longrightarrow \begin{array}{l}C_{411}H_{752}O_{227}N_{48} \\ + 1811CO_2 + 2981H_2O\end{array} \quad (6.39)$$

Equation 6.39 has an RQ of 1811/2787.8 = 0.65, which is what Yamagata[13] found experimentally. However, it is evident from Eq. 6.39 that only 411/(2 × 1111) × 100 = 18.5% of the carbon in the carbon source has become incorporated into cellular material. This seems like far too low a value. The difference of 13.2% between 31.7% incorporation from Eq. 6.33 and 18.5% from Eq. 6.39 is, however, in the right range for the reduction of nitrate to ammonia. This value is less than the 20% value found for the reduction of almost the same quantity of nitrate to ammonia in Eq. 6.36 where glucose is the carbon source because ethanol is considerably more reduced than glucose and contains more energy per unit carbon.

In addition to the general equations representing the construction of the fabric of a microorganism from a given carbon source, Tamiya[12] developed equations for calculating the weight in grams of any organic substance that is required to construct 1 g dry weight of microorganism, as follows:

$$g_S = \frac{n_C}{M_M} \cdot \frac{M_S}{\mathrm{n_C}} \cdot \frac{CQ_S}{CQ_M} \quad (6.40)$$

and

$$g_{S'} = \frac{n_C}{M_M} \cdot \frac{M_S}{\mathrm{n_C}} \quad (6.41)$$

where all the notations have been defined previously except g_S and g'_S, which represent the number of grams of carbon source having a hyperquotient or hypoquotient, respectively, and M_S and M_M, which represent the molecular weight and empirical formula weight, respectively, of the carbon source and the cellular material. As an example, using the data in Table 6.1 relating to the construction of *A. oryzae* using glucose as the carbon source, $\mathrm{n_C} = 6$, $n_C = 409$, $M_S = 180$, $M_M = 9{,}997$, $CQ_S = 1.000$, and $CQ_M = 0.935$. Inserting these values in Eq. 6.40 gives

$$g_S = \frac{409}{9{,}997} \times \frac{180}{6} \times \frac{1.000}{0.935} = 1.313 \text{ g of glucose}$$

which are required for the synthesis of 1 g dry weight of cells. The same value can be obtained from Eq. 6.18. If ethanol is the carbon source the amount required to synthesize 1 g of cellular material is only 0.947 g. Equations 6.40 and 6.41 are valid whether the nitrogen source is ammonia or nitrate, since they represent only that amount of carbon source which enters into the construction of the cellular material. However, if nitrate is the nitrogen source, an additional amount of carbon source will be required to reduce the nitrate to ammonia, which is the form of nitrogen actually used in growth.

6.11. TAMIYA'S CALCULATIONS OF THE HEAT OF GROWTH OF *ASPERGILLUS*

Not surprisingly, Tamiya was also interested in the energy exchanges accompanying growth. In 1932 he (Tamiya[7]) thus developed an indirect method of determining both the material and the energy efficiency accompanying growth processes. This was accomplished by measuring the amount of growth and the amount of respiration, rather than the amount of carbon source utilized. The research described in this extensive publication will not be covered in detail here because its inaccuracies make the conclusions little better than tentative suggestions. In 1932 Yamagata[11] had not yet carried out his accurate analyses of the elemental composition of *A. oryzae* mycelia, and Tamiya was still using the formula $C_{86}H_{160}O_{45}N_7$, which had been derived from the analyses done in 1902 by Mazé[9] of *Eurotyopsis Gayoni*. The use of this formula made his results questionable. This problem ceased to exist when Yamagata's[11] data became available, and in 1935 Tamiya[12] published a short summary of all his research on the material and energy balance during microbial growth. In it he condensed some 706 pages of research published by himself and his collaborators into 42 pages of summary, in itself no simple task and one which must have taken considerable courage. Few scientists are willing to admit that, with respect to any particular aspect of their published research effort, only 6% of what they have printed has any lasting value. It is here in Tamiya's 1935 article that his ideas on energy exchange are succinctly stated.

Although Tamiya and Yamamoto[16] had in 1933 described a microcalorimeter during the operation of which it was possible to determine the heat production, the O_2 consumption, and the CO_2 formation accompanying the growth of mold cultures, Tamiya[12] in 1935 used the values obtained by Algera[17] in his research on *A. niger* to confirm his own theories as to the production of heat. Tamiya[12,18] pointed out, however, that Algera was wrong in his interpretation that the amount of respiration could be determined by the amount of CO_2 produced. Algera grew *A. niger* on glucose in his calorimeter. This is a carbon source which, according to Tamiya's definition, has a hyperquotient. An inspection of Eq. 6.18 will show that with glucose, as with other substances having a hyperquotient, carbon dioxide is produced during the construction of the mold, this being the only method by which the level of reduction of the substance of the carbon source can be made to equal that of the cellular material comprising the mold. Thus, the total amount of carbon dioxide produced is greater than that which would be produced solely by respiration. As shown in Eq. 6.28, it is the oxygen consumed that indicates the extent of respiration. This was clearly recognized by Tamiya,[12] who corrected Algera's experimental values in this respect.

Using Eq. 6.40 the weight and from this the heat of combustion of the quantity of carbon source required to construct 1 g of cellular material can be calculated and compared with the experimentally determined heat of combustion of 1 g dry weight of cellular material. The results of such calculations and determinations made by four investigators are shown in Table 6.3. It is apparent that in most cases the heat of combustion for substances having a hyperquotient is greater than that of the mold formed from it. The difference would be even greater with respect to carbon sources having a hypoquotient. The exception is the datum of Molliard,[15] and it is difficult to know how to assess this or to know to what extent it represents an accurate analysis. In any case, Tamiya felt that cellular construction could be regarded as an exothermic reaction which produced heat quite apart from respiration.

Tamiya[12,18] used Algera's[17] data to produce a confirmation of his own ideas on mold energetics. Algera obtained the following values from the one experiment he carried out:

U = 9.08 kcal = total heat lost from the culture medium, as determined by indirect calorimetry (the difference between the heats of combustion of the medium at the beginning and at the end of the experiment)

U_M = 5.606 kcal = the heat of combustion of the mycelium formed

U_W = 3.299 kcal = the heat of growth of the mycelium, as determined by direct calorimetry

U_R = 2.937 kcal = the heat lost by the process of respiration (This was calculated by Tamiya on the basis of the oxygen consumed during Algera's experiment and not, as Algera did, on the basis of the total carbon dioxide production.)

U_B = 6.246 kcal = the heat of combustion of that fraction of the carbon source which physically enters into the construction of the mycelium (This could be calculated by Tamiya using Eq. 6.40 and the value of 4.757 kcal g^{-1} as the heat of combustion of glucose.)

TABLE 6.3. A comparison of the heat of combustion of the quantity of substance forming 1 g of *Aspergillus* sp. with the heat of combustion of 1 g of the mold which is formed

		Heat of Combustion of	
Aspergillus	Carbon Source	Substance Forming 1 g Dry Weight of Mold (kcal)	1 g Dry Weight of Mold Formed (kcal)
niger	*Glucose*	*4.91*	*4.8*[14]
niger	Glucose	4.91	5.19[15]
niger	Glucose	4.91	4.57[17]
oryzae	Galactose	4.91	4.6[19]

Source: Reprinted by permission from Reference 12.

Using modern values for the heats of formation of ammonium ion and of ammonia, the correction for the heat produced when ammonium ion becomes transformed into ammonia for utilization in the construction of the mycelium is negligible. Aside from direct measurement by difference, U can be calculated in two ways.

$$U = U_M + U_W = 5.606 \text{ kcal} + 3.299 \text{ kcal} = 8.905 \text{ kcal}$$

and

$$U = U_R + U_B = 2.937 \text{ kcal} + 6.426 \text{ kcal} = 9.363 \text{ kcal}$$

Accepting the potential analytical inaccuracies inherent in determining U, U_M, and U_W, the value of 9.363 kcal which is obtained by Tamiya's calculations is in good agreement with those of 9.08 and 8.905 obtained by thermal measurement, being within 4% of the average of these latter measurements.

6.12. THE CALORIMETRIC STUDIES OF TAMIYA AND YAMAMOTO, AND OF YAMAMOTO AND YAMAGATA

In 1933 Tamiya and Yamamoto[16] described the design and construction of a new microcalorimeter. It was basically a simpler instrument than that designed by Algera[17] in that it was not an isothermal calorimeter. The mold was grown in a nickel-plated, brass cylinder 80 mm long by 42 mm in diameter and closed at the upper end, to the bottom end of which was attached by means of a screw ring a hemispherical shell which could contain 5 ml of culture solution. This is illustrated in Fig. 6.5*a*. Two tubes through the cylinder head provided for the measurement of gas exchange. Calibration could be carried out by means of an electrical resistance wire made of manganin which dipped into

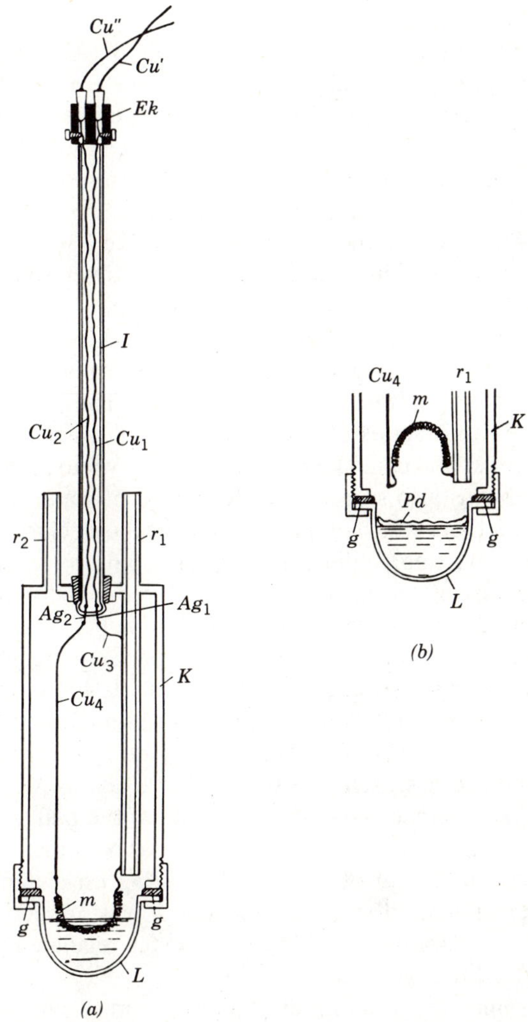

FIGURE 6.5. Diagram of the microcalorimeter vessel assembly used by Tamiya and Yamamoto. (*a*) Complete vessel assembly. (*b*) Lower end of the vessel assembly showing the calibration resistance wire withdrawn from the culture medium. *Cu'*, *Cu''*, copper leads to resistance wire; *Ek,* ebonite terminal connection; *I*, glass tube; Cu_1, Cu_2, copper leads; Ag_1, Ag_2, silver leads passing through glass of tube *I*; Cu_3, Cu_4, copper leads; *K*, nickel-plated metal cylinder; r_1, r_2, metal tubes for gas inlet and outlet; *m*, resistance wire for calibration; *g*, rubber washer; *L*, hemispherical, nickel-plated culture vessel; *Pd*, mycelial pad. Reprinted from Reference 16 with permission.

the culture solution, the two copper leads to which passed out of the cylinder through a third tube inserted into a hole in the cylinder head through a rubber stopper. After calibration, so that it would not interfere with the recovery of the mold pellicle after growth, the resistance wire could be lifted out of the culture solution by rotating the tube containing the leads. This is illustrated in Fig. 6.5*b*. The culture solution was inoculated with a spore suspension before being placed in the hemispherical shell, this latter being then attached to the metal cylinder. After the culture bomb was assembled it was placed in a round bottomed, cylindrical container made of polished silver which was 17.5 cm long and 5.8 cm in diameter and somewhat widened at the top. This was screwed to an ebonite sleeve, itself screwed to an ebonite cover through which were passed an ebonite stirring rod, a Beckman thermometer, the tube containing the resistance wire leads, and the two gas exchange tubes. The stirring rod was connected to a series of fingerlike projections that surrounded the culture bomb cylinder and could be moved up and down to accomplish stirring. The culture bomb was also surrounded by 200 ml of water contained within the silver receptacle, into which dipped the bulb of the Beckman thermometer. The whole of this assembly, illustrated in Fig. 6.6, constituted the actual calorimeter. The rest of the instrument constituted a thermal barrier which isolated the calorimetric system from the environment. For this purpose the calorimeter was placed in a Dewar flask, the upper edge being fitted into the ebonite cover of the silver receptacle. The Dewar flask was then placed in a cylindrical, sheet metal pail 38 cm high by 17.5 cm in diameter having a cork stopper 3 cm thick through which could be passed the stirring rod, the Beckman thermometer, and the gas exchange tubes. The leads to the resistance wire were connected to terminals sealed into the side of the pail. A double-walled mantle into which water from the thermostat could be drawn was placed over the pail, with the inner wall of the mantle resting on a flange around the outside of the pail. Finally, the whole instrument was placed in a water thermostat, the mantle evacuated, and the thermostat regulated at 30 ± 0.1°C. All this is illustrated in Fig. 6.7.

Tamiya and Yamamoto's calorimeter had the same operational difficulties as did Rubner's and those of others whose calorimeters were very well isolated thermally but from which losses of heat by conduction still took place during experiments of long duration. Thus, although the temperature of the calorimeter vessel never rose very much above that of its surrounding environment, it was always enough that some heat was lost by conduction and radiation which had to be corrected for by using Newton's Law of Cooling. Just as happened with Algera,[17] after all the work expended by Tamiya and Yamamoto in designing and constructing a microcalorimeter only one experiment seems to have been carried out with it. This was published in 1935 by Yamamoto and Yamagata.[19] They grew *A. oryzae* in the calorimeter on a nutrient medium containing galactose and ammonia, recording a corrected heat production by direct calorimetry of 311.6 cal. Of this quantity, 276.8 cal was calculated to be due to respiration, based on the amount of oxygen consumed during growth

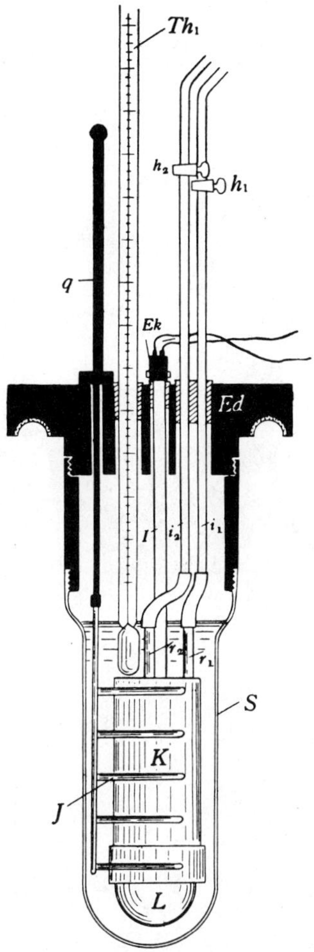

FIGURE 6.6. Diagram of the microcalorimeter vessel assembly in Fig. 6.5*a* mounted inside a silvered, copper container, *S*, filled with 200 ml water. Th_1, Beckman thermometer; h_1, h_2, stopcocks; *q*, stirring rod handle; *Ed*, ebonite upper enclosure; i_1, i_2, gas delivery tubes; *J*, stirring rod. Reprinted from Reference 16 with permission.

(galactose is a substance having a hyperquotient), leaving 34.8 cal as the quantity of heat produced during the construction of the mold. Since 0.055 g dry weight of mycelium was formed during growth, the quantity of heat produced in the formation of 1 g dry weight of mycelium was therefore 34.8/0.055 = 632.7 cal/g of mycelium formed according to the following equation.

$$\left.\begin{array}{l} 1.312\ \text{g}\ C_6H_{12}O_6 \\ +\ 0.0782\ \text{g}\ NH_3 \end{array}\right\} \longrightarrow \begin{array}{l} 1\ \text{g}\ (C_{409}H_{717}O_{233}N_{46}) \\ +\ 63.3\ \text{ml}\ CO_2 + 0.27\ \text{g}\ H_2O - 632.7\ \text{cal} \end{array} \qquad (6.42)$$

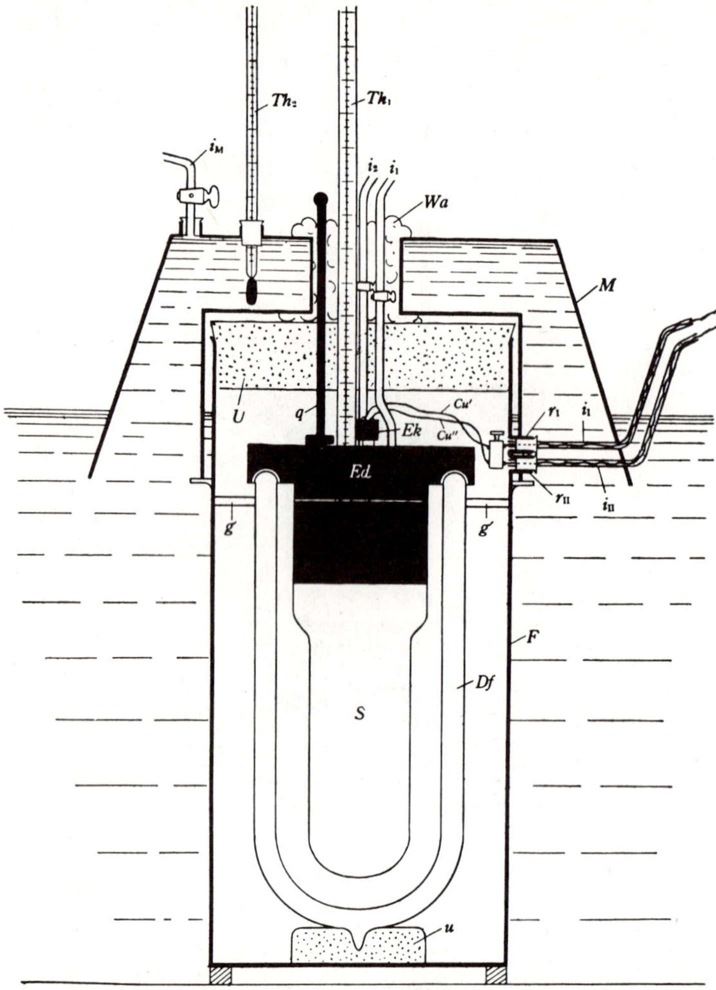

FIGURE 6.7. Diagram of Tamiya and Yamamoto's entire microcalorimeter. Th_2, Beckman thermometer; i_m, vacuum connections for evacuating hood so that the hood can be filled with water from the thermostat; *Wa*, cotton packing; *M*, hood; *U*, cork stopper; *g*, rubber washer; r_1, r_2, port and stopper through which pass calibration wire leads; i_1, i_2, calibration wire leads inside glass tubes; *Df*, Dewar flask; *u*, cork support; *F*, insulating can and hood support. Reprinted from Reference 16 with permission.

Although Yamamoto and Yamagata did not do so, Eq. 6.42 can be rewritten in terms of moles, using appropriate standard states.

$$\left.\begin{array}{l} 0.00728C_6H_{12}O_6(aq) \\ +\ 0.00459NH_3(aq) \end{array}\right\} \longrightarrow \begin{array}{l} 1/10{,}007C_{409}H_{717}O_{233}N_{46}(\text{cells}) \\ +\ 0.00282CO_2(aq)\ +\ 0.01474H_2O(l) \end{array} \qquad (6.43)$$

The equation representing the complete combustion of *A. oryzae* in a bomb calorimeter to carbon dioxide, water, and nitrogen gas can be represented as follows.

$$\left.\begin{array}{l} C_{409}H_{717}O_{233}N_{46}(\text{cells}) \\ +\ 471.75O_2(g) \end{array}\right\} \longrightarrow \begin{array}{l} 409CO_2(g) + 358.5H_2O(l) \\ +\ 23N_2(g) \end{array}$$

Using the heat of combustion of 4.6 kcal g^{-1} dry weight of cells which was determined by Yamamoto and Yamagata and the appropriate ΔH_f° values in Appendix 1, the enthalpy of formation of one empirical formula weight of cells is calculated to be $-16{,}969$ kcal, from which the enthalpy change accompanying the reaction represented by Eq. 6.43 is about -712 cal, using the $\Delta H_f^{\circ\prime}$ values in Appendix 1. This is within about 12% of the value of -632.7 kcal for the reaction represented by Eq. 6.43 which they determined from direct calorimetry. The difference, $-712 - (-632.7) = -79.3$ cal, represents only 1.4% of the total heat (-5.665 kcal/g of mycelium formed) produced during the growth process, and is probably well within the experimental error of direct calorimetric determinations. However, we do not really know this, since only one calorimetric measurement was made.

One possible reason why Yamamoto and Yamagata did not pursue further their studies on direct microbial calorimetry was that it was simply too difficult, even with the best of equipment then available. The values for the heat production accompanying the construction of the fabric of microbial cells from the carbon source are apparently quite small compared to the total heat produced in respiration. The sources of error lie in the determination of the quantity of oxygen consumed (when the carbon source has a hyperquotient) and in the total quantity of heat produced, both of which are extremely difficult to do with great accuracy. A small absolute error in the determination of either of these values would result in a large relative error in the determination of the heat of construction.

The last study by Tamiya and his collaborators on the material and energy exchanges accompanying the growth of *A. oryzae* appeared in 1939 (Tamiya and Usami[20]). In this research the mold was grown on 14 amino acids while determinations were made of the oxygen consumed, the carbon dioxide produced, the RQ, and the dry weight of mycelium produced. Equations were developed whereby RQ with respect to growth on these nitrogen-containing carbon sources could be calculated, much the same as this was done previously for carbon sources not containing nitrogen. The agreement between theoretical and experimental values was, in general, quite good. Their data showed up another problem in this sort of research. In 1934 Yamagata[11] had determined the empirical formula for *A. oryzae* cells grown on glucose and ammonia to be $C_{409}H_{717}O_{233}N_{46}$, the ash content of the cells to be 5.7%, the CQ to be 0.935, and the empirical formula dry weight to be 9997 daltons. Five years later Tamiya and Usami grew the same organism on the same medium and found the empirical formula to be $C_{397}H_{722}O_{241}N_{46}$, the ash content of the cells to be 11.0%, the CQ to be 0.940, and the empirical formula dry weight to be 9997 daltons. Although the data with respect to the organic part of the cell are remarkably similar, the ash content of the latter is nearly twice that of the former. This indicates the absolute necessity of using *only* data with respect to the organic part of the cells when making calculations on material and energy balances. If an ordinary dry weight is used, comparisons of growth on different carbon sources can result in errors of several per cent on the basis of ash alone.

It is obvious from reading Tamiya's work that what he regarded as the actual process of growth is the construction of cellular material from the carbon source. This in itself is a weakly exothermic reaction quite apart from respiration. Respiration serves two functions. Part of the respiratory energy is used for the growth process. The rest of it is used for maintenance.

Considering the magnitude of Tamiya's research it is not surprising that he held other views on growth and maintenance which will not be discussed in detail here. For example, adapting Terroine and Wurmser's approach to the calculation of the quantity of energy needed for maintenance (which has been shown earlier in Section 3.9 to be somewhat flawed), he came to the conclusion that respiration linked to maintenance diminishes progressively with the age of the mold, there then being an augmentation in the amount of respiration linked to growth. This would seem to have to do with the change in the RQ over time which a great many research workers have observed, and which may be related to the appearance of some condition which limits growth so that the carbon source is becoming assimilated but not used for cellular construction. Because of this Tamiya's interpretation may be wrong. However, there is little point in considering this further. The problem of a changing RQ can be avoided by using far smaller amounts of carbon source than Tamiya did in his nutrient media, so that the carbon source becomes limiting *before* anything else does. If this is done the RQ does *not* change.

Tamiya's final publication on the mass and energy exchanges associated with microbial growth was a monograph published in 1942 (Tamiya[21]) on the physiology of *Aspergillus sp.* in which his own work and that of his collaborators was included.

6.13. SUMMARY

The importance of Tamiya's work can hardly be overstated. He was years ahead of his time. He recognized that the organic part of the substance of cells could be represented by an empirical formula, and that the inorganic materials forming the ash which remains after combustion of the cells can be ignored from the point of view of the material and energy exchanges which accompany growth. He was fully aware of the idea of the "degree of reduction" of an organic substance. He realized that a relationship exists between the degree of reduction of a carbon source and that of the organic cellular material produced during growth, and was able to describe this quantitatively in mathematical terms. He was able to calculate the exact quantity of substrate necessary for the construction of an empirical formula weight or a unit mass of dried cells on an ash-free basis, and to use this information to formulate equations representing microbial growth. Knowing the RQ for the growth of a microorganism on a given substrate he was also able to calculate the exact quantity of substrate necessary to provide the energy for the growth process. This latter he regarded as respiration, something necessary for, but independent of, growth. His "theory of the CQ" was brilliant in its conception and elegant in its derivation. Unfortunately, it may also have been too complex to be comprehended by his own contemporaries. It was another 30 years before Mayberry, Prochazka, and Payne (see Section 9.1.3) conceived the idea of the "available electron" which was subsequently used by Minkevich and Eroshin (see Section 9.3) to develop a different method of constructing growth-reaction equations. This method is equally as elegant as that of Tamiya and possesses the advantage of being easier to use. For this reason it is presently the method of choice for establishing equations representing microbial growth. In using it, however, we ought not to forget that it was with Tamiya that most of these concepts originated.

REFERENCES

1. H. A. Barker, The oxidative metabolism of the colorless alga *Prototheca zopfii* (Krüger), *J. Cell. Comp. Biol.* **8,** 231–250 (1936).
2. H. Tamiya, Studien über die Stoffwechselphysiologie von *Aspergillus oryzae.* I, *Acta Phytochim.* **3,** 51–173 (1927).
3. H. Tamiya, Studien über die Stoffwechselphysiologie von *Aspergillus oryzae.* II, *Acta Phytochim.* **4,** 77–213 (1928–1929).
4. H. Tamiya, Studien über die Stoffwechselphysiologie von *Aspergillus oryzae.* III, *Acta Phytochim.* **4,** 227–295 (1928–1929).
5. H. Tamiya, Über die Verwendbarkeit von verschiedenen Kohlenstoffverbindigung im Bau- und Betriebsstoffwechsel der Schimmelpilze. Studien über die Stoffwechselphysiologie von *Aspergillus oryzae.* IV, *Acta Phytochim.* **6,** 1–129 (1932).
6. H. Tamiya, Zur Theorie des respiratorischen Quotienten nebst einer Bemerkung über den Einfluss der oxydoreduktiven Zellvorgänge auf den Gaswechsel der Zellen. Beiträge zur Atmungsphysiologie der Schimmelpilze. I, *Acta Phytochim.* **6,** 227–263 (1932).
7. H. Tamiya, Zur Energetik des Wachstums. Beiträge zur Atmungsphysiologie der Schimmelpilze. II, *Acta Phytochim.* **6,** 265–304 (1932).
8. H. Tamiya and S. Yamagutchi, Über die Aufbau und der Erhaltungsatmung. Beiträge zur Atmungsphysiologie der Schimmelpilze. III, *Acta Phytochim.* **7,** 43–64 (1933).
9. P. Mazé, Sur l'assimilation du sucre et de l'alcool par l'*Eurotyopsis Gayoni, C.R. Hebd. Acad. Sci.* **134,** 191–193 (1902).

10. R. Takata, An analysis of the ash composition of *Aspergillus oryzae*, *J. Soc. Chem. Indust. Japan* **32,** 544–547 (1929), plus several other references in the same volume (in Japanese).

11. S. Yamagata, Über die elementare Zusammensetzung des Schimmelpilzkorpers, *Acta Phytochim.* **8,** 107–116 (1934).

12. H. Tamiya, *Le Bilan Matériel et l'Energétique des Synthèses Biologiques*, Actualités Scientifiques et Industrielles No. 214, Série 1935, Hermann & Cie, Editeurs, Paris, 44 pp.

13. S. Yamagata, Über den Einfluss der Stickstoffquelle auf den Gaswechsel des Schimmelpilzes. Beiträge zur Physiologie der Nitratassimilation. I, *Acta Phytochim.* **8,** 117–155 (1934).

14. E. F. Terroine and R. Wurmser, Influence de la température sur l'utilisation du glucose dans le développement de l'*Aspergillus niger*, *C.R. Hebd. Séances Acad. Sci.* **173,** 482–483 (1921).

15. M. Molliard, Recherches calorimétriques sur l'utilisation de l'énergie respiratoire au cours du développement d'une culture de *Sterigmatocystis nigra*, *C.R. Soc. Biol.* **87,** 219–221 (1922).

16. H. Tamiya and A. Yamamoto, Über ein neues Mikrokalorimeter zur Messung der Wärmeabgabe von Schimmelpilzkulture, *Acta Phytochim.* **7,** 245–263 (1933).

17. L. Algera, Energiemessung bei *Aspergillus niger* mit Hilfe eines automatischen mikro-kompensations-Calorimeters, *Rec. Trav. Bot. Neerlandais* **29,** 47–163 (1932).

18. H. Tamiya, Über die Stoff- und die Energiebilanz bei dem Wachstumsvorgang des Schimmelpilzes. Bemerkungen zu der Arbeit von L. Algera, *Acta Phytochim.* **7,** 27–41 (1933).

19. A. Yamamoto and S. Yamagata, Thermochemische Untersuchungen über die Energiebilanz bei dem Wachstumsvorgang des Schimmelpilzes, *Acta Phytochim.* **8,** 245–254 (1935).

20. H. Tamiya and S. Usami, Über das Wachstum von *Aspergillus oryzae* bei Zugabe der Aminosäuren als alleinige Kohlenstoff- und Stickstoffquelle, *Acta Phytochim.* **11,** 261–298 (1939–1940).

21. H. Tamiya, Atmung, Gärung und die sich daran beteiligenden Enzyme von *Aspergillus*, *Adv. Enzymol.* **2,** 183–238 (1942).

duplicate samples taken from the control suspension contained in a flask immersed in the calorimeter thermostat.

The observed heat production was found to be 120.64 cal. This had to be corrected for the heat of neutralization of the nitrous acid resulting from the oxidation of ammonia. Hes's analyses showed that 25.5 mg of ammonia had been oxidized, resulting in the formation of 70.5 mg HNO_2 and the production of 15 cal as the heat of neutralization. The remaining heat, $120.64 - 15 = 105.64$ cal, thus represented the heat resulting from the metabolism of 25.5 mg of ammonium by *Nitrosomonas*. Hes then compared this value with that to be expected theoretically.

As an earlier part of this research he had determined that, in near agreement with the value of 35 obtained by Winogradsky[6] but in considerable disagreement with the values of 14.3 and 79 obtained by Nelson[7] and Engel,[8] respectively, the ratio of the weight of ammonia nitrogen oxidized over the weight of carbon dioxide carbon reduced to cellular material (the N/C ratio) during the growth of *Nitrosomonas* at 27°C averaged 33.3 for five experiments. Following the conclusions of Gaffon[9] and Van Niel[10] that the primary product of CO_2 fixation in the red nonsulfur bacteria and the red sulfur bacteria, respectively, had a formula about the same as that of crotonic acid ($C_4H_6O_2$), Hes felt that synthesis in *Nitrosomonas* could be represented by the equation

$$3NH_4OH + 4CO_2 \longrightarrow C_4H_6O_2 + 3HNO_2 + 3H_2O \tag{7.1}$$

Using the value of -498 kcal mole^{-1} for the heat of combustion of crotonic acid and the then-accepted values for the heats of formation of the other reactants and products, he calculated the heat of the reaction represented by Eq. 7.1 to be 264.33 kcal. For the same reaction, the N/C ratio was $(3 \times 14)/(4 \times 12) = 0.875$.

The equation representing the oxidation of ammonium during the respiration of *Nitrosomonas,* whereby this bacterium obtains energy for cellular synthesis is

$$NH_4OH + 2O_2 \longrightarrow HNO_2 + 2H_2O \tag{7.2}$$

for which Hes calculated a heat of reaction of -77.87 kcal. Since the total N/C ratio was 33.3, the N/C ratio with respect to the amount of nitrogen utilized in the reaction represented by Eq. 7.2 was therefore $33.3 - 0.875 = 32.425$, and the ratio of the total nitrogen used for the same reaction over that used in the reaction represented by Eq. 7.1 became 32.425/0.875. During Hes's experiment 21 mg of ammonium nitrogen disappeared from solution. Of this, $32.425/33.3 \times 21$ mg $= 20.499$ mg which was utilized according to Eq. 7.2, resulting in the production of $20.499 \times (-77.87/14) = -113.74$ cal. For the reaction represented by Eq. 7.1, $21.0 - 20.499 = 0.551$ mg of nitrogen were required. Therefore, with respect to this same reaction $0.551 \times 264.33/(3 \times 14) = 3.47$ cal were absorbed. From all these calculations, the consumption by *Nitrosomonas* of 21 mg of ammonium nitrogen should theoretically be accompanied by the production of $-113.74 + 3.47 = -110.27$ cal. Hes observed -105.64 cal, but did not consider this agreement to be especially good. However, he did make the point that quite certainly the oxidation of ammonium to nitrite was the only energy-producing process occurring during the growth of *Nitrosomonas*. He thus came to the same conclusion that Meyerhof[3] came to with respect to the oxidation of nitrite to nitrate during the growth of *Nitrobacter*.

Hes's study represents the last time anyone tried to carry out this kind of research with *Nitrosomonas* sp. In both Hes's and Meyerhof's[3] research, it is impossible to know if the bacterium were growing, assimilating, or simply oxidizing the substrate in the calorimeter vessel. There was no determination of the amount of carbon dioxide fixed in the one calorimetric experiment. In addition, the use of the formula for crotonic acid to represent the cellular material synthesized during the experiment was highly questionable, at least if microbial growth were being considered. Hes was in a position to carry out an ash-free, elemental analysis of *Nitrosomonas* cells since he made determinations of the N/C ratio with respect to their growth. For whatever reason he did not do so. Hes's[2] research along with that of Meyerhof[3] on the energy changes accompanying the metabolism of *Nitrosomonas* and of *Nitrobacter*, respectively, will be remembered more for the temerity of these investigators in choosing such difficult microorganisms with which to carry out their investigations than for the calorimetric results they obtained.

7.2. THE ADIABATIC CALORIMETER OF WINZLER AND BAUMBERGER

Prior to 1938 all calorimetric methods of determining the heat produced during microbial growth had either relied on extensive insulation around the calorimeter vessel for the purpose of eliminating or for the most part reducing heat loss to the environment, or had involved various mechanisms whereby heat production could be measured while at the same time the calorimeter vessel and its contents were maintained isothermal with respect to the environment. In 1938 Winzler and Baumberger[11] designed and constructed the first adiabatic calorimeter for use in microbial calorimetry. In an adiabatic calorimetric system the temperature of the environment, a liquid bath surrounding the calorimeter vessel, is raised at the same rate and to the same extent as that of the calorimeter vessel and its contents, so that there can be no heat loss between them.

Winzler and Baumberger's calorimeter consisted of a noninsulated calorimeter vessel suspended inside a highly polished, cylindrical container having a capacity of about 1 liter and a wide flange at the top to which a round, flat cover could be clamped. Leakage of water into the container was prevented by a vaseline seal. Four junctions of an eight-junction, copper–constantan thermopile were immersed in the suspension of microorganisms contained within the calorimeter vessel, the other four junctions being immersed in a water bath that surrounded the silver container. A temperature difference between the calorimeter vessel and the water bath caused an electric potential to be developed in the thermopile which activated a sensitive, mirror galvanometer, thereby causing a light beam reflected off the mirror to move off or onto a photoelectric cell that was located 3 m away at the end of a long tube (to reduce the effects of stray light). The voltage output of the photoelectric cell was amplified through a thyratron circuit, with the result that an electric current was sent through a naked resistance wire placed in the water bath, this latter being very vigorously stirred with a propeller. The galvanometer zero was adjusted so that a very small temperature difference of the calorimeter vessel above that of the water bath would cause the light beam to move such that it illuminated the photoelectric cell. This resulted in the water bath becoming heated until the galvanometer returned to the zero position, when the water bath and the calorimeter vessel and its contents were again at equal temperatures. At this time the electricity to the heating circuit was automatically cut off. With the light spot at the end of the light beam moving over the face of the photoelectric cell at a distance of 3 m, the sensitivity with which temperature differences could be detected was 0.0003°C mm^{-1} of lateral movement. Since the maximum variation of the position of the light spot was never more than 20 mm, the calorimeter vessel plus its contents and the water bath never differed in temperature more than 0.006°C, effectively preventing any heat transfer from one to the other. The temperature rise of the water bath was measured with a Beckman thermometer. The heat capacity of the calorimeter vessel plus its contents multiplied by the temperature rise gave the heat production of a microbial suspension. The calorimetric system could be calibrated electrically, and Winzler and Baumberger presented data showing close agreement between the amount of heat electrically introduced into the calorimeter vessel and the amount observed to be produced. A suspension of microbial cells contained in the calorimeter vessel could be agitated by bubbling gas through a tube inserted into the suspension. This also served to maintain aerobic or anaerobic conditions. The gas was first passed through 3 ft of copper tubing immersed in the water bath to bring the gas to the appropriate temperature, after which it was passed through a "saturation vessel," also immersed in the water bath, containing a solution identical to that in which the cells within the calorimeter vessel were suspended. In this manner the gas entering the calorimeter vessel became saturated with water vapor at the same temperature and pressure as that above the medium in the calorimeter vessel, thus avoiding any temperature changes due to evaporation. Oxygen-free gas for anaerobic experiments was obtained by bubbling nitrogen through Fieser's[12] reagent, and then through a wash bottle containing a lead acetate solution, prior to leading it into the calorimetric system. A diagram of Winzler and Baumberger's calorimeter is shown in Fig. 7.1.

In attempting to arrive at the true heat production from their experiments, as opposed to the observed heat production, Winzler and Baumberger corrected the latter using the data in Table 7.1. It is apparent that they regarded oxygen and carbon dioxide to be utilized biologically in their standard states,

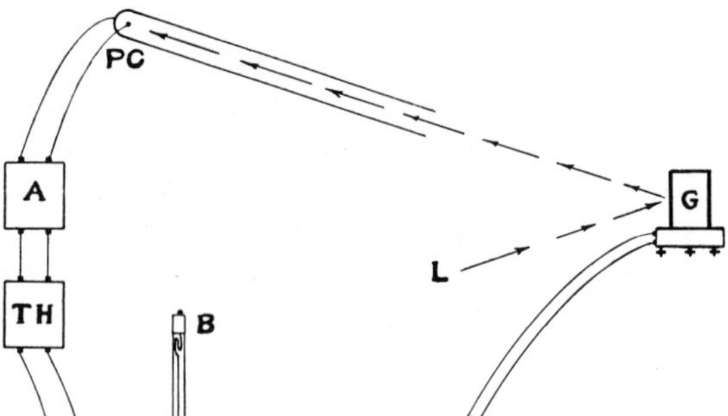

FIGURE 7.1. Diagram of Winzler and Baumberger's microcalorimeter. A, amplifying circuit; B, Beckman thermometer; b, bubbler; C, silver can; CT, copper tubing; G, galvanometer; H, heater; I, inlet, k, clamp; L. light source; O, outlet; PC, photoelectric cell; S, stirrer; SV, saturation vessel; T, thermopile; TH, thyratron tube; V, calorimeter vessel. Reprinted from Reference 11 with permission.

TABLE 7.1. Free energies and heats of formation at 298.15°K and 1 atm. ΔF and ΔH values are in cal mole^{-1}

Substance and State	Concentration	ΔF	ΔH
O_2(g)	0.2 atm	−950	0
CO_2(g)	0.0003 atm	−98,900	−94,240
H_2O(l)		−56,500	−68,310
NaOH(aq)	3×10^{-5} M	−106,200	−112,210
$C_6H_{12}O_6$(aq)	1×10^{-4} M	−222,460	−300,450
C_2H_5OH(aq)	2×10^{-3} M	−45,520	−67,230
CH_3COONa(aq)	3×10^{-5} M	−164,960	−175,100
$C_6H_{10}O_5$(aq)	Saturated solution	−158,580	−226,590

Source: Reprinted with permission from Reference 11.

rather than in aqueous solution. They did, however, recognize the importance of attempting to arrive at meaningful approximations of the values for the free energy of formation of various substances in aqueous solution, and used in these calculations the mean values for the concentrations of these substances. In doing so they pointed out that ". . . the reliability of the free energy data does not merit the taking into consideration of changing concentrations." (Winzler and Baumberger,[11] p. 185.) They had thus advanced further than any investigators in attempting to evaluate the thermodynamic properties of substances in the physical form in which microorganisms actually metabolize them.

Winzler and Baumberger used their calorimeter only for measuring heat production by yeast cells grown on a complex, solid medium, scraped off, and washed and suspended in 0.1 *M* phosphate buffer, pH 6.8. They measured the rate of endogenous heat production, the critical concentration of glucose, and the heat produced during oxidative and fermentative assimilation, no experiment lasting more than 2.5 hr. No experiments were carried out with growing cells. Thus, the importance of their work for future studies on the energetics of microbial growth was chiefly in their design and development of an adiabatic calorimeter.

7.3. THE STUDY BY HOOVER AND ALLISON ON THE GROWTH METABOLISM OF *RHIZOBIUM*

Apparently completely independently of the work of Tamiya and his associates, Hoover and Allison[13] in 1940 described an analogous method of representing the growth of microorganisms. They grew *Rhizobium* in a sucrose-inorganic salts medium containing iron and coenzyme R, using either ammonia or nitrate as a source of nitrogen. Aliquots of the culture medium were pipetted into Warburg vessels at various times during the growth of the organism and the RQ determined by the first method of Warburg. This method involved the manometric measurement of oxygen consumption at the same time that the carbon dioxide being produced was absorbed by a solution of barium hydroxide. The absorbed carbon dioxide was subsequently measured manometrically by acidifying the barium hydroxide solution after an experiment was terminated. Their report of the RQ values is inconsistent. They state that the highest values for the RQ were found in young cultures, after which it decreased as the amount of growth became maximal. This is illustrated by a graph of the RQ against time. On the other hand, they also show a graph of the Q_{CO_2} in mm^3 mg dry wt^{-1} hr^{-1} plotted against the Q_{O_2} in mm^3 mg dry wt^{-1} hr^{-1} (i.e., Q_{CO_2}/Q_{O_2} = RQ) in which the data clearly show the RQ to be remarkably constant. Hoover and Allison accepted this latter to be true for the purposes of constructing equations representing the growth of *R. meliloti* with the RQ for ammonia-grown cells being 1.20 and that for nitrate-grown cells being the extremely high value of 1.86. Analyses of the dried bacteria showed a somewhat lesser constancy in the percentages of carbon, hydrogen, oxygen, and nitrogen than were obtained by Yamagata[14] for *A. oryzae.* Hoover and Allison found the ash-free percentages of carbon, hydrogen, oxygen, and nitrogen to be 52.14, 7.42, 33.94, and 6.50, respectively. From these values they calculated the elementary composition based on a unit amount of nitrogen to be $C_{9.4}H_{16}NO_{4.6}$, which was rounded out to give a formula having subscripts which were whole numbers (again, this tendency to want to deal in terms of small, whole numbers) of $C_{19}H_{32}N_2O_9$. As was pointed out by Tamiya (see Chapter 6), if the RQ and the composition of the carbon source and of the cells are known, it becomes possible to write equations representing microbial growth. This was also recognized by Hoover and Allison. They did not, however, develop an elaborate "theory of the CQ" to do this, as did Tamiya (see Chapter 6) but rather, appear to have constructed their equations by a method of successive approximations. These equations are shown in Table 7.2. If it is assumed that the oxygen consumed is used solely for the purpose of respiration, then the amount of carbon source respired can be determined by measuring the oxygen consumption. This is illustrated by Eqs. 1 and 4 in Table 7.2. The difficulty comes in writing equations representing the conversion of the carbon source into cellular material, as illustrated by Eq. 2 in Table 7.2. If the empirical composition of both the carbon source and the cellular material are known, there is then only one equation that can represent the conversion of the former into the latter if the chemical nature of any other reactants and products is

TABLE 7.2. Hoover and Allison's equations representing the growth of *Rhizobium meliloti*

Growth using sucrose as the carbon source and ammonia as the nitrogen source.

$$\tfrac{10}{12}C_{12}H_{22}O_{11} + 10O_2 \longrightarrow 10CO_2 + \tfrac{110}{12}H_2O \qquad RQ = 1.00 \tag{1}$$

$$\tfrac{21}{12}C_{12}H_{22}O_{11} + 2NH_4OH \longrightarrow C_{19}H_{32}N_2O_9 + 2CO_2 + \tfrac{99}{12}H_2O \qquad RQ = \infty \tag{2}$$

$$\left.\begin{matrix}\tfrac{31}{12}C_{12}H_{22}O_{11} + 10O_2 \\ + 2NH_4OH\end{matrix}\right\} \longrightarrow C_{19}H_{32}N_2O_9 + 12CO_2 + \tfrac{209}{12}H_2O \qquad RQ = 1.20 \tag{3}$$

Growth using sucrose as the carbon source and nitrate as the nitrogen source.

$$\tfrac{7}{12}C_{12}H_{22}O_{11} + 7O_2 \longrightarrow 7CO_2 + \tfrac{77}{12}H_2O \qquad RQ = 1.00 \tag{4}$$

$$\tfrac{4}{12}C_{12}H_{22}O_{11} + 2HNO_3 + \tfrac{1}{3}H_2O \longrightarrow 2NH_4OH + 4CO_2 \qquad RQ = \infty \tag{5}$$

$$\tfrac{21}{12}C_{12}H_{22}O_{11} + 2NH_4OH \longrightarrow C_{19}H_{32}N_2O_9 + 2CO_2 + \tfrac{99}{12}H_2O \qquad RQ = \infty \tag{6}$$

$$\left.\begin{matrix}\tfrac{32}{12}C_{12}H_{22}O_{11} + 7O_2 \\ + 2\ HNO_3\end{matrix}\right\} \longrightarrow C_{19}H_{32}N_2O_9 + 13CO_2 + \tfrac{172}{12}H_2O \qquad RQ = 1.86 \tag{7}$$

Source: Data from Reference 14.

known. The only other reactant is the nitrogen source, which was chosen by Hoover and Allison to be NH_4OH (it could equally well have been represented by NH_3 or NH_4^+). The only other products of the reaction are CO_2 and H_2O. The problem is to find the coefficients. This can be done by fixing the quantity of cellular material, which then determines the quantity of nitrogen source. By successively approximating the quantity of carbon source one can eventually arrive at an equation which includes all the terms and which has the same elemental balance on both sides of the equation. This is a perfectly valid method of constructing such an equation, although it can be extremely tedious and time consuming. It obviously lacks the elegance and sophistication of the method developed by Tamiya for accomplishing the same thing in a far shorter time. The addition of Eqs. 1 and 2 in Table 7.2 gives Eq. 3 representing the growth of *R. meliloti* on sucrose when ammonium is the nitrogen source. When nitrate is the nitrogen source an additional equation (Eq. 5 in Table 7.2) must be included to represent the reduction of nitrate to ammonia by electrons from the carbon source. In general the equations resemble those of Tamiya's. No oxygen is involved in the actual synthesis of cellular material, and less oxygen gas is required if the nitrogen source is nitrate rather than ammonium. Everything about Hoover and Allison's equations is internally consistent and provides the required RQ values. However, equations such as these can be written to conform to any RQ value, and that of 1.86 seems excessively high. RQ values are extremely difficult to determine precisely. In addition, the use of barium hydroxide to absorb the carbon dioxide produced from respiration at the same time that the oxygen consumption is being measured manometrically is a very risky technique in that carbon dioxide is an absolute requirement for microbial growth.It is entirely possible that this became a limiting factor which affected the results of Hoover and Allison's RQ determinations. They did, however, also carry out an experiment in which *R. meliloti* was grown on a limiting amount of sucrose using either ammonium or nitrate as the nitrogen source, after which dry weights of the bacteria were determined. With these data and the empirical formulas for the bacteria it is possible to construct equations representing the growth of *R. meliloti* in a manner different from that employed either by Tamiya or by Hoover and Allison (discussed in Chapter 10).

7.4. STUDIES ON THERMOGENESIS

Following the investigation in 1929 by Bayne-Jones and Rhees[15] on the relationship of heat production to the phases of growth of bacteria grown on complex substrates, there seems to have been almost no research on thermogenesis for the next 10 years. There then began a whole series of studies more or less repeating what had been done previously (see Chapter 2) but with increasingly better heat measuring devices.

7.4.1. The Study by Wedberg and Rettger of Factors Influencing Microbial Thermogenesis

In 1940 Wedburg and Rettger[16] carried out an investigation similar to that of James et al.[17] using a modification of James' calorimeter. One result was that they recorded a maximum temperature of 67°C as a result of thermogenesis by unsterilized corn, this being 4.5°C higher than that obtained by James et al. Other experiments involved the measurement of thermogenesis caused by pure cultures of bacteria growing on corn-extract agar, or on cracked corn. None of these studies has any significance with respect to the energetics of microbial growth because these systems are manifestly undefined. This investigation is only mentioned here because it may have been the last of these qualitative efforts at thermogenesis measurement. Wedburg and Rettger also investigated the inhibition of various bacterial enzymes as a function of temperature, seeking to confirm the report of Edwards and Rettger[18] that the cessation of bacterial growth at certain temperatures was related to the thermal destruction of specific respiratory enzymes.

7.4.2. Tian's Compensating Microcalorimeter

One of the problems besetting earlier measurements of heat production by slow reactions was that of heat exchange between the calorimeter and the environment. Corrections for these exchanges can only be made with difficulty and are usually inadequate, especially when the magnitude of the correction approaches that of heat measurement in experiments where very little heat is absorbed or liberated. Such heat exchanges can be reduced to zero when the temperature difference between the calorimetric system and the environment is reduced to zero. One way of doing this is to create an *adiabatic* calorimetric system. The other way is to use an *isothermal* calorimetric system whereby the temperature inside the calorimeter is maintained exactly the same as that of the environment. Isothermal calorimeters for the measurement of biologically produced heat have been known since 1780, when Lavoisier and Laplace[19] used an ice calorimeter to measure the rate of heat production by a guinea pig. Such a calorimeter, while accurate and convenient, cannot be used to measure the heat production of poikilothermic organisms since most of these do not grow well at 0°C. Von Korosy's[20] isothermal calorimeter (see Section 2.6), in which the latent heat of vaporization of a liquid rather than the latent heat of fusion of a solid (as in the ice calorimeter) was used to keep the temperature constant, was an improvement over the ice calorimeter for the measurement of heat production by microorganisms. The disadvantage of Korosy's calorimeter was that it could only operate at the temperature of boiling pentane, which was 35.0°C. This disadvantage was eliminated in the design of the calorimeter constructed by Algera,[5] in which water was used as the means of evaporative cooling (see Section 3.11).

In 1924 A. Tian[21] described an isothermal calorimeter the operation of which was quite different from that of previous isothermal calorimeters in that the heat developed within the calorimeter could be removed electrically using the Peltier effect.

The Peltier effect can best be explained by first explaining the Seebeck effect, which is what is observed in the action of a thermocouple. All metals have more or less resistance to the flow of an electrical current which is a function of the density of so-called "free electrons" which can move about freely within a metal. These free electrons are loosely held in the outer orbit of any particular metal atom and can be easily displaced by other free electrons. When there is a net displacement of these electrons in one direction through a conductor the flow of electrons past a given location constitutes an electrical current. Different metals have different free-electron densities, those with the greater densities having the greater conductivities. If two metal wires having different free-electron densities are connected together in an electrical circuit, as illustrated in Fig. 7.2*a*, there will have to be two locations at which connections are made, each connection being called a "thermoelectric junction" and the whole circuit being called a thermoelectric couple, or "thermocouple," since there are two thermoelectric junctions. As an example, suppose that the two substances used in a thermocouple were copper and "constantan," a

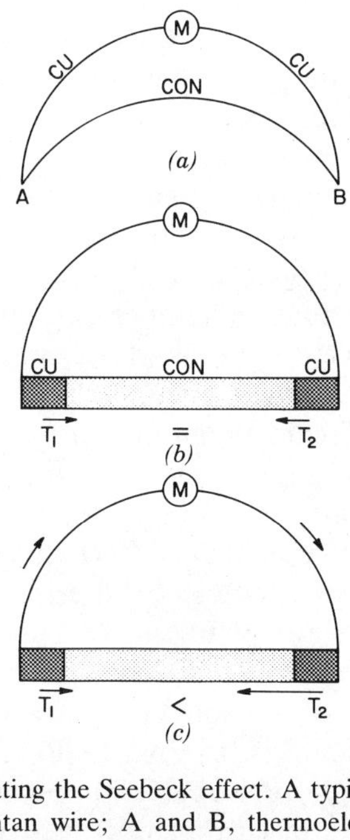

FIGURE 7.2. Diagrams illustrating the Seebeck effect. A typical thermocouple. M, voltmeter; Cu, copper wire; Con, constantan wire; A and B, thermoelectric junctions. Enlargement of thermoelectric junctions A and B illustrating the different free-electron densities in the copper and constantan wires. When $T_1 = T_2$ there will be no net tendency for electrons to flow around the circuit. Same as (*b*) except that if T_1 is lower than T_2 there will be a net tendency for electrons to flow in a clockwise direction around the circuit. This can be detected as a voltage difference that is proportional to the temperature difference.

copper–nickel alloy that is used in the form of electrical resistance wires to generate heat in electric heaters and stoves. The copper wire has a high conductivity and a low resistivity, and therefore a high density of free electrons. The constantan has a low conductivity and a high resistivity, and therefore a low density of free electrons. Free electrons act in a manner analogous to gas molecules in that the higher the temperature of the metal or alloy, the faster they will randomly move about within it. As illustrated in Fig. 7.2*b*, where the representation of the junctions has been enlarged, there will always be a tendency for the free electrons in a space (the wire) where there is a high free-electron density to move into a space where there is a lesser electron density, this tendency being a function of the temperature and being measurable in terms of voltage. Thus, there will always be a tendency for free electrons to pass from copper, which has a higher free-electron density, to constantan, which has a lower free-electron density. As shown in Fig. 7.2*b*, if the temperatures of the two junctions are identical, the tendency for electrons to pass from copper to constantan will be the same with respect to both junctions, and there will be no net tendency for electrons to flow through the circuit. On the other hand, if the temperature of one junction is higher than that of the other, there will be a greater tendency for electrons to pass from copper to constantan at the hotter junction than at the colder junction, as illustrated in Fig. 7.2*c*. This creates a voltage difference which can be detected and measured by a voltmeter inserted into the circuit. If one of the junctions of a thermocouple is kept at a known, constant temperature, the magnitude of the voltage developed in the circuit is a function of the difference in potential of the thermoelectric junctions resulting from the temperature difference between them. Individual thermocouples, or many of these connected in series called a "thermopile," can be used to measure temperature differences very accurately. The phenomenon of an electrical potential developed between two thermoelectric junctions at different temperatures is called the Seebeck effect. Related to this effect is the current flow which can be measured if an ammeter (which has a low resistance) rather than a voltmeter (which has a high resistance) is inserted into the circuit. When the total resistance in a thermocouple circuit is low, the flow of current results in the removal of heat from the environment at the hot junction and its transfer to the cold junction, where it is dissipated into the environment. Effectively, heat is transformed into electricity at the hot junction, the electrical energy then being dissipated as heat at various locations throughout the rest of the circuit because of its resistance (the Joule effect). The location having the most resistance is the cold junction, and therefore most of the electrical energy is converted into heat at this location. The Peltier effect has to do with the effect of passing an imposed electrical current (as from a battery) across a thermoelectric junction. Depending on the direction of current flow heat will either be absorbed or liberated at a junction. If the current flows in the same direction as does the current at a hot junction (from copper to constantan in Fig. 7.2*c*), heat will be absorbed. If it flows in the same direction as the current at a cold junction (from constantan to copper in Fig. 7.2*c*), heat will be liberated. Thus when, as a result of an imposed current flow, the free electrons are forced to pass from a region of high electron density (in the copper metal) to a region of low electron density (in the constantan alloy) heat is absorbed. The opposite phenomenon occurs when free electrons are forced to pass from a region of low electron density (in the copper alloy) to a region of high electron density (in the copper metal). The result of the Peltier effect is that heat can be removed from one location and released at another, and because of this it can be used as a heat pump. The only problem associated with the use of the Peltier effect is that all electrical conductors at ordinary temperatures, including thermoelectric junctions, have a finite resistance to the flow of current which results in the liberation of heat. Thus, the absorption of heat is equal to the Peltier coefficient (P) for a particular junction at a given temperature multiplied by the current (I), whereas the liberation of heat due to the Joule effect is I^2R, where R is the electrical resistance of the junction at the given temperature. When $PI = I^2R$, no further heat absorption is possible. For this reason, the use of the Peltier effect is limited to the transfer of only small amounts of heat because the effect can only be observed with small values of I.

Tian's calorimeter consisted of two concentric, cylindrical copper containers. One was a small diameter, inner cylinder mounted inside a much larger cylinder surrounded by a constant temperature water bath. The calorimeter vessel was a long, thin-walled glass tube in the bottom of which the experimental material (actually, the system to be studied) could be placed. The tube was sized so as to fit very snugly into the inner cylinder. The outer surface of the lower part of this tube was first ground and silvered, after which copper was plated over the silver coating. This helped to establish the best possible thermal contact between the calorimeter vessel and the wall of the inner cylinder. Two iron–constantan thermopiles, one of 7 thermocouples and one of 42 thermocouples, connected the outside wall of the inner cylinder with the inside wall of the outer cylinder. Each thermopile could be connected by means of mercury switches either to a sensitive galvanometer or to a battery. In the former case the thermopiles could serve as devices to sense the temperature difference between the calorimeter vessel and the water bath. This could be done with extreme sensitivity. Using an appropriate mirror galvanometer and a light beam of suitable length reflected onto a scale, the sensitivity of the 42-thermocouple thermopile was of the order of 1/12,000°C. If the reaction in the calorimeter vessel liberated heat, the thermopiles could be used as devices to remove this heat, using the Peltier effect, so that the temperature of the reacting system would remain isothermal with the water bath. If the reaction in the calorimeter vessel absorbed heat, an isothermal condition could be maintained by adding heat to it. This was done by passing a current through a resistance wire placed in the bottom of the calorimeter vessel, thus making use of the Joule effect. In either case, the amount of heat removed from or added to the calorimeter vessel in order to keep it and its contained system in temperature equilibrium with the water bath could be accurately measured.

Tian's calorimeter never seems to have been actually used for experimental

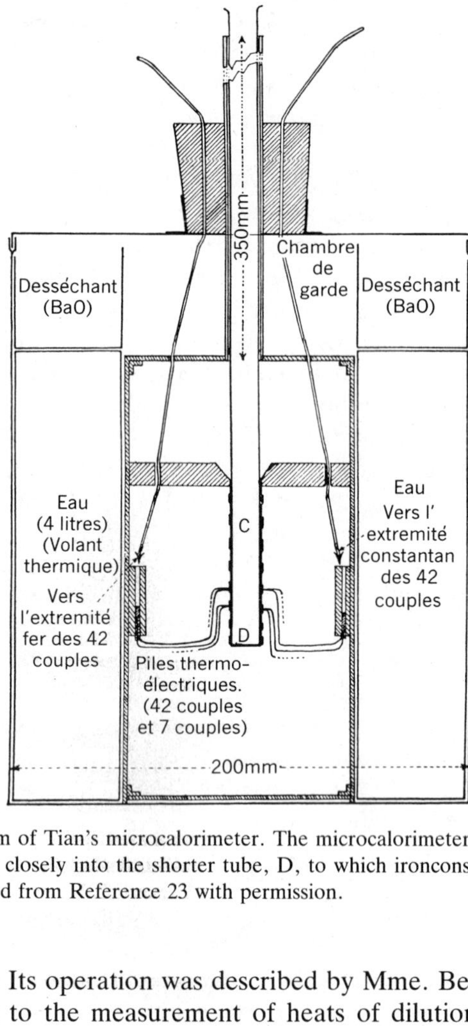

FIGURE 7.3. Diagram of Tian's microcalorimeter. The microcalorimeter vessel is a long tube, C, inserted and fitting closely into the shorter tube, D, to which ironconstantan thermocouples are attached. Reprinted from Reference 23 with permission.

purposes by Tian. Its operation was described by Mme. Berenger-Calvet[22] in 1927 with respect to the measurement of heats of dilution and of heats of hydrolysis and etherification. A diagram of the instrument shown in Fig. 7.3 appeared in a publication by E. Calvet[23] in 1933, in which he described both the operation of the calorimeter and its application for direct thermochemical measurement in organic chemistry. He also described the thermoelectrical calculations which are involved in the use of the Peltier effect.

Because of the electrical resistance of any thermoelectric junction, a Joule effect always accompanies the Peltier effect. By passing a sufficiently strong electric current through one of the thermopiles while using the other thermopile as a detector, it is possible to arrive at a value for the current, I, which results in the Peltier effect (PI) being exactly compensated by the Joule effect (I^2R). Under these conditions,

$$PI = I^2R$$

or

$$I = \frac{P}{R}$$

where I is here designated as the compensating current. The quantity of heat q, measured in g cal, which is absorbed in 1 hr by a thermopile of n couples through which a current i (which must be less than I) is passed is represented by the equation

$$q = \frac{3600}{4.185}(Pi - Ri^2) = \frac{3600\ Pni}{4.185}\left(1 - \frac{R}{P}i\right) = Ki\left(1 - \frac{i}{I}\right)$$

The constant, K, is obtained from a previous standardization and has the dimensions of $iRt/4.18$.

7.4.3. The First Biological Studies by Calvet and His Collaborators Using Tian's Calorimeter

The first biological use to which Tian's calorimeter was put was published by Prat and Calvet[24,25] in 1944 in a description of the thermal events accompanying the germination of cereal grains. In 1945 Calvet, Fricker, and Prat[26] published a short note on the measurement of thermogenesis by bacteria. They seem not to have been aware that considerable work of this kind had already been done during the preceding 40 years (see Sections 2.1 and 2.2) since they state, "The good results that we have obtained [with studies on the thermogenesis of germinations] have incited us to extend our research and to apply the same microcalorimetric method to the study of bacterial cultures, which had not been made until the present." (Calvet, Fricker, and Prat,[26] p. 797.) Eight milliliters of peptone broth were placed in the culture vessel and inoculated with 0.1 ml of a suspension of *Proteus vulgaris* containing 6×10^6 cells ml^{-1}. After a few hours of incubation at 17°C a slow thermogenesis was observed that increased to a maximum after about 24 hr. At that moment the culture was liberating 0.16 cal hr^{-1}. Following this the thermogenesis declined rapidly and became zero after about 18 hr. Under the same conditions but using a more concentrated peptone broth, the maximum thermogenesis was observed to occur later, after about 42 hr, when the intensity of the thermogenesis was 0.3 cal hr^{-1}. From this they concluded that the greater the quantity of nutrients, the more prolonged and intense was the thermogenesis, thus rediscovering what A. V. Hill[27] had originally discovered in 1912. Experiments with smaller inocula demonstrated that although the length of time required for a culture to reach maximum heat production was longer, the total quantity of heat produced was the same for a given quantity of nutrient as it was when the inocula were larger. The quantity of heat produced thus depended *only* on the quantity of nutrients, and not upon the rate at which heat was liberated.

Calvet et al.[26] suggested that microbial microcalorimetry would permit the study of the variations in activity of a given microorganism as a function of its environmental conditions and of the method of its inoculation. As such it might be of considerable service to bacteriological research. Possibly with this in mind Prat, Calvet, and Fricker[28] in 1946 published the results of some experiments which were carried out in a manner identical to those of Calvet et al.[26] but in which different bacteria were used. It is here (Prat et al.[28]) that we see one of their bacterial thermograms for the first time. This was a thermogram of *Bacterium proteus* (*Proteus vulgaris*?) grown on peptone broth at a temperature of 17°C. The reason for carrying out the experiments at this rather low temperature seems to have been that the location most suited to the operation of their calorimeter was in a cellar which happened to be at this temperature. The *B. proteus* thermogram shows a smooth rise in the rate of heat production which may have been exponential during the first 10 hr but which then became linear until the twenty-fourth hour when it reached a peak and subsequently declined in a nearly exponential manner. This surely indicates that the culture soon became oxygen limited after the first 10 hr, probably because the surface/volume relationship of the culture fluid in the calorimeter vessel was not particularly favorable for nonlimited growth in peptone broth. Experiments under identical conditions with strains of *Bacterium typhosum* (now *Salmonella typhi*) thermograms exhibited rhythmic pulses of thermal debt spaced 8 hr apart. Similar experiments with *Bacterium paratyphosum* (now *Salmonella paratyphi*), strains A and B, resulted in thermograms in the shape of an undulating curve quite unlike the smooth curve of *B. proteus*. Such data led them to the conclusion that each bacterial strain presents a specific thermogram that may be modified as a result of changes in the experimental conditions.

7.5. CALVET'S DIFFERENTIAL COMPENSATING MICROCALORIMETER

In Tian's[21] original descriptions of his compensating microcalorimeter he emphasized the necessity of maintaining the bath at a constant temperature. The use of ordinary thermostats, in which the bath is in direct contact with the thermoregulator, must be avoided. Instead, what ought to be used is a thermostat which Tian[29,30] had described earlier having multiple enclosures. This thermostat consisted of a water bath surrounded by one or more chambers having walls made of copper 2 mm thick which were separated from one another by a layer of felt or other material. The outermost enclosure had a mercury well attached to its inner surface into which an electrical heating element was placed. A thermoregulator was placed in an identical mercury well on the outermost enclosure wall diametrically opposite the heater well. Heating the mercury in the heater well caused the copper of the wall to which the well was attached to be heated around its entire circumference because of the high heat conductivity of copper metal. As heat became transmitted to the thermoregulator well, the thermoregulator deenergized the heater when the temperature reached the set point, whereupon the outermost wall cooled slightly, losing heat to the environment. In this manner the temperature of the outermost wall would oscillate slightly, these oscillations becoming successively damped with respect to each enclosure wall because of the felt insulation between them, to the extent that temperature oscillations became undetectable in the constant temperature water bath itself which was contained within the innermost wall. A diagram of one of Tian's multiple-enclosure thermostats is shown in Fig. 7.4.

In spite of Tian's ingenious efforts to obtain a constant reference temperature for the successful operation of his compensating microcalorimeter, at the end of his publication he states, "I hope to be able to continue my experiments in a cellar, in the best thermal conditions." (Tian,[21] p. 707.) Mme. Berenger-Calvet[22] carried out some of her experiments with Tian's calorimeter in an underground room in which the temperature varied in a regular manner

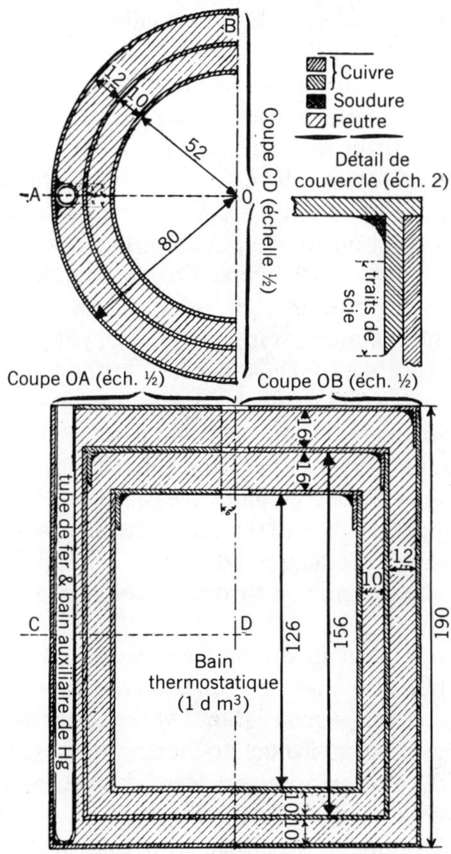

FIGURE 7.4. Diagram of Tian's multiple-enclosure thermostat. Reprinted from Reference 30 with permission.

no more than 0.02°C day^{-1}. Although this was of some benefit in maintaining the temperature of the thermostat constant, it was more essential in eliminating thermoelectric forces accidentally (and inevitably) produced in the galvanometric circuit by temperature changes, quite independently of the effects of these changes on other parts of the instrument. It was this latter problem that caused Prat et al.[28] to look for a location for their calorimeter which had an extremely constant temperature of only 17°C.

The difficulty of finding suitable locations for their calorimeter eventually led Calvet[31] to design a differential compensating microcalorimeter which could be operated at all physiological temperatures and be completely independent of the usual environmental temperature fluctuations. He pointed out that there were two causes of the variation in the experimental zero of a Tian microcalorimeter. One was the variation in the reference temperature (of the water bath) caused by small variations in the temperature of the environment, even in a cellar. This cause of experimental zero variation was thought not to be as important as the second cause which was due to the Thomson effect, this latter being the potential gradient along an electrical conductor which accompanies a thermal gradient along the same conductor. With Tian's microcalorimeter such gradients became established in the galvanometric circuit when the working temperature of the microcalorimeter was different from that of different parts of the circuit, as already emphasized by Berenger-Calvet.[22] In the experiments of Prat et al.[28] the calorimeter was buried in the ground, whereas the galvanometer and the commutators (which enabled one or other of the two thermopiles to be used) were mounted externally.

Calvet's brilliant solution to the above problem was to use two identical Tian compensating microcalorimeters, the thermopiles of which were connected in opposition to one another, which were placed *together* with the galvanometer and the commutator in a massive copper block which was suitably hollowed out to receive them. Thus, because of the high heat conductivity of the copper metal, the copper block, the reference galvanometer, and the galvanometric circuit including the galvanometer and the commutator were all at exactly the same temperature, and the Thomson effects were thereby eliminated. The copper block containing the calorimeter was itself placed inside another massive copper block, and this latter was in turn placed in a thermostat. The calorimetric assembly is diagramed in Fig. 7.5. Calibration could be carried out by passing an electric current through a resistance wire inserted into the calorimeter vessel.

The advantage of Calvet's calorimeter was not only that Thomson effects were eliminated, but also that any variation in the reference temperature affected both calorimeters equally. Since these latter were connected in opposition, any effects of an environmental temperature variation were canceled out provided that the two calorimeters were, in fact, perfectly identical. Any change in the potential of the galvanometric circuit could be due only to the difference in temperature between the two calorimeters, thus making this instrument a differential, compensating microcalorimeter. Once the accuracy of the instrument was assured, Calvet proceeded to increase its sensitivity by

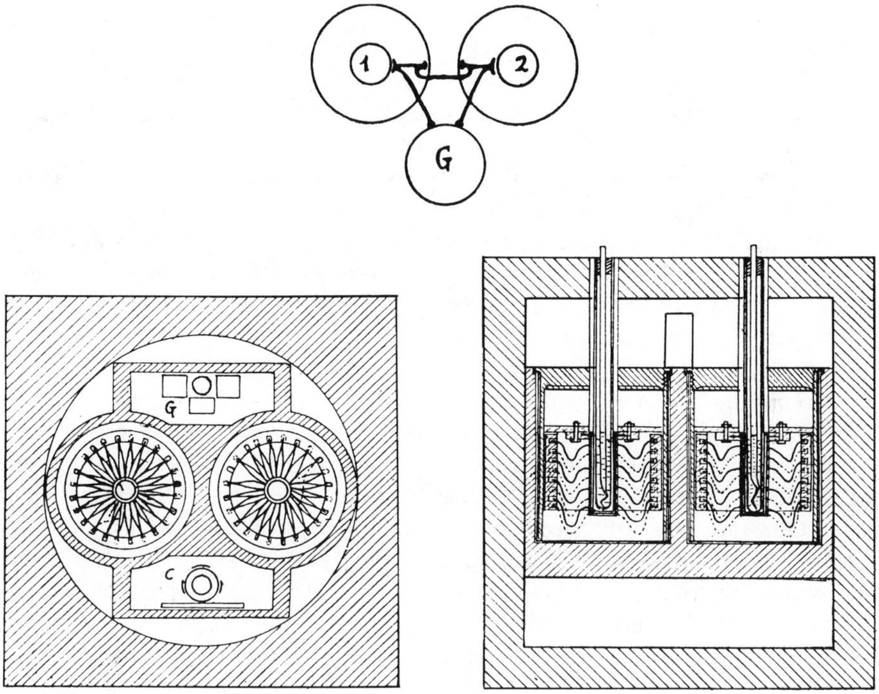

FIGURE 7.5. Diagrams of Calvet's microcalorimeter. The lower left and right diagrams are top and side views, respectively, of the copper block containing twin Tian microcalorimeters. The upper diagram illustrates how the output leads from the thermopile circuits of the twin microcalorimeters, 1 and 2, are connected in opposition to a galvanometer. Reprinted from Reference 31 with permission.

increasing the size of the largest thermopile to 192 iron–constantan thermocouples for each calorimetric cell. The calorimeter was equipped with a Zernike mirror galvanometer having a 1 m light lever. A deviation of 1 mm on the galvanometer scale corresponded to a heat production of 0.001 cal hr^{-1}. The deviations of the galvanometer were registered photographically by means of a periscopic device which reflected the light beam to and from the mirror of the galvanometer inside the copper block containing the calorimeter. The constancy of Calvet's calorimeter was such that after an experiment of 10 to 15 days during which bacterial thermogenesis was measured, the pointer of the galvanometer light spot would return exactly to its initial zero. Thus, although the idea of a differential calorimeter was not new, being first used by Hill[27] in 1912 for measuring the thermogenesis accompanying the lactic fermentation of milk (see Section 2.5), it is clear that Calvet's differential compensating calorimeter was by far the most sensitive and accurate of all the calorimeters constructed through 1948. Its principal limitation was therefore not physical, but microbiological in that the small diameter tube serving as the calorimeter vessel provided for only a limited surface area through which oxygen could pass into a liquid culture medium contained in the tube. Thus, many microbial reactions taking place in the calorimeter vessel might sooner or later become limited by the lack of oxygen unless some provision were made to supply this latter. Otherwise, the calorimeter would best be suited for measuring the thermogenesis of anaerobic microbial reactions, for which no oxygen is required.

The first experiments with Calvet's new differential calorimeter were published in 1948 by Calvet and Fricker.[32] These involved the growth of *B. typhosum* and *B. paratyphosum* (the same organisms used by Prat et al.,[28] for which a thermogram was not published) in peptone broth. Calvet and Fricker's thermograms are shown in Fig. 7.6. With the new instrument it was possible not only to determine the rate of thermogenesis by the bacteria at any given time, but also to integrate the thermogram to obtain the total amount of heat produced. Thus, with *B. typhosum* they observed that at 37.4°C the amount of heat produced from 10 ml of peptone broth varied between 60 and 70 cal at the end of 190 hr (at which time, according to Fig. 7.6, the thermogram had also not yet returned to the baseline), whereas Prat et al.[28] had previously found that only 2 cal were produced during the same time when this microorganism was grown at 17°C. Calvet and Fricker[32] pointed out that calorimetry represented another method of following the development of a population of microorganisms which had some advantage over photocolorimetry or nephalometry in that it was independent of the optical variables associated with cell suspensions. They also suggested that thermograms such as those in Fig. 7.6 might have some use in the identification of different types or strains of microorganisms.

Although Calvet's microcalorimeter was designed to be capable of being operated through the use of the Peltier effect to compensate for the heat produced as a result of bacterial metabolism, there is little evidence that it was actually used in this manner. To accomplish this a feedback device would be needed to pass current through the thermopile proportional to the amount of heat produced by the bacteria, thereby removing this heat by means of the Peltier effect. Otherwise, an operator would have to sit by the instrument to monitor constantly the deviation of the galvanometer light beam from the

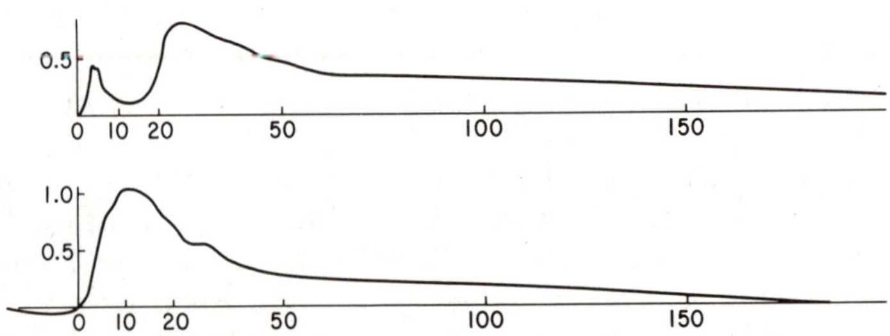

FIGURE 7.6. Calvet and Fricker's thermograms of the growth of *Bacterium typhosum* (upper) and *Bacterium paratyphosum* (lower) in a peptone medium. The *y* axis is in cal hr^{-1} and the *x* axis is in hr. Reprinted from Reference 32 with permission.

initial zero position, and to compensate for this by manually passing current through the large thermopile to return the galvanometer light beam to zero. Neither of these methods is mentioned by Calvet. For an operator to constantly monitor the instrument for up to 190 hr would have been extremely tedious. What seems to have happened is that although Calvet's calorimeter was admirably designed as an isothermal instrument, it was never used in this manner. It was used instead as a heat-flow calorimeter, much as Hill's[27] heat-flow calorimeter was used. The thermograms are therefore direct traces of the movement of the galvanometer light spot as recorded on a moving, photographic film, the position of this spot being related to the temperature difference between the experimental calorimeter vessel and the control. This was a far easier procedure than to measure the total quantity of current per unit time used to produce either an analogue or digital compensation for the heat produced. The cooling of the experimental calorimeter vessel when it was not compensated was thus accomplished partly by the conduction of heat through the thermopile wires (192 couples per calorimeter cell) and through the calorimeter vessel receiver tube; and partly by air convection (since the space between the calorimeter vessel receiver tube and the outer wall of the calorimeter cell was not evacuated). Nevertheless, the instrument was still essentially an isothermal calorimeter in that the temperature difference between the experimental calorimeter vessel and the reference temperature was never greater than a few thousandths of a degree Celsius. A treatise covering the theoretical principles and the physico-chemical and biological applications of their microcalorimeter was published by Calvet and Prat[33] in 1956, including a diagram of the whole calorimetric apparatus. This is reproduced in Fig. 7.7. A monograph on microcalorimetry was published in 1958 by Calvet and Prat[34] and subsequently updated in an English version[35] published in 1963. Here they presented a more recent modification of the microcalorimetric assembly, shown in Fig. 7.8. This seems to be the end point of the evolution of the Tian–Calvet microcalorimeter as developed by these investigators. However, a further evolution occurred in 1985 when Dermoun et al.[36] changed the shape of the calorimeter vessel to one in which the surface/volume ratio was more favorable for the growth of aerobic microorganisms.

The last of a series of investigations on bacterial thermogenesis was published in 1953 by Prat.[37] Using a Calvet microcalorimeter he carried out experiments similar to those previously carried out by Calvet and his collaborators[26,28,32] and also made comparisons of thermograms with respect to bacterial growth, germination of *Zea mays* seeds, the metabolism of an isolated orthopteran (*Melanoplus mexicanus*), and the metabolism of a population of fruit flies (*Drosophila* sp.), emphasizing differences between them. Prat[33] emphasized that thermograms were related not only to the amount of biologically produced heat, but also to strictly physical effects such as evaporation, dilution, hydration, solution, absorption, and so on. In summarizing, he stated that bacterial thermograms are characteristic of a particular bacterial strain and are modified by temperature, composition of the culture medium,

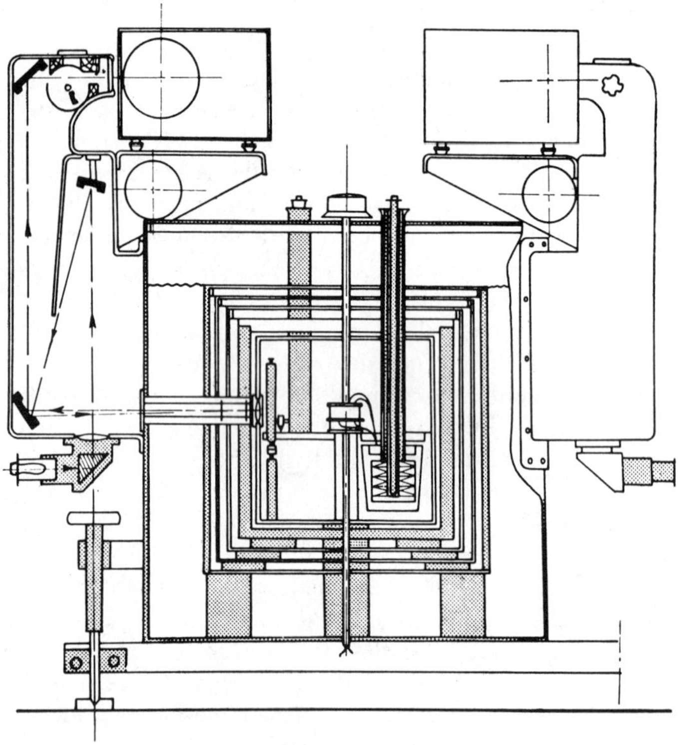

FIGURE 7.7. Diagram showing two half-views of the entire microcalorimeter assembly of Calvet and Prat. The twin microcalorimeters are encased in a massive, copper heat sink, the whole apparatus being isolated from the environment in one of Tian's multiple-enclosure thermostats. The device projecting to one side of the thermostat is used for recording photographically to position of the light beam reflected from the mirror galvanometer. Light enters the microcalorimetric system through the prism at the base of the optical reflecting chamber. Light-lever galvanometric amplification systems were used because of their lack of drift, and were later replaced by electronic instruments when stable amplifiers became available. Reprinted from Reference 33 with permission.

the initial number of microorganisms that are inoculated into the medium, and by the presence of various drugs. In these respects the production of heat by bacterial cultures closely parallels the measurement of bacterial growth as followed by optical methods. Microcalorimetry thus offers an alternative method for such studies.

In general, the idea that the thermogram of a particular microorganism was a characteristic that could be applied to its identification has not proved useful, any more than has the growth curve for a particular microorganism obtained by other methods. In special cases, however, it may be useful. Because optical methods of measuring the extent of cellular growth are far more simple than

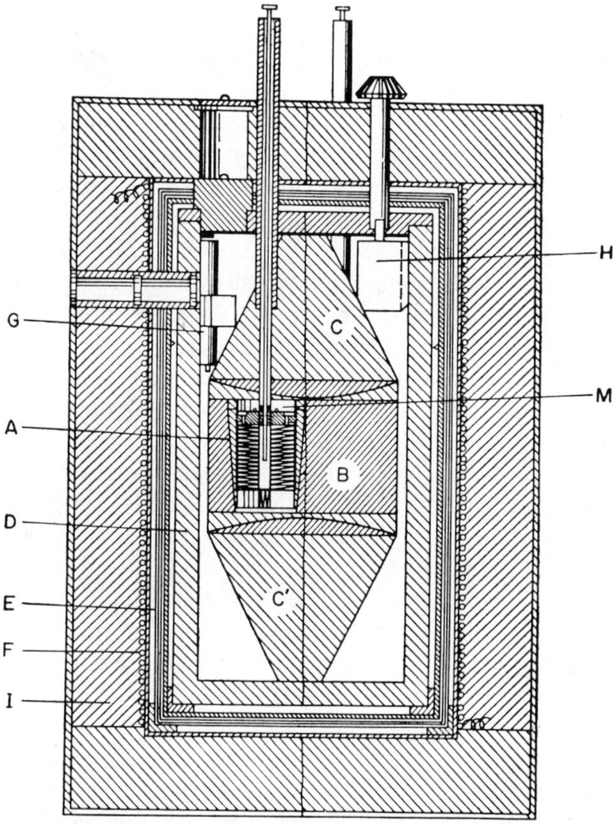

FIGURE 7.8. Diagram of a later modification of Calvet and Prat's microcalorimeter. Here metallic cones have been introduced to provide for an equipartition of thermal perturbations. Tian's multiple-enclosure thermostat is still present, but the whole instrument has been encased in dry thermal insulation rather than immersed in a water bath. A, conical socket cut into block B; B, metallic block; C, C′, metallic cones designed to give equipartition of the thermal perturbations; D, thick metal cylinder; E, thermostat consisting of several concentric cannisters; F, electrical heat controlled by the thermoregulator; G, galvanometer; H, switch; I, thermal insulation; M, microcalorimetric vessel assembly (placed inside its cavity A). Reprinted with permission from *Recent Progress in Microcalorimetry,* by E. Calvet and H. Prat, 1963, Pergamon Press, Ltd.

calorimetric methods, these latter are not often employed for this purpose. Another difficulty associated with Calvet's calorimeter was that it was not designed to be used with cultures of microorganisms which had to be grown aerobically. All of the thermograms presented by Calvet and his collaborators were of bacteria grown anaerobically in peptone broth. Thus, in spite of its obvious elegance of design and its capability of accurate and precise measurements, Calvet's calorimeter was not much used except by Senez, Belaich, and their collaborators (see Section 9.4.3).

7.6. THE CALORIMETRIC STUDIES BY KORTE AND ENGEL ON THE ENERGETICS OF GROWTH OF *HYPHOMICROBIUM VULGARE*

The sources of energy and carbon for the growth of "chemolithoautotrophic" bacteria such as *Nitrosomonas* and *Nitrobacter* are known with certainty. This is not necessarily true of those bacteria such as *Hyphomicrobium vulgare*, for which there is at least some evidence that although the energy source is an organic substance, the carbon source is carbon dioxide, thus making such bacteria "chemoorganoautotrophs." In 1955 Korte and Engel[38] published the results of attempts to determine the extent to which the quantity of heat produced when *H. vulgare* was grown on organic substrates equaled that which was expected theoretically if the same substrates were completely oxidized. Their calorimetric measurements were carried out with an instrument of the same design as that used by Meyerhof[35] (see Section 3.6) for measuring heat production by *Nitrobacter*. Their method of experimentation was also physically the same. A great many difficulties were expected, amongst which was the long duration of their experiments. A suspension of *Hyphomicrobium* required a full week to oxidize 40 mg of formic acid. Just as with the research of Meyerhof[39] with *Nitrobacter* and of Hes[2] with *Nitrosomonas*, because a large quantity of the test organism was inoculated into the calorimeter vessel, it was impossible to know if the hyphomicrobia were growing, assimilating, or simply oxidizing the substrate. The thermograms presented by Korte and Engel do not seem to show exponential growth. In addition, no carbon values were determined. Only the heat production was measured, the observed values having to be extensively corrected for various sources of error. Their measurements showed that about 90–95% of the heat theoretically expected from the complete combustion of formic acid was produced during its utilization by *H. vulgare*, whereas the value for oxalic acid was about 85–95%. On the basis of these values they concluded that this bacterium probably utilized formic and oxalic acids solely as sources of energy, not carbon.

These experiments of Korte and Engel appear to be the last in which a simple Dewar flask immersed in an accurately regulated thermostat was used as a calorimeter. It is surprising that they did so. Calvet's calorimeter would have been far better suited for the purpose, especially in that *H. vulgare* is such a slow-growing microorganism. Calorimetric experiments with autotrophic bacteria have always been very difficult to carry out meaningfully and successfully.

7.7. THE THEORETICAL ENTROPY CALCULATIONS OF MOROWITZ

In 1955 H. J. Morowitz[40] published an imaginative approach to the calculation of entropy changes accompanying cellular synthesis that involved order disorder considerations with respect to atomic configurations within a cell.

Of the very large number of atomic configurations, N, that are possible within a system consisting of a single cell, a much smaller number, L, corresponds to the possible configurations existing when the cell is alive. Assuming that all states (configurations) have equal probability, the quantity L/N represents the probability of the system being a living cell. The idea that this latter can be represented in terms of some function of the state of the system suggested the use of entropy and information to make quantitative evaluations of cellular order. Entropy and information can be represented as follows:

$$\Delta S = k \log N - k \log L \tag{7.3}$$

and

$$I = \log_2 N - \log_2 L \tag{7.4}$$

where ΔS represents the entropy, k the Boltzmann constant, and I the information. Using the concept of "information" has the advantage that no assumptions have to be made about equilibrium distributions of the stable states within the system.

Morowitz pointed out that certain experimental findings can simplify the system being studied. These involve the removal of water from the cells comprising the system. The practice of lyophilization removes over 95% of cellular water from microbial cells without killing them, and has long been used as a method of preservation and storage. The addition of water to lyophilized cells to reactivate them can be regarded as a random process which does not add information to the system, and it can be concluded that it is the nonaqueous part of living cells that constitutes the basic information unit. In addition, Morowitz reported his own experimentation together with W. Hansen in which dry *Bacillus subtilis* spores were demonstrated to survive exposure to a temperature of about 1.3°K for 5.5 hr, showing, in their opinion, that the basic information unit is independent of the rotational and vibrational states of the system. From these two observations Morowitz concluded that at least for computational purposes the living system can be specified as the nonaqueous portion of a cell at absolute zero.

The assumption was also made that $\log_2 L$ in Eq. 7.4 is negligible in comparison to $\log_2 N$, which is equivalent to assuming that there is only one state of the system in which it (the cell) is alive. Equation 7.4 then takes the form

$$I = \log_2 N \tag{7.5}$$

Because Morowitz regarded the system to be at absolute zero, only the coordination and chemical bonds of each atom needed to be considered, and a further assumption could then be made that each atom occupies the same volume, so that the cell can be considered as consisting of a number of boxes representing unit atomic volumes, each box containing one atom. As a first approximation the number of states of the system is then equal to the number of permutations of the number of atoms in the system contained in the same number of unit atomic volumes.

Morowitz then carried out a series of mathematical manipulations that are not at all clear in which he calculated the information in a structure the size of a small bacterium, assuming the mass of a single cell to be of the order of 10^{-13} g. Knowing the atomic composition of this single cell (source unidentified) in terms of the kinds and numbers of atoms it contains, the value if I was found to be 4.6×10^{10} bits. This can be compared with the limiting value of 10^{13} bits for a bacterial cell calculated in 1953 by Linschitz[41] by a completely different method making use of estimated entropy values. However, Linschitz did comment that if it is estimated that even under conditions of optimal growth a cell wastes 99% of its entropy supply in maintenance and metabolic activity, the value of I per cell might be reduced to 10^{10}–10^{11} bits. This latter range of values is similar to that also calculated in 1953 by Dancoff and Quastler[42] for a germ cell, based on structural models and genetic characteristics of cells.

Returning to Eqs. 7.3 and 7.5 Morowitz noted that entropy was related to information in the following manner:

$$\Delta S = -kI \ln 2 \tag{7.6}$$

He then considered the growth of bacteria in an isothermal calorimeter (not, this time, at absolute zero) and stated that accompanying the production of each bacterium there is a local increase in information I. There must therefore be a corresponding increase of at least ΔS somewhere else in the system, according to The Second Law of Thermodynamics. Provided that heat is transferred slowly from the growing bacteria to the sink (not specified), the entropy change will be of the order of $\Delta Q/T$. From this the maximum information content will then be

$$\frac{\Delta Q}{T} = -kI \ln 2 \tag{7.7}$$

Values for ΔQ had been measured previously with respect to *Bacterium coli* by Bayne-Jones and Rhees,[15] and Morowitz used their data to calculate that at 37.5°C ΔQ for a single cell had a value of 4×10^{-10} cal. Using Eq. 7.5, the minimum value for ΔQ based on the value for I obtained theoretically and corrected for the mass of a single *B. coli* cell (about 4×10^{-13} g dry weight) is about 1.2×10^{-10} cal. This indicated that the heat given off during the production of a single cell is about three times that required to account for the decrease in entropy. It could also be interpreted to mean that only about 30% of the heat of growth is the result of a local decrease in entropy, the remaining 70% being the result of other processes of metabolism, motility, or the maintenance of concentration gradients. Since the order of magnitude of the theoretically calculated value of ΔQ was the same as that obtained experimentally, Morowitz felt that the former was within a reasonable range.

Morowitz[43,44] subsequently published in 1960 a continuation of his theoretical studies in which he attempted to calculate the entropy changes accompanying cell production, making use of data accumulated by Roberts et al.[45] on biosynthesis in *E. coli*. Here he described the growth of a single *E. coli* cell by means of the following equation:

$$\begin{aligned}&(3.22 \times 10^9)C_6H_{12}O_6 + (1.94 \times 10^9)NH_3 + (1.69 \times 10^8)H_3PO_4\\&+ (4.8 \times 10^7)H_2SO_4 + (9.40 \times 19^9)O_2 \longrightarrow (1)\ E.\ coli\\&+ (9.55 \times 10^9)CO_2 + (1.48 \times 10^9)CH_3COOH\\&+ (2.11 \times 10^9)\text{end products} + (13.63 \times 10^9)H_2O \qquad (7.8)\end{aligned}$$

where the coefficients in parentheses represent numbers of atoms of the substance they precede. Physically, during the reaction represented by Eq. 7.6: (a) nutrients are taken in by a parent cell to form a daughter cell; (b) these nutrients are either anabolized into cellular material or catabolized and excreted as waste products; and (c) heat is given off during the whole process of metabolism and transferred to the isothermal reservoir. During the same reaction, if the Second Law of Thermodynamics is to hold, the entropy change for the whole system including the reservoir must be greater than or equal to zero. Morowitz considered cell production to consist of two linked processes, one being anabolism and the other catabolism. This linkage can be expressed by a parameter f, which represents the number of grams of waste product produced per gram of cells produced. Thus, if the mass of a single cell is m, fm g of material (unspecified) will be given off during the process of catabolism.

Four types of entropy changes were considered, these being:

1. The entropy decrease accompanying the condensation of nutrient molecules in the cell, according to the following approximation

$$\Delta S_c = -mR/M \log Cc/Cm$$

where ΔS_c represents the entropy decrease, m the mass of a single cell, R the gas constant, M the average molecular weight of the nutrients, Cc the concentration of the nutrient in the medium, and Cm the concentration of cells.

2. The entropy decrease accompanying the ordering of the nutrient materials into a cellular structure using a modification of Eq. 7.6, as follows:

$$\Delta S_0 = -kmNi \ln 2$$

where ΔS_0 represents the entropy decrease at absolute zero, k the Boltzmann constant, and i the ratio I/mN. This last Morowitz defined as a "normalized information density, where N is Avagadro's number."

3. The entropy change, which is probably positive, accompanying the conversion of nutrients to waste products during catabolism. If this entropy change per gram is designated as ΔS_w, then the total entropy change during the growth of a single cell will be $fm\Delta S_w$.

4. The entropy change, $\Delta Q/T$, accompanying the transfer of heat to or from the reservoir and represented by the following equation

$$\frac{\Delta Q}{T} = -\frac{(m\,\Delta Hc + fm\,\Delta Hw)}{T}$$

where ΔHc represents the enthalpy change per gram for cell synthesis and ΔHw that per gram of waste produced. Adding the four entropy changes gives

$$\Delta S = -\frac{mR}{M}\log\frac{Cc}{Cm} - mRi\ln 2 + fm\,\Delta S_w - \frac{(m\,\Delta Hc + fm\,\Delta Hw)}{T}$$

from which

$$f \geq -\frac{1}{\Delta Fw}\frac{RT}{M}\log\frac{Cc}{Cm} + RTi\ln 2 + \Delta Hc \tag{7.9}$$

According to Morowitz, the quantity *f* in Eq. 7.9 is a measure of the "minimum price that must be paid" to convert unorganized nutrient material into highly organized cellular material. To calculate *f* it is necessary to obtain some idea of the value of ΔHc, which Morowitz at this point identifies as the "enthalpy of formation" of a cell. To do this he split Eq. 7.8 into two equations, one representing anabolism (cellular synthesis) and the other metabolism (metabolic formation of waste products). The equation for anabolism is as follows:

$$\left.\begin{array}{l}(0.97\times 10^9)C_6H_{12}O_6 + (1.73\times 10^9NH_3 \\ +\ (0.196\times 10^6)H_3PO_4 + (0.048\times 10^9)H_2SO_4\end{array}\right\}\longrightarrow \begin{array}{r}1\ \textit{E. coli}\ \text{cell} \\ +\ (4.03\times 10^9)H_2O\end{array} \tag{7.10}$$

Again, the coefficients in parentheses represent numbers of atoms of the substance they precede. Although he did not include a formula representing the *E. coli* cell in Eq. 7.8, water was added to the right side to balance the equation. From data provided by Roberts et al.[41] on the macromolecular composition of *E. coli* cells, Morowitz[40] was able to determine both the type and number of individual covalent bonds within a single cell. This was also done with respect to the nutrients that form the cell. Then, using the available data on the heats of formation of these bonds he was able to calculate ΔH_c for the reaction represented by Eq. 7.10 as -682 cal g^{-1}. The formation of water as represented on the right side of this equation was considered by Morowitz to be a linked reaction, and by an "approximate method" that is completely obscure because it was not stated, he then dropped the value of ΔH_c to -125 cal g^{-1}. What he was apparently trying to do was to arrive at a system in which there were nutrients on the left side of an equation and only one *E. coli* cell on the right, although no such equation was presented nor could it have been written in a usual chemical form. What was probably used was an equation containing only types and numbers of covalent bonds.

Various values were then substituted into Eq. 7.9. *Cc* and *Cm* were taken as 250 g liter^{-1} and 5 g liter^{-1}, respectively, *M* as 100 g, and *T* as 300°K, making the term $RTM^{-1} \log CcCm^{-1}$ 23.4 cal g^{-1}. Based on his previously determined[34] information content of a single 10^{-13} g bacterial cell as 4×10^{10} bits, the term $RTi \ln 2$ equals 159 cal. ΔHc was taken as -124 cal g^{-1}, but it was pointed out that this quantity is very dependent on the nature of the nutrient medium and in a very rich nutrient medium probably approaches zero. As a first approximation Morowitz assumed that ΔFw is the free energy of glucose oxidation, equal to $-691{,}000$ cal mole^{-1} or -1860 cal g^{-1}. The result of this calculation was

$$f \geq \frac{(58.4)}{1860} = 0.03 \tag{7.11}$$

Because the quantity in parentheses in Eq. 7.11 is positive, *f* becomes positive, and this demonstrated to Morowitz the existence of linked reactions. Although the calculated value of *f* seems small, it was thought to be a minimum, the assumed value for ΔFw being high because the ΔF values for the other linked reactions are lower than that for glucose. Experimentally, *f* had a value of 4.95 rather than 0.03, in that for each gram of cell material which was synthesized 4.95 g of waste product (mostly CO_2 and H_2O) were given off. However, thermodynamically the minimum value for *f* is of the order of 0.01 of this value (see Eq. 7.11), indicating that "the process of *coli* duplication in minimal medium produces far more entropy than is required to satisfy the second law." (Morowitz,[40] p. 343.)

Morowitz's moderate conclusion from the results of his theoretical considerations was that such thermodynamic analyses can be useful, but sufficient data were not available. His method of counting up the number and distribution of covalent bonds in a single *E. coli* cell and in the quantities of glucose, ammonia, phosphoric acid, and sulfuric acid needed to synthesize such a cell, for the purpose of calculating total bond energies which could be used in his calculations, was most ingenious. On the other hand, the large number of assumptions which are made with respect to other calculations, some of them uncertain even as to the order of magnitude of the figures involved, gives one pause to wonder what the calculations really mean. Nevertheless, such calculations cannot be disregarded, and it may be that with better data and more accurate assumptions the methods Morowitz used might prove to be indeed useful.

7.8 SUMMARY

It is apparent from the discussion in this chapter that the most important developments during this period were technical.

Winzler and Baumberger's[11] adiabatic microcalorimeter was simple in design and construction and easy to operate, and as such had an elegance all its own. Large quantities of culture medium could be employed (up to about 100 ml), whereby suitable analyses could be carried out on the culture and on the medium after the experiment was terminated. Further, either aerobic or anaerobic experiments could be conducted. Finally, these investigators demonstrated clearly that their calorimetric system was easy to calibrate, and accurate to use. It did lack the sensitivity of Calvet's[32] compensating microcalorimeter, but this problem was eliminated by the ability of the experimenter to use larger quantities of experimental material, resulting in larger quantities of heat produced and a lesser need for a high sensitivity on the part of the calorimeter. Inherent in any adiabatic calorimeter is the change in the temperature of the system which occurs during an experiment. Correcting for this may be important if the heat production is used to calculate heats of formation at a standard temperature. However, for most biological measurements the temperature change is less than one degree, and any errors caused by this are likely of no consequence. They can be made negligible if the standard temperature (for example, 25°C) is bracketed by the experimental temperature change.

Calvet's[31] compensating microcalorimeter was not simple in design and in construction, although it was not difficult to operate. It possessed the advantages of an isothermal calorimeter. It was also extremely sensitive, making it useful for the measurement of small quantities of heat, and it was a remarkably accurate and stable instrument. However, it is in the nature of the Peltier effect that only small quantities of heat production can be compensated, and for this reason only small quantities of experimental material could be used. It may thus be more difficult to carry out analyses of the experimental material after an experiment has been terminated, simply because of the small amounts of substances involved. For example, Prat[37] used only 4 ml of culture suspension, about 25 times less than Winzler and Baumberger[11] used.

In retrospect, these technical advances have only served as stepping stones. Present-day microcalorimeters are not adiabatic, although they can easily handle the quantities of culture suspensions used by Winzler and Baumberger.[11] On the other hand, they do not employ the Peltier effect to compensate for the heat produced during a reaction. Rather, modern calorimeters seem to have evolved from the use of Calvet's microcalorimeter as a heat-flow instrument, rather than as a compensating instrument. This was because of the later availability of "thermoelectric devices" which simply drain off the heat produced in a calorimeter vessel as fast as it is produced, so that the actual temperature rise above the heat sink is never more than a few thousandths of a degree. In doing so they act as massive thermocouples, the electrical signal from which is proportional to the amount of heat transferred from the source to the sink. These calorimeters will be described in Chapter 9.

REFERENCES

1. M. Stephenson, *Bacterial Metabolism*, 2nd ed., Longmans, Green, and Co., London, 1939.
2. J. W. Hes, Zur Stoffwechselphysiologie von *Nitrosomonas, Rec. Trav. Bot. Neerland.* **34,** 233–277 (1937).
3. O. Meyerhof, Untersuchungen über den Atmungsvorgang nitrifizierender Bakterien. I. Die Atmung des Nitratbildners, *Pfluger's Archiv gesamte Physiol. Menschen Tiere* **165,** 353–427 (1916).
4. T. Y. Kingma-Boltjes, Untersuchungen über die nitrifizierender Bakterien, *Arch. Mikrobial.* **6,** 79–138 (1934).
5. L. Algera, Energiemessungen bei *Aspergillus niger* mit Hilfe eines automatischen mikro-kompensations-Calorimeters, *Rec. Trav. Bot. Neerland.* **29,** 47–163 (1932).
6. S. Winogradsky, Recherches sur les organismes de la nitrification, *Ann. Inst. Pasteur* **4,** 760–771 (1890).
7. D. H. Nelson, Isolation and characterization of *Nitrosomonas* and *Nitrobacter, Zentralbl. Bakteriol. Pasasitenkd. Infektionskr. Hyg. Abt. II,* **83,** 280–311 (1931).
8. H. Engel, Die Kohlenstoffassimilation des Nitritbildners, *Planta* **8,** 423–426 (1929).
9. H. Gaffron, Über den Stoffwechsel der schwelfrien Purpurbakterien, *Biochem. Z.* **260,** 1–17 (1933).
10. C. B. Van Niel, On the metabolism of the Thiorhodaceae, *Arch. Mikro.* **7,** 323–358 (1956).
11. R. J. Winzler and J. P. Baumberger, The degradation of energy in the metabolism of yeast cells, *J. Cell. Comp. Biol.* **12,** 183–211 (1938).
12. L. F. Fieser, A new absorbant for oxygen in gas analysis, *J. Am. Chem. Soc.* **46,** 2639–2647 (1924).
13. S. R. Hoover and P. E. Allison, The growth metabolism of *Rhizobium*, with evidence on the interrelations between respiration and synthesis, *J. Biol. Chem.* **134,** 181–192 (1940).
14. S. Yamagata, Über die elementare Zusammensetzung des Schimmelpilzkorpers, *Acta Phytochim.* **8,** 107–116 (1934).
15. S. Bayne-Jones and H. Rees, Bacterial calorimetry II. Relationship of heat production to phases of growth of bacteria, *J. Bact.* **17,** 23–40 (1929).
16. S. E. Wedburg and L.F. Rettger, Factors influencing microbial thermogenesis, *J. Bact.* 725–743 (1941).
17. L. H. James, L. F. Rettger, and C. Thom, Microbial thermogenesis. II. Heat production in moist organic materials with special reference to the part played by microorganisms, *J. Bact.* **15,** 117–141 (1928).
18. O. F. Edwards and L. F. Rettger, The relation of certain respiratory enzymes to the maximum growth temperatures of bacteria, *J. Bact.* **34,** 489–515 (1937).

19. A. L. Lavoisier and P. S. LaPlace, Mémoire sur la Chaleur, *Mémoires de l'Academie Royale*, 355 (1780).
20. K. von Körosy, Mikrokalorimeter zur Bestimmung der Wärmeproduktion von Bakterien, *Z. physiol. Chem.* **86,** 383–400 (1913).
21. A. Tian, Microcalorimètre à compensation par effets Peltier et Joule, *C. R. Hebd. Séances Acad. Sci.* **178,** 705–707 (1924).
22. Mme. Berenger-Calvet, Calorimétrie par compensation à l'aide des effects Peltier et Joule: Application à l'étude de chaleurs de dissolution et à la calorimétrie des réactions lentes de faible thermicité, *J. Chim. Phys.* **24,** 325–345 (1927).
23. E. Calvet, Mesures thermochimiques directes en chimie organique. Première Partie. Appareils et méthodes utilisés pour la mesure des chaleurs de saponification des amides et pour la mesure des constantes de vitesse de saponification. Chapitre Premier. Mesure des chaleurs de saponification des amides. A. Le microcalorimètre de *M.* A. Tian et son emploi, *J. Chim. Phys.* **30,** 5–12 (1933).
24. H. Prat and E. Calvet, Application de la microcalorimétrie à l'analyse des stades initiaux de la germination, *C. R. Séances Soc.Biol. Fil.* **138,** 662–663 (1944).
25. H. Prat and E. Calvet, Sur le dégagement de chaleur manifeste au cours des premiers stades de l'hydration des graines, *C. R. Hebd. Séances Acad. Sci.* **218,** 944–945 (1944).
26. E. Calvet, J. Fricker, and H. Prat, Mesure de la thermogénèse des bactéries, *C. R. Hebd. Séances Acad.Sci.* **220,** 797–798 (1945).
27. A. V. Hill, A new form of differential microcalorimeter, for the estimation of heat production in physiological, bacteriological, or ferment actions, *J. Physiol.* **43,** 261–285 (1911–12).
28. H. Prat, E. Calvet, and J. Fricker, Etude microcalorimétrique du développement des cultures bactériennes et de la germination des graines, *Rev. Canad. Biol.* **5,** 247–250 (1946).
29. A. Tian, Thermostats à enceintes multiples, *C. R. Hebd. Séances Acad. Sci.* **174,** 1453–1455 (1922).
30. A. Tian, Recherches sur les thermostats. Contribution à l'étude du ŕeglage—Thermostats à enceintes multiples, *J. Chim. Phys.* **20,** 132–166 (1923).
31. E. Calvet, Nouveau micro-calorimètre différentiel et à compensation, *C. R. Hebd. Séances Acad.Sci.* **226,** 1702–1704 (1948).
32. E. Calvet and J. Fricker, Sur la thermogénèse des Bactéries, *C. R. Hebd. Séances Acad. Sci.* **226,** 1846–1847 (1948).
33. E. Calvet and H. Prat, *Microcalorimétrie: Applications physico-chimiques et biologiques*, Masson et Cie., Paris, 1956.
34. E. Calvet and H. Prat, *Récents Progrès en Microcalorimétrie*, Dunod, Paris 1958.
35. E. Calvet and H. Prat, *Recent Progress in Microcalorimetry.*, ed. and trans. by H. A. Skinner, Pergamon Press, Ltd., Oxford, 1963.
36. Z. Dermoun, R. Bloussand, D. Cotten, and J. P. Belaich, A new batch calorimeter for aerobic growth studies, *Biotech. Bioeng.* **27,** 996–1004 (1985).
37. H. Prat, Observations sur la thermogénèse bactérienne, *Rev. Canad. Biol.* **12,** 19–34 (1953).
38. I. Korte and H. Engel, Energiemessung an *Hyphomicrobium vurgare* Stutzer and Hartleb, *Arch. Mikrobiol.* **21,** 248–254 (1955).

39. O. Meyerhof, Untersuchungen über den Atmungsvorgang nitrifizierender Bakterien. I. Die Atmung des Nitratbildners, *Pfluger's Archiv für gesammte Physiologie* **164,** 353–427 (1916).
40. H. J. Morowitz, Some order-disorder considerations in living systems, *Bull. Math. Biophys.* **17,** 81–86 (1955).
41. H. Linschitz, The information contents of a bacterial cell, in H. Quastler, Ed., *Information Theory in Biology.*, Univ. of Illinois Press, Urbana, 1953, pp. 251–262.
42. S. M. Dancoff and H. Quastler, The information content and error rate of living things, in H. Quastler, Ed., *Information Theory in Biology.*, Univ. of Illinois Press, Urbana, 1953, pp. 263–273.
43. H. J. Morowitz, Cells: Living; Molecular Physics., in O. Glasser, Ed., *Medical Physics*, Vol. III, The Yearbook Publishers, Chicago, 1960, pp. 111–113.
44. H. J. Morowitz, Some consequences of the application of the second law of thermodynamics to cellular systems, *Biochim. Biophys. Acta* **40,** 340–345 (1960).
45. R. B. Roberts, P. H. Ableson, D. B. Cowie, E. T. Bolton, and R. J. Britten, *Studies of Biosynthesis in Escherichia coli.*, Carnegie Institute of Washington Pub. 607, Washington, D.C., 1957.

8

THE STUDIES BY BATTLEY ON THE GROWTH OF *SACCHAROMYCES CEREVISIAE*

Following the 1931 review by Wilson and Peterson[1] on "The Energetics of Heterotrophic Bacteria," nothing appeared in the literature for the next 30 years in which an investigation was made into the free-energy changes accompanying the growth of a microorganism on a single substance as the sole source of carbon and energy. This may have been because during this same period a vast amount of work was being done with respect to the biochemistry of intermediary metabolism, and $\Delta G^{\circ\prime}$ values were calculated for the many enzymatically catalyzed reactions that were discovered as soon as these latter became known. It was natural that some attempt should have been made to sum up the sequential chemical events and the accompanying energy exchanges which comprised biological processes at a higher level of integration. This culminated in the classical paper by Burton and Krebs[2] (see also Burton[3]) in which they were able to list all the sequential reactions of glycolysis plus the tricarboxylic acid cycle and of alcoholic fermentation together with their accompanying free-energy changes. Although they did not actually do so, it is possible to use their data to sum up the free-energy changes of these individual reactions almost exactly to equal the free-energy change accompanying glucose oxidation or fermentation minus the energy conserved in ATP. This was a remarkable achievement, utilizing the biochemical contributions provided by a great many investigators during the preceding 50 years. It clearly demonstrated that $\Delta G'$ values calculated using thermodynamic equilibrium constants (K_{eq}) and oxidation-reduction potentials (E_0') for the many sequential reactions of glycolysis plus the tricarboxylic acid cycle and of fermentation could be added together to give the same figures obtainable with the completely different method of using the free energies of formation of the reactants and products. This was a triumph of biochemical thermodynamics which surely comes close to that of being able to calculate absolute entropies using either thermal or statistical methods. On the other hand, no one could believe that it would ever become possible to add up the free-energy changes accompanying the individual reactions comprising the process of cellular growth, beside the complexities of which those of glycolysis and the tricarboxylic acid cycle faded into insignificance. Obviously, if the free-energy change accompanying the process of growth were to be determined, it would have to be by some more simple means than the addition of all the component reactions.

8.1. THE INITIAL EXPERIMENTS

The studies published in 1960 by E. H. Battley[4–6] on the thermodynamics of growth of *Saccharomyces cerevisiae* were motivated initially by the chance observation of the discrepancy between the amounts of growth obtained aerobically or anaerobically when this facultative yeast was grown on glucose. The oxidation of glucose to carbon dioxide and water and its fermentation to carbon dioxide and ethanol are represented by the following equations.

Glucose oxidation:

$$C_6H_{12}O_6(aq) + 6O_2(aq) \longrightarrow 6CO_2(aq) + 6H_2O(1) \qquad (8.1)$$

Glucose fermentation (*alcoholic*):

$$C_6H_{12}O_6(aq) \longrightarrow 2CO_2(aq) + 2CH_3CH_2OH(aq) \qquad (8.2)$$

Using the values for the thermodynamic properties listed in Appendix 1, the free-energy changes accompanying the reactions represented by Eqs. 8.1 and 8.2 are -698 and -53 kcal mole^{-1}, respectively. It could thus reasonably be expected that were this yeast to be grown aerobically on 1 mole of glucose there would be observed approximately $-698/-53$, or 13.2 times more growth aerobically than when the same organism were grown anaerobically on the same quantity of glucose. However, when this experiment was actually carried out, on the basis of turbidimetric measurements there was observed only 3.4 times more growth aerobically than anaerobically. The reason for this was subsequently investigated.

8.2. GROWTH-REACTION SYSTEMS

The application of classical, closed-system thermodynamics to the phenomenon of microbial growth requires that a reaction system be established having an initial and final state. This is necessary in order to have some basis for determining any change in the thermodynamic properties of the system that may occur as it passes from one state to the other. The initial and final states chosen are arbitrary, the only criterion being that if growth is to be studied it must be included as something which happens within the system.

The word "growth" is a generalized term used to symbolize a process consisting of a great many interrelated phenomena. It is only when the concept of growth is restricted to some small set of these phenomena, such as growth in the absence of cellular and tissue differentiation, that the variables included with them can be reduced to a minimum and an arbitrary definition of growth can be made. This is possible when unicellular microorganisms are used as tools with which to carry out a study. In this case, growth can be considered as an increase in the mass of cellular material, exclusive of storage products, so that only the structural part of each cell is included. Battley[7] more recently called this latter the "fabric" of a cell, by analogy with the architectural term denoting the structural part of a building *exclusive* of the inside furnishings. As was pointed out by Duclaux[8] (p. 328) in 1900, storage products such as glycogen constitute an internal substrate as opposed to an external substrate. Since this former can also be utilized as a source of carbon and energy, it can disappear from the cell and thus does not constitute a part of the permanent cellular structure, or fabric.

The above consideration enables growth to be regarded in molecular terms in that the mass of cellular substance produced during growth can be represented as a formula in a chemical reaction equation, as was first done by Tamiya (see Chapter 6). With equations of this kind it ought to be possible to use the principles of chemical thermodynamics to determine the free-energy, enthalpy, and entropy changes accompanying the microbial growth process. To express a cellular mass as a chemical formula it must be ascertained that it represents a constant entity. Ethyl alcohol has the same molecular formula under all physiological conditions. However, it is not necessarily true that a microbial cell has the same empirical composition under different environmental conditions and as Spoehr and Milner[9] have shown with *Chlorella*, may vary widely depending on how a microorganism is grown. This empirical composition is a function largely of the genetics of the microorganism, the chemical composition of the environment, and the physical factors that affect the environment such as temperature, osmolality, and radiant energy. For example, during conditions unfavorable for growth microbial cells may accumulate storage products as a result of oxidative or fermentative assimilation (see Chapter 5). Depending upon the nature of the storage products such an accumulation may result in an ash-free, empirical cellular composition which is different from that of exponentially grown cells. Similarly, growth under unfavorably high temperature conditions may give rise to involution forms which have a different appearance and composition from that of normal cells. Thus, although the initial and final states chosen for the study of microbial growth are theoretically arbitrary, there is the practical consideration that the constancy of composition of the cellular substance should be preserved. As emphasized by Monod,[10] this constancy can only occur during the exponential phase of microbial growth. If the logarithm of the increase in cellular mass during growth is plotted against time an approximately linear relationship is observed. As long as this relationship does not change the cells may be regarded as remaining constant in their average chemical composition for any length of time. It is not practical, however, to carry out experiments of great length and so the duration of an experiment must be limited. This can best be done by limiting the source of energy for growth. It can reasonably be expected that under conditions where exponential growth is possible it will, in fact, occur and will continue until all the substrate (i.e., the source of carbon and energy) is utilized. In other words, the constancy of composition of the cellular material will persist until the substrate becomes exhausted, at which point the slope of the line representing the (logarithm of the increase in mass)/time relationship ceases to be positive and falls to zero. This point is the practical limit in time for a microbial growth experiment.

8.3. INITIAL AND FINAL STATES OF GROWTH-REACTION SYSTEMS

With the above considerations in mind it is possible to establish the initial and final states of what Battley[4] called a *growth-reaction system*, or more simply, a *growth reaction*. He defined the initial state as one in which all the potentialities for unlimited growth exist with respect to energy and materials. The final state was defined as one in which the potentiality for growth no longer exists because of the exhaustion of the substrate. In order to study a growth-reaction system it must be separated intellectually and physically from its environment. The most convenient way of doing this is to establish a closed system within which growth occurs. A batch culture for the most part represents such a system provided that the substrate concentration is kept small and that only small amounts of gas are produced or absorbed. Intellectually, if it can be conceived that the concentrations of all soluble reactants and products are at "hypothetical infinite dilution" there are no volume changes, and thus no thermodynamic problems with pressure–volume energy changes. Practically, these latter are usually insignificant.

Battley further described the initial state of a growth reaction as the culture medium plus at least one living cell. The amount of substrate that is included in the initial state is an arbitrary matter. Since the duration of an experiment is of practical importance, there is good reason to keep the quantity of substrate as small as possible. This confers the additional advantage that, with a small quantity of substrate in a correspondingly low concentration, the values of its thermodynamic properties in a defined aqueous standard state are very closely approximated. Except for the nitrogen source the minerals and trace elements that form a part of the culture medium, while necessary for growth, do not enter significantly into the material exchange of the growth process. The initial state may also include vitamins and organic growth factors as a part of the culture medium, but their concentrations compared to that of the substrate are so low as to make their material contributions to the total mass of substance involved in the growth process completely negligible. Similarly, if the inoculum is kept low the initial amount of cellular material can be neglected. The final state was further defined as the residual culture medium devoid of any substrate but including the products of the growth reaction, namely, the microbial cells which have been grown and the by-products of the reaction such as carbon dioxide, water, and/or organic substances, as is the case with fermentations.

In order to carry out studies on the energetics of microbial growth it is not sufficient to define the initial and final states as they have been defined above, in terms of the material balance of the process of growth. It is further necessary to be able to define the initial and final states thermochemically. The most convenient way of doing this is to use a single, chemically pure substance as a source of carbon and of energy, the thermodynamic properties of which are known (see Chapter 1). The thermodynamic properties of the oxygen which is a reactant in aerobic growth reactions are also known, as are those of ammonia and nitrate, the two most common nitrogen sources. The other substances that are present at the beginning of a growth reaction in the form of minerals, trace elements, or vitamins either do not have any net effect on its energetics or are present in quantities so small that their mass and thus their thermodynamic properties can be ignored. This latter is also true for the cells that are used as an inoculum, provided that their quantity is kept sufficiently small that no detectable growth occurs for the first few hours of an experiment. The final state must also be thermochemically definable. Thus, except for the cells that are grown it is necessary to know the thermodynamic properties of the products of the reaction and for this, of course, it is necessary to know what these latter are. In the cases of many aerobic growth reactions it is known that the only products of the reaction other than the cells are carbon dioxide and water, which have been thermochemically described. On the other hand, many fermentative growth reactions have a number of organic substances as products of the reaction, for some of which, even if they could be detected and analysed, the thermodynamic properties have never been determined. In such cases it is impossible to define the final state. For these reasons, the microbial tools with which to conduct experiments on the energetics of growth must be carefully chosen so that the thermodynamic properties of all the reactants and products of the growth reaction other than the cells are known. Unfortunately, this somewhat limits the potentialities for study to a restricted number of microorganisms.

8.4. THE NUTRIENT CULTURE MEDIUM

The composition of any given culture medium depends on the nature of the microorganism to be grown in it. The initial question Battley asked was why the ratio of aerobic/anaerobic yeast growth on glucose was so low, and to attempt to answer this it was necessary to devise a strictly defined medium that could be used both under aerobic and anaerobic conditions. By 1960 the nutritional requirements for yeasts had been well established, and the defined culture medium originally used by him contained all the minerals, trace elements, and vitamins thought to be necessary. However, when this medium was used with glucose as the sole source of carbon and energy and inoculated with a very small quantity of cells so as not to introduce any appreciable extraneous quantity of organic material into the medium, *S. cerevisiae* grew well aerobically, but not at all on the same substrate under rigorously anaerobic conditions. In addition to the above-mentioned vitamins, anaerobically-grown yeasts also have a requirement for steroids. This was first discovered by Devloo[11] and further investigated by Andraesen and Stier.[12] The need for steroids seems to apply only to anaerobic growth, as made evident from the data of Maguigan and Walker[13] which indicate that steroid synthesis in yeasts takes place only under aerobic conditions. When Battley incorporated ergosterol into the culture medium it became possible to obtain strictly anerobic growth.

Since the intention of Battley's investigations was to find out something about the energetics of microbial growth, in addition to a substrate such as glucose which could be metabolized both aerobically and anaerobically two other substrates were chosen, one of which was more and the other less reduced than carbohydrate. These were acetic acid, supplied to the medium as sodium acetate, and ethanol, both of which could only be metabolized aerobically. The choice of all three substrates was also predicated on the fact that they had been previously defined thermodynamically and that the products of their metabolism, which were carbon dioxide and ethanol for glucose fermentation, and carbon dioxide and water for the aerobic oxidation of glucose, acetic acid, and ethanol, had also been previously thermochemically defined. The nitrogen source in the culture medium was ammonia, and oxygen gas served as the final electron acceptor for the aerobic growth-reaction studies. For both of these latter substances the thermodynamic properties were also known.

One of the variables usually controlled by means of the culture medium is the pH. To minimize pH changes to ±0.1 pH unit Battley incorporated 0.3% phosphate into the medium, considerably more than required for growth. It is possible that this had some effect on the growth of *S. cerevisiae.* However, since the same medium was used for all growth reactions the amounts of growth observed on the three substrates could still be considered relative.

8.5. THE NECESSITY OF ESTABLISHING THAT THE AMOUNT OF GROWTH IS LIMITED ONLY BY THE QUANTITY OF SUBSTRATE

Because growth represents the interaction of an organism with its environment, it is useful to know as much as possible about the nature of this interaction. Ideally, it should be established that nothing in the environment is limiting the amount of growth that can be obtained from a given quantity of substrate. This is usually the case when microorganisms are growing at their maximum rate. In general, there are two kinds of limitations to the rate of growth. One of these is temperature. Microorganisms have an optimal temperature at which they grow the fastest on a given substrate, all other conditions being equally near optimal. Even if all other conditions are optimal, at other temperatures growth will proceed at a slower rate the farther the temperature is from the optimum until at the maximal and minimal temperatures for growth there will be none at all. Under temperature-restricted conditions microorganisms may still be able oxidatively or fermentatively to assimilate the substrate and to form storage materials, but in these cases an increase in the mass of the cells will not be due entirely to the synthesis of cellular fabric. Concentration effects also limit the rate of growth in many ways. For example, as a lactic acid bacterium ferments glucose the formation of lactic acid will lower the pH, causing the organism to grow more slowly until a pH is reached at which growth will stop completely. Conversely, a bacterium growing aerobically on sodium acetate will utilize the organic part of this substrate, leaving the sodium ion in solution with a resultant rise in pH which can be equally growth-rate limiting if not controlled. If a micronutrient essential for growth becomes exhausted from the medium by incorporation into cellular material, growth will cease but fermentative or oxidative assimilation may continue. This will result in the growth rate dropping to zero, however, the mass of cellular material will continue to increase due to the accumulation of storage products. The growth rate of an aerobic microorganism may also become limited by the rate at which oxygen enters the medium after a certain cell density has been achieved. If oxidative assimilation takes place faster than the cells can utilize the substrate in the process of growth, accumulation of storage products can be expected. Even the substrate can be growth-rate limiting. This commonly occurs in chemostat experiments in which a given microorganism is cultured which has a low affinity for the substrate and thus a saturation constant, K_s, which has a high value. For this reason batch cultures rather than chemostat cultures may be preferable for studies on microbial energetics. On the other hand, if the concentration of substrate were growth-rate limiting, would fermentative or oxidative assimilation not be expected, and therefore possibly no accumulation of storage products? Even at the present time there seem to be no data in the literature indicating to what extent, if at all, the composition of the fabric of the cells changes with increasing substrate concentration limitation. Lacking such knowledge, it seems more prudent to use batch rather than continuous cultures.

Battley checked for the possibility of growth limitation by adding increasing concentrations of individual substrates to a unit amount of medium at an optimal temperature of 30°C and an optimal pH at 6.5, and inoculating each of these with a small number of cells so that the initial concentration of these latter was effectively zero. Each culture then grew to a density limited by the initial concentration of substrate. His results are shown in Fig. 8.1. If growth, as determined by whatever criterion, is plotted against substrate concentration a linear relationship establishes that the only thing limiting the amount of growth is the quantity of substrate. Optical density was used to estimate the relative quantity of cells. Figure 8.1 shows that provided Battley's experiments were carried out within the range of substrate concentrations exhibiting this linear relationship to the quantity of growth, nothing would limit the amount of growth except the quantity of substrate. All the requisite experimental conditions are met in experiments of this type. The optimum temperature for growth can be maintained constant by means of a thermostat. The flasks are

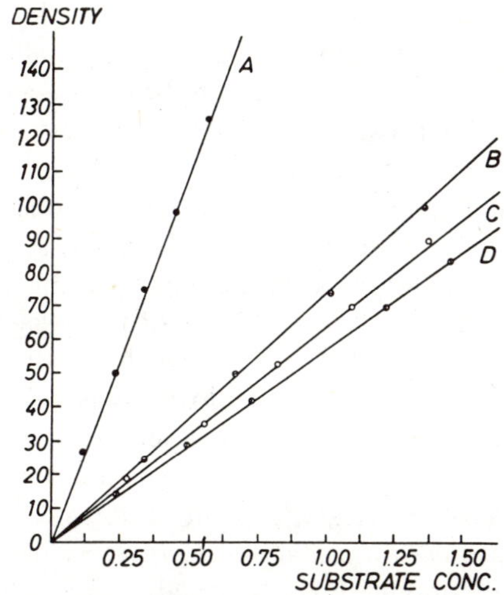

FIGURE 8.1. Graph of cell suspension optical density in Klett units plotted against substrate concentration in c*M* for the growth of *Saccharomyces cerevisiae*, illustrating the nonlimiting nature of the culture medium. A, aerobic growth on glucose; B, aerobic growth on ethanol; C, anaerobic growth on glucose; D, aerobic growth on acetic acid. Reprinted from Reference 4 with permission.

not sealed and thus the experiment runs at constant atmospheric pressure. The use of a very small inoculum ensures that by the time growth begins to be detected the cells will be in the exponential phase of growth and will remain there until the substrate has become exhausted. The optimum pH can be satisfactorily maintained by a buffer system within the culture medium. And finally, as illustrated by Fig. 8.1, whether or not the culture medium is in any way limiting can be checked.

8.6. THE ANALYSIS OF REACTIONS IN WHICH *S. CEREVISIAE* WAS GROWN ANAEROBICALLY ON GLUCOSE AND AEROBICALLY ON GLUCOSE, ETHANOL, AND ACETIC ACID

Whereas Tamiya (see Chapter 6) constructed equations representing cellular synthesis and respiration by means of his ingenious "theory of the combustion quotient," Battley analyzed the gas exchange and the production and composition of cells and organic substances formed from a given quantity of a known substrate, all other reactants or products being determinable by difference. The concentrations of substrates used were not more than 0.007 *M*. The oxygen consumed and the carbon dioxide produced during growth were determined by inoculating a small number of cells into 2 ml of sterile culture medium contained in "Warburg" vessels (microrespirometer flasks) and measuring the exchange of these two gases by constant volume manometry as described in general by Umbreit, Burris, and Stauffer,[14] but using the method of Schatz[15] as modified by Baalsrud and Baalsrud[16] and by Clayton.[17] This latter was necessary so that the cells could be grown in the absence of KOH in the center wells of the Warburg vessels. Carbon dioxide is essential for the growth of microorganisms. In the presence of KOH, when very small inocula are used carbon dioxide absorption prevents growth. During the initial stages of such manometric experiments no gas exchange could be observed because of the small number of cells which were beginning to grow, so that by the time that gas exchange was detectable the cells were presumably already in the exponential phase of growth, as shown by a plot of the logarithm of gas pressure against time. This is shown in Fig. 8.2, in which the first part of the curve is due only to the production of carbon dioxide gas. The amount of cellular material formed during growth was determined gravimetrically, and the ash-free, empirical formula for the cells by combustion analyses for carbon, hydrogen, nitrogen, and ash, with oxygen being determined by difference. Little difficulty was encountered with analyses involving aerobic growth reactions and good balances were obtained. However, after anaerobic growth on glucose it was discovered that only 77.0% of the amount of carbon dioxide predicted on the basis of a complete alcoholic fermentation of the available glucose was produced, and only 65.0% of the ethanol. When the amount of carbon represented by these quantities was added to that found in the cells that were also produced during the growth reaction, there was still a marked deficit in the carbon balance. This could only mean that there were fermentation products still unaccounted for. A series of analyses finally showed that the missing carbon was present in solution in the culture medium as a mixture of glycerol

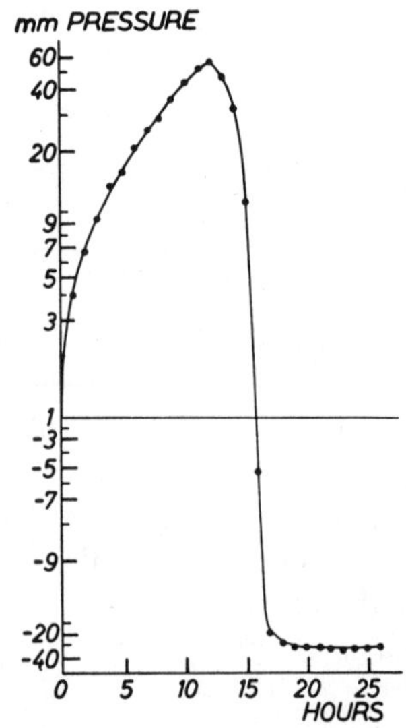

FIGURE 8.2. Graph showing total gas pressure in mm Brodie's solution plotted against time for the growth of *Saccharomyces cerevisiae* aerobically in a Warburg vessel on 5.555 μ*M* of glucose. Reprinted from Reference 4 with permission.

and α-glycerophosphate. Which substance was actually the product of the fermentation was uncertain. Neuberg and Karczag[18] had shown that fresh suspensions of yeast are able to catalyze the hydrolysis of α-glycerophosphate, and the immunological evidence of Sevag[19] indicated that glycerophosphatase is located, at least in part, at the surface of yeast cells. Further, the studies of Meyerhof and Green[20] demonstrated that glycerophosphatase activity is freely reversible. In either case, when the amount of carbon from these two substances was added to that of the cells, carbon dioxide, and ethanol which were produced, a good carbon balance was also obtained for the growth reaction during which glucose was fermented.

During the experiments with the aerobic glucose growth reaction it was found in complete confirmation of Brown's[21] original observation in 1892 and Swanson and Clifton's[22] subsequent observation of growing cells in 1948, that in the presence of oxygen the initial stage of the aerobic utilization of glucose by growing cells still consists of a fermentative growth reaction. This was true even in the presence of a gas phase consisting of pure oxygen. The anaerobic growth reaction was then followed by the aerobic utilization of the products of glucose metabolism which were formed anaerobically. This is shown in Fig. 8.2.

8.7. GROWTH REACTION EQUATIONS

Constructing equations representing microbial growth, which Battley called *growth-reaction equations*, required arbitrary decisions as to how to represent the various reactants and products. In what form, for example, does a cell receive the nitrogen source at the cell surface, when it first enters into the reaction? Since $NH_4^+(aq)$ and $NH_3(aq)$ both exist in an equilibrium determined by the pH of the culture medium, is it $NH_4^+(aq)$ or $NH_3(aq)$ which should be used in the growth-reaction equation? Does oxygen gas enter the cell as $O_2(g)$, or as $O_2(aq)$? Does carbon dioxide interact with the cell as $CO_2(g)$, $CO_2(aq)$, or $HCO_3^-(aq)$? The general rule Battley followed was that except for the cells all substances participating in a growth reaction should be considered to be in their aqueous standard states that are more likely to pass through the cell membrane in an uncharged condition. For this reason $O_2(aq)$, $CO_2(aq)$, and $NH_3(aq)$ are the forms of these substances that were used in his growth-reaction equations. Such equations are thus slightly less complex and long. It was also for this reason that the mixture of glycerol and α-glycerophosphoric acid found in the culture medium after anaerobic growth on glucose is represented simply as glycerol. The representation of the cellular material formed during a growth reaction is also arbitrary. Cells are not substances in true solution. For the most part they behave as insoluble precipitates which cannot reversibly go into solution. On the other hand, protoplasm consists at least in part of highly hydrophilic colloids, so that for the living cell there is the absolute necessity for the presence of water. Thus, there is no satisfactory standard state for cells. In growth-reaction equations the cellular "state" is probably best represented by the suffix (cells).

The cellular material produced during a growth reaction was originally represented by Tamiya (see Chapter 6) as the relative proportions of carbon, hydrogen, oxygen, and nitrogen that contribute to its *ash-free* dry weight. The only other element that contributes appreciably to the ash-free dry weight is sulfur, and this is usually present in considerably less than 1%. There is little reason to change Tamiya's original representation except in the actual formulation. Tamiya represented cellular material by dividing the ash-free percentages of carbon, hydrogen, oxygen, and nitrogen by their respective atomic weights and multiplying by 100. An example is the representation of *Aspergillus oryzae* cells grown aerobically on glucose with ammonium chloride as the nitrogen source, taken from Table 6.1, which has the formula $C_{409}H_{717}O_{233}N_{46}$. Hoover and Allison[23] represented the cellular substance of *Rhizobium meliloti* by the formula $C_{19}H_{32}N_2O_9$. In doing so they sacrificed a greater accuracy of representation to the convenience of dealing with small, whole numbers. Battley adopted a different form of representation in that the subscripts for the respective elements in the empirical formula were divided through by the subscript for carbon, giving a formula with only one carbon atom. Thus, if the formula above for *A. oryzae* were divided through by 409, the formula $CH_{1.753}O_{0.569}N_{0.112}$ results. Battley called this a *unit-carbon formula*. It is equally accurate a representation as that of Tamiya, and possesses the advantage that quantities of growth can easily be compared simply by comparing the coefficients in front of these formulas, especially for the same microorganism grown on different substrates. This is possible because carbon forms the greatest percentage by weight of any of the four elements in the empirical formula, and because this percentage is very nearly the same for all cells of the same kind.

Battley's equations representing the growth of *S. cerevisiae* anaerobically on glucose and aerobically on acetic acid and on ethanol are given in Table 8.1.

TABLE 8.1. Equations representing the growth of *Saccharomyces cerevisiae* anaerobically on glucose (1), and aerobically on glucose (2), ethanol (3), and acetic acid (4)

1.	$C_6H_{12}O_6(aq) + 0.12NH_3(aq)$	$\longrightarrow$	$0.59CH_{1.75}O_{0.45}N_{0.20}(cells) + 1.54CO_2(aq) + 0.43C_3H_8O_3(aq) + 1.30CH_3CH_2OH(aq)$
2.	$C_6H_{12}O_6(aq) + 0.33NH_3(aq) + 3.84O_2(aq)$	$\longrightarrow$	$1.95CH_{1.71}O_{0.46}N_{0.17}(cells) + 4.09CO_2(aq) + 4.73H_2O(l)$
3.	$CH_3CH_2OH(aq) + 0.15NH_3(aq) + 1.82O_2(aq)$	$\longrightarrow$	$1.03CH_{1.72}O_{0.41}N_{0.15}(cells) + 0.97CO_2(aq) + 2.31H_2O(l)$
4.	$CH_3COOH(aq) + 0.09NH_3(aq) + 1.35O_2(aq)$	$\longrightarrow$	$0.62CH_{1.62}O_{0.44}N_{0.15}(cells) + 1.38CO_2(aq) + 1.65H_2O(l)$

Source: Reference 4.

8.8. THE ADDITIVITY OF GROWTH-REACTION EQUATIONS

An inspection of Eq. 1 in Table 8.1 shows a most unusual aspect of the reaction represented by this growth-reaction equation. It is usually thought that the alcoholic fermentation of glucose results only in the production of carbon dioxide and alcohol, although small amounts of other substances were detected by Pasteur[24] in his efforts to achieve a complete fermentation balance. When, for example, he fermented a solution containing 9.998% pure sugar with 2.026 g of yeast paste, he recovered 0.34 g of glycerol and 0.065 g of succinic acid amounting to 3.4 and 0.6% of the weight of the initial sugar, respectively, these values being common for an experiment of this kind. Thus, Pasteur concluded that the Gay-Lussac equation for the alcoholic fermentation of sugar by yeast did not exactly hold, and that glycerol and succinic acid were normal products of the fermentation. However, since the medium in which the yeast was suspended contained nothing but sugar, Pasteur was not working with growing yeast cells in these experiments. If any growth did occur, it could only have done so using resources other than sugar present in the yeast cells at the time of inoculation, such as an amino acid pool, minerals, trace elements, vitamins, and so on. As shown in Eq. 1 in Table 8.1, during anaerobic growth on glucose a far greater quantity of glycerol (and/or α-glycerophosphoric acid) is formed than was ever observed previously except when suspensions of yeast were incubated in the presence of sugar and sodium bisulfite (Connstein and Lüdecke,[25] Fulmer, et al.[26]). Under these latter, extremely nonphysiological conditions the amount of glycerol produced is 20–30% of the sugar used. On the other hand, the conditions for growth of *S. cerevisiae* were selected by Battley as being optimal. Why is so much glycerol produced during normal growth? The reason is probably that, given the restraints with respect to the kinds of molecules that can be produced during glycolysis, glycerol comes closest to having the degree of reduction required to balance that of the cells also produced, the composition of which can be regarded as being fixed. If this is so, it should be possible to "ferment" the cells plus the glycerol to produce the quantities of carbon dioxide and ethanol which are not produced during the growth reaction, relative to the nonconservative reaction. That this can be done is demonstrated by the following equation.

$$\left.\begin{array}{l} 0.59CH_{1.75}O_{0.45}N_{0.20}(\text{cells}) \\ +\ 0.43C_3H_8O_3(aq) \end{array}\right\} \longrightarrow \begin{array}{l} 0.46CO_2(aq) + 0.12NH_3(aq) \\ +\ 0.70CH_3CH_2OH(aq) \end{array}$$

The material balance of this equation is satisfactory considering that the coefficients and subscripts are expressed to only two places.

It is reasonable to expect, if we are to believe in Hess's Law, that it should be possible to add the equation representing the growth of *S. cerevisiae* anaerobically on glucose and the equations representing aerobic growth on the organic products of the anaerobic growth reaction other than the cells, to equal the equation representing aerobic growth on glucose. Equation 3 in Table 8.1 shows that *S. cerevisiae* can grow aerobically on ethanol, but attempts to grow it aerobically on glycerol were only successful after a long period of incubation, during which the yeast apparently became slowly adapted to this substrate. It is evidently not simultaneously adapted to *growth* both on ethanol and on glycerol, preferentially growing on the former while simply oxidizing the latter. Figure 8.2 illustrates the gas exchange accompanying the growth of *S. cerevisiae* aerobically on glucose. The initial period of fermentation during which only carbon dioxide gas is produced is followed by an uptake of oxygen gas at a far greater rate than the production of carbon dioxide gas. This conversion from an anaerobic to an aerobic type of metabolism can be considered a true diauxy, since the yeast is effectively using two different substrates, that is, glucose (fermentatively) and ethanol (aerobically). There is no actual growth aerobically on glucose. But, there is also no diauxy observed with respect to gas exchange as a result of growth on ethanol and glycerol. Thus, although both substances are being rapidly oxidized after fermentative growth has ceased, it is apparently only the energy from the oxidation of ethanol that is used for aerobic growth. The energy from the oxidation of glycerol seems to be completely lost. This is made evident by the addition of the equation representing the anaerobic growth of *S. cerevisiae* on glucose (1), the equation representing aerobic growth on the ethanol produced during the anaerobic growth reaction (2), and the equation representing the complete combustion to carbon dioxide and water of the glycerol produced during the anaerobic growth reaction (3), to equal the equation representing aerobic growth on glucose, as shown in Table 8.2. The agreement of Eq. 4 in Table 8.2 with Eq. 2 in Table 8.1 is not perfect, but can be considered to be within about 2%. In the former equation the unit-carbon formula for the cells was taken as that which was determined experimentally rather than that representing a weighted average of formulas

TABLE 8.2. The addition of the equations representing (1) the anaerobic growth of *Saccharomyces cerevisiae* on glucose, (2) aerobic growth on the ethanol produced during the anaerobic growth reaction, and (3) the total oxidation of the glycerol produced during the anaerobic growth reaction, to equal the equation (4) representing aerobic growth on glucose[a]

1.	$C_6H_{12}O_6(aq) + 0.12NH_3(aq)$	$\longrightarrow$	$0.59CH_{1.75}O_{0.45}N_{0.20}(\text{cells}) + 1.54CO_2(aq) + 0.43C_3H_8O_3(aq) + 1.30CH_3CH_2OH(aq)$
2.	$1.30CH_3CH_2OH(aq) + 0.20NH_3(aq) + 2.38O_2(aq)$	$\longrightarrow$	$1.34CH_{1.72}O_{0.41}N_{0.15}(\text{cells}) + 1.26CO_2(aq) + 3.00H_2O(l)$
3.	$0.43C_3H_8O_3(aq) + 1.50O_2(aq)$	$\longrightarrow$	$1.29CO_2(aq) + 1.72H_2O(l)$
4.	$C_6H_{12}O_6(aq) + 0.32NH_3(aq) + 3.88O_2(aq)$	$\longrightarrow$	$1.93CH_{1.71}O_{0.46}N_{0.17}(\text{cells}) + 4.09CO_2(aq) + 4.72H_2O(l)$

[a]Equation 4 should be compared with Eq. 2 in Table 8.1.

Source: Reference 4.

for the cells grown anaerobically on glucose and aerobically on ethanol. The ratio of the unit-carbon formula coefficient for the cells grown aerobically on glucose over that for cells grown anaerobically on glucose is 1.93/0.59 = 3.27. This is close to the ratio of 3.4 observed with turbidimetric measurements as shown in Fig. 8.1. It is very close to the ratio of cellular dry weights (not ash free), which has a value of 3.21.

The useful aspect of the set of four equations in Table 8.2 is that the additivity of Eqs. 1, 2, and 3 to equal Eq. 4, which is almost identical with a similar equation which has been determined experimentally, shows that the data are internally consistent. This in turn gives some degree of confidence that these equations are good representations of the mass balance with respect to the individual growth reactions that are represented. The main question that then arises is whether the ash-free, empirical composition of yeast cells grown on different substrates is the same, or different. The analyses seemed to indicate that these compositions were slightly different, but perhaps this is an artifact of preparation. The preparation of cells for elemental analysis has not been sufficiently standardized to warrant any firm conclusion. For example, how many times should the cells be washed, and how should they be dried to get the most accurate and consistent results?

8.9. THE MEASUREMENT OF THE ENTHALPY CHANGES ACCOMPANYING THE GROWTH OF *S. CEREVISIAE* ANAEROBICALLY ON GLUCOSE AND AEROBICALLY ON GLUCOSE, ETHANOL, AND ACETIC ACID

During the middle 1950s the extremely accurate and sensitive calorimeters of the type designed by Calvet[27] were not readily available. For this reason the instrument used by Battley[5] for measuring the heat of growth of yeast was essentially of the same design as the adiabatic calorimeter of Winzler and Baumberger[28] (see Fig. 7.1), except for a few modifications. The calorimeter vessel of Winzler and Baumberger was replaced by an ordinary wide-mouth vacuum bottle having a capacity of one quart. The top of the vessel was closed with a large rubber stopper through which passed various tubes and leads into the space within the vessel. The whole vessel assembly could be autoclaved. The culture medium within the vessel was agitated by means of a Teflon-coated stirring bar actuated with a rotating permanent magnet mounted beneath the vessel. A 50-junction copper–constantan thermopile was used to sense the temperature difference between the inside of the vessel and the water-filled thermostat within which the vessel was mounted. A mirror galvanometer with a coil having a sensitivity of 0.5 μV mm^{-1} at a distance of 3 m was used. Since the thermal e.m.f. of copper against constantan is 43 μV °C^{-1}, the sensitivity of the temperature-sensing system was 0.00023°C mm^{-1} of deflection of the light lever. During an experimental run the deflection of the light lever seldom varied more than 1 mm, and thus it was possible for the temperature within the calorimeter vessel to differ from that of the water bath by no more than this amount. In practice, however, it was necessary to drain off the heat produced by the mechanical action of the magnetic stirring bar by holding the temperature of the water bath slightly lower than that of the culture medium in the calorimeter vessel. If this balance were not accurately adjusted or if the speed of the stirring motor varied, an experiment was doomed. For this reason the calorimeter had to be operated for several hours prior to inoculation of the culture medium, until there was no detectable temperature change within the vessel. The temperature rise of the culture medium was read with a Beckman thermometer placed in a glass thermometer well that passed through the rubber stopper and dipped into the culture medium. A similar well was used to contain one set of thermopile junctions. The calorimeter could be calibrated electrically by means of a resistance wire also immersed in the culture medium. Calibration was carried out after the calorimeter had come to a temperature balance with the water bath. The gas phase within the calorimeter vessel could be varied by flushing the appropriate gas through an inlet and outlet tube that also passed through the rubber stopper. Anaerobic experiments were carried out with 200 ml of the same culture medium used for the growth-reaction studies, the gas phase being pure nitrogen. For aerobic experiments 100 ml of culture medium were used, the gas phase being pure oxygen. Gas was not passed through the calorimeter after it had achieved a stable temperature. All experiments were conducted at an initial temperature between 29 and 30°C, the assumption being made that the thermal exchanges taking place within the calorimeter vessel would not be significantly different from those at 25°C. Thus, values for the thermodynamic properties of the reactants and products at 25°C were used without temperature correction. During inoculation 1 ml of a sufficiently small quantity of cells was added so that no detectable rise in temperature of the calorimeter vessel would occur until the cells were well into the exponential phase of growth. This was also necessary because the inoculum was never at exactly the same temperature as the contents of the calorimeter vessel, and it took a period of time before the temperature of the calorimeter was again balanced with respect to that of the water bath.

Battley paid considerable attention to calorimetric correction factors. Corrections for side reactions such as the vaporization of volatile products, the solubilization of gases, and the neutralization of bases are of extreme importance if completely ambiguous results are not to be obtained. Corrections were made for the following chemical events, which are listed together with the appropriate reaction equations.

1. The formation of ammonia, under the assumption that this was the form in which this nitrogen source was actually taken up by the cells.

$$NH_4^+(aq) + OH^-(aq) \longrightarrow NH_3(aq) + H_2O(l)$$

2. The heat of solution of the oxygen gas used in the aerobic experiments.

$$O_2(g) \longrightarrow O_2(aq)$$

3. The formation of HCO_3^- from $CO_2(aq)$ at pH 6.55.

$$CO_2(aq) + H_2O(l) \longrightarrow HCO_3^-(aq) + H^+(aq)$$

4. The movement of $CO_2(g)$ into the gas phase from the $CO_2(aq)$ produced by the growth reactions.

$$CO_2(aq) \longrightarrow CO_2(g)$$

5. The vaporization of ethanol in the anaerobic glucose growth reaction.

$$CH_3CH_2OH(aq) \longrightarrow CH_3CH_2OH(g)$$

6. The solution of ethanol from the gas phase in the aerobic ethanol growth reaction.

$$CH_3CH_2OH(g) \longrightarrow CH_3CH_2OH(aq)$$

7. The neutralization of the $OH^-(aq)$ produced by the utilization of the sodium acetate in the acetic acid growth reaction at a pH of 6.55.

$$H^+(aq) + OH^-(aq) \longrightarrow H_2O(l)$$

Although he did not, Battley should have included the enthalpy change accompanying the formation of acetic acid from acetate, since the acid was assumed to be a reactant in the acetic acid growth reaction. The correction for the phosphorylation of the glycerol produced during the anaerobic glucose growth reaction was also not made, but can probably be neglected. The values for the corrections were calculated using the ΔH_f° and $\Delta H_f^{\circ\prime}$ values in Appendix 1 for the substances involved and their quantities as they were used or produced during the growth reactions or as they existed in equilibrium between the gas and liquid phases in the calorimeter vessel at the end of the growth reactions. Although several corrections were made for each of the four reactions, it is difficult to know what others might also have been made. The corrections varied from 1 to 8% for the aerobic growth reactions, to as high as 48% for the anaerobic glucose growth reaction in the experiments in which the culture medium was stirred. While it might not be thought necessary to stir anaerobic cultures because there is no need to distribute oxygen throughout the culture medium, experiments were conducted both with and without stirring so as to have identical physical conditions with respect to anaerobic and aerobic experiments. It developed that stirring the medium resulted in a considerably lower observed heat of reaction. This was apparently due to a far greater amount of carbon dioxide and ethanol entering the gas phase above the culture medium than occurred with the nonstirred cultures, due to a constant renewal of the gas–liquid interface. In the experiments in which the medium was not stirred carbon dioxide gas and ethanol vapor, being heavier than the other gases in the system, probably evolved slowly and evenly from the aqueous phase to form a thin layer at the interface. Under these circumstances very little carbon dioxide and ethanol would be expected to leave the aqueous phase, and the observed heat of reaction would approach closely the true heat of reaction. For this reason the observed heats of reaction of the anaerobic experiments that were *not* stirred were accepted.

The thermograms of the enthalpy changes accompanying the four growth reactions are about what might be expected, as illustrated in Fig. 8.3. The initial slope of this graph is probably due to the fermentative growth reaction, after which the cells change to growth on ethanol and the oxidation of glycerol resulting in a rate of heat production which is slightly lower than that of fermentative growth. The rate of heat production then begins to fall off as the concentration of substrate become rate limiting toward the end of the experiment. One important aspect of Fig. 8.3 with respect to the energetics of microbial growth is that in experiments of this kind there is no demonstration

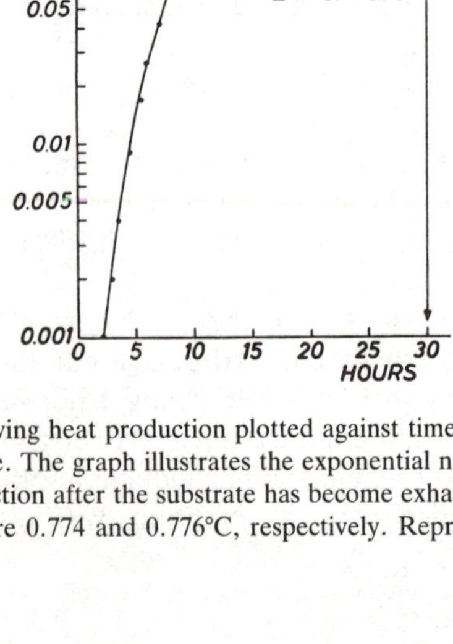

FIGURE 8.3. Graph showing heat production plotted against time for the growth of *Saccharomyces cerevisiae* on glucose. The graph illustrates the exponential nature of the thermogram and the absence of heat production after the substrate has become exhausted. The values for the last two points on the curve are 0.774 and 0.776°C, respectively. Reprinted from Reference 5 with permission.

of further production of heat after the substrate has become exhausted. This must surely mean that there are no stored sources of energy within the cells, and that all of the substrate has been either catabolized or converted into the fabric of the cells. It is also evidence that only growth takes place, and not growth plus oxidative assimilation. Battley[5] determined the enthalpy change accompanying the growth of yeast anaerobically on glucose to be −23 kcal $mole^{-1}$, in excellent agreement with the average (−23.1) of the values obtained by Dubrunfaut[29] (−22.7), Bouffard[30] (−23.5), A. J. Brown[31] (−21.4), M. Rubner[32] (−25.6), and H. T. Brown[33] (−22.5). The corrected values for aerobic growth on glucose, ethanol, and acetic acid were determined to be −479, −204, and −162 kcal $mole^{-1}$, respectively.

8.10. CONSERVATIVE AND NONCONSERVATIVE REACTIONS. CONSERVATION THEORY

It was shown by Clifton[34–35] and by Winzler and Baumberger[28] that chemical inhibitors such as dinitro phenol (DNP) or sodium azide in low concentrations can cause an inhibition of the processes of assimilation and growth in bacteria and yeasts without appreciably hindering catabolic processes. Battley[5] pointed out that it was possible to conceive of experiments identical to the growth-reaction studies he carried out except for the inclusion of DNP in the growth-reaction systems. Under these conditions such systems would then be capable of undergoing a maximum change in free energy, enthalpy, and entropy. As an example, suppose that DNP were included as a component of the initial state in the reaction during which *S. cerevisiae* would otherwise be grown aerobically on glucose. Although the yeast cells would no longer be able to grow on or to assimilate the glucose, they would still be capable of oxidizing the glucose completely to carbon dioxide and water. The DNP does not enter into the reaction system since there is as much present at the end of the experiment as there is at the beginning. Under these circumstances all of the free energy that is available from the oxidation of glucose would become converted into heat and entropy, and the enthalpy change for the system would be at its maximum value of −684.0 kcal $mole^{-1}$ (see Table 8.3). Battley defined such reactions as being "non-conservative." If DNP is not present in this particular growth-reaction system growth will occur. However, even though the glucose becomes completely consumed during the growth reaction it is evident from the measurement of heat produced that this latter is less than maximum, in this case −479 kcal $mole.^{-1}$ Battley reasoned that some of the free energy that is potentially available from the substrate is therefore not converted into heat and entropy, but remains somewhere within the growth-reaction system. He defined such reactions as being "conservative" and concluded that ". . . the [enthalpic] energy which has been conserved must reside in the products of the growth-reaction which are different from those of the relative non-conservative reaction." (Battley,[5] p. 635.) It is apparent that the

TABLE 8.3. Equations representing the nonconservative reactions that are related to the conservative reactions (growth reactions) listed in Table 8.1, together with their accompanying changes in thermodynamic properties at 298.15°K and 1 atm

Reaction	$\Delta G°$, $\Delta H°$, and T $\Delta S°$ (kcal mole^{-1})
1. Glucose fermentation (alcoholic): $C_6H_{12}O_6(aq) \longrightarrow 2CO_2(aq) + 2CH_3CH_2OH(aq)$	$\Delta G° = -52.82$ $\Delta H° = -33.57$ $T\,\Delta S° = 19.25$
2. Glucose oxidation: $C_6H_{12}O_6(aq) + 6O_2(aq) \longrightarrow 6CO_2(aq) + 6H_2O(l)$	$\Delta G° = -698.52$ $\Delta H° = -684.04$ $T\,\Delta S° = 14.48$
3. Ethanol oxidation: $CH_3CH_2OH(aq) + 3O_2(aq) \longrightarrow 2CO_2(aq) + 3H_2O(l)$	$\Delta G° = -322.84$ $\Delta H° = -325.44$ $T\,\Delta S° = -2.60$
4. Acetic acid oxidation: $CH_3COOH(aq) + 2O_2 \longrightarrow 2CO_2(aq) + 2H_2O(l)$	$\Delta G° = -210.91$ $\Delta H° = -212.73$ $T\,\Delta S° = -1.82$

adjective "conservative" is somewhat similar to the adjectives "hypoosmotic," "isoosmotic," and "hyperosmotic" in that a reaction is conservative only when compared or related to an analogous reaction that is not. A given reaction is neither conservative nor nonconservative when considered by itself. The nonconservative reactions relative to the conservative reactions (growth reactions) represented by Eqs. 1–4 in Table 8.1 are shown in Table 8.3, together with the values of $\Delta G°'$, $\Delta H°'$, and $T\,\Delta S°'$ for each that has been calculated using the values for the thermodynamic properties listed in Appendix 1. The enthalpy change accompanying a growth reaction (which is one type of conservative reaction) can be represented by the symbol ΔH_C and that accompanying its relative nonconservative reaction by the symbol ΔH_{NC}. Figure 8.4 shows that if the ΔH_C values for the four growth reactions Battley studied are plotted against the ΔH_{NC} values of their relative nonconservative reactions, a close but not exact proportionality exists in which the slope has a value of 0.712. Whether this proportionality would be more exact if the calorimetric data were more accurate is purcly conjectural. It is not necessary that all the points lie exactly on a straight line. This could be simply the way in which the data naturally occur.

The percent molar efficiency with which enthalpy is conserved within a growth-reaction system was defined by Battley through the use of the following equation.

$$\begin{array}{l}\text{percent molar efficiency}\\ \text{of enthalpy conservation}\end{array} = \frac{\Delta H^{\circ}_{NC} - \Delta H_C}{\Delta H^{\circ}_{NC}} \times 100 \qquad (8.3)$$

where ΔH°_{NC} and ΔH_C represent the enthalpy changes accompanying the relative nonconservative and conservative reactions (growth reactions), respectively. If the only product of the growth reaction that is different from those produced during the relative nonconservative reaction happens to be cells, then the efficiency as calculated with Eq. 8.3 is a direct measure of the enthalpy conserved in the substance of the cellular material. However, this is not the case with respect to the anaerobic glucose growth reaction (Eq. 1 in Table 8.1), during which both cells and glycerol are formed as products that are different from those (carbon dioxide and ethanol) of the relative nonconservative reaction. Thus, the efficiency with which enthalpy is conserved in the cells must be separated from that conserved in the glycerol. Battley was able to accomplish this by carrying out a theoretical oxidation of the cellular substance and of the glycerol and by comparing the enthalpy changes accompanying each reaction. Using Eq. 1 in Table 8.1, the enthalpy change (-23 kcal mole^{-1}) which was determined calorimetrically for this reaction, and the appropriate $\Delta H_f^{\circ\prime}$ values from Appendix 1, the ΔH_f (cells) for the cellular material formed during anaerobic growth on glucose was calculated to be:

$$\begin{aligned}0.59\ \Delta H_f\ (\text{cells}) &= \Delta H - 0.43\ \Delta H_f^{\circ\prime}\ C_3H_8O_3 - 1.54\ \Delta H_f^{\circ\prime}\ CO_2 \\ &\quad - 1.30\ \Delta H_f^{\circ\prime}\ CH_3CH_2OH + \Delta H_f^{\circ\prime}\ C_6H_{12}O_6 \\ &\quad + 0.12\ \Delta H_f^{\circ\prime}\ NH_3 \\ &= -23 + 69.53 + 152.31 + 89.18 - 301.88 - 2.30 \\ &= -16.16\ \text{kcal}\end{aligned}$$

from which

$$\Delta H_f\ (\text{cells}) = -27.39\ \text{kcal unit-carbon formula}^{-1}$$

This value is slightly different from that actually calculated by Battley because of the use of the more accurate $\Delta H_f^{\circ\prime}$ values in Appendix 1. The following equation was then used to calculate the heat of combustion of the cells that were grown.

$$\left.\begin{array}{l}0.59CH_{1.75}O_{0.45}N_{0.20}(\text{cells}) \\ +\ 06.3O_2(aq)\end{array}\right\}\longrightarrow \begin{array}{l}0.59CO_2(aq) + 0.34H_2O(l) \\ +\ 0.12NH_3(aq)\end{array} \tag{8.4}$$

from which, using the appropriate $\Delta H_f^{\circ\prime}$ values in Appendix 1,

$$\begin{aligned}\Delta H &= 0.59\ \Delta H^{\circ\prime}\ CO_2 + 0.34H_f^{\circ}\ H_2O + 0.12\ \Delta H_f^{\circ\prime}\ NH_3 \\ &\quad - 0.59\ \Delta H_f\ (\text{cells}) - 0.63\ \Delta H_f^{\circ\prime}\ O_2 \\ &= -58.35 - 23.23 - 2.30 + 16.16 + 1.82 \\ &= -65.90\ \text{kcal}\end{aligned}$$

Using a similar calculation the heat of combustion of 0.43 moles of glycerol (aq) was found to be −171.42 kcal. The sum of these two values was therefore −65.90 + (−171.42) = −237.09 kcal as the enthalpy change accompanying the oxidation of both products. Of this, (−65.90/−237.09) × 100 = 27.79% is the energetic contribution of the cellular material in terms of enthalpy. The total enthalpy conserved during the anerobic glucose growth reaction was that given with Eq. 1 in Table 8.3 minus the enthalpy change accompanying the anaerobic glucose growth reaction, or −33.57 − (−23) = −10.57 kcal. The actual quantity of enthalpic energy conserved within the cells is then (0.2779 × −10.57 × 100)/−33.57 = 8.75% of the total available. This also represents the percent efficiency of enthalpic energy conservation in the cells. The rest of the conserved enthalpic energy is found in the glyercol, which contains 22.73% of the total available. The remaining quantity, 68.52%, is what is observed as the heat of reaction (i.e., 0.6852 × −33.57 kcal = −23.0 kcal).

The efficiencies of energy conservation within the substance of the cells produced during the growth reaction cannot be compared directly unless the same substrates are involved. This happens to be the case with respect to the reactions in which *S. cerevisiae* is grown anaerobically or aerobically on glucose. Using Eq. 8.3 it is apparent that with aerobic growth on glucose the efficiency of enthalpy conservation is 100 × [−684.0 − (−479)]/−684.0 = 30.0%. One would therefore expect to obtain 30.0/8.8 = 3.4 times more growth on glucose aerobically than anaerobically, which is exactly what one does observe on the basis of the slopes representing aerobic and anaerobic growth in Fig. 8.1. However, it must be remembered that a comparison such as this that has been derived from enthalpy data can only be considered approximate. What should actually be used for such a comparison is free-energy data, since it is the potential free-energy change and not the enthalpy change of a chemical process that is the true measure of the amount of work which can be accomplished.

8.11. AN INITIAL ATTEMPT AT CALCULATING THE FREE-ENERGY CHANGE ACCOMPANYING A GROWTH REACTION

Whereas the enthalpy change accompanying a growth reaction can be directly measured, the free-energy change cannot. This latter can be calculated if the free energies of formation of all the reactants and products of a reaction are known. This was not the case with Battley's growth reactions, as represented by the growth-reaction equations in Table 8.1, because there were no data available at that time with respect to the free energy of formation of cellular material. Growth reactions are not reversible, and because cells may be regarded as precipitates there is no chemical activity that can be ascribed to them. It is thus impossible to determine a thermodynamic equilibrium constant for the purpose of calculating $\Delta G^{\circ\prime}$ by means of Eq. 1.47. In addition, there are no electrode potentials that can be directly related to the free-energy changes accompanying growth reactions and thus $\Delta G^{\circ\prime}$ cannot be calculated in this manner. A free-energy change *can* be calculated using the Gibbs free-energy equation *provided* that both the related enthalpy and entropy changes can be thermally measured. However, whereas the former can be accomplished readily, no data had been accumulated on the absolute entropy of cells. It was therefore impossible to calculate the entropy of formation of cells as one of the values that would have to be used to calculate the entropy change accompanying a growth reaction.

The first approach that Battley[6] used to calculate the free-energy change accompanying a growth reaction was little more than an approximation. On the basis of his original hypothesis that the energy conserved during a growth reaction must reside in the products of the growth reaction that are different from those of the relative nonconservative reaction, he then reasoned that the extent to which energy is conserved probably bears some relationship to the total free energy that is initially available for expenditure. This latter is suggested by the proportionality that appears to exist between the enthalpy changes of the nonconservative and the conservative reactions, as illustrated by Fig. 8.4. He then further hypothesized that ". . . the free-energy accompanying a growth-reaction is proportional to the extent to which its relative non-conservation reaction has proceeded to completion." (Battley,[6] p. 676.)

As an example, if a quantity of viable yeast cells were suspended in nutrient medium containing glucose as a sole carbon and energy source, and if assimilation and growth were prevented from occurring by the addition of a metabolic poison such as DNP or NaN_3 in the appropriate concentration, the catalytic mechanisms of the yeast would still allow the complete oxidation of

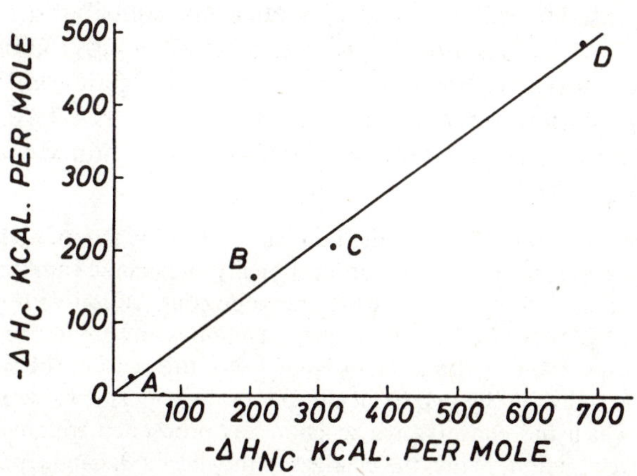

FIGURE 8.4. Graph showing heat production accompanying the growth of *Saccharomyces cerevisiae* on different substrates plotted against the enthalpy changes accompanying the relative nonconservative reactions. A, anaerobic growth on glucose; B, C, and D, aerobic growth on acetic acid, ethanol, and glucose, respectively. The average slope of the line is 0.712. Reprinted from Reference 5 with permission.

the glucose as represented by the following equation:

$$C_6H_{12}O_6(aq) + 6O_2(aq) \longrightarrow 6CO_2 + 6H_2O(l) \qquad (8.5)$$

From Table 8.3 the reaction represented by Eq. 8.5 is seen to be accompanied by a maximum loss in free energy to the extent of −698.52 kcal mole^{-1}, of which −684.04 kcal become converted into thermal energy and 14.48 kcal become reversibly exchanged with the environment because of the change in entropy of the reaction system as it passes from its initial to its final state. However, during growth on this substrate there is less energy lost than in the reaction represented by Eq. 8.5. The difference is conserved in the cellular material represented in the following growth-reaction equation from Table 8.1.

$$\left.\begin{array}{l} C_6H_{12}O_6(aq) + NH_3(aq) \\ + 3.84O_2(aq) \end{array}\right\} \longrightarrow \begin{array}{l} 1.95CH_{1.71}O_{0.46}N_{0.17}(\text{cells}) \\ + 4.09CO_2(aq) + 4.73H_2O(l) \end{array} \qquad (8.6)$$

According to Battley, "If the hypothesis is made that the relationship in the energetics of the reactions expressed by equations (2) and (3) [in this text Eqs. 8.5 and 8.6] can be determined by a proportion, then this proportion must be calculable from the chemical data contained in these equations." (Battley,[6] p. 677.)

The most simple method by which to make the above calculation was to assume that the extent to which the reaction represented by Eq. 8.5 (the nonconservative reaction relative to that represented by Eq. 8.6) proceeds to completion is proportional to the extent to which its products appear in the reaction represented by Eq. 8.6. Thus, the energetic contribution with respect to the formation of $CO_2(aq)$ in the growth reaction is 4.09/6.00 = 0.681 of that of the nonconservative reaction. Similarly, the energetic contribution with respect to the formation of $H_2O(l)$ in the growth reaction is 4.73/6.00 = 0.787 of that of the nonconservative reaction. Battley then reasoned that:

> Since there are two products of the non-conservative reaction, and since energy levels tend to balance themselves proportionally in dynamic equilibrium systems, there is no reason to believe that the energetic contribution with respect to the formation of $CO_2(aq)$ in the conservative reaction is any greater or any less than that with respect to the formation of $H_2O(l)$. For this reason, the sum of the two proportions is divided by two to give 0.734, or 73.4%. This value represents the extent to which the conservative reaction has proceeded, relative to the non-conservative reaction. When the changes in the thermodynamic quantities which accompany the non-conservative oxidation of glucose are multiplied by this percentage, values are obtained which represent the changes in the thermodynamic quantities which accompany the aerobic glucose growth-reaction. [Battley,[6] p. 677.]

The results of such calculations are shown in Table 8.4, and represent the first attempt at actually calculating the changes in thermodynamic properties which accompany growth reactions. A plot of these *calculated* free-energy and enthalpy changes (ΔG_C and ΔH_C) accompanying the four growth reactions represented by the equations in Table 8.1 against those accompanying the relative nonconservative reactions ($\Delta G^{\circ\prime}_{NC}$) and $\Delta H^{\circ\prime}_{NC}$) as given in Table 8.3, is shown in Fig. 8.5. It is apparent that the value of 0.707 for the slope of the graph in Fig. 8.5 is almost identical with that of 0.712 for the slope in Fig. 8.4.

The percent molar efficiency with which free energy and enthalpy can become conserved during growth reactions in the mass of substances that are produced other than those produced by the relative nonconservative reactions can only be compared by making all substrates comparable through a calculation of their average potential free energy or enthalpy per unit carbon atom. This is because not all substrates have equal numbers of carbon atoms per molecule. The percent molar efficiency with respect to free energy was calculated by means of an equation similar to Eq. 8.3, as follows:

$$\begin{array}{l}\text{percent molar efficiency of} \\ \text{free-energy conservation}\end{array} = \frac{\Delta G^{\circ}_{NC} - \Delta G_C}{\Delta G^{\circ}_{NC}} \times 100 \qquad (8.7)$$

where ΔG°_{NC} and ΔG°_{C} represent the free-energy changes accompanying the relative nonconservative and conservative reactions (growth-reactions), respectively. The nonconservative enthalpy and free-energy changes per unit carbon atom of the substrate were calculated by means of the following equations.

$$\begin{array}{l}\text{nonconservative enthalpy} \\ \text{change per unit carbon atom}\end{array} = \frac{\Delta H^{\circ}_{NC}}{n\text{C}} \qquad (8.8)$$

and

$$\begin{array}{l}\text{nonconservative free-energy} \\ \text{change per unit carbon atom}\end{array} = \frac{\Delta G^{\circ}_{NC}}{n\text{C}} \qquad (8.9)$$

where nC represents the number of carbon atoms in a given substrate. The result of plotting the percent molar efficiencies as calculated by means of Eqs. 8.3 and 8.7 against the nonconservative enthalpy and free-energy changes per unit carbon atom of the substrate as calculated using Eqs. 8.8 and 8.9 is shown in Fig. 8.6. From this Battley concluded that even though the method of calculating energy changes for the growth reactions that he used was at best an approximation, the efficiency with which energy is conserved during a growth reaction was a function of the nonconservative energy change per unit carbon atom of the substrate, that is, to the average degree of reduction of

TABLE 8.4. Calculated and experimentally determined values for the changes in thermodynamic properties accompanying the nonconservative and conservative reactions relative to the growth of *Saccharomyces cerevisiae* anaerobically on glucose, and aerobically on glucose, ethanol, and acetic acid. The proportion of 73.4% for glucose fermentation is based on the assumption that the efficiency of this reaction is the same as that for glucose oxidation

		Thermodynamic Quantities (kcal mole^{-1})				
Reaction	Proportion	ΔF (calc.)	ΔH (calc.)	ΔH (exp.)	$T\,\Delta S$ (calc.)	S (calc.)
Glucose fermentation	71.0%					
Nonconservative		−52.2	−33.0		19.2	0.064
Conservative		−37.1	−23.4	−23	13.7	0.045
Glucose fermentation	73.4%					
Nonconservative		−52.2	−33.0		19.2	0.064
Conservative		−38.3	−24.4	−23	14.1	0.047
Glucose oxidation	73.4%					
Nonconservative		−698.2	−677.2		21.0	0.070
Conservative		−512.5	−497.1	−479	15.4	0.052
Ethanol oxidation	62.7%					
Nonconservative		−323.0	−322.1		0.9	0.003
Conservative		−202.5	−202.0	−204	0.5	0.002
Acetic acid oxidation	75.7%					
Nonconservative		−210.3	−209.7		0.6	0.002
Conservative		−159.4	−158.9	−162	0.5	0.002

Source: Reprinted with permission from Reference 6.

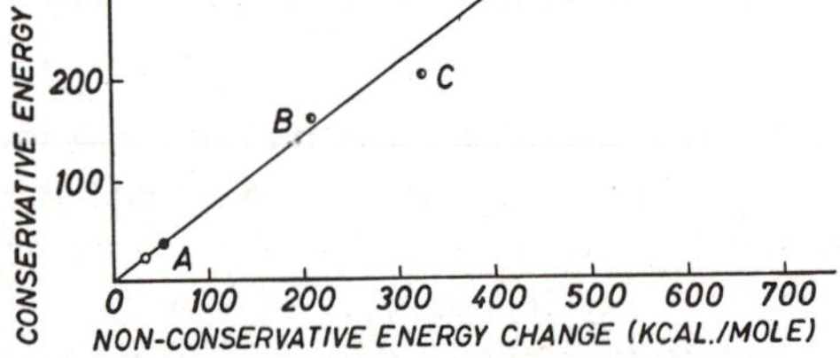

FIGURE 8.5. Graph showing calculated conservative energy changes accompanying the growth of *Saccharomyces cerevisiae* on different substrates plotted against those accompanying the relative nonconservative reactions. A, anaerobic growth on glucose; B, C, and D, aerobic growth on acetic acid, ethanol, and glucose, respectively. Filled circles, free-energy changes; open circles, enthalpy changes; half-filled circles, nearly identical free-energy and enthalpy changes. The average slope of the line is 0.707. Reprinted from Reference 6 with permission.

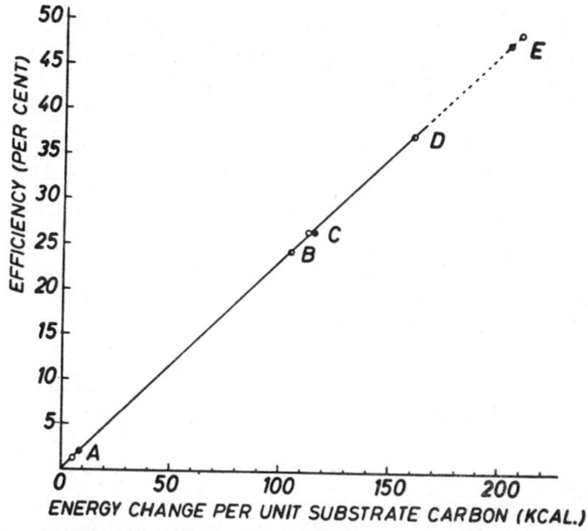

FIGURE 8.6. Graph showing percent molar efficiency of the growth of *Saccharomyces cerevisiae* on different substrates plotted against the average energy changes per unit-carbon atom of the substrate. A, anaerobic growth on glucose; B, C, and D, aerobic growth on acetic acid, glucose, and ethanol, respectively. Filled circles, free-energy changes; open circles, enthalpy changes, half-filled circles, nearly identical free-energy and enthalpy changes. Reprinted from Reference 6 with permission.

the substrate carbon. From Fig. 8.6 it is also evident that

$$\begin{matrix}\text{percent molar efficiency of} \\ \text{enthalpy conservation}\end{matrix} = K\frac{\Delta H^\circ_{NC}}{n\text{C}} \tag{8.10}$$

and

$$\begin{matrix}\text{percent molar efficiency of} \\ \text{free-energy conservation}\end{matrix} = K\frac{\Delta G^\circ_{NC}}{n\text{C}} \tag{8.11}$$

where $K = 0.230$ for *S. cerevisiae*. The percent molar efficiencies as calculated using Eqs. 8.10 and 8.11 should have the same values as those calculated using Eqs. 8.3 and 8.7, and do. Figure 8.6 shows further that when the free-energy change (-205.3 kcal mole^{-1}) accompanying the nonconservative oxidation of methane to carbon dioxide and water is taken as the maximum per unit carbon atom of the substrate (methane is 100% reduced), the maximum molar efficiency with which *S. cerevisiae* can conserve free energy would extrapolate to about 47.5%, *if* this yeast were capable of growing on methane (which it is not). Although it was impossible to draw any firm conclusions on the basis of experiments with only one microorganism, it was Battley's opinion that a molar efficiency of 50% might well be a maximum. This was because it seemed unlikely to him that more energy could be conserved in a reaction than was used up in causing it to proceed. If anything, the maximum molar efficiency

TABLE 8.5. Calculated values for the thermodynamic properties of the cellular material formed during the growth of *Saccharomyces cerevisiae* anaerobically on glucose and aerobically on glucose, ethanol, and acetic acid

	Energies of Formation from the Elements, at 298.15°K and 1 atm, Expressed in kcal $mole^{-1}$ of Substrate Consumed or in kcal Unit-Carbon Formula $Weight^{-1}$					
Growth Reaction	ΔF_f	ΔF_f^a	ΔH_f	ΔH_f^a	$T\,\Delta S_f$	$T\,\Delta S_f^a$
Glucose fermentation	−8.3	−14.0	−16.9	−28.7	−8.6	−14.7
Glucose fermentation[b]	−9.5	−16.2	−17.7	−30.0	−8.2	−13.8
Glucose oxidation	−86.1	−44.1	−93.3	−47.8	−7.2	−3.7
Ethanol oxidation	−19.3	−18.7	−27.1	−26.3	−7.8	−7.6
Acetic acid oxidation	−29.3	−47.2	−33.6	−54.2	−4.3	−7.0

[a]Values for the cellular material as represented by a unit amount of carbon.
[b]Values calculated on the assumption that the efficiency of the anaerobic glucose growth reaction is the same as that of the aerobic glucose growth reaction.
Source: Reprinted by permission from Reference 6.

might well be expected to be slightly less than 50%, due to inefficiences accompanying the transfer of molecules about the enzyme systems within the cells.

Using his approximate method of calculating the energy changes accompanying growth reactions Battley then calculated values for the thermodynamic properties of the cellular material formed during these reactions. These are found in Table 8.5. It would be thought that yeast cells are yeast cells, no matter on what substrate they are exponentially grown, and that therefore their composition and energy content per unit quantity of cells would be about the same no matter on what they are grown. It is apparent from Table 8.5 that ΔF_f for a unit-carbon formula weight of cells does *not* have roughly the same value with respect to growth on the four substrates. One would suppose that it should. The same is true for ΔH_f and $T\,\Delta S_f$. In fact, these values are sufficiently disparate that Battley's approximate method must be seriously questioned in spite of the niceness of fit of the points on the line in Fig. 8.6 which is apparently only fortuitous.

8.12. THE HYPOTHESIS THAT THE FREE-ENERGY CHANGE ACCOMPANYING THE CONSTRUCTION OF CELLS FROM A SINGLE SOURCE OF CARBON AND ENERGY IS EQUAL TO ZERO

The inadequacy of Battley's initial method of calculating the free-energy, enthalpy, and entropy changes accompanying growth reactions caused him to search for other methods of doing this which were more thermodynamically sound. In 1960 Bauchop and Elsden[36] described a series of experiments on the growth of microorganisms in relation to the quantity of ATP that was made available from the anaerobic metabolism of several substrates. It had been previously observed by a number of investigators that when fermentative microorganisms were grown in the presence of extraneous carbon sources such as peptone and yeast extract, all of the sugar became catabolized, as evidenced by 100% or close to 100% carbon balances that did not include the carbon in the cells carrying out the fermentation (see Section 4.9). Bauchop and Elsden did not actually determine carbon balances in their growth experiments. However, in order to ascertain to what extent the substrate was becoming incorporated into the material of the cells in their experiments, they grew *Streptococcus faecalis* on 150 mg of D-(^{14}C)-glucose in a culture medium that also contained minerals, casein hydrolysate, and several amino acids and vitamins. When growth had ceased the dry weight of the cells was found to be 27.8 mg, and 0.92% of the glucose carbon had become incorporated into cellular substance. Since the carbon content of the cells was taken as 50% of the dry weight, this meant that the fermented substrate had supplied about 4% of the cell carbon, and that all the rest of the cellular carbon originated in the casein hydrolysate or in the individual amino acids that were also added as a part of the culture medium. In effect, the glucose was used by these cells only as a source of energy for growth.

Bauchop and Elsden were interested solely in relating the quantity of bacterial growth to the quantity of ATP formed during the anaerobic catabolism of substrates along known biochemical pathways. It became apparent to Battley[37–38] that quite apart from the interesting data they presented in this respect, their data on the uptake of radioactive glucose by *S. faecalis* explained completely why 100% fermentation balances had been observed where no accounting had been made of the carbon in the cells that had grown while fermenting. As stated in Section 4.9, in *all* cases where this was observed peptone, yeast extract, casein hydrolysate, or other organic carbon sources were present in the culture media in addition to the sugar which was being fermented. The logical conclusion had to be that a 100% fermentation balance could only have occurred if the material for cellular construction came from the organic substances other than the source of energy which were present in the culture media. The apparently complete biochemical separation of the source of energy used for the process of heterotrophic growth from the source of organic material used for the construction of the cells was recognized as representing a unique thermodynamic system. *If* the source of energy were to be completely oxidized or fermented, *all* the free energy would become converted into enthalpy and entropy and no free energy from the substrate could become conserved in the substance of the cells. This being the case, the reaction whereby the extraneous organic material also in the culture medium became converted into cellular material must have occurred with *no net change in free energy*, since the cells could not grow on this organic material in the absence of an energy source. Bauchop and Elsden[36] also grew *S. cerevisiae* anaerobically on glucose in a medium which contained in addition casein hydrolysate, vitamins, and ergosterol, and found that 21 g dry weight of yeast cells were produced per mole of glucose fermented. What Battley felt must be happening energetically in this system is shown in Table 8.6. If glucose were to be completely fermented to ethanol and carbon dioxide this *must* be accompanied by the conversion of all the available free energy (-52.82 kcal mole^{-1}) into enthalpy (-33.57 kcal mole^{-1}) and entropy (19.25 kcal mole^{-1}), and none of the free energy from this reaction could be conserved in the cells. Nevertheless, 21 g dry weight of cells were still produced, the substance of which, if appropriately reacted, could be made to yield free energy. Since the yeast cells would not grow anaerobically on casein hydrolysate in the absence of glucose, the *only* source of free energy in Bauchop and Elsden's system had to be the glucose which, however, was probably all fermented to carbon dioxide and alcohol. Thus, with this organism as well the only possible conclusion was that various organic substances present in the casein hydrolysate were also converted into cellular substance with no *net* change in free energy. As a check on this Battley[38] grew *S. cerevisiae* anaerobically on glucose and found that whereas in a defined medium 1.54 moles of carbon dioxide and 1.30 moles of ethanol were produced, under the same conditions but with increasing concentrations of yeast extract the molar amounts of these products approached 2, as predicted by the Gay-Lussac equation.

It should be noted that Battley's hypothesis applies only to the free-energy changes, and not to enthalpy or entropy changes. Enthalpy changes can be measured, and if free-energy changes can be calculated, the entropy changes can be determined by means of the Gibbs free-energy equation.

TABLE 8.6. Theoretical representation of the thermochemical events accompanying the growth of *Saccharomyces cerevisiae* anaerobically on glucose in the presence of casein hydrolysate, vitamins, ergosterol, and minerals

Initial State		Final State		
$C_6H_{12}O_6$(aq)	⟶	$2CH_3CH_2OH$(aq)	+	$2CO_2$(aq)
$\Delta G^{\circ\prime}$	⟶	$\Delta H^{\circ\prime}$	−	$T\,\Delta S^{\circ\prime}$
-52.02 kcal mole^{-1}	⟶	-33.57 kcal mole^{-1}		-18.45 kcal mole^{-1}
nonthermal chemical energy	⟶	thermal energy	+	absorbed thermal energy
casein hydrolysate	⟶	21 g dry wt cells		(Ref. 36)
$\Delta G'$	⟶	$\Delta H'$	−	$T\,\Delta S'$
0 kcal mole^{-1}	⟶	? kcal mole^{-1}		? kcal mole^{-1}
$C_6H_{12}O_6$(aq) + organic substances	⟶	$2CH_3CH_2OH$(aq)	+	$2CO_2$(aq) + cells
$\Delta G^{\circ\prime}$	⟶	$\Delta H^{0\prime}$	−	$T\,\Delta S^{0\prime}$
-52.02 kcal mole^{-1} + 0 kcal mole^{-1}	⟶	-33.57 kcal mole^{-1} + ? kcal mole^{-1}		-18.45 kcal mole^{-1} + ? kcal mole^{-1}

The criticism can be leveled that a reaction in which the ΔG is zero cannot proceed since it is not spontaneous, but it should be remembered that whatever processes are used to convert organic materials in casein hydrolysate into cellular material are coupled to other processes which provide energy enabling cellular syntheses to take place. Another method of representing free-energy changes that occur with respect to this reaction system is given in Table 8.7 It is apparent that the free-energy change for the coupled reaction is the same for the two methods of representation. However, perhaps the best way of regarding the production of 21 g dry weight of cells from casein hydrolysate in Table 8.6 is not as if this reaction represents a condition of dynamic equilibrium (growth reactions are completely irreversible), but as a mass balance between the initial and final states of cellular construction in which $\Sigma\ \Delta G_f$ of the reactants is exactly equal to $\Sigma\ \Delta G_f$ of the products, the net ΔG being zero. These reactions are thus neither exergonic nor endergonic.

On the basis of the above reasoning Battley[39] came to the conclusion that the same thermochemical events might take place in the absence of substances other than the energy source, such as casein hydrolysate, in which case some the substrate would be anabolized in cellular construction and the remaining part would be catabolized for the production of energy to make anabolism possible. His then recently published equations representing the growth of *S. cerevisiae* anaerobically on glucose, and aerobically on glucose, ethanol, and acetic acid provided useful tools for applying his hypothesis. Each growth-reaction equation was divided into two parts. One part represented a reaction, called a "work-reaction," during which part of the substrate that is present in the initial state is oxidized or fermented to accomplish the work of converting the remaining part into cells and other products of the growth reaction. This

TABLE 8.7. A theoretical representation of the function of ATP during the growth of *Saccharomyces cerevisiae* anaerobically on glucose in the presence of peptone and yeast extract

Initial State	Final State
$C_6H_{12}O_6(aq) + 2ADP + 2P_i \longrightarrow$	$2CH_3CH_2OH(aq) + 2CO_2(aq) + 2\ ATP$
$\Delta G = -35.0$ kcal	
yeast extract, peptone, casein hydrolysate, etc. $+ 2ATP \longrightarrow$	21 g dry wt yeast cells $+ 2ADP + 2P_i$
$\Delta G = -17.8$ kcal (see source note below)	
$C_6H_{12}O_6(aq) +$ organic substances $\longrightarrow$	$2CH_3CH_2OH(aq) + CO_2(aq) +$ cells
$\Delta G = -52.82$ kcal	

Source: Reference 3, Table 13, giving the $\Delta G'$ for the hydrolysis of ATP^{4-} to ADP^{3-}, P_i^{2-}, and H^+ at pH 7.5 as -8.9 kcal $mole^{-1}$. The actual value for the free-energy change accompanying the hydrolysis of ATP within the cells may actually be closer to -13 kcal $mole^{-1}$. Whatever it is, the values for ATP synthesis and hydrolysis cancel out.

latter reaction was called a "conversion-reaction" and could also be represented by an equation. Battley's work reaction could equally well be called "catabolism" and his conversion reaction "anabolism," the two reactions adding up to constitute "metabolism." The equation sets he constructed to represent these ideas are shown in Tables 8.8–8.11. In each case the ΔG for the conversion reaction is equal to zero.

The conversion reactions represented in Tables 8.8–8.11 are equivalent to what Tamiya (see Chapter 6) regarded as "growth," the aerobic work reactions in the same tables being what Tamiya regarded as "respiration." However, Battley's method of constructing conversion-reaction equations differs from that of Tamiya in one respect. Tamiya fully recognized that if the substrate were less reduced than the average degree of reduction of the cellular material, in order to become incorporated a part of it would have to become internally oxidized (an internal dismutation, somewhat like alcoholic fermentation) so that the remaining part could become reduced to the level of the cellular material. This would result in carbon dioxide being a product in the conversion-reaction equation, and can be seen in Eq. 3 in Tables 8.8, 8.9, and 8.11. On the other hand, if the substrate were more reduced than the average degree of reduction of the cellular material, Tamiya felt that it should become externally oxidized by gaseous oxygen to the level of reduction of the cells. This would result in oxygen being a reactant in a conversion-reaction equation. Battley felt that this was not the case, since oxygen is usually considered to function in metabolism only as a terminal electron acceptor for catabolic reactions. It was shown by Müller[40] in 1933 that when photosynthetic bacteria were grown anaerobically in a medium containing only minerals and low concentrations of a single source of organic carbon, the organic substrate disap-

TABLE 8.8. Equations representing the anaerobic growth of *Saccharomyces cerevisiae* on glucose. $\Delta G^{\circ\prime}$ values were calculated using the $\Delta G_f^{\circ\prime}$ values in Appendix 1.

Non-conservative reaction: $\Delta G^{\circ\prime} = -52.82$ kcal mole^{-1}(calc.)

$$(1)\ C_6H_{12}O_6(aq) \longrightarrow 2CO_2(aq) + 2CH_3CH_2OH(aq)$$

Conservative reaction:

Work reaction: $\Delta G^{\circ\prime} = -34.2$ kcal (calc.)

$$(2)\ 0.896C_6H_{12}O_6(aq) + 0.240H_2O(aq) \longrightarrow 1.490CO_2(aq) + 0.430C_3H_8O_3(aq) + 1.300CH_3CH_2OH(aq)$$

Conversion reaction: $\Delta G' = 0$ (hypothesis)

$$(3)\ 0.104C_6H_{12}O_6(aq) + 0.120NH_3(aq) \longrightarrow 0.590CH_{1.75}O_{0.45}N_{0.20}(\text{cells}) + 0.045CO_2(aq) + 0.288H_2O(l)$$

Total conservative reaction: $\Delta G' = -34.2$ kcal

$$(4)\ C_6H_{12}O_6(aq) + 0.12NH_3(aq) \longrightarrow 0.59CH_{1.75}O_{0.45}N_{0.20}(\text{cells}) + 1.55CO_2(aq) + 0.43C_3H_8O_3(aq) + 1.30CH_3CH_2OH(aq) + 0.05H_2O(l)$$

Source: Reference 39.

TABLE 8.9. Equations representing the aerobic growth of *Saccharomyces cerevisiae* on glucose. $\Delta G^{\circ\prime}$ values were calculated using the $\Delta G_f^{\circ\prime}$ values in Appendix 1.

Non-conservative reaction: $\Delta G^{\circ\prime} = -698.52$ kcal mole^{-1}(calc.)		
(1) $C_6H_{12}O_6(aq) + 6O_2(aq)$	$\longrightarrow$	$6CO_2(aq) + 6H_2O(l)$
Conservative reaction:		
Work reaction: $\Delta G^{\circ\prime} = -450.54$ kcal (calc.)		
(2) $0.645C_6H_{12}O_6(aq) + 3.870CO_2(aq)$	$\longrightarrow$	$3.870CO_2(aq) + 3.870H_2O(l)$
Conversion reaction: $\Delta G' = 0$ (hypothesis)		
(3) $0.355C_6H_{12}O_6(aq) + 0.290NH_3(aq)$	$\longrightarrow$	$1.950CH_{1.72}O_{0.44}N_{0.15}$(cells) $+ 0.200CO_2(aq) + 0.890H_2O(l)$
Total conservative reaction: $\Delta G' = -450.54$ kcal		
(4) $C_6H_{12}O_6(aq) + 0.29NH_3(aq) + 3.87O_2(aq)$	$\longrightarrow$	$1.95CH_{1.72}O_{0.44}N_{0.15}$(cells) $+ 4.07CO_2(aq) + 4.76H_2O(l)$

Source: Reference 39.

peared completely from the medium. When he used acetic or lactic acid as the substrate there was always a production of carbon dioxide gas during the growth of these bacteria, but when he used substrates more reduced than these carbon dioxide gas was always absorbed. These phenomena were explained by Van Niel[41] in 1941 in a consideration of the relative degrees of reduction of the substrate and the cellular material being grown. If a substrate more oxidized than cellular material becomes incorporated into cellular substance there will be some carbon remaining which must be eliminated as carbon dioxide. Conversely, under the anaerobic conditions in which the photosynthetic bacteria live, if the substrate is more reduced than the cells there will have to be an uptake of carbon dioxide so that the average degree of reduction

TABLE 8.10. Equations representing the aerobic growth of *Saccharomyces cerevisiae* on ethanol. $\Delta G^{\circ\prime}$ values were calculated using the $\Delta G_f^{\circ\prime}$ values in Appendix 1.

Nonconservative reaction: $\Delta G^{\circ\prime} = -322.84$ kcal mole^{-1} (calc.)		
(1) $CH_3CH_2OH(aq) + 3O_2(aq)$	$\longrightarrow$	$2CO_2(aq) + 3H_2O(l)$
Conservative reaction (growth reaction):		
Work reaction: $\Delta G^{\circ\prime} = -199.19$ kcal (calc.)		
(2) $0.617CH_3CH_2OH(aq) + 1.851O_2(aq)$	$\longrightarrow$	$1.235CO_2(aq) + 1.851H_2O(l)$
Conversion reaction: $\Delta G' = 0$ (hypothesis)		
(3) $0.383CH_3CH_2OH(aq) + 0.150NH_3(aq) + 0.264CO_2(aq)$	$\longrightarrow$	$1.030CH_{1.72}O_{0.41}N_{0.15}$(cells) $+ 0.488H_2O(l)$
Total conservative reaction: $\Delta G' = -199.19$ kcal		
(4) $CH_3CH_2OH(aq) + 0.15NH_3(aq) + 1.851O_2(aq)$	$\longrightarrow$	$1.030\ CH_{1.72}O_{0.41}N_{0.15}$(cells) $+ 0.97CO_2(aq) + 2.34H_2O(l)$

Source: Reference 39.

TABLE 8.11. Equations representing the aerobic growth of *Saccharomyces cerevisiae* on acetic acid. $\Delta G^{\circ\prime}$ values were calculated using the $\Delta G_f^{\circ\prime}$ values from Appendix 1.

Nonconservative reaction: $\Delta G^{\circ\prime} = -210.91$ kcal mole^{-1} (calc.)

$$(1)\ CH_3COOH(aq) + 2O_2(aq) \longrightarrow 2CO_2(aq) + 2H_2O(l)$$

Conservative reaction (growth reaction):

Work reaction: $\Delta G^{\circ\prime} = -140.35$ kcal (calc.)

$$(2)\ 0.668CH_3COOH(aq) + 1.336O_2(aq) \longrightarrow 1.336CO_2(aq) + 1.336H_2O(l)$$

Conversion reaction: $\Delta G' = 0$ (hypothesis)

$$(3)\ 0.332CH_3COOH(aq) + 0.090NH_3(aq) \longrightarrow 0.620CH_{1.62}O_{0.44}N_{0.15}(cells) + 0.044CO_2(aq) + 0.303H_2O(l)$$

Total conservative reaction: $\Delta G' = -140.35$ kcal

$$(4)\ \left.\begin{array}{l} CH_3COOH(aq) + 0.09NH_3(aq) \\ + 1.34O_2(aq) \end{array}\right\} \longrightarrow \begin{array}{l} 0.62CH_{1.62}O_{0.44}N_{0.15}(cells) \\ + 1.38CO_2(aq) + 1.64H_2O(l) \end{array}$$

Source: Reference 39.

of both of these substances taken together will be equivalent to that of the cellular substance. Battley adopted this idea with respect to constructing conversion-reaction equations when the substrate is more reduced than the cells. Thus, in these cases it is not oxygen, as Tamiya felt, but carbon dioxide which is the oxidant that brings the level of reduction of the substrate to that of the cells. This can be seen in Eq. 3 in Table 8.10, where carbon dioxide is represented as a reactant. This is also a more efficient process in that more cells are produced for less substrate carbon, and explains why there is a greater percentage efficiency of substrate energy conservation in the cells if the substrate is more reduced. A plot of the free-energy changes of the conservative (growth) reactions as calculated using Battley's hypothesis, ΔG_C, against those of the relative nonconservative reactions, $\Delta G^{\circ\prime}_{NC}$, is given in Fig. 8.7 and shows a close relationship which can be represented by the equation

$$\Delta G_C = K\,\Delta G^{\circ\prime}_{NC} \tag{8.12}$$

where K was defined by Battley as being a "thermodynamic growth constant" having a value of 0.645. This is a considerably different value from that of 0.707 obtained from Fig. 8.5 when the initial method of free-energy calculation was used. Battley emphasized that whether K could be considered a true constant or varies depending upon the kind of microorganism being grown remained to be decided by future experimentation.

The hypothesis that the conversion reactions in Tables 8.8–8.11 are accompanied by a zero change in free energy was suggested by at least one other biological system. This is the one described by Mechalas and Rittenberg[42] in 1960, in which *Desulfovibrio desulfuricans* was grown anaerobically on hydrogen as a source of energy and yeast extract as the source of carbon for cellular

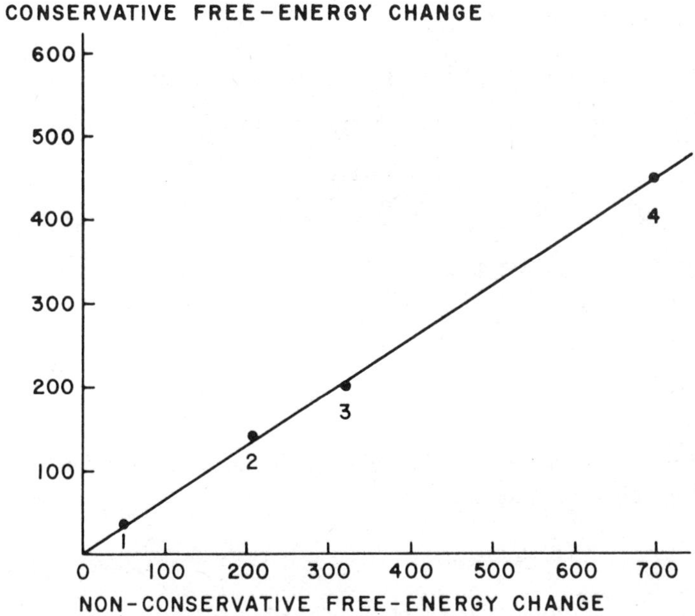

FIGURE 8.7. Graph showing calculated conservative free-energy changes accompanying the growth of *Saccharomyces cerevisiae* on different substrates plotted against those of the relative, nonconservative reactions. 1, anaerobic growth on glucose; 2, 3, and 4, aerobic growth on acetic acid, ethanol, and glucose, respectively. Reprinted from Reference 39 with permission.

construction. Here, the only source of free energy is an inorganic substance, $H_2(aq)$, the oxidation of which provides energy for the incorporation of organic material in the form of yeast extract into cellular material. The cells would not grow anaerobically on the yeast extract alone. Mechalas and Rittenberg carried out a similar experiment in which a nonassimilable organic substrate, isobutanol, was anaerobically oxidized by a growing culture of *D. sulfuricans* to provide energy for the conversion of substances in yeast extract into cells. Thus, this must be another example in which the free energy of the energy source is completely converted into enthalpy and entropy, and the net free-energy change accompanying the conversion of organic substances into cellular material can be considered to be zero.

The amount of free-energy conserved within the fabric of the cells which have been formed during the anaerobic and aerobic growth of *S. cerevisiae* on glucose can be calculated using conservation theory in exactly the same manner as has been done for enthalpy changes in Section 8.9. Battley did not actually make this calculation, but it is worth doing for purposes of demonstration. Thus, using Eq. 1 in Table 8.1, the calculated value for the free-energy change accompanying the anaerobic growth of *S. cerevisiae* on 1 mole of glucose of -34.2 kcal from Table 8.8, and the appropriate values for

$\Delta G_f^{\circ\prime}$ from Appendix 1,

$$\begin{aligned}0.59\ \Delta G_f(\text{cells}) &= \Delta G - 0.43\ \Delta G_f^{\circ\prime}\ C_3H_8O_3 - 1.54\ \Delta G_f^{\circ\prime}\ CO_2\\ &\quad - 1.30\ \Delta G_f^{\circ\prime}\ CH_3CH_2OH + \Delta G_f^{\circ\prime}\ C_6H_{12}O_6\\ &\quad + 0.12\ \Delta G_f^{\circ\prime}\ NH_3\\ &= -34.2 + 51.13 + 142.08 + 56.47 - 218.58 - 0.76\\ &= -3.86\ \text{kcal}\end{aligned}$$

from which

$$\Delta G_f(\text{cells}) = -6.54\ \text{kcal}$$

Using Eq. 8.2 to calculate the free-energy change accompanying the combustion of the cells which were grown

$$\begin{aligned}\Delta G &= 0.59\ \Delta G_f^{\circ\prime}\ CO_2 + 0.34\ \Delta G_f^{\circ}\ H_2O + 0.12\ \Delta G_f^{\circ\prime}\ NH_3\\ &\quad - 0.59\ \Delta G_f(\text{cells}) - 0.63\ \Delta G_f^{\circ\prime}\ O_2\\ &= -54.43 - 19.27 - 0.76 + 3.86 - 2.46\\ &= -73.06\ \text{kcal}\end{aligned}$$

Using a similar calculation the free-energy change accompanying the combustion of 0.43 mole of glycerol(aq) was found to be -171.19 kcal. The sum of these two values is therefore $-73.06 + -171.19 = -244.25$ kcal as the free-energy change accompanying the oxidation of both products. Of this, $(-73.06/-244.25) \times 100 = 29.91\%$ is the energetic contribution of the cellular material in terms of free energy. The total free energy conserved during the anaerobic growth reaction is that given with Eq. 1 in Table 8.3 minus the free-energy change accompanying the anaerobic glucose growth reaction, or $-52.82 - (-34.2) = -18.62$ kcal. The actual quantity of free energy conserved within the cells is then $(0.2991 \times -18.62 \times 100)/-52.82 = 10.54\%$ of the total available. The rest of the conserved free energy is found in the glycerol, which contains 24.70% of the total free energy available. The remaining quantity, 64.76%, represents the free-energy change of the growth reaction (i.e., $0.6476 \times 52.82 = -34.2$ kcal). The free energy conserved in the aerobic glucose growth reaction is that of the relative nonconservative reaction given with Eq. 2 in Table 8.3 minus that of the growth reaction (since there are no products of this latter other than the cells) from Table 8.9. The efficiency of free-energy conservation is therefore $100 \times [(-698.52 - (-450.54)]/-698.52 = 35.50\%$. The ratio of the free-energy efficiency of growth aerobically on glucose over that obtained anaerobically should be similar to that found with other methods such as cell dry weights or turbidities. This ratio is calculated to be 35.50/10.54 = 3.4. The ratio found using cell turbidity as a criterion or growth is seen from Fig. 8.1 to be 3.4. The ratio of the dry weights of cells produced during the two growth reactions is 3.25 on an "ashes present" basis and 3.26 on an "ash-free" basis. In principle, it is the ratio of the efficiencies of cellular free-energy conservation in this comparison that would be expected to be the most accurate. An addition of the free-energy changes accompanying the reactions represented by Eqs. 1, 2, and 3 in Table 8.2 gives −34.2 + (−258.94) + (−171.19), respectively, = −464.33 kcal as the free-energy change accompanying the aerobic glucose growth reaction. The calculated value found in Table 8.9 is −450.54 kcal $mole^{-1}$ of glucose oxidized, the value found by summation agreeing with this latter by 3.0%, which is quite good.

8.12. SUMMARY

The work of Battley described in this chapter is important in that he recognized the distinction between growth and oxidative or fermentative assimilation and how this latter might interfere with measurements involving the former, resulting in values for growth that were too high. With respect to this he emphasized the necessity of establishing the nonlimiting nature of the culture medium and of working only with cells which were in the exponential phase of growth. He was the first to work with a facultative microorganism so that the energetics of aerobic and anaerobic growth could be compared, and to use unit-carbon formulas to represent the cellular material formed during growth. He conceived the idea of nonconservative and conservative reactions as applied to growth-reaction systems, and originated the concept that the energy conserved during a growth reaction was a function of the extent to which the relative nonconservative reaction had not proceeded to completion. He established the additivity of growth-reaction equations, thereby demonstrating that in spite of their internal complexity they could be treated just as reaction equations representing more simple reactions. Hess's Law was demonstrated to apply also to growth reactions, as it would be expected to do. Battley was the first to attempt to calculate ΔG, ΔH, and ΔS values for growth reactions. His initial attempts were not particularly successful. His subsequent hypothesis that the anabolic part of a growth reaction is accompanied by a zero change in free energy was based on more sound thermodynamic reasoning and may well prove to be true. He was the first to demonstrate that the energy change accompanying the growth of a given microorganism on several substrates is directly proportional to the extent to which the relative nonconservative reaction has proceeded to completion, and also that the efficiency with which energy is conserved is a function of the degree of reduction of the substrate. Finally, he emphasized the necessity for carrying out experiments on the energetics of microbial growth using low concentrations of substrate, of the order of 0.01 M or less.

REFERENCES

1. P. W. Wilson and W. H. Peterson, The energetics of heterotrophic bacteria, *Chem. Rev.* **8,** 427–480 (1931).
2. K. Burton and H. A. Krebs, The free-energy changes associated with the individual steps of the tricarboxylic acid cycle, glycolysis, and alcoholic fermentation and with the hydrolysis of the pyrophosphate groups of adenosine triphosphate, *Biochem. J.* **54,** 94–107 (1953).
3. K. Burton, Free-energy data of biological interest, *appended to* H. A. Krebs and H. L. Kornberg, Energy transformations in living matter. A survey, *Ergeb. Physiol. Biol. Chem. Exp. Pharmokol.* **49,** 212–298 (1957).
4. E. H. Battley, Growth-reaction equations for *Saccharomyces cerevisiae*, *Physiol. Plant.* **13,** 192–203 (1960).
5. E. H. Battley, Enthalpy changes accompanying the growth of *Saccharomyces cerevisiae* (Hansen), *Physiol. Plant.* **13,** 628–640 (1960).
6. E. H. Battley, A theoretical approach to the study of the thermodynamics of growth of *Saccharomyces cerevisiae* (Hansen), *Physiol. Plant.* **13,** 674–686 (1960).
7. E. H. Battley, Alternate method of calculating the free-energy change accompanying the growth of *Saccharomyces cerevisiae* (Hansen) on three substrates, *Biotech. Bioeng.* **21,** 1929–1961 (1979).
8. E. Duclaux, *Traité de Microbiologie.*, III, Masson et Cie., Editeurs, Paris, 1900.
9. H. A. Spoehr and H. W. Milner, The chemical composition of *Chlorella*; effect of environmental conditions, *Plant Physiol.* **24,** 120–149 (1949).
10. J. Monod, *Recherches sur la Croissance des Cultures Bactériennes*, Herman & Cie., Paris, 1942.
11. W. R. Devloo, A sterol indispensable to yeast, *Arch. Physiol.* **46,** 157–188 (1938).
12. A. A. Anderson and T. J. B. Stier, Anaerobic nutrition of *Saccharomyces cerevisiae*; Ergosterol requirement for growth in a defined medium, *J. Cell. Comp. Physiol.* **36,** 159–164 (1950).
13. W. H. Maguigan and E. Walker, Sterol metabolism of microorganisms, I. Yeast, *Biochem. J.* **34,** 804–813 (1940).
14. W. W. Umbreit, R. H. Burris, and J. F. Stauffer, *Manometric Techniques and Tissue Metabolism*, Burgess Publishing Co., Minneapolis, 1949.
15. A. Schatz, Uptake of carbon dioxide hydrogen, and oxygen by *Hydrogenomonas facilis*, *J. Gen. Microbial.* **6,** 329–335 (1952).
16. K. Baalsrud and K. S. Baalsrud, The role of phosphate in CO_2 assimilation of thiobacilli, in W. D. McElroy and H. B. Glass, Eds., *Phosphorous Metabolism*, Johns Hopkins Press, Baltimore, 1952.
17. R. K. Clayton, Photosynthesis and respiration in *Rhodospirillum rubrum*, *Arch. Mikrobiol.* **22,** 180–194 (1955).
18. C. Neuberg and K. Karczag, Über zuckerfrei Hefegärung. III; 1. Quantitative Versuche über die Zerstörung von Brenztraubensäure, Weinsäure, and Glycerinphosphosäure durch Hefen, *Biochem. Z.* **36,** 60–67 (1911).
19. M. G. Sevag, M. D. Newcomb, and R. E. Miller, Phosphatase-antiphosphatase reaction. Competition between specific substrate and antiphosphatase for phosphatase, *J. Immunol.* **72,** 1–11 (1954).

20. O. Meyerhof and H. Green, Synthetic action of phosphatase, *J. Biol. Chem.* **178,** 655–667 (1949).

21. A. J. Brown, Influence of oxygen and concentration of alcoholic fermentation, *J. Chem. Soc., Trans.* **61,** 369–385 (1892).

22. W. H. Swanson and C. E. Clifton, Growth and assimilation in cultures of *Saccharomyces cerevisiae*, *J. Bact.* **56,** 115–124 (1948).

23. S. R. Hoover and F. E. Allison, The growth metabolism of *Rhizobium*, with evidence on the interrelations between respiration and synthesis, *J. Biol. Chem.* **134,** 181–192 (1940).

24. L. Pasteur, Mémoire sur la fermentation alcoolique, *Ann. Chim. Phys., 3è série,* **58,** 323–426 (1860).

25. W. Connstein and L. Lüdecke, U.S. Patent 1511754, 1924.

26. E. I. Fulmer, L. A. Underkofler, and R. J. Hickey, U.S. Patent 241645, 1947.

27. E. Calvet, Nouveau microcalorimètre différentiel et à compensation, *C.R. Hebd. Séances Acad. Sci.* **226,** 1702–1704 (1948).

28. J. W. Winzler and J. P. Baumberger, The degradation of energy in the metabolism of yeast cells, *J. Cell. Comp. Physiol.* **12,** 183–211 (1938).

29. M. Dubrunfaut, Note sur la chaleur et le travail méchanique produit par la fermentation vineuse, *C.R. Hebd. Séances Acad. Sci.* **42,** 945–948 (1856).

30. A. Bouffard, Détermination de la chaleur dégagée dans la fermentation alcoolique, *C.R. Hebd. Séances Acad. Sci.* **121,** 357–360 (1895).

31. A. J. Brown, The heat of fermentation, *Inst. Brew. London, J.* **7,** 93–103 (1903).

32. M. Rubner, Die Umsetzungswärme bei der Alkolgärung, *Arch. Hyg.* **49,** 355–418 (1904).

33. H. T. Brown, Some studies on yeast, *Ann. Bot.* **28,** 197–226 (1914).

34. C. E. Clifton, Prevention of assimilation in respiring cells, *Nature* **140,** 318 (1937).

35. C. E. Clifton, On the possibility of preventing assimilation in respiring cells, *Enzymologia* **4,** 246–253 (1937).

36. T. Bauchop and S. R. Elsden, On the growth of microorganisms in relation to their energy supply, *J. Gen. Microbial.* **23,** 457–469 (1960).

37. E. H. Battley, On the calculation of the thermodynamic changes accompanying the growth of microorganisms, Abstracts of Contributed Papers, International Biophysics Congress, Stockholm, Proceedings, p. 55 (1961).

38. E. H. Battley, A calculation of the enthalpy change accompanying glucose fermentation, *Bact. Proc.* **P47** (1962).

39. E. H. Battley, A calculation of the free-energy changes accompanying the growth of *Saccharomyces cerevisiae* on several substrates, Proc. First European Biophys. Congr., V.IV, Baden, Austria, pp. 299–305 (1971).

40. F. M. Müller, On the metabolism of the purple sulfur bacteria in organic media, *Arch. Mikrobiol.* **4,** 131–166 (1933).

41. C. B. Van Niel, The bacterial photosyntheses, *Adv. Enzymol.* **1,** 263–328 (1941).

42. B. J. Mechalas and S. C. Rittenberg, Energy coupling in *Desulfovibrio desulfuricans, J. Bact.* **80,** 501–507 (1960).

9

RECENT DEVELOPMENTS, 1961–1985

Studies related to the energetics of microbial growth during the period 1961–1985 can be divided roughly into four groups. One group consists of largely theoretical investigations into the nature of entropy changes as these take place during microbial growth. A second group is represented by an extensive series of studies on the relationships between the amount of microbial growth obtained from a given substrate and various energetic parameters. A third group is made up of an increasing number of studies dealing with bioengineering of one kind or another intended to lead toward more efficient industrial process design. The literature in this group is far too extensive, and most of the mathematical or instrumental methods much too intricate to describe in detail. The fourth group includes studies in microcalorimetry and in methods of estimating the changes in free energy, enthalpy, and entropy which accompany microbial growth. Chapter 9 is intended to cover the various developments leading to the comprehensive description of the energetics of microbial growth in Chapter 10.

9.1. STUDIES ON ENTROPY

As studies of cellular biochemistry progressed through the past 85 years the almost incomprehensible structural complexity of cells became increasingly evident. Concurrently, the idea developed that because this complexity was so much greater than that of the far more simple substances serving as substrates and nutrients for cellular growth, the state of being a cell was highly improbable, and therefore the formation of a cell must be accompanied by a considerable decrease in entropy. The confirmation of this idea has not been readily forthcoming. For whatever reasons, no one has yet attempted by direct thermal measurements to determine experimentally the absolute entropy of a unit mass of dried microbial cells. Such a determination could be made indirectly if the free energy and the enthalpy of formation of this unit mass of cells were known, along with its empirical composition. This method also has not been used. The only remaining method is that first investigated by Morowitz[1–3] (see also Section 7.7) in which the calculation of entropy is made partly through the use of thermal physics and partly through the use of information theory. It is these last techniques which have in general prevailed in entropy studies.

9.1.1. The Comments of Passynsky and of Morrison

In 1961 A. Passynsky[4] made some comments on the limitations in the biological role of entropy. He pointed out (as others had done before him) that the local decrease in entropy of an organism is overcompensated by the entropy increase which occurs in other parts of the system. There is thus no contradiction with respect to the second law. However, he also made the following statements.

> "The change of the entropy or of the intensity of its production (as considered in the thermodynamics of irreversible processes) is not the driving force of the processes of evolution or ontogenesis. The change of the negative entropy in organisms cannot serve as the measure of the ordering of biological structures and biochemical processes.", and "The thermodynamic entropy in its physical sense is different from the information function and therefore the connection between the entropy and disordering in biological systems must be very carefully considered."(Passynsky,[4] p. 53.)

Passynsky may have been the first to emphasize that informational entropy and physical (structural) entropy are not the same thing. This distinction was also emphasized by Chambadal[5] in 1963. That the probabilistic concept of informational entropy and the physical concept of structural entropy are in many ways similar but not identical was seemingly ignored by Morowitz[1–3] in the late 1950s and again in 1968 when he (Morowitz[6]) published a book on the general subject of energy flow in biology. Here the entropy decrease which takes place when nutrient molecules are ordered into a cellular structure was calculated by means of the following equation.

$$\Delta S_0 = -kI \ln 2 \tag{9.1}$$

where k represents Boltzmann's constant and I the increase in information of the cell over that of the unordered clusters of atoms comprising the component molecules of the cell. A more restricted attempt to estimate the structural entropy of a cell was made in 1962 and 1969 by Lwoff,[7,8] who related it to the logarithm of the number of permutations of the bases comprising the DNA of the cell.

Passynsky's comments with respect to entropy changes at higher levels of integration were amplified by P. Morrison in the following profound statement.

> "Nearly all the manifest visual and mechanical intricacy of organisms, like their apt behavior, turns out to be without quantitative thermodynamic importance. Morphology and ecology are . . . only small secondary properties of a fundamentally thermodynamic system." (Morrison,[9] p. 520.)

Passynsky's and Morrison's comments were intuitive, and based on theoretical reasoning rather than experimental finding. But what they suggest is that whatever structural entropy is in living cells, its quantity is related largely to the lowest level of cellular integration, which is that of the atomic building blocks of which cells are composed.

9.1.2. The Theoretical Investigations by Gorski

Beginning in 1966 a series of articles was published by F. Gorski[10–15] in which he developed his ideas on the structural entropy of organisms. Essentially, his technique is similar to that employed by Morowitz[1] (p. 83) in that a cell can be considered to be made up of a number of boxes each of an average unit atomic volume. According to Gorski, "The space V occupied by an organism is divided into n very small but very numerous space elements $\Delta \nu$ (cubes); n is the sum of all the numbers of C, H, O, N, etc. atoms and water molecules entering into the composition of the organism. The atoms and water molecules may be distributed in space V in N ways, N being a very large number." (Gorski,[13] p. 263.) Following the mathematical development of his theory and making a few simplifications, he came to the conclusion that, "The structural entropy of an organism is equal to the logarithm of the number of permissible distributions of its atoms and H_2O molecules preceded by a negative sign." (Gorski,[13] p. 268.) This conclusion can be represented by the following equation:

$$\Delta S_s = -k \ln N \tag{9.2}$$

where ΔS_s represents the structural entropy, k is Boltzmann's constant, and N the number of distributions of atoms and water molecules within the space occupied by a cell (or an organism). Morowitz[6] extended his own argument somewhat further by relating N to the number of bits of information contained within a cell, thus arriving at Eq. 9.1.

Neither Morowitz's nor Gorski's ideas with respect to structural entropy have been the subjects of rigorous experimental verification. Seemingly, the only way to obtain valid entropy values would be to deal with balanced chemical reaction equations representing systems in which one of the products of the reaction is a given mass of cells. Then, provided that the free-energy and enthalpy changes accompanying such reactions were known, and that the free energies and enthalpies of formation of all the reactants and products except the cells were also known, the entropy of formation of the cells from the elements could be calculated using the Gibbs free-energy equation. This could be considered as the most standardized form of structural entropy. The entropy of formation from preformed organic substances as nutrients could also be calculated *if* the nature, thermodynamic properties, and quantities of these nutrients were accurately known in such a way as to be able to include them in a balanced equation. This is presumably what Morowitz[2,3] did in his compilations of the types and quantities of chemical bonds in the nutrient medium and in the cells which were formed from the nutrients. Calculations of ΔS_f values for cells will be considered in Chapter 10.

9.2. YIELD STUDIES AND ENERGETICS

The work of Monod[16] established clearly that the amount of growth obtained from a microorganism was related directly to the quantity of a single source of carbon and energy on which it was grown, and that a given microorganism grown on a given carbon source under a given set of physiological conditions would always produce the same yield of cells. This yield, designated by Y or Y_{sub}, is usually expressed in terms of grams dry weight of cells produced per unit quantity of substrate utilized, the most common being grams dry weight of cells per gram of substrate, or per mole of substrate. The latter is called the molar growth yield. When Bauchop,[17] and Bauchop and Elsden[18] conceived the idea of a Y_{ATP} and demonstrated that the amount of microbial growth was proportional to the amount of ATP generated during catabolism, a whole new field of microbial energetics came into view, and both biochemists and microbiologists entered it with enthusiasm. It is at this point that energetics seems to depart from thermodynamics. Classical thermodynamics is a rigorous discipline dealing with substances the thermodynamic properties of which have been very accurately determined, and with energy exchanges which can be closely predicted. Biological energetics, on the other hand, while it also deals with energy exchanges, often cannot be considered particularly rigorous, partly because of what appears to be either inherent biological variability or insufficiently controlled experimentation, and partly because we may not know how to define the systems.

9.2.1. The Review by Senez

In 1962 J. Senez[19] published a review in which he considered what was then known about the energetics of bacterial growth. In it he cited earlier literature and pointed out that calculations of free-energy efficiencies of growth were for the most part made using data derived from experiments using autotrophic bacteria. Thus, Meyerhof[20] and Baas-Becking and Parks[21] defined free-energy efficiency according to the following relationship.

$$\text{free-energy efficiency (\%)} = 100 \times \frac{E_c}{E_s} \qquad (9.3)$$

where E_s represents the free-energy change accompanying the oxidation of the inorganic substrate utilized as a source of energy, and E_c represents the heat of combustion of the cells that are grown. This mixture of free-energy and enthalpy changes, which is not particularly good thermodynamics, was made because of the general observation that the free-energy and enthalpy changes accompanying the oxidation of organic substances to carbon dioxide and water are often not greatly different from one another, and thus that one value could reasonably replace the other. The experiments of Lees[22] with short-term cultures had demonstrated that free-energy efficiencies calculated by means of Eq. 9.3 ranged from 25 to 50% of the total free-energy change accompanying substrate oxidation.

Senez went on to consider certain aspects of the energetics of heterotrophic metabolism, stating that:

> . . . for a heterotrophic organism growing at the expense of an organic compound or some of its intermediary products as the carbon source, it is clear that the quantity of energy incorporated into cellular material is the difference between the heat of formation of the cells and the heat of formation of the form under which carbon is assimilated. In most cases, this form is not known, so that the free-energy efficiency cannot be estimated. [Quoted by permission from Reference 19, p. 96.]

The reaction referred to can be represented as follows:

$$\left.\begin{array}{r}\text{assimilated substances from}\\ \text{which cells are constructed}\end{array}\right\} \longrightarrow \text{cells} + \text{products?} \qquad (9.4)$$

However, even if the reaction represented by Eq. 9.4 could be better represented by a stoichiometrically balanced equation, the difference between the sum of the heats of formation of the cells and the products (if any) and those of the assimilated substances from which the cells are constructed would be the heat of the reaction represented by Eq. 9.4. It is the *biological* heat of oxidation of the cells which indicates the quantity of enthalpic energy incorporated in the cells. This latter would still bear no relationship, per se, to the free-energy efficiency.

Senez then went on to state:

> . . . it must be stressed that if a heterotroph initiates directly the biosynthesis of the cell material from glucose or from another substance in which the carbon is at the same level of oxidation as in the cell constituents (CH_2O), then no energy is required for the reduction of the carbon source and the concept of the free-

energy efficiency [Eq. 9.3] as defined above would be meaningless. A similar case is provided by an organism growing on a medium in which all the monomers of the cellular macromolecules are provided and assimilated as such. [Quoted by permission from Reference 19, p. 96.]

As a basis for this latter statement he cited the studies of Morowitz[3] (see also Section 7.7) in which the minimal variation of entropy was calculated for the biosynthesis of *Escherichia coli* in a synthetic medium containing glucose as the sole source of carbon and energy. The calculated value of only 58.4 cal g^{-1} (dry weight) represented less than 1% of the free energy made available by the accompanying oxidation of the glucose in the culture medium, and Senez assumed that theoretically the minimal energy required for the synthesis of bacterial cells from their preformed monomers is of the same order of magnitude. Both Morowitz[3] and Senez[19] believed that the expenditure of free energy for cellular biosynthesis was considerably more than that represented by Morowitz's theoretical minimum, and Senez was of the opinion that:

the measurements of free energy efficiency have restricted application and significance. They only take into consideration the free energy incorporated into the cell constituents for the reduction of the nutrients to the cellular level and do not provide information on the requirement and efficiency of free energy for the other steps of the biosynthesis. [Quoted by permission from Reference 19, p. 97.]

It is clear that Senez's views were considerably different from those of Battley[23] (see also Section 8.12) who had expressed the idea one year earlier that the free-energy change accompanying the conversion of substrate substances into cells must be very small (Battley hypothesized that it was zero). The principal difference between the two approaches to the study of energy exchanges accompanying microbial growth was Senez's apparent belief that the subject of greatest importance to consider was the conversion of ATP energy into cellular energy, whereas Battley was concerned only with initial and final states of processes involving or including growth which could be represented by equations. Inherent in Battley's approach was that since ATP and ADP plus P_i are recycled during metabolism, the energy conserved in ATP synthesis becomes converted into heat and entropy upon hydrolysis, and thus these energy exchanges can be considered to be part of the overall energy exchanges accompanying catabolism. The free-energy change accompanying anabolism is effectively zero. Senez's approach and that of all who studied the proportionality of microbial growth to anything other than free energy (such as the Y_{ATP}) was obviously that of energetics. Battley's was more properly that of thermodynamics. Ideally, the two should give the same answers. That they often do not is because of the simplicity of thermodynamics in dealing with an overall phenomenon, as opposed to the complexities of energetics in which what is studied is often only indirectly related to energy exchange.

At the time Senez wrote his review the work carried out previously appeared to suggest that the Y_{ATP} could be considered a constant. It was his opinion that:

> The recent discovery that the growth-yield per mole of ATP is identical for very different microorganisms growing from the preformed monomers of cellular macromolecules is of special importance. It provides a quantitative estimation of the actual energy requirements of the polymerization process and shows that these steps of cell biosynthesis are the principal consumers of energy under the form of phosphate-anhydride bonds. Moreover, some of the experimental observations reported in this discussion suggest that this new biological constant may provide a most valuable tool for the elucidation of unknown metabolic pathways. [Quoted by permission from Reference 19, p. 106.]

9.2.2. Studies on the Y_{ATP}

The enthusiastic predictions of Senez in 1962 did not come true. In 1963 Hungate[24] carried out a study with *Ruminococcus albus* indicating either that sources of energy exist for the growth of these bacteria other than those found in the usual fermentation reactions, or that the Y_{ATP} is not constant. The same conclusion was made by Hopson[25] in 1965 with respect to *Selenomonas ruminantium.* In 1967 after studying anerobic growth yields with *Aerobacter cloacae* and *Escherichia coli* Hernandez and Johnson concluded that "although Y_{ATP} may have a maximal value, it cannot be considered to have a constant value." (Hernandez and Johnson,[26] p. 994). As seen in subsequent reviews by Stouthamer[27] in 1969, Payne[28] in 1970, Decker, Jungermann, and Thauer[29] in 1970, Forrest and Walker[30] in 1971, Stouthamer and Bettenhaussen[31] in 1973, and Stouthamer[32] in 1976, the results of a great deal of experimental investigation involving both biochemists and microbiologists made it abundantly evident that the Y_{ATP} is *not* a constant. Even in anaerobic microorganisms during the metabolism of which the number of ATPs generated by substrate-level phosphorylation can be ascertained with considerable accuracy, the Y_{ATP} has been demonstrated to vary from 5.9[33] to 20.9.[34] The variation in Y_{ATP} values is even worse in aerobic organisms because of the difficulty in determining P/O ratios representing oxidative phosphorylation during the metabolism of growing cells. Many ingenious methods have been devised to estimate this ratio, all of them having to make use of one assumption or another. It finally became apparent that good generalizations with respect to the Y_{ATP} are hard to make, and that probably each microorganism should be treated on its own merits as regards a Y_{ATP} value if it is important to know this. At the present time interest in Y_{ATP} determinations seems to have waned, much as it did in fermentative and oxidative assimilation during the early 1950s, although the occasional article on determining P/O ratios sometimes still appears.

9.2.3. The Yield Factor Determinations by Payne and His Collaborators

The intensive studies on the Y_{ATP} would seem to reflect a need on the part of interested investigators to relate the quantity of microbial growth to some energetic parameter such as, for example, the amount of ATP which was formed during catabolism. This ought not to be surprising since these relationships represent a higher level of integration than that represented by the more simple biochemical reactions and pathways that participate in catabolism. It is one of several "next steps" which must be taken to enable us to understand more about cellular synthesis. Thus, it is equally not surprising that other yield relationships should be considered. In fact, microbial growth can be demonstrated to be proportional to quite a large number of things. It had been pointed out many years previously by Karasch and Scher[35] and by Karasch[36] that any organic compound can be characterized in part by its constituent number of electrons, which is the number of electrons capable of being transferred from a unit mass of a given substance to oxygen gas. Normally the unit mass is represented by 1 mole of substance if its composition and molecular weight are known; otherwise, the mass of the substance can be expressed in grams. All organic substances that can be oxidized have a certain number of transferrable electrons. For example, the combustion of glucose in a bomb calorimeter is an oxidation–reduction reaction which can be represented as follows.

$$C_6H_{12}O_6 + 6H_2O \longrightarrow 6CO_2 + 24H^+ + 24e^- \quad \text{oxidation}$$

$$24H^+ + 24e^- + 6O_2 \longrightarrow 12H_2O \quad \text{reduction}$$

$$C_6H_{12}O_6 + 6O_2 \longrightarrow 6CO_2 + 6H_2O \quad \text{oxidation–reduction}$$

Because the oxidation of glucose involves the transfer of 24 electrons to oxygen, glucose is said to be a 24 electron substance, or to be able to transfer 24 electron equivalents to oxygen. Extending the ideas of Karasch and Sher, and Karasch, it is apparent that substances containing nitrogen can be treated similarly, except that here attention must be paid to the oxidation state of the nitrogenous product of combustion. The product of the biological respiration of nitrogen-containing organic substances by microorganisms is usually ammonia. However, if such substances are combusted in a bomb calorimeter the organic nitrogen becomes oxidized to nitrogen gas. The number of electrons transferred will be less in the former than in the latter case. Thus, the combustion of glycine in a bomb calorimeter can be represented in the following manner.

$$H_2NCH_2COOH + 2H_2O \longrightarrow 2CO_2 + 0.5N_2 + 9H^+ + 9e^-$$
oxidation

$$9H^+ + 9e^- + 4.5O_2 \longrightarrow 4.5H_2O$$
reduction

$$H_2NCH_2COOH + 4.5O_2 \longrightarrow 2CO_2 + 0.5N_2 + 2.5H_2O$$
oxidation–reduction

Under these conditions of oxidation glycine is a 9 electron substance, because 9 electron equivalents have been transferred to oxygen. This is *not* what happens biologically. If glycine were to be oxidized by the process of biological respiration, instead of combustion as in a bomb calorimeter, a different reaction would take place which can be represented as follows:

$$H_2NCH_2COOH + 2H_2O \longrightarrow 2CO_2 + NH_3 + 6H^+ + 6e^-$$
oxidation

$$6H^+ + 6e^- + 1.5O_2 \longrightarrow 3H_2O$$
reduction

$$H_2NCHCOOH + 1.5O_2 \longrightarrow 2CO_2 + NH_3 + H_2O$$
oxidation–reduction

Under the conditions of biological oxidation, where the nitrogenous product is ammonia rather than nitrogen gas, the number of electrons transferred from glycine is 6 rather than 9. This difference between the number of electrons transferred during bomb calorimetry and during biological respiration has not always been recognized. It should also be emphasized that because *all* organic substances have transferrable electrons, substances containing C, H, O, and N for which only an empirical formula is known, whether or not the subscripts for these atoms are expressed as integers, the same (or other) methods of determining the number of transferable electrons per unit mass of substance can be used equally well.

In 1967 Mayberry et al. conceived the idea of an "available electron" as one which is available "either for transfer to oxygen in respiration, or for incorporation and non-respiratory reductions involved in synthesis." (Mayberry et al.,[37] p. 1334.) They then used this concept to define a "$Y_{av\ e^-}$" as the molar growth yield in grams divided by the number of available electrons from 1 mole of substrate, finding for a number of soil pseudomonads that this value fell near 3.0, being specifically 3.14 ± 0.11 for their particular study. Their experimental method was to grow the cells in Warburg respirometer flasks in a medium containing a single source of carbon and energy, and to monitor the oxygen uptake as a function of time. Similar experiments were carried out in 1969 with *Pseudomonas aeruginosa* by Mackechnie and Dawes,[38] who found that after the consumption of oxygen had ceased to become exponential it still continued although at a steadily decreasing rate. Accompanying this latter was an inversely related decrease in the molar growth yield (see Reference 38, graph on p. 346). These authors state that, "The oxygen consumption which occurs at the end of exponential growth is presumably due to the oxidation of endogenous materials and therefore not a reliable indication of the 'growth curve' of the culture." (Mackechnie and Dawes,[38] p. 345.) Mayberry et al.[37] apparently faced similar problems, and Payne[28] was careful to point out that:

> . . . for precision in studies of this sort a number of restrictions must be observed. The dry weight determinations of yield must be made as soon as oxygen uptake in the culture no longer increases logarithmically with time and exact accounting

for the quantity of substrate consumed must be obtained at that time to avoid weight loss due to endogenous respiration. [Payne,[28] p. 34.]

It is thus abundantly evident that the yields which Mackechnie and Dawes and Mayberry et al. obtained may have been those of cells containing considerable quantities of storage materials. On the other hand, their results may also have been due to the degradation of some of the fabric of the cells following the exhaustion of the external substrate. In either case, the growth yield would be less if the experiments had been run until there was no further change in the weight of cells per unit volume of culture medium, until there was no further gas exchange, or until no further heat was produced within the medium. Similar difficulties may have been encountered in other studies of this sort which have been summarized by Payne. It is apparent that from a total of 79 experiments carried out by different investigators from the data of which $Y_{\text{av } e^-}$ values could be calculated, an average value of 3.07 was obtained. However, these values ranged from 1.20 for *Salmonella typhimurium* growing on citrate to 3.75 for *Aerobacter aerogenes* growing on sorbitol, both on minimal medium. Thus, just as with the Y_{ATP} which was initially thought to be constant but for which it was later proven otherwise, the average value of 3.07 for the $Y_{\text{av } e^-}$ may also not be constant, and if it is important to determine this value for a given microorganism it would be better to treat the latter on its own merits. However, perhaps the value of 3.07 is in fact nearly constant, the large scatter of the data being due to nonexacting experimental assays. In either case, although averages of Y_{ATP} and $Y_{\text{av } e^-}$ can be calculated, it must be questioned what they mean, especially when it is frequently apparent that the data from which they were obtained included those with respect to cells which were both growing and incorporating storage products, so that it was not just growth that was being measured. The quantity Y is a weight measurement in which no accounting is made of what is being weighed. If the cells have transformed some of the substrate into a storage carbohydrate which contains a high percentage by weight of oxygen, or if the cells have simply stored a carbohydrate substrate in a polymerized form such as starch or glycogen, the yield will appear to be high. This is because the overall composition of the cells will then be less reduced than it would be were only structural cellular fabric to be formed. The quantity Y may well mean the weight of cells produced during a period of growth, but it can also mean either (1) the weight of cellular structure excluding all storage products, in which case it represents a true growth yield or (2), the weight of cellular structure plus assimilated storage products, in which case it does *not* represent a true growth yield. All other conditions for growth being equal, the former value for Y will always be less than the latter. This distinction is almost never made, and the failure to make it has had a considerable effect on the interpretation of experimental results involving yield values. All yield values are not necessarily growth-yield values.

Mayberry et al.[37] also considered the relationship of the molar growth yield to the quantity of energy removed from the medium during growth. For this purpose a Y_{kcal} was defined as the grams dry weight of cells (not ash free) produced during growth divided by the total quantity of enthalpic energy disappearing from the medium. Taking from the literature the heats of combustion of 18 selected compounds, they calculated that the heat of combustion per available electron had a mean value of 26.53 kcal eq^{-1}, in good agreement with the value of 26.05 kcal eq^{-1} as determined by Karasch[36] for the transfer of electrons from a "methane-type" bond to oxygen. Dividing the value of 3.14 which they had obtained for $Y_{av\ e^-}$ by 26.53 kcal eq^{-1} gave a value for Y_{kcal} of 0.118 g dry wt $kcal^{-1}$ of enthalpic energy utilized.

In a continuation of their studies published in 1968 Mayberry et al.[39] carried out an elemental analysis of the two bacteria used in their previous studies, finding mean values of: carbon, 45.6%; hydrogen, 7.6%; nitrogen, 12.1%; ash, 3.0%; and oxygen (by difference), 31.7%. On an ash-free basis the empirical formula for the cells was therefore $C_{3.914}H_{7.773}O_{2.042}N_{0.890}$. This was rounded off to $C_4H_8O_2N$, the formula weight then becoming 102 daltons. The mean heat of combustion of these cells was 5.27 kcal g^{-1}. By theoretically burning one empirical formula weight of these cells in a bomb calorimeter to CO_2, H_2O, and N_2 20 electron equivalents are transferred, or $20 \times 102^{-1} = 0.196$ eq g^{-1}. Multiplying this value by 26.53 kcal eq^{-1} gave 5.20 kcal g^{-1} as a calculated value for the heat of combustion of the bacterial cells used, compared to 5.27 as the mean experimentally measured value. However, Mayberry et al. tended to favor the value of 0.199 eq g^{-1} obtained by dividing 5.27 kcal g^{-1} by 26.53 kcal eq^{-1}. Multiplying the molar growth yield from a given substrate by 0.199 should give the number of available electrons incorporated into the cellular material. Similarly, multiplying the number of moles of oxygen consumed during the utilization of the same substrate by four should give the number of electron equivalents transferred to oxygen during catabolism. The sum of the two should equal the total number of electrons originally available, that is, for a microorganism grown on glucose as the sole source of carbon and energy this number should be 24. These calculations were made for the growth of *Pseudomonas* $C_{12}B$ and bacterium TEG-5, and in general the calculated values were found to be from 0 to 27.3% too high. Although these authors conjectured that this may have been due to the synthesis of amino acids, it was more probably because their value for the mean heat of combustion of 1 g of cells was uncorrected for the enthalpy change accompanying the oxidation of amino (ammonia) nitrogen to nitrogen gas as this occurs in a bomb calorimeter. This correction should have been made because the nitrogen source for growing the cells was ammonia.

In 1970, Prochazka, Payne, and Mayberry[40] published the results of an extensive study in which they used bomb calorimetry to determine the heat content of a number of dried bacteria, yeasts, and molds. Combining the values they obtained without regard to whether the microorganisms were grown aerobically or anaerobically, or on minimal or complex media, they found an overall average of 5383 cal g^{-1} of ash-free dry weight. Although the precision of their measurements was better than 1% with the same microorganisms, the range among different microorganisms varied from 6370 cal g^{-1} for *A. aerogenes* grown on nutrient broth to 4627 cal g^{-1} for *Clostridium pasteurianum* grown on minimal medium. They concluded that a value rounded to 5.3 kcal g^{-1} represented an "acceptable approximation" of the average heat of combustion of bacteria. A continuation of this study published in 1973 by the same authors[41] resulted in values of 5411 cal g^{-1} ash-free dry weight (range 5089 to 5683) for 15 species of bacteria, 5607 (range 5083 to 6425) for 9 species of septate and nonseptate filamentous fungi, and 5324 (range 5069 to 5762) for 8 species of yeasts, the mean value for 485 samples representing 32 species of microorganisms being 5455 cal g^{-1} ash-free dry weight. Only one of these microorganisms was grown in a minimal medium, all the others being grown in nutrient broth, tryptic soy broth, or Sabouraud dextrose broth. The value of 5.41 kcal g^{-1} for the combustion of bacteria thus represented a slight upward revision from that of 5.27 kcal g^{-1} obtained earlier by Mayberry et al.[39] An interesting close approximation to this former value was obtained by Mennett and Nakayama[42] in 1971. These authors used the percent dry weight values for protein, nucleic acids, and carbohydrate obtained by Tempest and Hunter[43] for *A. aerogenes* cells and, assuming that the ash content was 7% and that the lipid content is the difference between 100% and the sum of the percentages of all the other macromolecular constituents, were thus able to arrive at percentages for all four macromolecular constituents. The heat content of the cells was estimated by using the commonly accepted values for the heats of combustion of proteins as 5.4 kcal g^{-1}, of lipids as 9.3 kcal g^{-1}, and of carbohydrates as 4.2 kcal g^{-1}, and of experimentally determining a value for nucleic acids as 3.5 kcal g^{-1}. In this manner the heat of combustion of bacterial cells was calculated as 5.3 kcal g^{-1}. On the other hand, Mennett and Nakayama[42] used similar data from a study on the composition of *Candida utilis* by Brown and Rose[44] to calculate a value of 5 kcal g^{-1} for this yeast.

The interest of Payne and his collaborators in microbial energetics culminated in 1979 with a publication by Ho and Payne[45] on the assimilation efficiency and the heat content of several prototrophic bacteria. A calculation using the data from Table II of this article shows that the average heat of combustion for 11 species of bacteria was 5195 kcal g^{-1} ash-free dry weight, a value somewhat lower than that of 5411 obtained earlier by Prochazka et al.[41] From the variation of data of this kind it can only be concluded that for critical studies in microbial energetics in which the heat of combustion of a given bacterium (or other microorganism) is important, it would be better not to use average values and to determine the needed heat of combustion directly. One of the major aspects of this work, however, was an attempt to calculate the heat of combustion of bacteria on the basis of the elemental composition of the cells. In this respect Ho and Payne probably performed more elemental analyses of cellular material than anyone since Yamagata[46] did this in 1934. Determinations of the heat content of 75 samples provided an average value of 26.8 kcal available electron^{-1} on an ash-free, dry weight basis, in excellent agreement with the value of 26.9 kcal available electron^{-1}

found by Minkevich and Eroshin[47] for *Pseudomonas* and *Candida* cells. Although Ho and Payne's method of calculation made use of this value, their data showed a considerable variation of heat content on either side of 26.8 kcal available electron^{-1} with respect to individual species. Because of this they introduced a correction factor based on the carbon content of the cells of individual species. Their method of calculation was not particularly successful, the average of the calculated values for the heat of combustion being 6.45% lower than that for the experimentally determined values, with a range of +6 to −16%.

In their 1970 publication Prochazka et al.[40] reasoned that the total energy consumed by a growing culture of microorganisms, E_c, was equal to the amount of energy assimilated by the growing cells, E_a, plus that dissimilated by the oxidation of the substrate, E_d. E_a could be calculated by multiplying the yield of cells in g mole^{-1} of substrate consumed (presumably on an ash-free dry weight basis, although this was not stipulated) by the energy content of that yield, which was taken as 5.3 kcal g^{-1}. The energy dissimilated during the process of catabolism is related to the number of electrons transferred to oxygen, four electrons being transferred for every mole of oxygen consumed during growth. Since Mayberry et al. had previously shown that the enthalpic energy produced when one electron equivalent is transferred to oxygen is 26.5 kcal, E_d can be closely approximated by multiplying this value by four times the number of moles of oxygen consumed. The yield of cells per kcal of energy taken in any way from the medium is thus

$$Y_{\text{kcal}} = \frac{Y}{E_c} = \frac{Y}{E_a + E_d} \tag{9.5}$$

It is apparent that Prochazka et al. did not take into account that whereas combusting dried cells in a bomb calorimeter does indeed give a value for the heat content of the cells, as applied to the energetics of microbial growth this value is too great by an amount equal to the enthalpy change accompanying the oxidation of the nitrogen in the cells, which is largely amino nitrogen, to nitrogen gas. As an example, using the data from Mayberry et al.[39] for the growth of *P.* $C_{12}B$ in minimal medium on 1 mole of benzoate, the total number of available electrons removed from the medium and transferred to the cells is given as 17.3 and that transferred to oxygen as 13.8. This gives a total of 31.1 available electrons which, when multiplied by −26.5 kcal available electron^{-1} gives −824 kcal as the total enthalpic energy removed from the culture medium. However, the complete oxidation of 1 mole of benzoate results in the transfer of 30 electrons which, when multiplied by −26.5 kcal available electron^{-1} gives only −795 kcal. Thus, by using values obtained from bomb calorimetry without correcting for the heat of combustion of cellular nitrogen to nitrogen gas, more enthalpic energy is recovered from the medium than was put into it, which is obviously impossible. If this correction had been made the number of available electrons in one formula weight of cellular material would drop from 20 to 17, and the value for E_a from -458 to -389 kcal, a difference of 15%. This would cause an increase in the value of Y_{kcal} in this particular example from 86.8/824 = 0.105 g $kcal^{-1}$ to 86.8/755 = 0.115 g $kcal^{-1}$.

Payne[28] summarized the then-prevailing ideas on energy yields and growth of heterotrophic microorganisms in a review published in 1970. In this he suggested that there might be limits imposed upon bacteria to prevent them from going beyond the production of a certain mass of cells per unit quantity of substrate, and that therefore there might exist an "investment-expenditure" budget for the available energy for growth. Finding it reasonable to assume that such limits might be genetically predetermined, he postulated the existence of a gene which controls the energy budget. However, since that time no such gene has been discovered. Payne's ideas on investment expenditure were extended in 1976 when Payne and Williams[48] published the results of a study in which they grew *Enterobacter aerogenes* in a minimal medium containing succinate as the sole source of carbon and energy, and *Pseudomonas perfectomarinus* in a minimal medium containing asparagine. They then supplemented each medium in separate experiments with casamino acids and with yeast extract. Both bacteria were found to assimilate 56–57% of the carbon removed from the minimal media during growth. However, in minimal medium plus casamino acids or yeast extract *E. aerogenes* assimilated a remarkably high 85 and 75%, respectively, of the carbon removed from the medium. For *P. perfectomarinus* these values were 75 and 65%, respectively. Thus, in the control experiments in which there was only one source of carbon and energy roughly 40% of the substrate is expended to accomplish an investment of about 60% of the available energy in cellular material. This ratio of expenditure to investment is very much lowered when preformed organic materials in complex mixtures are present in the environment in addition to what can still be regarded as a basic source of carbon and energy, being 15:85 in the case just cited of *E. aerogenes*. The exact reasons for the observed results are not indicated from the data accumulated.

In 1978 Payne and Wiebe[49] published a later review on the subject of growth yield and efficiency in chemosynthetic microorganisms. Here they extended the use of Eq. 9.5 by accepting as a constant the value of Y_{kcal} which they had previously predicted to be 0.118 kcal g^{-1}. E_c then becomes equal to $8.47 \times Y_{sub}$.

The two publications by Mayberry et al.[37,39] and the review by Payne[28] (which contains some original material) are extremely interesting in that they represent an evolution of thought on the energetics of microbial growth from the more abstracted (the Y_{ATP}) through a middle position (the $Y_{av\ e^-}$) to the more directly related (the Y_{kcal}), at least as far as energy, per se, is involved. Unlike the uncertainties involved in determining the quantities of ATP generated from unit quantities of substrate or the efficiencies with which ATP energy becomes incorporated (if it does) into cellular material, there are no uncertainties involved in determining the number of electron equivalents transferred from an initial to a final state if these states can be defined. The only difficulty with the use of electron equivalents is that not all electrons are equivalent, and so the tendency is, as always, to take averages. The use of forms of energy is better in this respect, since all calories (or all joules, if preferred, which it really is not) are equivalent, and finally we begin to realize the possibility of some reasonably accurate determinations *provided* the appropriate biological precaution is taken that the molar growth yield is a function of cellular growth only, and not cellular growth plus the accumulation of storage products, especially those which have a high percentage of oxygen. The wide range of experimental results which have been described must be due at least in part to this. The difficulty with yield studies related to energy is what energy to use. Mayberry et al. made use of enthalpy, because it can be directly measured. The tendency of microbiologists (and ecologists) has always been to rationalize this with the knowledge that $\Delta G°$ and $\Delta H°$ values accompanying the combustion of organic compounds are often quite close, and that for all practical purposes one can be replaced by the other. This is in many cases true, although it is not at all true for fermentations. The point is that whether or not $\Delta G°$ and $\Delta H°$ values for a particular reaction happen to be close, we will not understand the energetics of microbial growth until we are able to calculate the accompanying changes of free energy and enthalpy separately.

The definition of what constitutes an available electron is subject to interpretation. If it is defined as the number of electron equivalents transferred from an organic substance to oxygen, then the definition is perfectly clear, at least with respect to aerobic reactions or combustions for which definite initial and final states can also be defined. However, Mayberry et al. refined the definition so that, ". . . those electrons in a compound not involved in orbitals with oxygen are considered 'available' for (i) transfer to oxygen or (ii) synthetic reductive reactions." (Mayberry et al.,[39] p. 1424). Again, category (i) is perfectly clear. With respect to category (ii), these authors believed that the "synthesis of cellular material is a process entailing a dimunition of entropy" (Mayberry et al.,[39] p. 1425) and that the reactions comprising cellular synthesis are endergonic. However, if what Battley[23] suggested is true, that the free-energy change accompanying the formation of cells from substrate molecules or preformed organic monomers is zero or at least very small, there will then be no net electron transfer and thus no electrons which need to be available for synthetic reductive reactions. In this respect, although electrons are, indeed, removed from the medium in the form of substrate material that is used for cellular synthesis, this removal will *not* make electrons "available" in the sense that their transfer provides energy. Under these conditions the only electrons which are available to provide energy are those transferred to oxygen. Y_{kcal} then becomes Y_{sub} divided by the number of kcal of heat energy resulting from the oxidation of that part of the substrate which becomes converted to carbon dioxide and water during catabolism. This gives quite different values for Y_{kcal}.

The importance of the studies of Mayberry et al.[37,39] and of Payne[14] seems to be largely in their consideration of the energetics of microbial growth in terms of electron transfer and the accompanying enthalpy changes, and *especially* in their development and use of the concept of "available electrons." Mayberry et al.[39] appear to have been the first to calculate the number of electron equivalents incorporated within the biomass of cells grown on 1 mole of substrate, although they used these values plus those for electron equivalents transferred to oxygen during growth on 1 mole of substrate only to check on whether the sum of these two values with respect to a given substrate would equal that for 1 mole of substrate consumed (since molar growth yields were used).

The values obtained by Prochazka et al.[40,41] for the heats of combustion per unit ash-free dry weight of various microorganisms are of no direct use in calculating the enthalpy changes accompanying microbial growth. In most cases the microorganisms used to obtain these values were grown on complex media containing many kinds of organic compounds, and it is not known to what extent (if any) the cells were loaded with storage substances which would alter their overall composition and energy state from that of the fabric of the cells alone. Even in the cases of cells grown on a single substrate in minimal media it is usually not certain that it was only growth that occurred, and not growth plus oxidative or fermentative assimilation resulting in the formation of storage products. In addition, no correction was made for the oxidation of the nitrogenous portion of the cells to nitrogen gas which happens during combustion in a bomb calorimeter. Where heat of combustion data *could* be useful is in determining the enthalpy of formation of a unit mass of cells provided, of course, that it was only the fabric of the cells that was being burned. Such values could then be utilized in appropriate equations representing the growth of microorganisms, along with values for the enthalpy of formation of the reactants and the other products.

Just as studies on fermentative and oxidative assimilation largely ceased in the early 1950s, so have studies involving the determination of the Y_{ATP}, $Y_{av\ e^-}$, and Y_{kcal} at the present time, although articles occasionally still appear on some new way of determining P/O ratios in growing microorganisms. And, although average values for these quantities can certainly be computed, their use is dubious considering the considerable range of values around these means, suggesting that if such information is important, it would be safer to determine it specifically for a given microorganism.

Payne's[28] idea that microorganisms live within an "energy budget" and must "decide" on how much of the available energy can be "expended" for the purpose of "investing" the rest in the mass of cells and other products that are formed as a result of growth has not been further investigated. It seems difficult to relate this idea directly to anything done previously or since. No "budget genes" have been discovered. It might even be asked, "What is the budget?" This appears to be a ratio of energy expended:energy conserved in cellular substance. Whether such a ratio has any predictive value is purely conjectural at the present time. With thermochemically defined systems

Battley[23,50] has shown that the free-energy change accompanying what he called a "growth reaction" is directly proportional to the free-energy change of the relative nonconservative reaction. This being so, if the proportionality constant is determined for growth on one substrate, the amount of growth obtainable on any other substrate can be calculated. This suggests a sort of "budget." On the other hand, the free-energy change accompanying the formation of cells from organic substances during anabolism (investment) appears to be zero, or nearly so, and there is thus no information to indicate exactly how much catabolic energy (expenditure) is required to bring this about. Perhaps much of this latter is wasted, or unnecessary.

9.2.4. Bell's Analysis of Yield Factors

In 1972 G. H. Bell[51] published an analysis of microbial yield factors (such as Y_{kcal} and $Y_{\text{av } e^-}$) and their significance. As a consequence of this he introduced yet another yield factor, the Y_{carbon}, which was defined as the grams dry weight of biomass produced per gram of carbon metabolized. Previously Payne[28] had shown that mean $Y_{\text{av } e^-}$ for sugars and polyhydric alcohols were identical (3.28 and 3.21, respectively), with those for organic acids being somewhat lower (2.84) and those for *n*-paraffins being much lower (2.06). What Bell found using mean Y_{carbon} values calculated from the same data used by Payne was that those for sugars and *n*-paraffins were identical (1.07 and 1.04, respectively), whereas those for polydydric alcohols were considerably higher (1.22) and those for organic acids being almost equally lower (0.80). On the basis of these observations he came to the conclusion that it might be reasonable "to consider the substrate primarily as a source of carbon and secondarily as a source of energy." (Bell,[51] p. 23.) Assuming that it is the conversion of carbon from substrate into cellular material that is the primary controlling influence, he further concluded that "the total energy available for the formation of a unit-quantity of biomass is thus proportional to the heat of combustion of a quantity of substrate containing 1 g of carbon (heat of combustion, kcal/g of substrate carbon)." (Bell,[51] p. 23.) Calculations of these latter values for different substrates showed that the heat of combustion of 13.3 kcal g^{-1} for *n*-paraffins is very much higher than that for sugars of about 9.3 kcal g^{-1} of substrate carbon. However, the Y_{carbon} values for these two classes of substances were identical. Bell therefore reasoned that "the paraffins must be oxidized to a level similar to that of the sugars before the energy produced can be coupled to the metabolism." (Bell,[51] p. 23.) This was made plausible in that *n*-paraffins had been shown to be oxidized via the corresponding primary alcohols to the corresponding carboxylic acids, being then subjected to beta-oxidation with the formation of acetic acid. This last has the same oxidation state as glucose and, when metabolized from dilute solution, gives yields comparable to those on sugars. Bell felt that no utilizable energy could be obtained from hydrocarbons until they had become oxidized to the level of acetic acid or other small molecular weight intermediates at a similar level of reduction, the energy released from this initial oxidation being wasted as heat.

The finding that the use of Y_{carbon} was preferable to the use of Y_{kcal} or $Y_{av\ e^-}$ in that growth on hydrocarbons, carbohydrates, and polyhydric alcohols gave approximately constant Y_{carbon} values enabled Bell to propose a "unified concept" of substrate utilization. The empirical formulas for hydrocarbons, carbohydrates, and polydydric alcohols can be generalized by the formulas C_xH_{2x+2}, $C_x(H_2O)_y$, and $C_x(H_2O)_yH_z$. In all three cases the amount of oxygen is less than or equal to the amount required to form water from the hydrogen these substances contain. Low energy substances, on the other hand, can be generalized by the formula $C_x(H_2O)_yO_w$, characterizing them as containing an excess of oxygen. With respect to these latter if the assumption is made that the quantity of carbon in the substance equivalent to O_w is *not* available for metabolism, the fraction of carbon that *is* available is equal to $(x - 0.5w)/x$. The yield of cells on this carbon that is available can then be designated as $Y_{av\ carbon}$. Bell was able to show that substrates which are organic acids containing an excess of oxygen gave $Y_{av\ carbon}$ values that fell within the same range as those for Y_{carbon} with respect to other substances. Organic acids such as acetate and lactate seemed to be exceptions to the generalization, probably because they are low-energy substances having the empirical formula of carbohydrates. The conclusion was then made that $Y_{av\ carbon}$ values are much more constant than those for $Y_{av\ e^-}$ and that therefore the most significant factor must be the energy derived from the carbon atoms. The electrons available from the hydrogen atoms in a substance do *not* have a major role in providing energy for growth. Both aspects of this conclusion are unique in their conception.

The ideas Bell presented represent an advance over those of Payne and his collaborators (see Section 9.2.3) in that $Y_{av\ carbon}$ values with respect to microbial growth on dilute solutions of most classes of substances, including *n*-paraffins, appear to fall within a somewhat more restricted range. However, this range is still wide and the predictive value of $Y_{av\ carbon}$ still limited. Attempts to relate growth as a weight measurement to specific factors such as kcal, available electrons, or available carbon per mole of substance metabolized have not been particularly successful, probably because the molar growth yield is not related to only one of these factors but to a combination of all of them. On the other hand, the large data scatter may be obscuring an appropriate conclusion. As McCarty[52] has pointed out, the differences in the values obtained by various workers for a given Y factor are greater than those obtained for different substrates.

Although he did not specifically mention it, it is apparent that in his unified concept Bell was considering some measure of the degree of reduction of an organic substrate and its relationship to the process of microbial growth. His conclusions that energy can be considered as being obtained separately from the transfer of electrons from the carbon or from the hydrogen atoms in a substrate, and that the electrons available from the hydrogen atoms did not play a major role in providing metabolic energy for growth, were highly original and have not been developed since. The usual concept of electron transfer then, as now, is that the available electrons from any metabolizable organic substance can be used for any biological purpose, irrespective of their intramolecular source. The idea that substrates that are more reduced than cellular material must be oxidized to the level of reduction of the cells before they can be utilized during growth was first clearly elucidated by Tamiya.[53] The idea that the energy released by this oxidation was wasted as heat was original to Bell, and has not been subjected to experimental verification. An alternate possibility is the one suggested by Battley,[50] that the oxidant for oxidizing highly reduced substrates to the level of the cells could be carbon dioxide which, becoming reduced to an appropriate level, is then incorporated into cellular material. In this case there would be no loss of energy at all.

9.3. ENGINEERING STUDIES ON MICROBIAL GROWTH

To sanitary engineers "microbial growth" means "increase in biomass of activated sludge." Activated sludge is a complex mixture of many different communities of microorganisms which together remove organic substances from raw sewage and industrial wastewaters. In doing so they feed on these substances and utilize them in the process of growth. This results in an increase in the biomass of activated sludge in proportion to the efficiency with which the waste organic substances become incorporated into the cell bodies of the activated sludge microorganisms. Thus, while dissolved waste organic substances are indeed removed from raw sewage and industrial wastewaters so that these become purified of pollutants, some disposition has to be made of the continuously accumulating activated sludge. It therefore was (and is) useful to know something about the relationship between the energy provided to activated sludge and the biomass produced from it as a result. This led to studies on the kinetics and thermodynamics of activated sludge production. Whereas microbiologists tended to use enthalpy changes as a measurable function of the energetics of microbial growth, engineers immediately recognized that the most important parameter to study was free-energy change. Since this cannot be directly measured with respect to the systems with which they worked, their approach to the energetics of activated sludge production was largely theoretical, with data occasionally being borrowed from the microbiological literature. This latter was the source of certain difficulties of interpretation, because whereas engineers studied the increase in biomass of mixed communities of many different kinds of microorganisms including bacteria and protozoa which grew on substrates of undefined composition (raw sewage), microbiologists almost universally worked with pure cultures growing on a single organic substance as the substrate. The impossibility of thermochemically defining their systems made it necessary for engineers to rely on indirect methods of measuring energy content and from this to develop theories that could account for what they empirically observed.

One of the early empirical relationships established by sanitary engineers was that activated sludge production is related to oxygen consumption. In 1947 Placak and Ruchoft[54] used biochemical oxygen demand (BOD) and suspended solids (activated sludge) to compute oxygen consumption and sludge production in experiments designed to study the metabolism of pure organic substances by activated sludge. They found that whereas all the substances they tested were used, the amount of sludge synthesis and the accompanying oxygen utilization varied considerably depending on the nature of the substrate. One immediate uncertainty is whether the increase in biomass was due to growth, to the synthesis of storage products, or both. In fact, it represents exactly the same problem faced by chemists such as Pasteur nearly 80 years previously, when they found differing results with respect to alcoholic fermentation depending on quantity of yeast that was incubated with sugar. A different interpretation from that of Placak and Ruchoft was made by Helmers et al.,[55] who incubated activated sludge with peptone, glucose, cotton kiering liquor, ragrope waste, and brewery waste (i.e., in this case all complex substances), concluding that the increase in biomass of the activated sludge was directly related to the BOD loading. However, these investigators also concluded that the proportion of organic substance that was converted to sludge was the same, regardless of the nature of the complex substrates that were used.

The interpretations of the work in the two articles cited above are examples of what by the early 1960s had developed into two opinions held by engineers regarding what they called "oxidative assimilation" relationships. It should be noted that this term represented to them the increase in activated sludge biomass, including growth plus the intracellular storage of organic substances, and is quite different from the same term as originally used by Barker[56] in 1936, which came to mean the intracellular accumulation of storage substances by resting cell suspensions. One group of engineers thought that the amount of sludge produced during wastewater treatment varied with the type of wastewater being treated, that is, with the types of substrate used by the activated sludge microorganisms. The other group felt that activated sludge production was independent of the nature of the wastewater being treated, and thus of the substrates being used. At the same time considerations were being given to the relationship between thermodynamics and the overall exchange of mass which took place when the oxidizable substances in wastewater became converted into activated sludge microorganisms. These resulted in three general concepts advanced by McKinney, by Servizi and Bogan, and by McCarty.

9.3.1. The Studies by McKinney of Activated Sludge

In 1962 R. E. McKinney[57,58] proposed that the nonthermal energy made available to microorganisms from their substrates is proportional to the heat liberated during their metabolism. He defined this energy as "biological energy," as differentiated from any established thermodynamic function, largely because no study had been made up until that time with respect to free-energy, enthalpy, and entropy changes as they actually occur in systems involving activated sludge. However, the basic equation he used

$$\text{``biological energy''} = K\,\Delta H$$

indicated that he considered the heat liberated during growth to be synonymous with the closed system state variable, enthalpy. There is nothing basically wrong in this. Given a repeatable reaction within a defined system, a change in any characteristic of the system (such as enthalpy) can be related to any variable (such as biomass). However, if the reaction within the same system is changed, for example, by using a different substrate, or by studying fermentation of the same substrate rather than its oxidation, then the free-energy change would be the better function to use.

McKinney felt that the net increase in biomass depended upon the rate of endogenous metabolism of the microorganisms being studied. This opinion is completely opposite to that of Battley,[59] who emphasized in 1960 that for studies of microbial growth it was extremely important to establish that it was only growth that was occurring, and not also an increase in biomass due to the intracellular accumulation of endogenous substrates.

McKinney concluded that there was a fixed relationship between activated sludge synthesis and all types of substrates being metabolized. However, in 1968 Burkhead and McKinney[60] demonstrated that this was not the case, and that the rate and amount of sludge synthesis depended on the nature of the substrate.

9.3.2. The Studies by Servizi and Bogan on Activated Sludge

Although they did not specifically use Y_{ATP} values, in 1963 J. A. Servizi and R. H. Bogan[61] adopted the conclusion of Bauchop and Elsden[18] that the quantity of energy available for cellular synthesis was more precisely related to the quantity of ATP synthesized than to the free-energy change accompanying substrate oxidation. This latter relationship was expressed by the following equation.

$$Y = k_1 N_{ATP} \tag{9.6}$$

Here Y represents the molar growth yield, k_1 a proportionality constant having the dimensions of grams dry weight of cells per mole of ATP generated (actually, Y_{ATP} as Bauchop and Elsden defined it), and N_{ATP} the number of moles of ATP generated per mole of substrate metabolized. Servizi and Bogan were careful to point out that what they meant by cell synthesis was gross production of cell mass exclusive of endogenous metabolism (i.e., production of storage materials). They then reasoned that if the production of ATP is proportional to the free-energy change accompanying the oxidation of the substrate, then

$$N_{ATP} = -k_2 \Delta F_{ox} \tag{9.7}$$

where $-k_2$ represents a proportionality constant having the dimensions of moles of ATP $kcal^{-1}$, and ΔF_{ox} the free energy of oxidation. Combining Eqs. 9.6 and 9.7 gives the following equation.

$$Y = -k_1 k_2 \Delta F_{ox} \tag{9.8}$$

To test whether or not Eq. 9.8 represented a reasonable hypothesis they then calculated values for k_2 using values for N_{ATP} assuming a P/O ratio of 3 and standard values for the free energies of oxidation of nine common organic substrates, finding it to have the remarkably constant value of 0.0542 ± 0.0008 mole of ATP $kcal^{-1}$. Whatever the values for k_1 and k_2, the dimensions of moles ATP cancel out and Y becomes Y_{kcal}, as defined in 1967 by Mayberry et al.[37] Servizi and Bogan then proceeded to test whether the growth of activated sludge was proportional to the free energy of oxidation of the substrate. Their initial experiments involved incubating sludge with varying quantities of a single organic substrate and following growth as a function of time. They found that the growth curve for this mixture of microorganisms was very similar to that of pure cultures in that it could be demonstrated to be exponential, and in that the quantity of growth was proportional to the quantity of substrate. This was done for 10 different substrates with the result that when Y was plotted against $-\Delta F^{\circ}_{ox}$, a nearly perfect fit was found with respect to a line having a slope ($k_1 k_2$) of 0.104 g $kcal^{-1}$. This slope represented an average of the $k_1 k_2$ values for all of the many different kinds of microorganisms found in the mixed population constituting activated sludge. However, Servizi and Bogan also used the data of Siegel and Clifton[62] for carbon converted into cell mass versus carbon utilized with respect to the growth of *E. coli* in pure culture on several carbon sources. Calculating Y values from these data they also found a direct relationship between Y and ΔF°_{ox} with $k_1 k_2$, however, being slightly greater in value. The respirometric data of McKinney et al.[63] were also used, molar growth yields being calculated with the assumption that the oxygen consumed represented that portion of the utilized substrate converted to carbon dioxide and water, the rest having been converted into protoplasm. Similar graphic results were obtained, except that with the same activated sludge, different $k_1 k_2$ values were obtained depending on whether the sludge were grown on carbohydrates or Krebs cycle intermediates, or on aromatic substances or aliphatic acids. Servizi and Bogan concluded that the increase in biomass of activated sludge and of bacteria is directly proportional to the free-energy change accompanying the oxidation of the substrate. Prior to this time Hoover and Porges[64] and Symons and McKinney[65] had published formulas representing the empirical composition of several different activated sludges. These were summarized by Servizi and Bogan in a table which indicated clearly that different sludges could have considerably different compositions. One such formula, $C_5H_7O_2N$, determined by Hoover and Porges came to be widely used to represent the general composition of bacteria. A subsequent study in 1964 by Servizi and Bogen[66] expanded the ideas they had developed to include the growth of activated sludge on multicomponent substrates, with the assumption that the total growth yield is a summation of the yields from each individual component.

9.3.3. The Theoretical Study by McCarty of the Energetics of Microbial Growth

In 1965 P. O. McCarty[67] published an extensive article on the thermodynamics of biological synthesis and growth in which he made use of essentially the same approach as did Servizi and Bogan in assuming a constant production of biomass per unit amount of ATP generated. However, he also believed that the best indicator of the energy actually obtained by the cells was the chemical oxygen demand (COD), which is the oxygen equivalent of the organic material dissolved or suspended in a unit volume of sample and is usually expressed in milligrams of oxygen per liter. Thus, the equation he used to describe biological synthesis is

$$-k_1 A\ \Delta F_r = \Delta F_s + B\ \Delta F_m \tag{9.9}$$

where k_1 represents the transfer efficiency of substrate energy to ATP energy, A the grams COD of substrate converted for energy per net gram COD of cells synthesized, B the total grams COD of active cells requiring maintenance energy per gram COD of cells synthesized, ΔF_r the free-energy per gram COD of substrate converted for energy, ΔF_s the ATP free energy required for the net synthesis of 1 g COD of cells, and ΔF_m the ATP free energy required for the maintenance of 1 g COD of active cells. It is clear that by this time the treatment of the energetics of microbial growth by engineers was becoming more and more sophisticated, and has become even more so up to the present time. McCarty postulated that the ATP free energy required for synthesis, ΔF_s, is the sum of two quantities. One of these is the ATP energy which is either required or liberated, depending on the degree of reduction of the substrate, when the material of the substrate is brought to an intermediate level which can be used by all microorganisms. The intermediate chosen was pyruvate. The other is the ATP energy required to convert pyruvate into cells. The energy of maintenance, F_m, was analyzed but was eventually neglected, because his calculations of transfer efficiencies were for the most part made from cell yields representing relatively "young" cultures, in which cases the maintenance energy would be expected to be small or nonexistent. ΔF_s in Eq. 9.9 was further represented as follows:

$$\Delta F_s = \frac{\Delta F_p}{k_2^n} + \Delta F_c \tag{9.10}$$

where k_2 represents the transfer of ATP energy for the conversion of the carbon source to pyruvate, n, a constant having a value of $+1$ if ΔF_p is positive, and -1 if ΔF_p is negative, ΔF_p, the free-energy change accompanying the conversion of 1 g COD of carbon source to 1 g COD of pyruvate, and ΔF_c, the free-energy change accompanying the conversion of 1 g of pyruvate to 1 g COD of cells. ΔF_c was evaluated by means of the following equation:

$$\Delta F_c = \frac{\text{free energy of hydrolysis of ATP}}{Y_{\text{ATP}} \times \begin{pmatrix}\text{\% organic composition} \\ \text{of cells}\end{pmatrix}\begin{pmatrix}\text{COD of organic} \\ \text{material in cells}\end{pmatrix}} \tag{9.11}$$

McCarty adopted the value of 10.5 g dry wt of cells per mole of ATP generated[17,18] and the value for the free-energy of hydrolysis of ATP under physiological conditions of -12.5 kcal mole^{-1}.[68,69] He considered bacteria to be 90% organic material, and that this oxidizable material represents a COD of 1.42 g g^{-1}. Using Eq. 9.11 ΔF_c then becomes:

$$\Delta F_c = \frac{12{,}500}{(10.5)(0.90)(1.42)} = 932 \text{ cal g}^{-1} \text{ cell COD} \tag{9.12}$$

Combining Eqs. 9.10, 9.11, and 9.12 and neglecting the term representing maintenance, McCarty's energy transfer equation then became

$$-k_1 A \, \Delta F_r = \frac{\Delta F_p}{k_2^n} + 932 \tag{9.13}$$

Here, k_2 from Eq. 9.10 was set equal to k_1, the transfer energy then obtained being an average for the transfer of energy from the substrate to ATP and from ATP to pyruvate.

Equation 9.13 can be used with respect to either autotrophic or heterotrophic microorganisms, depending on the sign of n, and data were calculated for both groups. However, one of most important data sets that McCarty calculated has to do with aerobic heterotrophs in the form of activated sludge cultures, in which for 24 experiments involving 10 different substrates no energy transfer efficiency was greater than 44% (glucose) or less than 11% (formate), while most ranged from 20 to 40%. These data obtained from experiments of others, some of which were with pure cultures and some with mixed cultures, provide strong evidence that the energy transfer efficiency may be considerably lower than is frequently thought, and that efficiencies of over 50% might well be suspect.

McCarty's publication is important in that it represents the first attempt to calculate free-energy exchanges during growth which is based upon the formation of a biologically utilizable intermediate organic compound. His whole approach to the calculation of energy transfer efficiencies is based on the assumption that ATP energy is, in fact, transferred to the cells, an assumption which had the authority of dogma in 1965. On the other hand, if the energy for *net* cell synthesis from a carbon source is zero, as Battley[23,50] has proposed, then the energy from ATP would only be used to accomplish the chemical mechanics of cell production, being converted into heat and entropy in the process, and *none* of this energy would become incorporated into the substance of the cells themselves. These two approaches are very much a matter of interpretation.

McCarty later came to the conclusion that his original proposal that COD be used to evaluate the portions of the substrate which are converted into end products for energy production and into cells was confusing because of the units involved, which are used only by sanitary engineers. In 1971, four years after the introduction of the "available electron" concept into the biological literature by Mayberry et al.,[37] McCarty[70] proposed that electron equivalents should be used instead of COD units, one electron equivalent being equal to 8 g of COD, the major portion of his publication being the reiteration of his previous ideas[67] in terms of the new unit. It is thus apparent that the available electron concept was beginning to become accepted by the engineering fraternity. The subsequent development of the use of this concept by Minkevich and Eroshin (see Section 9.4) is an example of how important it has become.

9.3.4. The Objections of Schroeder and Busch

The publications of McKinney,[57,58] of Servizi and Bogan,[61,66] and of McCarty[67,70] initiated considerable discussion in the literature more or less culminating in 1968 in a publication by E. D. Schroeder and A. W. Busch,[71] who questioned the validity of energy change as a growth parameter. One of their criticisms was that the thermodynamic considerations made by these authors applied only to closed systems, and were valid only for conditions of constant temperature and pressure where the only work done is that caused by a change in volume. They conceded that a BOD bottle might be considered a closed system, but that a single microorganism in the bottle could not. Similarly, sewage treatment systems, which are continuous flow processes, also cannot be considered to be closed. The lack of boundary considerations by McKinney, by Servizi and Bogan, and by McCarty therefore invalidated their conclusions. This criticism notwithstanding, Schroeder and Busch then went on to compare the growth of an unidentified bacterium on glucose using either aerobic respiration on oxygen or anaerobic respiration on nitrate. Table 1 in their article shows clearly that an average of almost exactly twice as much growth was obtained per mole of glucose utilized during growth on nitrate as on oxygen gas. The theoretical yields they calculate using their own arguments seem to indicate that the opposite should happen, since the $Y_{NO_3^-}/Y_{O_2}$ ratio they obtained was 0.65. In spite of their contention that closed system thermodynamics ought not to be applied to open systems, many investigators since that time have continued to study the energetics of microbial growth as it occurs in closed containers.

9.3.5. The Sludge Production Equations of Burkhead and McKinney

A publication in 1969 by Burkhead and McKinney[72] represented the first systematic approach to representing activated sludge production as a balanced chemical reaction, the concept of which had originated in 1952 with Hoover and Porges[64] in a study of the assimilation of dairy wastes by activated sludge. Burkhead and McKinney[60] had previously acclimated batch cultures of activated sludge to pure substrates representing the different food classes, after which samples were placed in Warburg respirometers at 25°C. Analyses were made at frequent intervals for substrate removal measured as soluble COD, total and volatile suspended solids, oxygen uptake, and total COD. Total suspended solids (TSS) represent the grams dry weight of substances not in solution per unit volume of suspending fluid, the volatile suspended solids (VSS) being that portion of the TSS which can be oxidized (volatilized) to carbon dioxide and water and which thus represents the organic part of the TSS. From the data collected theoretical oxidative assimilation equations were constructed which included a term representing the mass of activated sludge. The empirical formula representing the sludge was based on whatever would give the best chemical balance for the following general equation.

$$\left.\begin{matrix}\text{substrate formula} + O_2 \\ \pm\ NH_3\end{matrix}\right\} \longrightarrow \begin{matrix}\text{protoplasm formula} + CO_2 \\ + H_2O\end{matrix}$$

The data needed for use in the above equation were (1), the chemical formula of the substrate to be tested; (2), the moles of oxygen required and the VSS produced per mole of substrate removed; and (3), the protoplasm composition of $C_5H_{7\pm2n}O_{2\pm n}N$ whose value for n gave the best chemical balance. This last was determined by a trial and error method. For example, in a particular experiment the quantity of oxygen required per mole of glucose utilized was calculated as follows:

$$\begin{aligned}\frac{\text{moles } O_2}{\text{mole glucose}} &= \left(\frac{\text{mg liter}^{-1}\ O_2}{\text{mg liter}^{-1}\ \text{COD}}\right)\left(\frac{\text{wt } O_2}{\text{wt glucose}}\right) \\ &\quad \times \left(\frac{\text{g mole glucose}}{\text{mole glucose}}\right)\left(\frac{\text{mole } O_2}{\text{g mole } O_2}\right) \qquad (9.14) \\ &= (0.21)(1.07)(180)\left(\frac{1}{32}\right) = 1.26\end{aligned}$$

For the above value of 1.26 it was therefore necessary to know the quantity of oxygen consumed (by respirometric determinations) and the quantity of glucose utilized (calculated as reduction in the soluble COD). Volatile suspended solids were calculated in the following manner.

$$\frac{\text{moles VSS}}{\text{mole glucose}} = \left(\frac{\text{mg liter}^{-1}\text{ VSS}}{\text{mg liter}^{-1}\text{ COD}}\right)\left(\frac{\text{wt } O_2}{\text{wt glucose}}\right) \times \left(\frac{\text{g mole glucose}}{\text{mole glucose}}\right)\left(\frac{\text{mole } C_5H_7O_2N}{\text{g mole } C_5H_7O_2N}\right) \quad (9.15)$$

To use Eq. 9.15 it is necessary to measure the mg liter^{-1} increase in VSS (cells) as well as the quantity of glucose utilized (calculated as reduction in the soluble COD). For the growth of activated sludge on glucose the experimentally measured value for VSS/soluble COD was 0.74. For the empirical composition of activated sludge they initially used that obtained by Hoover and Porges[64] of $C_5H_7O_2N$, the "molecular weight" of this sludge thus being 113 daltons. Inserting these data into Eq. 9.15 gives

$$\frac{\text{moles VSS}}{\text{mole glucose}} = (0.74)(1.07)(180)\left(\frac{1}{113}\right) = 1.26$$

The following trial equation was then set up with respect to growth on glucose, for which the chemical formula for the substrate, the quantity of oyxgen consumed per mole of glucose (from Eq. 9.14), and the quantity of activated sludge per mole of glucose (from Eq. 9.15) are known. Since a unit-nitrogen formula for the activated sludge cells is used, the molar yield coefficient also becomes the coefficient for the ammonia used as a nitrogen source.

$$C_6H_{12}O_6 + 1.26O_2 + 1.26NH_3 \longrightarrow 1.26C_5H_7O_2N + XCO_2 + YH_2O$$

X then becomes $6 - (1.26 \times 5) = -0.3$, which is not a good balance. Therefore, new compositions of $C_5H_{7\pm 2n}O_{2\pm n}N$ must be tried, effectively adding or subtracting the elements of water to or from the empirical formula representing the cells. Burkhead and McKinney found that the best chemical balance was obtained when $n = 2$. Thus,

$$C_6H_{12}O_6 + 1.26O_2 + 0.96NH_3 \longrightarrow 0.96C_5H_{11}O_4N + 1.20CO_2 + 2.16H_2O$$

where the coefficients for carbon dioxide and water were calculated by difference.

This method of constructing equations representing microbial growth used by Burkhead and McKinney is just the opposite from that used by anyone previously. By assuming an empirical composition for the activated sludge of $C_5H_7O_2N$ from Hoover and Porges,[64] they adjusted the latter by adding or subtracting the elements of water (always permitted in balancing organic oxidation–reduction reactions) by a series of successive approximations until a balanced equation was achieved. This is ingenious, and within the limits of accuracy acceptable to most sanitary engineers. The two publications of Burkhead and McKinney were important for the development of studies on the energetics of microbial growth not so much for their results as for their systematic and comparative use of equations to represent growth processes. These equations were used to calculate the molar growth yields of activated sludge cells grown on different substrates, from which plots could be made of these yields versus the ΔF°_{ox} for the respective substrates. They appear to have been the only authors to have used this method of calculation. All others have used either (1) the RQ for the oxidation of the substrate *plus* an experimentally determined empirical formula for the cells or (2) the empirical formula for the substrate, the empirical formula for the cells (or other products), and the molar growth yields of the cells and products.

9.4. THE MASS-BALANCE EQUATIONS OF MINKEVICH AND EROSHIN

In the early 1930s H. Tamiya (see Chapter 6) became interested in characterizing the growth of *Aspergillus oryzae* and realized that the process of growth could be considered as an oxidation–reduction reaction. He was the first to construct equations representing the growth of microorganisms on organic substrates, and to do so developed his "theory of the CQ," the combustion quotient (CQ), this being the respiratory quotient (RQ, CO_2/O_2) for the complete combustion of an organic substance to carbon dioxide, water, and in the case of nitrogenous substances, ammonia. The CQ is therefore related to the degree of reduction of an organic substance, and values for this quantity were calculated in the following manner.

$$\mathrm{CQ} = \frac{4n_C}{4n_C + n_H - 2n_O - 3n_N} \tag{9.16}$$

where n_C, n_H, n_O, and n_N represent the relative numbers of carbon, hydrogen, oxygen, and nitrogen atoms, respectively, in an organic substance. Using Eq. 9.16 methane has a degree of reduction of 0.5, carbohydrate of 1.0, and carbon dioxide of infinity. The concept of the CQ was developed in this particular manner largely because of the interest in the 1930s in the importance of the RQ in respiring tissues, and because respirometric methods existed for determining RQ values. Knowing the degrees of reduction for a substrate and a microorganism grown on it, by using his theory of the CQ Tamiya could predict the RQ for growth and the quantity of substrate needed to form the cells. He could also predict the quantity of substrate needed for respiration by knowing the RQ. For whatever reason he never actually put the two together to represent a complete growth reaction. For him growth represented simply the formation of the cells, apart from respiration. In some respects his theory of the CQ was indirect and cumbersome, though no less so than the use of COD values, which also represent degrees of reduction, by sanitary engineers for calculating the energetics of activated sludge production. The remarkable thing is that there is nothing that can be done at the present time using the available electron concept to construct equations representing microbial growth that either was not done or could not have been done using Tamiya's methods. The former is simply a more elegant refinement of the latter.

Other methods of representing degrees of reduction of organic substances have been defined, one of the most interesting being the percentage method of Spoehr and Milner,[73] in which methane is taken as being 100% reduced, carbon dioxide as 100% oxidized, and other organic substances falling somewhere in between on a unit-carbon basis. This method, however, has never been used for the purpose of constructing equations representing microbial growth, by means of which studies on the energetics of growth could be made.

The concept of the "available electron" came into use through the studies of Mayberry et al.,[37] Prochazka et al.,[74] and Payne et al.,[75] which were directed toward finding significant parameters that could be used to relate the quantity of microorganisms produced during growth to the quantity of a single organic substance that was used as a source of carbon and energy (see Section 9.1.3), and resulted in their definition of the $Y_{\text{av } e^-}$. However, the concept of the degree of reduction of an organic substance did not enter into the considerations of these investigators except in that more reduced substrates have greater numbers of available electrons and that these, in turn, yield a greater number of kcal when transferred to oxygen.

In 1972 I. G. Minkevich and V. K. Eroshin[76] and Minkevich et al.[77] began what was to be the first of an extended series of publications on the development and use of what they called "mass balance equations" in predicting various aspects of microbial growth in industrial fermenters or continuous cultures. This was followed by a short article in 1973 by Minkevich and Eroshin[78] in which microbial growth was generalized as follows:

$$\begin{aligned} CH_mO_l + aNH_3 + bO_2 &\longrightarrow \\ &yCH_pO_nN_q + zCH_rO_sN_t + (l - y - z)CO_2 \end{aligned} \tag{9.17}$$

where CH_mO_l, $CH_pO_nN_q$, and $CH_rO_sN_t$ represent the substrate, cells (biomass), and product, respectively, m, l, p, n, q, r, s, and t represent the subscripts of their respective atoms in the substrate, cells, and product, and y and z are the coefficients with respect to the unit-carbon formulas representing the cells and products, respectively. Appearing in this article is their definition of the "reductance degree," (γ), of an organic substance, where γ is the number of gram-equivalents of oxygen required for the complete combustion of one gram-atom of carbon in the substance, or, the number of available electrons *per unit carbon atom* of the substance. Thus, the reductance degrees of the substrate, γ_S, and that of the cells (biomass), γ_B, are calculated from Eq. 9.17 in the following manner:

$$\gamma_S = 4 + m - 21 \tag{9.18}$$

and

$$\gamma_B = 4 + p - 2n - 3q \tag{9.19}$$

Using Eqs. 9.18 or 9.19 methane has a reductance degree of 8, carbohydrate of 4, and carbon dioxide of zero. It is apparent that compared to Tamiya's degree of reductance

$$\gamma = \frac{4}{\mathrm{CQ}}$$

and that for whatever γ can be used, CQ can be used also. The point to be considered is, however, that the approach of Minkevich and Eroshin in using available electrons as a measure of reductance is far more direct and elegant in its simplicity. In 1973 these latter investigators[79] repeated the experiments of Karasch and Scher,[35] Karasch,[36] and Mayberry et al.,[37] confirming the results of these latter that the heat of combustion of many organic compounds per equivalent of available electrons is essentially constant, varying between 25 and 27 kcal eq^{-1}. For dry biomass they obtained an experimental value of 25.36 kcal eq^{-1}, which can be compared to the value which Mayberry et al. adopted of 25.53 kcal eq^{-1}. From considerations of this kind Minkevich and Eroshin in 1975 made the statement, ". . . it follows that the equivalent of available electrons is the measure of substance quantity which gives the best reflection of its chemical store." (Minkevich and Eroshin,[80] p. 45.) *It is this statement relating the quantity of available electrons to the quantity of substance in which they are incorporated which provides the means for constructing with the greatest simplicity equations representing microbial growth.*

Subsequently, many publications involving the available electron concept appeared in the engineering literature, principally from the laboratories of Minkevich and Eroshin[81–92] and Erickson[93–109] and their collaborators. These investigations deal largely with rate process equations which can be of general industrial applicability, for which some of the parameters are weighted averages. These data are far beyond the scope of this book and will not be covered here. Rather, a more simplistic approach will be adopted requiring the fewest possible variables, the aim being only to construct equations representing microbial growth and with these to calculate the free-energy, enthalpy, and entropy changes accompanying this process as it proceeds from an initial to a final state. This is what Schroeder and Busch[71] called the "BOD bottle" approach, as opposed to the "open system" approach which they felt was necessary for considering the energetics of sewage treatment plants and continuous industrial fermentations. Presumably the energetics should be the same provided that the same thing is being studied.

For microbiologists used to molar growth yields, the use of total available electron values is easier than computing γ values, although if the method originally devised by Battley[110] of representing cells in terms of unit-carbon formulas is used, the total available electron values are the same as the γ_B values of Minkevich and Eroshin. Otherwise, the method is much the same. It is necessary to know the empirical formulas for all the reactants and products, which latter will include the cells that have grown. Any organic product for which a specific molecular formula is not known, such as cells or biomas in general, should be represented chemically in terms of its ash-free composition, although it is not absolutely imperative to use unit-carbon formulas. It is also necessary to know the molar yield coefficients for all organic products. Once the mole fractions of the substrate used to form the organic products of the growth process have been determined, the coefficients for the inorganic reactants and products, such as ammonia as a nitrogen source, oxygen if the growth process is aerobic, carbon dioxide, and water, can all be determined by difference. This is easiest to do simply by systematically counting up atoms, in the time-honored way learned by students of general chemistry. It is sometimes less easy to do this by hand using mathematical formulations which can, of course, be written for the purpose of calculating each coefficient other than the molar yield coefficients for the organic products. All this can also be computerized, but since the calculations can so easily be done by hand, why bother? Inherent in the method of constructing these equations is the remarkably simple observation by Minkevich and Eroshin[80] cited previously, that the number of available electrons in a given mass of organic substance is also a measure of the *quantity* of the substance. This had been known ever since Karasch and Scher[35] and Karasch[36] had related available electrons to the heats of combustion, but it had never before been used in quite this manner. Minkevich and Eroshin were effectively saying that the available electrons in the organic products can only have come from the substrate, and because of this it becomes possible to calculate precisely the quantity of substrate that is used to form a given organic product. Thus, *if*, for example, during the growth of *Saccharomyces cerevisiae* on 1 mole of glucose 1.30 moles of ethanol are formed, we can easily determine what quantity of glucose *must* have been used for this purpose (since the ethanol could not have come from any other source). The empirical (and molecular) formulas for glucose and for ethanol are $C_6H_{12}O_6$ and CH_3CH_2OH, respectively. The total number of available electrons (AE) with respect to the empirical formula of any organic substance containing C, H, O, and N can be calculated using the following equation.

$$\mathrm{AE} = 4n_{\mathrm{C}} + n_{\mathrm{H}} - 2n_{\mathrm{O}} - 3n_{\mathrm{N}} \tag{9.20}$$

Using Eq. 9.20 the numbers of available electrons for glucose and for ethanol are 24 and 12, respectively. The molar yield coefficient ($\mathrm{MYC_{eth}}$) for ethanol is 1.30. The mole fraction of substrate, S_{eth}, used to form this quantity of ethanol is therefore $(MYC_{eth} \times AE_{eth})/AE_{glu}) = (1.30 \times 12)/24 = 0.650$. Since it is a *molar* yield that is being used in this calculation, this fraction then becomes 0.650 mole of glucose required for the formation of the ethanol. This can be represented as follows:

$$\begin{array}{ccc} 0.65C_6H_{12}O_6 & \longrightarrow & 1.30CH_3CH_2OH \\ 15.6AE & \longrightarrow & 15.6AE \end{array} \tag{9.21}$$

Equation 9.21 represents an electronic balance, but it does not represent a material balance, which can be made up by difference. This can be done by the addition or subtraction of formulas representing substances that have no available electrons, such as carbon dioxide or water or their equivalent ions if these are required. Thus, the appropriate equation for the formation of 1.30 moles of ethanol from 0.65 mole of glucose becomes

$$0.65C_6H_{12}O_6 \longrightarrow 1.30CH_3CH_2OH + 1.30CO_2 \tag{9.22}$$

The same type of calculation and use of equations can be applied for any other organic products, such as cells. Eventually, the addition of equations such as 9.22 or others in which the substrate is oxidized during respiration must sum up to a total quantity of 1 mole of substrate, since it is molar yield coefficients that are being used. It is important to point out that although an available electron is defined as one that can be transferred to elemental oxygen during the process of complete combustion to carbon dioxide, water, and ammonia, this is only intended to serve as a basis for comparison. Thus, AE values can be used whether or not an equation in which a given substance is included contains oxygen as a reactant. As an example, the number of electrons transferred from substrate to product in Eq. 9.22 can be calculated, even though no oxygen is involved. It is thus evident that the number of available electrons in a substance can be used with respect to that substance whether or not the reaction is aerobic or anaerobic. Examples of the construction of aerobic and anaerobic growth-process equations are given in Sections 10.5.1–10.5.5. It should be emphasized that given the requisite basic data for each there are several other methods of arriving at growth-process equations (growth-reaction equations, mass-balance equations, etc.) all of which, if carried out properly, give exactly the same results. This is simply a reflection of the fact that there may be several ways of proceeding mathematically, chemically, or thermodynamically from the same initial to the same final state. However, with respect to convenience of manipulation the method of using available electrons is far superior.

Once growth-process equations have been constructed, they may be used to calculate changes in energy provided that the thermodynamic properties of all the reactants and products of the growth processes represented by the equation are known except those of the cells. The use of available electrons is not of very much use here, although average values for enthalpy changes in kcal AE^{-1} have in fact been used by Mayberry et al.[37] and Payne[28] and by many investigators who have contributed since 1967 to the engineering literature. This method may be used if results are acceptable to within ±5%, which is usually satisfactory for most practical purposes. However, for academic purposes the better to hope to understand what is going on, a more accurate treatment is preferable. The problem is that although the number of available electrons is constant for all substances having an equivalent composition on a unit-carbon basis (i.e., γ values) the amount of energy per available electron is usually not, at least for small molecules of the kind that serve as substrates. Thus, formaldehyde (HCHO), acetic acid (CH_3COOH), and glucose ($C_6H_{12}O_6$) all have γ values of 4.000, respectively, but the ΔH° values per available electron are −33.59, −26.11, and −27.90 kcal, respectively, for the combustion of these substances. For this reason it is better to use the values for the thermodynamic properties of each substance rather than average values for energy changes per available electron, especially for small molecular weight reactants and products. Examples of the calculation of energy exchanges accompanying growth-process reactions are also given in Sections 10.5.1–10.5.5.

9.5. MICROCALORIMETRY OF MICROBIAL GROWTH

The enthalpy change is the only one of the three changes in energy (ΔG, ΔH, and ΔS) accompanying microbial growth that can be practically measured, and beginning with the early 1960s several such measurements have been made. This was partly because of the increasing availability of good instrumentation and partly because of the interest of a few investigators in studying this particular aspect of microbial energetics. No attempt will be made here to cover all the publications involving the microcalorimetry of microbial growth. Many of these are not pertinent because they describe the use of thermal measurements for the same purpose that the optical density of cell suspensions can be used. Thermograms have been thought to be useful in characterizing microorganisms, both with respect to the shape of the curve and to the rate of growth. However, with respect to microbial energetics, the observed heat of growth is not meaningful unless the nature of the reactions causing the heat production is also known. It is only then that the "true" heat of growth can be determined. By 1970 microcalorimeters had become more available, and the use of these instruments to study various aspects of microbial growth increased. Unfortunately, many of these studies are of no use to a consideration of the energetics of microbial growth because an appropriate thermodynamic system was not described having a defined initial and final state. The observed thermogram or measured heat of growth therefore had a limited meaning. To attempt to gain some idea of what is really taking place thermally relative to microbial growth it is necessary to work with some form of chemical equation (growth-reaction equation, mass-balance equation, etc.) which indicates what sorts of things are happening inside the calorimeter vessel. For this reason, in Section 9.5 for the time period after 1970 only those studies will be described involving batch-culture microcalorimetry of microorganisms grown in a minimal medium on a single source of carbon and energy in which the kinds and quantities of reactants and products are known. These are the simplest sorts of thermodynamic systems, and the ones most likely to give interpretable results.

9.5.1. The Studies of Forrest and His Collaborators

In 1961 W. W. Forrest[111] published the first of several articles dealing with microbial heat exchange, including a description of a twin microcalorimeter specifically designed for this purpose. This latter consisted of two enclosed 200-ml aluminum vessels constructed to be as identical as possible, each of which was contained inside a convection shield which was in turn contained inside an isothermal enclosure jacket. A diagram of a vessel assembly is shown in Fig. 9.1. The vessel in one of these served as a reaction vessel, and contained 175 ml of culture medium or cell suspension. The other vessel served as a control vessel and contained 175 ml of water. Temperature differences between the two vessels were sensed by thermistors cemented to the outside of the bottom of each vessel. These served as two arms of a Wheatstone bridge, in the circuit of which was a mirror galvanometer. This latter operated an optical-

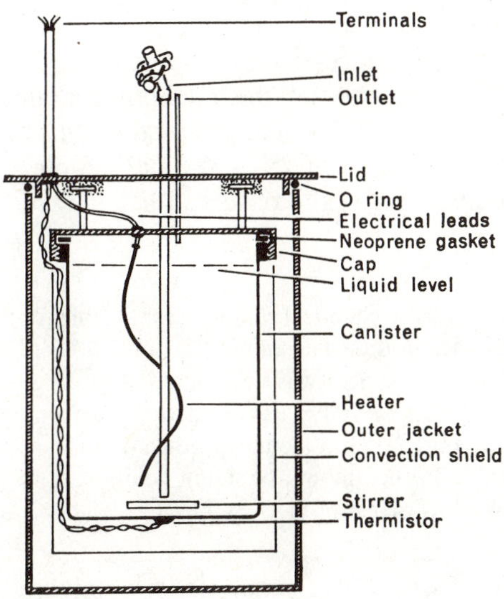

FIGURE 9.1. Diagram of Forrest's microcalorimeter. Reprinted from Reference 111 with permission.

lever-photocell amplifier controlling a thyratron which supplied current to a resistance wire immersed in the water of the balance vessel, causing the water to become heated. Thus, as a microbial culture or suspension gave off heat its temperature would rise, the temperature difference being sensed by the thermistors. The temperature of the water in the control vessel would then become automatically heated until it equaled that of the reaction vessel, the quantity of heat generated electrically in the former being proportional to the quantity of heat generated by the microorganisms in the latter. Any heat losses to the environment by the two vessels would then be equal, and the quantity of microbial heat generated could be precisely determined. The great advantage of Forrest's calorimeter was that the reaction vessel contained a quantity of suspending medium that was similar to that used in ordinary bacteriological experiments in the laboratory, and that it could be easily sterilized. Calibration of the calorimetric system was carried out electrically by means of a resistance element immersed in the contents of the reaction vessel. A microbial culture could be stirred by means of a magnetic stirrer on the bottom of the reaction vessel. During tests of their instrument intermittently over several weeks, they found a variation of ±3% with respect to their metering circuits. The sensitivity was 0.4 cal hr^{-1}. The operation of the calorimeter was demonstrated by an experiment in which a suspension of *Streptococcus lactis* fermented glucose to lactic acid. They observed that heat generation almost exactly matched the consumption of glucose, so that heat production ceased when the glucose had become exhausted.

During the same year Forrest et al.[112] published their experiments in which *Streptococcus faecalis* was grown within the calorimeter vessel. A defined medium was not used. In one set of experiments the medium contained yeast extract, casamino acids, and glucose, resulting in a heat production during growth of 33 cal $mmole^{-1}$ of glucose with a range of 31.3–35.6 cal $mmole^{-1}$ for six experiments. In a second set the medium contained yeast extract, tryptone, and glucose, resulting in a heat production during growth of 28 cal $mmole^{-1}$ of glucose with a range of 25.2–29.8 cal $mmole^{-1}$ for four experiments. These two are averages of the observed heats of reaction and, as the authors state:

> . . . in neither case could the chemical changes responsible for the heat production be defined. The lactate produced accounted for only 50% of the glucose carbon that disappeared. The lactic acid was accompanied by large amounts of steam-volatile acids, ranging up to 100% of the lactic acid formed. These were not identified. In the second set, the lactate produced corresponded only to 60 to 70% of the glucose, with the amount of steam-volatile acids being up to 90% of the lactic acid formed. [Forrest et al.,[112] p. 687.]

It is thus evident that with these experiments involving growing cells no conclusions could be drawn relative to the quantity of heat produced. This is because neither the initial nor the final states of these systems were thermochemically defined. Forrest et al. did, however, demonstrate clearly that during growth the heat production very closely matches the increase in microbial biomass. This is shown in Fig. 9.2. The data in this graph also confirm the observations of Winzler and Baumberger[113] and of Battley[114] with *S. cerevisiae* grown anaerobically on glucose that after the fermentation of the substrate has ceased there is no further heat production. Characteristic of all previous investigators who attempted to study the heat of growth, Forrest and his collaborators never again carried out another experiment in which cells were grown in a calorimeter vessel, although Forrest and Walker[115] did attempt to measure calorimetrically what they regarded as the energy of maintenance in *S. faecalis*. Forrest[116,117] published two articles on the subject of microbial calorimetry, and one[118] on the energetic aspects of microbial growth. Much of this last was devoted to yield studies, but for the first time since 1960 when Battley[59] recognized and emphasized the importance of its consideration, the ease of polymerization of substrate substances into storage materials was mentioned. Forrest[118] had found that substances such as cellobiose and other car-

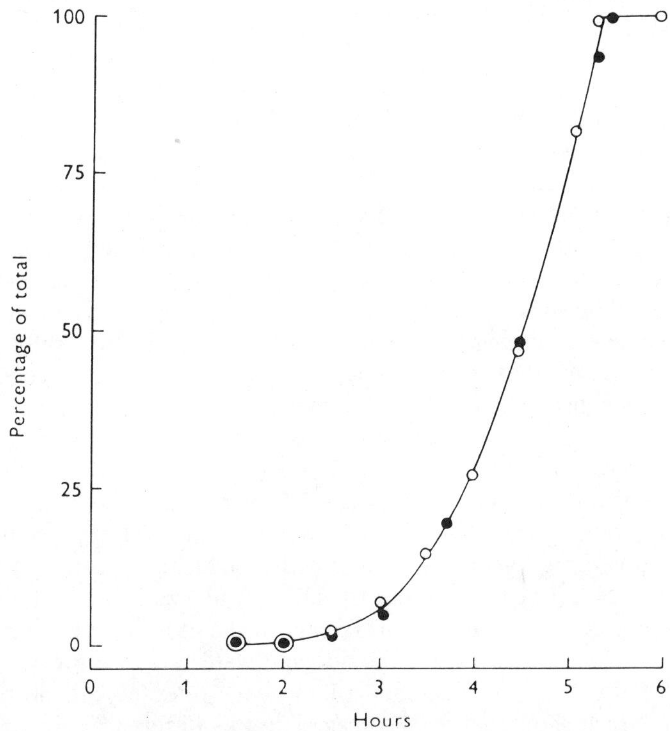

FIGURE 9.2. Heat production by *Streptococcus faecalis* growing anaerobically on a limited quantity of glucose as the energy source. Glucose fermented 1.8 m*M*; final bacterial dry weight, 27 mg; total heat produced, 49 cal. Filled circles, dry weight; open circles, heat production. Reprinted from Reference 112 with permission.

bohydrates could be polymerized via glucose-6-phosphate to storage polysaccharides, the attachment of each glucose residue requiring the expenditure of one ATP molecule. This process appeared to be stoichiometric, giving an apparent Y_{ATP} of 180. Such a polymerization would be accompanied by little entropy change, and in his opinion:

> . . . it is a reasonable speculation that reserve storage in general where little structural organization is involved may give high efficiencies of use of ATP. For this reason analysis of the cells produced under conditions of high growth yield is of considerable importance. It is perhaps significant that the reported cases of high growth yields occur concurrently with unusual carbon assimilation, either as storage polysaccharides or by CO_2 fixation [Forrest,[118] p. 71.]

Forrest made an interesting observation with respect to the enthalpy change accompanying anabolism. Prigogine[119] had previously stated that the biological heat production that is experimentally observed during growth results from a summation of all the exergonic and endergonic reactions which take place, and that the enthalpy changes of the endergonic reactions that occur during the synthesis of cellular material, ΔH_b, will subtract from the exergonic reactions of catabolism, ΔH_c. This implies that cellular-synthetic (anabolic) processes are, in fact, endergonic. However, Forrest pointed out that for the anaerobic growth of *S. lactis* (Boivinet[120]), *Zymomonas mobilis* (Belaich and Senez[121]), and *E. coli* (Senez and Belaich[122]) on glucose, and for the aerobic growth of *A. aerogenes* (Grangetto[123]) on succinate, in nutritionally adequate media in which the carbon source was limiting, the heat production agreed within experimental error (about ±2%) with that calculated using data from tables of thermodynamic properties for the enthalpy change accompanying the conversion of the substrate to the products of catabolism. This means that the enthalpy change accompanying anabolism, ΔH_b, must be too small to measure (whether or not it results from an endergonic reaction). This abundantly confirmed the conclusion made 47 years earlier by Molliard (see Section 3.10) that it was respiratory (or catabolic) energy ". . . that is uniquely associated with the production of heat . . ." Forrest was clear in his opinion that the difference between the sum of the heats of formation of the materials assimilated in whatever form they are assimilated and the heat of formation of the cells that are formed as a result of growth would be expected to be small. With respect to this last he apparently failed to realize that it is not only cells that are the products of anabolism. Except in the cases of substrates that are more reduced than cells, carbon dioxide and water are also formed, and it is the sum of the heats of formation of the cells *plus* the other products of anabolism which should be used to calculate this difference. An example he offered was that of glucose as a substrate being converted into cellular material at the same level of oxidation, which he considered to be represented by (CH_2O). This was not a good example in that Tamiya (see Chapter 6) had demonstrated some 30 years earlier that cells were, in fact, more reduced than carbohydrates and required a nitrogen source as a substrate material. Nevertheless, the idea had become established that the enthalpy change accompanying anabolism is likely to be very small. The connection between the hypothesis made in 1961 by Battley[23] that the free-energy change accompanying anabolism was zero and the demonstrable fact that the corresponding enthalpy change must be very small was never made.

Forrest's[118] publication reflects the opinions prevailing at the time with respect to the utilization of ATP energy. He pointed out that although the free-energy change accompanying the alcoholic fermentation of glucose by *S. cerevisiae* and *Saccharomyces rosei* is considerably greater than that accompanying the homolactic fermentation of glucose by *S. lactis*, *S. faecalis*, and *Lactobacillus plantarum*, the number of moles of ATP (2) generated in both fermentations is the same. Because the average yield (Y_{ATP}) for the growth of these microorganisms is the same for either fermentation, it was thought that the "quantized" production of ATP did not allow for the greater quantity of free energy provided to the alcoholic fermenters to be used. The implication is, of course, that ATP energy is transferred into cellular material in proportion to the amount which is provided.

On the other hand, two years later in 1971 Forrest and Walker[30] produced a far more extensive description even than that of Forrest[118] on all aspects of the energetics involved in the generation and utilization of energy during microbial growth. It is one of the most exhaustive treatments of this subject yet written. Most of the article is devoted to the metabolism and utilization of ATP in the synthesis of intermediate metabolites or in the polymerization of preformed organic substances into cells. With respect to this latter the data of Mandelstam and McQuillen[124] and Morowitz[6] were used to calculate that only 0.036 to 0.037 mole of ATP is required for the monomer polymerizations involved in the synthesis of 1 g of bacterial cells. Gunsalus and Schuster[125] had previously obtained a value of 0.03 mole of ATP per gram of cells, the lower value being due to the use in their calculation of only three ATP molecules for the incorporation of one amino acid into protein rather than the five used by Forrest and Walker. Using the remarkably complete analysis of the composition of bacterial cells made by Morowitz,[3] Forrest and Walker calculated that only 7.64×10^{-4} moles of ATP were required for the synthesis of the monomers present in 1 g of bacterial cells, concluding that such syntheses require little or no energy on the part of the cell. The total quantity of ATP required to synthesize 1 g of bacterial cells is therefore $0.0365 + (7.64 \times 10^{-4}) \approx 0.037$ mole of ATP. These values would give a value for Y_{ATP} of $1/0.037 = 27$. Forrest and Walker calculated in addition that autotrophic microorganisms would require about 0.17 mole of ATP to synthesize 1 g of glucose from carbon dioxide. To this must then be added the 7.64×10^{-4} moles of ATP required for monomer synthesis and the 0.036 mole required for polymerization of the monomers, to give a total of 0.207 mole of ATP for the synthesis of 1 g of autotrophic bacterial cells. Although not calculated by Forrest and Walker, this would give a Y_{ATP} value of about 5. A comparison of the data from these calculations with those experimentally found showed that whereas a Y_{ATP} value with respect to heterotrophic growth of about 27 is to be expected, what is actually observed in anaerobically grown microorganisms in which the quantity of ATP generated by substrate-level phosphorylation can often be quite accurately known is a Y_{ATP} value of about 10. From this they concluded that, clearly, most of the catabolically produced ATP is utilized in ways that are not chemically understood. No comparisons of this latter kind were made with respect to autotrophic microorganisms.

Whatever the use of ATP in synthesis, Forrest and Walker reiterated Forrest's[118] opinion that because the heat of growth, ΔH_b, is too small to measure, and because the theoretical calculations of Morowitz[3] indicated that even in a minimal medium the maximum value of ΔH_b should be only about 3% of the enthalpy change accompanying catabolism, therefore "*Both the experimental and theoretical approaches then show that only a small fraction of the energy available from catabolism is actually incorporated into the organisms as energies of chemical bonds.*" (Forrest and Walker,[30] p. 266, italicized by the present author to emphasize its importance). When Bauchop[17] and later Bauchop and Elsden[18] first described a Y_{ATP}, their conclusion was that growth was *proportional* to the quantity of ATP generated, and their studies indicated that this proportionality was nearly constant at about 10.5 g dry weight of cells per mole of ATP generated by catabolism. In the linkage of monomers into polymers in biochemical reactions involving ATP some of the energy of the ATP molecules becomes conserved within the substance of the polymer which is equivalent to that which would be released were these polymers to be enzymatically hydrolysed back to the constituent monomers at the same temperature and pressure. Thus, the transfer of ATP energy into polymers such as proteins, glycogen, lipoproteins, and so on, has long been known and well understood, and it was not unreasonable to think that the same sort of energy transfer might take place with respect to the synthesis of cells. For example, with respect to growth Senez states that, ". . . an effective coupling between the energy-yielding metabolism and the energy-consuming reactions of cell biosynthesis must necessarily exist . . ." (Senez,[19] p. 102), implying that free energy is transferred during growth from ATP to the cells as they are synthesized. Explicit statements to this effect are also to be found in Servizi and Bogan[61,62] and McCarty.[67] By the time Forrest and Walker[30] published their studies there was a common understanding that energy from ATP was transferred to the substance of growing cells with an efficiency that was not always easy to estimate but which was certainly less than 100%. This has been described previously by Eq. 9.6 from the work of Servizi and Bogan. McCarty states that ". . . the concept of ATP as a common energy *intermediate* [italicized by the present author] in all microorganisms appears valid and should permit formulation of a general relationship between substrate energy and growth." (McCarthy,[67] p. 171.) He found energy transfer efficiencies for fermentative microbial growth to be from 45–66%, and those for aerobic, heterotrophic growth to be from 20 to 40%. Thus, the quantities of energy transferred from

ATP into the substance of growing cells were not considered to be insignificant. How, then, does one reconcile the still presently accepted idea of the transfer of ATP energy to the substance of cells with the italicized statement of Forrest and Walker at the beginning of this paragraph? In fact, there can be no reconciliation. The two ideas are completely different because they represent two different interpretations of the same observable phenomenon. Forrest and Walker simply made their statement as a logical conclusion drawn from one experimental and one theoretical observation, and let it go at that. In retrospect we can see the dichotomy. The observation of Bauchop[17] that growth is proportional to the quantity of ATP generated has two interpretations, only one of which was made. These are:

1. Growth is proportional to the quantity of ATP generated because a certain proportion of ATP free energy becomes incorporated into cellular substance (i.e., the fabric, or structure of the cells).
2. Growth is proportional to the quantity of ATP generated because ATP energy is used to effect cellular syntheses as a net energy source only, that performs the "work" of transforming some of the substrate material into cells and in so doing becomes completely converted into heat and entropy. According to this interpretation there would be no net ATP free-energy incorporation in the cells, that is, no energy transfer. Whatever free energy is present in the cells has come only from that part of the substrate that is used as the carbon source.

Although they themselves did not consider it, it is this second interpretation that completely accounts for what is implied in Forrest and Walker's statement and in Battley's[23] hypothesis made 10 years previously, that the free-energy change accompanying cell synthesis (anabolism) is equal to zero. This will be discussed further in Chapter 10.

Considering that he had a well-designed calorimeter at his disposal and wrote two articles on the subject of bacterial microcalorimetry, it is indeed unfortunate that Forrest could not have carried out more extensive microcalorimetric investigations on microbial growth.

9.5.2. Further Developments in Microcalorimetry

In 1963 E. Calvet and H. Prat[126] published what was partly an English edition of their two previous books[127,128] on the subject of microcalorimetry and partly an updating through 1963 of work which they had done in perfecting and using the Tian-Calvet microcalorimeter. By this time calorimeters of this type had come to be called "heat-flow," or "conduction" calorimeters, because a considerable portion of the heat generated within them becomes transferred from the source to the sink by means of thermopiles, whereby at the same time an electrical signal is generated which is proportional to the amount of heat transferred. If two such microcalorimetric elements are used together within

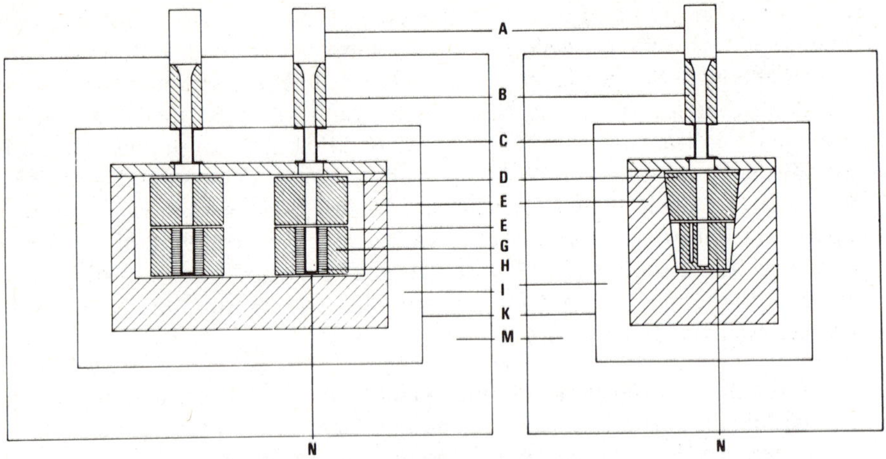

FIGURE 9.3. Diagram of Wadsö's microcalorimeter. Longitudinal and transverse sections on left and right, respectively. A, steel tube; B, copper constriction; C, steel tube; D, aluminum block; E, main heat sink; F, air space; G, aluminum block; H, thermocouple plate; I, air space; K, steel container; M, water thermostat; N, aluminum plate with hole for ampoule (microcalorimeter vessel). Reprinted from Reference 139 with permission.

a heat sink the calorimetric assembly is called a "twin-calorimeter," one calorimeter serving as an active element in which a heat-exchanging reaction takes place, and the other as a reference element that compensates for thermal changes in the environment (the heat sink). With conduction calorimeters it is not necessary that all of the heat generated within a reaction cell be transferred to the heat sink. However, it is essential for meaningful results that whatever heat is transferred be a constant fraction of the total heat generated. The quantity of heat capable of being transferred determines the speed of response of a conduction calorimeter, and the extent to which the calorimetric system is actually isothermal. The Tian-Calvet calorimeter has a limited number of thermoelectric junctions and is best suited to studies involving reactions accompanied by slow heat exchange, such as microbial growth. That designed by Benzinger and Kitzinger[129–132] has a much larger number of junctions (10^4), so that more heat is transferred in a shorter time and the speed of response is correspondingly increased. Such a calorimeter is useful for reactions during which there is a "heat burst," as when a solution of an enzyme is mixed with that of a substrate. More recently the thermoelectric junctions made of soldered or plated wire that were used in the Tian-Calvet or Benzinger-Kitzinger microcalorimeters, respectively, have been replaced by "thermoelectric devices." These latter consist of a flat array of massive thermocouples in the form of small blocks of alloy having a high thermoelectric power, connected in series, and sandwiched between two plates of thermally conducting but electrically insulating material. When such plates are placed between the heat

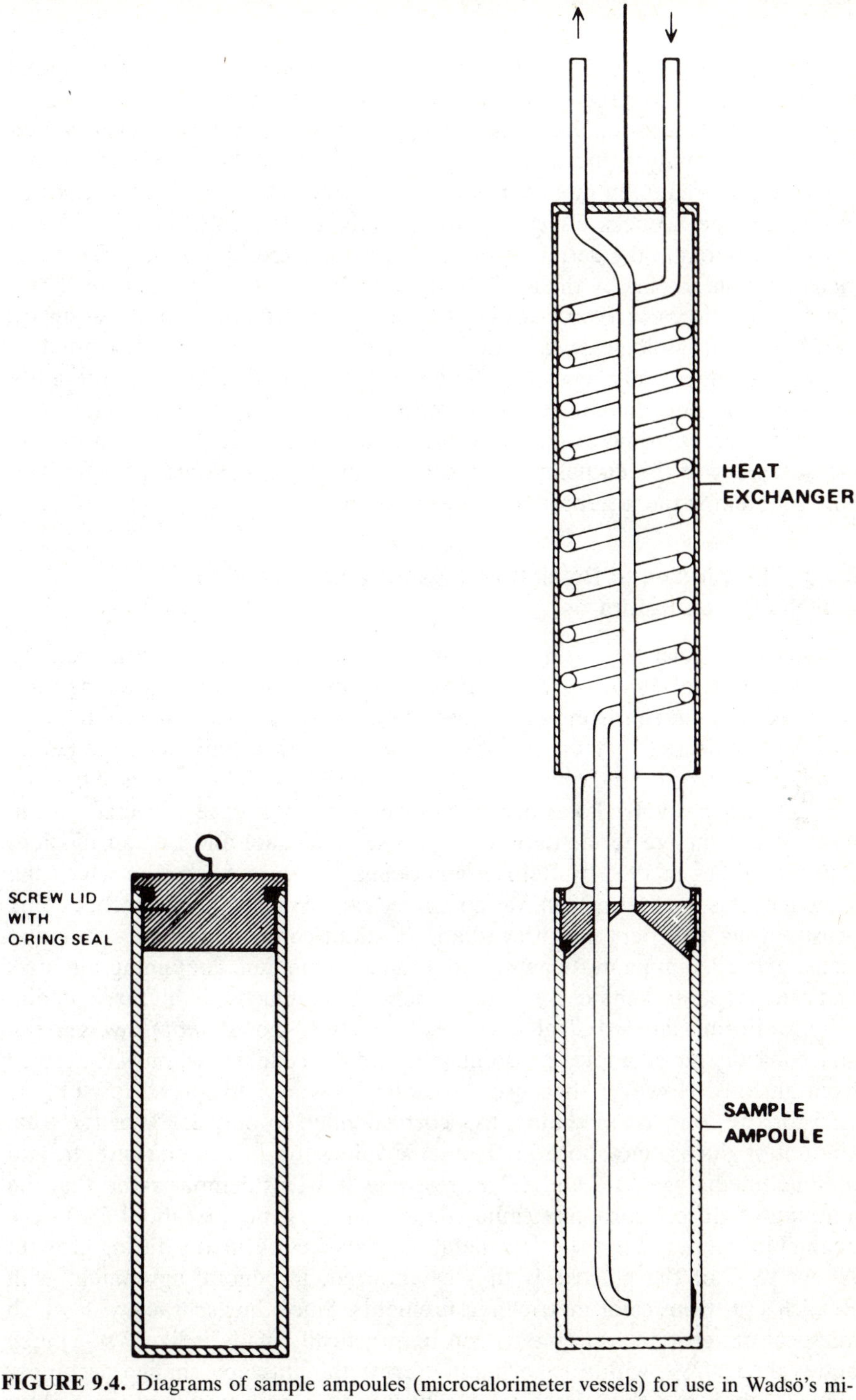

FIGURE 9.4. Diagrams of sample ampoules (microcalorimeter vessels) for use in Wadsö's microcalorimeter. Simple, closed ampoule and flow cells for gas or liquid flow on left and right, respectively. Reprinted from Reference 139 with permission.

source and the heat sink most of the heat generated is transferred almost totally (there has to be some small temperature difference, otherwise there is no electrical signal) from the former to the latter, making calorimeters in which they are used nearly isothermal. Originally devised to cool various components in electrical circuits by means of the Peltier effect, they have been used in reverse for the Seebeck effect (the thermocouple effect) by Wadsö[133–138] and his collaborators in the design of several types of microcalorimeters. This was made possible in part by the availability of highly stable amplifiers which did not exist in the early 1960s. A diagram of a twin microcalorimeter designed by Wadsö[139] for biological use is shown in Figs. 9.3 and 9.4. These calorimeters are now commercially available, and it is apparent that there is presently absolutely no impediment except the cost of purchase to the acquisition and use of a microcalorimeter that is sensitive enough and stable enough for the measurement of the enthalpy changes accompanying microbial growth. The only problem is the interpretation of the results.

9.5.3. The Studies of Belaich and Senez and of Belaich and His Collaborators

In 1963 J. P. Belaich,[139] using a Tian-Calvet microcalorimeter, measured the thermogenesis of *Pseudomonas lindneri* (*Zymomonas mobilis*) growing on a complex medium consisting of glucose, peptone, and yeast extract. His thermograms indicated that this microorganism grew in a typically exponential fashion, and that it used glucose as its only source of energy for growth. The amount of heat evolved was nearly equivalent to that to be expected on the basis of a complete fermentation of glucose to alcohol and carbon dioxide. Further studies in 1965 by Belaich and Senez[140] were published in which the growth of this microorganism was compared on a complex medium containing yeast extract and peptone, a synthetic medium containing minerals, amino acids, and calcium pantothenate, and a minimal medium containing minerals and calcium pantothenate, the energy source being glucose in all three media. Using uniformly labeled C^{14} glucose they found that when *Z. mobilis* was grown anaerobically on a complex medium 48.1% of the cellular carbon was derived from glucose. However, because the molar growth yield was so low (Y_G = 7.95 for the complex medium), this corresponded to only 2.1% of the total amount of glucose metabolized. The same values for growth on synthetic and minimal media were 1.7 and 2.3%, respectively. This demonstrated that the amounts of glucose carbon assimilated into cellular structure (about 2%) were negligible compared to the other metabolic balances occurring during growth. About 98% of the glucose is thus catabolized, in general agreement with Belaich's previous calorimetric measurements. Since the accuracy with which microcalorimetric measurements can be practically made is likely not much better than ±2%, within experimental error the enthalpy change accompanying catabolism is thus effectively the heat of the whole growth process. Also in 1965 Belaich and Senez[141] published an article describing in general the use of microcalorimetric methods for the study of the energetics of bacterial growth. In 1968 Belaich et al.[142] undertook a study of glucose permeation in microbial cells, during the course of which they measured the enthalpy change accompanying the growth of *Z. mobilis* and *S. cerevisiae* on complex media. Once again they found that the corrected enthalpy changes were very close to those expected on the basis of a complete fermentation of the glucose, showing that in accord with their previous findings:

> . . . the enthalpy incorporated into covalent bonds of biological macromolecules [in the structure of the cells which were grown] is negligible compared to the enthalpy of catabolic processes. In other words, under these conditions the corrected experimental enthalpy change (K_c) is practically equal to the enthalpy change corresponding to the "non-conservative reaction" as defined by Battley[114]. [Belaich et al.,[142] p. 1752.]

Their conclusion in this respect is very similar to the statement of Morowitz that, "In a very enriched medium [the conditions of growth of Belaich et al.[112]] ΔH_c [the enthalpy change accompanying anabolism] probably approaches zero." (Morowitz,[3] p. 343.) Belaich et al. also found that after exponential growth of *Z. mobilis* and *S. cerevisiae*, when the concentration of glucose had become growth-rate limiting, the microcalorimetric curves returned to the base line as the glucose became exhausted. In common with the findings of Winzler and Baumberger,[113] Battley,[114] and Forrest et al.,[112] this indicated that when growth is limited *only* by the energy source there is no accumulation of stored reserves and therefore no endogenous metabolism. In fact, this sort of experimentation with an exceedingly sensitive calorimeter such as that used by Belaich et al. is undoubtedly the best proof that exponentially growing microbial cells do not store reserve materials. The studies of Belaich et al.[142] were continued by Murgier and Belaich[143] with similar experiments involving galactose and maltose as limiting energy sources.

The microcalorimetric studies of J. P. Belaich continued in 1976 with two publications together with his wife, Anne.[144–145] The first of these was a microcalorimetric study of the anaerobic growth of *E. coli* in a synthetic medium with the intention of investigating the thermal behavior of a microorganism having a complex fermentation pattern. They found that *E. coli* K-12 exhibited a different thermogram at pH 6.5 than at pH 7.8 when grown with glucose as the sole source of carbon and energy. Two peaks were observed at the lower pH which seemed to be due at least partly to formic hydrogen lyase activity which removed the formic acid formed during the fermentation after this latter had ceased. However, a calculation showed that not sufficient heat could have been generated during this reaction to account for all the heat produced. This activity was not observed at the higher pH, and the formic acid concentration remained constant after the glucose substrate had become exhausted. Their analysis of the metabolic products formed during the fermentation of glucose was almost identical to that previously reported by Stokes.[146] The quantities of these products indicated that 21% of the substrate was unaccounted for. A molal growth yield was not determined. The second study involved measurements of the affinity of whole *E. coli* cells for various substrates. Here they reported the enthalpy changes accompanying the fermentation of 1 mole of several substrates, as well as the yield values in grams per mole of substrate. Their correction of the observed heats of fermentation was made using the following formula:

$$\Delta H_c = \frac{K}{1 - \left(\frac{Y}{2} \times \frac{30}{12} \times \frac{1}{P_m}\right) - x\,\Delta H_n} \tag{9.23}$$

where ΔH_c represents the true heat of fermentation; K, the observed heat of fermentation; Y, the molar growth yield; P_m, the molecular weight of the substrate; x, the number of protons generated per mole of substrate fermented; and ΔH, the heat of neutralization of one proton by the phosphate buffer used. The constants 30 and 12 represent the molecular weight of (CH_2O) and the atomic weight of carbon, respectively. $Y/2$ represents percentage of carbon in dried *E. coli* cells, taken as 50. Belaich and Belaich assumed that the cellular carbon and carbohydrates are at the same oxidation level. Their use of Eq. 9.23 represents an interesting departure from the more conventional method of using the enthalpies of formation of the reactants and products, if these can be determined. The form of Eq. 9.23 had changed by 1980 when Belaich[147] published a review article on the subject of microcalorimetry. ΔH_c became ΔH_{met} as a more general symbol to represent the heat exchange associated with the metabolism of a substrate. This latter could then be calculated as follows:

$$\Delta H_{\text{met}} = (1 - \alpha)\,\Delta H_{\text{cat}} + Y_s\,\Delta h_{\text{an}} \tag{9.24}$$

where α represents the fraction of cellular carbon derived from the substrate; ΔH_{cat}, the enthalpy change corresponding to catabolism; Y_s, the molar growth yield; and Δh_{an}, the "enthalpy variation" associated with the synthesis of 1 g (dry weight) of bacteria. This last value was shown by Belaich to be of the order of 0.011 kcal g^{-1} for microorganisms grown in complex media.

Equation 9.24 was used by Traore et al.[148] as part of a microcalorimetric study of the growth of sulfate-reducing bacteria. Because a complex medium is required for the growth of *Desulfovibrio vulgaris*, it was difficult to know exactly how much of the organic substance present other than the substrate was incorporated into cellular material. Sorokin Yu[149] had previously shown that when *Desulfovibrio desulfuricans* was grown in a complex medium with glucose only 30% of the cellular carbon was derived from lactate. However, Belaich[150] had previously shown that when *Z. mobilis* was grown anaerobically on glucose in a complex medium 50% of the cellular carbon was derived from this substrate. Traore et al. accepted this latter value. Elemental analysis of the Hildenborough strain of *D. vulgaris* grown on lactate and pyruvate, respectively, gave unit carbon formulas of $CH_{1.64}O_{0.33}N_{0.23}S_{0.01}P_{0.014}$ and $CH_{1.64}O_{0.31}N_{0.25}S_{0.07}P_{0.012}$, and fractional dry weights of carbon of 0.4892 and 0.4839, respectively. With these data α_{lact}, which is the quantity of lactate entering into the formation of cellular material per mole of substrate consumed, could be calculated in the following manner.

$$\begin{aligned} \text{lact} &= Y_{\text{lact}} \times 0.4892 \times \frac{1}{12} \times \frac{1}{3} \times \frac{50}{100} \\ &= 6.7 \times 0.4892 \times 0.0139 \qquad (9.25) \\ &+ 0.045 \text{ mole} \end{aligned}$$

In Eq. 9.25, 12 is the atomic weight of carbon, 3 is the number of carbon atoms in the substrate (lactate and pyruvate), and $\frac{50}{100}$ is the fraction of cellular carbon originating in the substrate. A similar calculation with respect to pyruvate metabolism gave 0.067 mole. When these values were inserted into Eq. 9.24 together with those calculated for ΔH_{cat} using standard enthalpies of formation and for Δh_{an}, the values for ΔH_{met} were found to be -6.83 and -16.77 kcal for growth on 1 mole of lactate and pyruvate, respectively. These should be compared with the microcalorimetrically determined values of -8.69 and -16.78 kcal, respectively, per mole of substrate consumed. Although the calculated and experimentally determined ΔH_{met} values did not agree particularly well with respect to growth on lactate, Traore et al. emphasized that both values show that growth on lactate is considerably less than that on pyruvate. This was thought to be explainable by a large loss of energy due to the production of hydrogen gas during growth on lactate, which did not occur during growth on pyruvate. Traore et al.[151] carried out additional studies involving the determination of energetic growth parameters for three other *Desulfovibrio* sp. In general, the agreement of their experimentally determined and theoretical H_{met} values for growth on lactate and pyruvate was good except for *D. desulfuricans* Norway, for which this agreement was not good at all. The explanation of these investigators was that clearly, all of the available electrons from lactate or pyruvate oxidation were not used to reduce sulfate. It should be remembered that it is difficult to account for all the reactants and products in any complex fermentation or anaerobic respiration. In the case of *D. sulfuricans* Norway butanol was produced in addition to the acetate usually produced by this genus. When this was taken into account in the theoretical calculation of ΔH_{met}, the agreement of this corrected value with that which experimentally determined was somewhat improved, although still not good.

Possibly the formation of stored, intracellular products of assimilation was responsible for the lower experimental heat production, but this was not investigated.

The experiments of Belaich and his collaborators made use of the Tian-Calvet microcalorimeter almost entirely for studies involving anaerobic microbial growth, which seemed to indicate that this particular instrument is best suited in this respect, and less suited for studies with aerobically grown microorganisms. This was confirmed by Dermoun and Belaich[152] who, in 1979, published the results of experiments making use of a new Tian-Calvet type of microcalorimeter in which the original instrument was modified by changing the shape of the calorimeter vessel and by using a different arrangement of thermopile plates. The original calorimeter vessel was a tube containing 5 ml of culture fluid, giving this latter a low surface/volume ratio. This tube was replaced by a flat vessel somewhat resembling a Petri dish, in which 5 ml of culture fluid would form a thin layer having a high surface/volume ratio. This served to increase the oxygen transfer efficiency to a point where it would only become limiting at higher concentrations of substrate. Thus, the new calorimeter vessel was a thin, inox cylinder 15 mm high by 42 mm in diameter, which could be closed with a screw-on Teflon cover. To facilitate oxygen transfer the thin layer of culture medium could be stirred by means of a Teflon paddle, the shaft of which passed through the cover. Two thin rubber tubes were also sealed into the cover, so that oxygen transfer could be increased by passing pure oxygen over the surface of the medium. A syringe attachment was also sealed into the cover, so that an inoculum could be injected into the culture medium. The whole calorimetric vessel assembly could be dismantled and autoclaved at 120°C. After reassembling the parts, 5 ml of sterile culture medium were pipetted into the sterile calorimeter vessel, after which a syringe containing 0.05 ml of inoculum was attached to the cover and the whole assembly placed in the calorimeter for temperature equilibration. The stirring system was started immediately, as was aeration with pure oxygen. The heat of stirring was constant, and amounted to 35 μW. Evaporative cooling due to aeration was shown to be negligible. When the baseline on the recording apparatus became stable, microbial growth was initiated by injecting into the medium the inoculum contained in the syringe by means of a thin, rigid plastic rod passing down through the calorimeter to the top of the syringe plunger. Figures 9.5 and 9.6 show details of the calorimeter block and the calorimeter vessel published by Dermoun et al.[153] in 1985.

The experiments of Dermoun and Belaich[152] resulted in what are probably the most accurate measurements yet made of the observed heat of microbial growth. Figure 9.7 shows the thermogram obtained when *E. coli* was grown in 5 ml of culture medium containing 0.0085 *M* succinic acid. This is what might be regarded as a theoretically ideal, simple thermogram, in that the power output is seen to rise exponentially with time until the substrate becomes limiting, after which it declines precipitously to zero. The slight shoulder at the bottom of the curve may be indicative of the utilization of metabolic

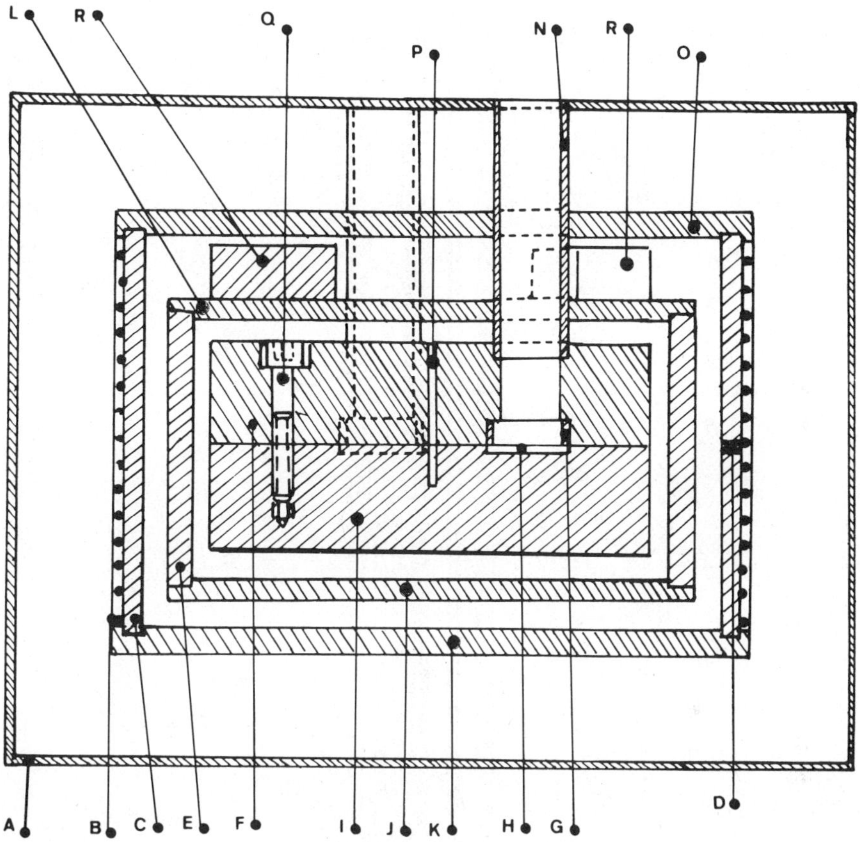

FIGURE 9.5. Diagram of Dermoun et al's microcalorimeter block assembly. A, tank; B, check heater; C, second box; D, temperature sensor; E, first box; F, first cylinder block; G, thermopile plate assembly screw; H, thermopile plate; I, second cylinder block; N, plastic sheath; P, temperature sensor; Q, block assembly screw; R, electronic amplifier. Reprinted from Reference 153 with permission.

intermediates remaining after the substrate has become exhausted and exponential growth has ceased. It can also indicate that something volatile, such as CO_2 gas, is slowly leaving the aqueous phase where it had been dissolved as $CO_2(aq)$, $H_2CO_3(aq)$, or $HCO_3^-(aq)$. In any case, 26 hr after inoculation no further heat output is evident, indicating that no storage products have accumulated within the cells, and that metabolism has effectively stopped. Figure 9.8 shows the thermogram obtained when *E. coli* was grown in 5 ml of a culture medium containing 0.034 *M* succinic acid. Here the power output levels off at a point where the transfer of oxygen into the medium becomes limiting and remains stationary until the substrate becomes exhausted, after which it also drops rapidly to zero. Interestingly, Dermoun and Belaich were able to demonstrate that growth occurred during this stationary phase of the

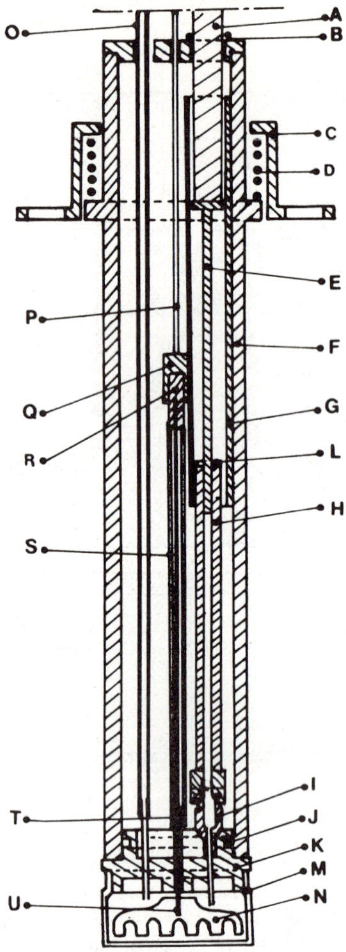

FIGURE 9.6. Diagram of the vessel assembly used in the microcalorimeter of Dermoun et al. A, plastic stick used for injection; B, withholding Teflon ring; C, removable cap; D, spring; E, syringe piston; F, plastic sheath; G, piston guide; H, syringe; I, truncated syringe needle; J, sheath-vessel assembly screw; K, screw-on Teflon cover; L, withholding Teflon ring; M, calorimeter vessel; N, Teflon stirring blade; O, fluxing gas tube; P, stirring stick; Q, stirring stick junction; R, stirring guide junction; S, waterproof sheath; T, stirring guide. Reprinted from Reference 153 with permission.

thermogram in Fig. 9.8 in that the molar growth yield obtained here was the same as that in the experiment illustrated by Fig. 9.7, and that the heat output was the same in both cases. Figure 9.9 shows the straight line obtained when the heat produced by *E. coli* grown on succinic acid is plotted against substrate concentration. Such experiments are absolutely necessary to establish reliable values for the observed heat of growth. One or two determinations at a single substrate concentration may not give reliable data. The value Dermoun and

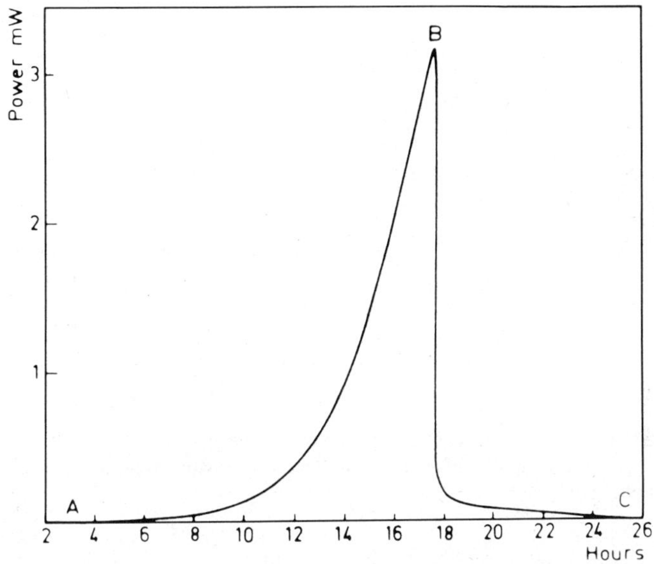

FIGURE 9.7. Dermoun and Belaich's thermogram of the growth of *Escherichia coli* in 5 ml of a defined medium containing 0.0085 *M* succinic acid. Reprinted from Reference 152 with permission.

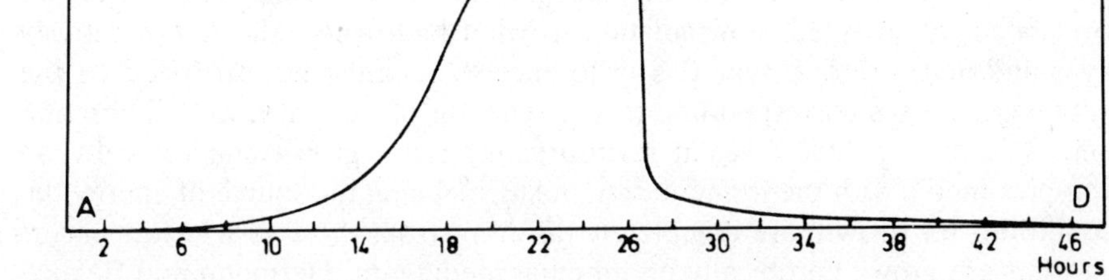

FIGURE 9.8. Dermoun and Belaich's thermogram for the growth of *Escherichia coli* in 5 ml of a defined medium containing 0.034 *M* succinic acid. Here oxygen has become limiting, causing the thermogram to level off. Reprinted from Reference 152 with permission.

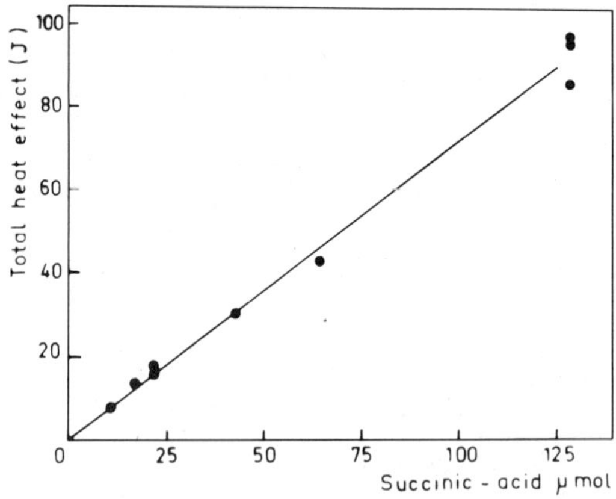

FIGURE 9.9. Graph of heat production accompanying the growth of *Escherichia coli* on succinic acid plotted against the quantity of substrate utilized for growth. Reprinted from Reference 152 with permission.

Belaich found for the observed enthalpy change accompanying the growth of *E. coli* on succinic acid was -176.72 ± 7.49 kcal mole^{-1} (-739.40 ± 31.35 kJ mole^{-1}). The unit-carbon formula for *E. coli* cells grown on succinic acid was $CH_{1.56}O_{0.36}N_{0.23}$. The molar growth yield, Y, was 37.5 g dry wt mole^{-1}, making the molar yield coefficient 1.66. Dermoun and Belaich were then able to represent the phenomenon of *E. coli* growth on this substrate in the following manner.

$$C_4H_6O_4 + 0.38NH_4OH + 1.77O_2 \longrightarrow \begin{array}{l} 1.66CH_{1.56}O_{0.36}N_{0.23} \\ + 2.34CO_2 + 2.65H_2O \end{array} \tag{9.26}$$

According to Eq. 9.26, 58.5% of the substrate is catabolized, with a corresponding enthalpy change which they calculated to be -209.44 kcal (-876.32 kJ) mole^{-1}. The difference between this last value and that experimentally measured, which is $-209.44 - (-176.72) = -32.72$ kcal or $-876.32 - (-739.40) = -136.92$ kJ is too great to be ignored and cannot be assigned to some error associated with the calorimetric measurement. They therefore concluded, in complete contradiction to what had been indicated previously by many investigators, that this difference ". . . must be attributed to the enthalpy changes corresponding to the reaction of biosynthesis." (Dermoun and Belaich,[151] p. 380.) When microorganisms are grown anaerobically on complex media with the fermentation of sugars being the source of energy the conditions for growth are completely different from those in which microorganisms are grown aerobically on minimal media, and Dermoun and Belaich felt that under these latter conditions the enthalpy change accompanying cellular synthesis may be significant.

In a subsequent article published in 1980, Dermoun and Belaich[154] attempted to account for the enthalpy change which was measured during the synthesis of *E. coli* cells from succinic acid. After having been grown in the same medium as that used for the previous study, the cells were harvested at the beginning of the stationary phase, washed twice with cold, distilled water, and lyophilized. Samples of 1 g each were then burned in a bomb calorimeter and the heat of combustion calculated taking into account the residual water content of the lyophilized cells. This latter was found to be -5.47 kcal (-22.9 kJ) g^{-1} dry wt of cells. The weight of ash in the dried cells was apparently not taken into consideration in this calculation, since this was not specifically mentioned. The formula weight corresponding to the empirical formula representing the dried cells, $CH_{1.56}O_{0.36}N_{0.23}$, was 22.54 daltons. Multiplying this value by -5.47 kcal (-22.9 kJ) gave -123.29 kcal (-516.17 kJ) as the enthalpy change accompanying the reaction represented by the following equation.

$$CH_{1.56}O_{0.36}N_{0.23} + 1.21O_2 \longrightarrow CO_2 + 0.78H_2O + 0.115N_2 \quad (9.27)$$

Equation 9.27 could then be used to calculate the ΔH_f of the cells as -23.82 kcal (-99.68 kJ) for one empirical formula weight of cells. Dermoun and Belaich considered the reaction whereby *E. coli* cells were synthesized from succinic acid to be represented as follows.

$$CH_{1.5}O + 0.23NH_4OH \longrightarrow CH_{1.56}O_{0.36}N_{0.23} + 0.54H_2O + 0.16O_2 \quad (9.28)$$

Inserting the appropriate data for the heats of formation of the reactants and products gave 14.41 kcal (60.31 kJ) as the enthalpy change accompanying the reaction represented by Eq. 9.28, or 0.64 kcal (2.67 kJ) g^{-1} of cells synthesized. Multiplying this value by Y_s gave 0.64 kcal $\times$ 37.5 g $=$ 24 kcal (100.12 kJ) as the enthalpy change for the synthesis of *E. coli* cells from 1 mole of succinic acid. The enthalpy change, ΔH_{met}, for the whole growth process can be calculated most easily by means of Eq. 9.26, and was found by Dermoun and Belaich using a slightly different method to be -185.52 kcal (-776.2 kJ) $mole^{-1}$ of succinic acid metabolized. This compares more favorably with the calorimetrically measured value of -176.72 kcal (-739.4 kJ) than does that of -209.44 kcal (-876.32 kJ) calculated previously on the assumption that the anabolic change was negligible.

Dermoun and Belaich[154] also calculated Δh^1_{an} values with respect to the synthesis from succinic acid of all the monomers in 1 g of dried *E. coli* cells. The values of Morowitz[1] were used for the quantities of individual amino acids, nucleic acid bases, lipid components, and sugars. To these were added Δh^2_{an} and Δh^3_{an} values corresponding to the formation of polymers from monomers, and the folding of polymers into tertiary structures, respectively. The summation of Δh^1_{an}, Δh^2_{an}, and Δh^3_{an} values gave Δh_{an} values for the proteins, nucleic acids, lipids, and polysaccharides found in 1 g of dried *E. coli* cells. The summation of these then gave a total value of 0.581 kcal (2.43 kJ) g^{-1} for Δh_{an} with respect to cellular synthesis during growth, in remarkably good agreement, considering all the uncertainties, with that of 0.640 kcal (2.67 kJ) found through experimental determination of the heat of combustion of the cells. According to Dermoun and Belaich, this ". . . demonstrated that the anabolic processes were indeed endothermic." (Dermoun and Belaich,[154] p. 745.) The enthalpy change accompanying anabolism was therefore not zero, as had previously been thought.

As always, the conclusions arising from any experimental investigation are the results of the use of the data obtained and the interpretations of the investigators. A different approach may result in a different interpretation. There can be no doubt that Dermoun and Belaich's measurement of the heat of growth of *E. coli* was made with great accuracy. The problem arises as to how to equate what happens in the calorimeter with the quantity of heat given off, and it is here that differences in interpretation can take place.

Determining the ΔH_f for the cells formed during growth by measuring the heat of combustion of a unit weight of dried cells and using this in Eq. 9.27 has the disadvantage that the cells are no longer in their native state when the measurement is carried out. The biological heat of combustion of the cells might better be determined by difference by calculating the enthalpic energy conserved within the substance of the cells during growth in the manner described by Battley (see Section 8.10) using conservation theory. Thus *if* the quantity of cells synthesized during growth on 1 mole of succinic acid is represented by $1.66CH_{1.56}O_{0.36}N_{0.23}$, *then* the quantity of substrate necessary for this process using the methods of Minkevich and Eroshin (see Section 9.3) is $(1.66 \times AE_c)/AE_S = 0.49$ mole. The remainder of the substrate, $(1 - 0.49) = 0.51$ mole, must be catabolized since there are no products formed during growth other than the cells. Catabolism can therefore be represented by the following equation.

$$0.51C_4H_6O_4(aq) + 1.78O_2(aq) \longrightarrow 2.04CO_2(aq) + 1.53H_2O(l) \qquad (9.29)$$

from which, using the appropriate values for $\Delta H_f^{0'}$ found in Appendix 2, the enthalpy change for the catabolic reaction represented by Eq. 9.29 is calculated to be −189.92 kcal (−794.64 kJ). However, were 1 mole of succinic acid to be completely oxidized under biological conditions, the enthalpy change would be −372.40 kcal (−1558.12 kJ). The difference, −182.48 kcal (−878.46 kJ), is the enthalpic energy that has been conserved within the substance of the cells in their living, hydrated state. This enthalpic energy would be released as heat if the cells were to be oxidized biologically in the following manner, using the quantity of cells from Eq. 9.26.

$$\left.\begin{matrix}1.66CH_{1.56}O_{0.36}N_{0.23}\\ + 1.72O_2(aq)\end{matrix}\right\} \longrightarrow \begin{matrix}1.66CO_2(aq) + 0.72H_2O(l)\\ + 0.38NH_3(aq)\end{matrix} \qquad (9.30)$$

The value of -182.48 kcal divided by 37.5 g dry weight of cells gives -4.87 kcal (-20.38 kJ) as compared with -5.47 kcal (-22.9 kJ) obtained by bomb calorimetry. This considerable difference is due largely to the oxidation of NH_3 to N_2 in the bomb calorimeter. Using Eq. 9.30 and the appropriate values for $\Delta H'$ and $\Delta H'_f$, the value of ΔH_f for the cellular material is calculated to be -20.00 kcal (-83.68 kJ) for one formula weight of cells. The equation representing anabolism is as follows:

$$0.49C_4H_6O_4(aq) + 0.38NH_3(aq) \longrightarrow \begin{array}{l} 1.66CH_{1.56}O_{0.36}N_{0.23} \\ + 0.30CO_2(aq) \\ + 0.75H_2O(l) \end{array} \tag{9.31}$$

Using the value of -20.00 obtained for the cells and the other appropriate values for $\Delta H_f^{0\prime}$ from Appendix 2, the enthalpy change accompanying the anabolic reaction is calculated from Eq. 9.31 to be -0.020 kcal (-0.084 kJ), or practically zero. Adding Eqs. 9.29 and 9.31 gives an equation representing the complete growth process.

$$\left.\begin{array}{l} C_4H_6O_4(aq) + 0.38NH_3(aq) \\ + 1.78O_2(aq) \end{array}\right\} \longrightarrow \begin{array}{l} 1.66CH_{1.56}O_{0.36}N_{0.23} \\ + 2.34CO_2(aq) + 2.28H_2O(l) \end{array} \tag{9.32}$$

The enthalpy change accompanying the reaction represented by Eq. 9.32 is then $\Delta H_{cat} + \Delta H_{an} = -189.92 + (-0.020) = -189.94$ kcal (-794.71 kJ), and can be considered the "true" heat of growth. The difference between this value and that of -176.74 kcal (-739.4 kJ) measured by Dermoun and Belaich[151] must therefore be due to physical reactions external to the process of catabolism and anabolism. The most probable one is the passage of CO_2 generated as a result of metabolism from the aqueous phase of the culture medium to the gas phase above it in the calorimeter vessel, since the latter was constantly being flushed with O_2 gas during a calorimetric experimental run. This would tend to remove CO_2 completely from the culture medium by the time the thermogram had returned to the base line. There are thus *two* physical reaction systems which must be considered here. One is the growth process itself, which takes place entirely within the aqueous phase and has its own heat of reaction. The other is the reaction whereby $CO_2(aq)$ leaves the aqueous phase and passes into the gas phase within the calorimeter vessel according to the following equation.

$$2.34CO_2(aq) \longrightarrow 2.34CO_2(g) \tag{9.33}$$

The reaction represented by Eq. 9.33 is accompanied by its own enthalpy change of 11.35 kcal (48.48 kJ). The addition of Eqs. 9.32 and 9.33 gives an equation representing more correctly than Eq. 9.32 alone what takes place within the calorimeter vessel during the growth of *E. coli*.

$$\left.\begin{array}{l} C_4H_6O_4(aq) + 0.38NH_3(aq) \\ + 1.77O_2(aq) \end{array}\right\} \longrightarrow \begin{array}{l} 1.66CH_{1.56}O_{0.36}N_{0.23} \\ + 2.34CO_2(g) + 2.65H_2O(l) \end{array} \quad (9.34)$$

Correspondingly, the addition of the enthalpy changes accompanying the reactions represented by Eqs. 9.32 and 9.33 gives $-189.94 + 11.35 = -178.59$ kcal (-747.22 kJ), which is within 1% of the value experimentally measured by Dermoun and Belaich.[151] This is an excellent agreement which must be regarded as confirming both Battley's theoretical approach to calculating enthalpy changes using conservation theory and Dermoun's and Belaich's microcalorimetric measurements. However, it does not confirm the recent conclusion of these latter authors that anabolism is endothermic. The enthalpy change accompanying anabolism as represented by Eq. 9.31 is for all practical purposes zero.

When Dermoun et al.[152] published a technical description of their new microcalorimeter in 1985, they also included studies of the enthalpy changes occurring when *E. coli* was grown aerobically on acetate and on glucose. For this purpose they used the equation developed by Dermoun and Belaich[153] for calculating the mole fraction of substrate involved in catabolism and anabolism, as follows:

$$\Delta H_{met} = (1 - \alpha)\,\Delta H_{cat} + \Delta H_{an} \quad (9.35)$$

where ΔH_{met} is the total enthalpy change accompanying the growth process (metabolism), ΔH_{cat} and ΔH_{an} the enthalpy changes accompanying the catabolic and anabolic processes, respectively, which go to make up metabolism, and α is the mole fraction of substrate used for the construction of cellular material. The value for α was calculated according to the following equation.

$$\alpha = \frac{y}{100} \times Y \times \frac{1}{12} \times \frac{1}{a} \quad (9.36)$$

where y is the quantity of carbon contained in 100 g dry weight of cells, Y represents the molar growth yield, 12 is the atomic weight of carbon, and a the number of carbon atoms contained in 1 mole of the carbon and energy source. Dermoun et al. found this value to be 0.41, 0.29, and 0.54 for the growth of *E. coli* on succinic acid, acetic acid, and glucose, respectively. The problem with using Eq. 9.36 is that this equates the carbon content of the substrate and cells, but does not address the degree of reduction of either. Thus, the equation representing the synthesis of *E. coli* from succinic acid using 0.41 as the value for α is as follows.

$$0.41C_4H_6O_4 + 0.38NH_3 \longrightarrow \begin{array}{l} 1.66CH_{1.56}O_{0.36}N_{0.23} + 0.50H_2O \\ + 0.27O_2 \end{array} \quad (9.37)$$

In order to balance Eq. 9.37 it is necessary to introduce 0.27 mole of O_2 as a product in order to achieve a material balance, something which is not known to happen biochemically except during the photolysis of water in photosynthesis. However, there is no reason why this balance cannot be made theoretically. The difficulty is that the available electrons in the cells could only have come from the substrate, as pointed out by Minkevich and Eroshin (see Section 9.3). The quantity of cells in Eq. 9.37 contains 6.89AE, whereas the quantity of substrate contains only 5.74AE. Thus, although Eq. 9.37 is materially balanced, it is not electronically balanced. A more correct representation is found in Eq. 9.31, wherein the value for α is 0.49 rather than 0.41. Using the same technique for constructing Eqs. 9.29 through 9.34 with respect to the growth of *E. coli* on molar quantities of acetic acid and on glucose, and the cellular compositions and values for Y found in Dermoun et al., the observed enthalpy changes for growth on acetic acid and glucose are calculated to be -142.0 kcal (-594 kJ) and -286.1 kcal (-1196 kJ), respectively. These values should be compared with -113.7 kcal (-476 kJ) and -300.4 kcal (-1257 kJ), respectively, which were obtained by Dermoun et al. experimentally. The agreement between the theoretical and experimental values for acetic acid-grown cells is only slightly better than that obtained by Dermoun et al. The agreement between the same values for glucose-grown cells is much better, and perhaps quite good considering that the thermogram was biphasic, indicating possible uncertainties in the description of what is happening within the calorimeter vessel during growth.

It is apparent that even with the extremely fine calorimetric instrument that Dermoun et al. used, it is necessary to know exactly the nature of the growth-process phenomena, and what corrections to apply with respect to reactions that take place within the calorimeter vessel apart from the growth-process system itself. Nevertheless, one of the most important developments to come out of the work of Dermoun and Belaich[151] is that with their yield data and empirical analysis of *E. coli* cells grown aerobically on succinic acid, and by using Battley's conservation theory, it is possible to predict almost exactly the observed heat of growth on succinic acid which they obtained with this bacterium. This must surely be more than coincidence, and may indicate that we are on the right track toward conceptualizing microbial growth correctly.

9.5.4. The Microcalorimetric Experiments of Ishikawa, Nonoyama, and Shodo

At about the same time that Dermoun and Belaich[152] began working with their modified Tian-Calvet microcalorimeter, Ishikawa et al.[155] built a similar elegant calorimeter making use of wire thermopiles rather than thermoelectric modules to measure the heat transferred from the calorimeter vessel to the heat sink. The calorimeter vessel volume was 30 ml. The calorimetric system also had the added advantage that it was designed to be operated either with batch or continuous cultures within the calorimeter vessel. In a subsequent publication

Ishikawa et al.[156] described experiments in which *E. coli* K-12 was grown aerobically in a minimal medium under conditions of glucose, nitrogen, and oxygen limitation. It is only the first of these that is of interest here, in that under the second two conditions oxidative assimilation would be likely to occur. In fact, this is demonstrated by the difference in the empirical formulas for the cells grown under the three conditions, which are $CH_{1.74}O_{0.464}N_{0.260}$, $CH_{1.76}O_{0.460}N_{0.253}$, and $CH_{1.80}O_{0.470}N_{0.225}$ for flucose, oxygen, and nitrogen limited conditions, respectively. The relative degrees of reduction (γ, using Eroshin and Minkevich's notation; see Section 9.4) are 4.032, 4.081, and 4.185, respectively. This indicates that the carbon in cells grown under substrate limitation is not as reduced as that in cells grown under oxygen or nitrogen limitation, and that therefore, these latter may contain storage materials which are more reduced than carbohydrates, such as, for example, poly β-hydroxy butyric acid.

With respect to the growth of *E. coli* aerobically on glucose Ishikawa et al.[156] determined the molar growth yield to be 74.34 g, in good agreement with that obtained by Dermoun et al.[153] of 75.72 g. Further, the same biphasic thermogram was observed by both sets of investigators, indicating that glucose was first metabolized to acetic acid, which was subsequently used as a carbon source for growth. However, the heat of growth measured by Ishikawa et al. was -257 kcal mole^{-1} of glucose consumed, considerably less than that of -300.4 kcal measured by Dermoun et al. In addition, the empirical formula and the heat of combustion of cells grown on glucose were determined by Ishikawa et al. to be $CH_{1.74}O_{0.464}N_{0.260}$ and -5.572 kcal g^{-1}, respectively, where the same values determined by Dermoun et al. were $CH_{1.54}O_{0.37}N_{0.24}$ and -5.406 kcal g^{-1}. Thus, the experimental results of the former were for the most part different from those of the latter. The following equation was used to calculate ΔH_f for the cells.

$$CH_{1.74}O_{0.464}N_{0.26} + 1.2O_2 \longrightarrow CO_2 + 0.87H_2O + 0.13N_2 \qquad (9.38)$$

Using Eq. 9.38 ΔH_f for the cells was calculated to be -15.14 kcal for one empirical formula wt of cells. This value was used together with other appropriate values for ΔH_f of the reactants and products in an equation representing cell synthesis.

$$\tfrac{1}{6}C_6H_{12}O_6 + 0.26NH_3 \longrightarrow CH_{1.74}O_{0.464}N_{0.26} + 0.52H_2O + 0.008O_2 \qquad (9.39)$$

The value for the enthalpy change accompanying cell synthesis as represented by Eq. 9.39 was calculated to be 7.90 kcal. The equivalent to Eq. 9.39 which was used by Dermoun et al. is as follows:

$$\tfrac{1}{6}C_6H_{12}O_6 + 0.24NH_3 \longrightarrow CH_{1.54}O_{0.34}N_{0.24} + 0.60H_2O + 0.02O_2 \qquad (9.40)$$

The enthalpy change calculated for the reaction represented by Eq. 9.40 is −8.42 kcal. It is apparent that although Eqs. 9.39 and 9.40 are similar, the results of the calculations are greatly different.

Ishikawa et al. described the total process of the growth of *E. coli* aerobically on glucose in the following manner.

$$\left.\begin{matrix} C_6H_{12}O_6 + 0.777NH_3 \\ + 2.98O_2 \end{matrix}\right\} \longrightarrow \begin{matrix} 2.99CH_{1.74}O_{0.46}N_{0.26} \\ + 3.01CO_2 + 4.56H_2O \end{matrix} \qquad (9.41)$$

The enthalpy change accompanying growth as represented by Eq. 9.41 was calculated to be −314.8 kcal, which is a value not far from that of −337.5 kcal calculated by Dermoun et al. for the same phenomenon. When using conservation theory, the ΔH_C (ΔH_G of Ishikawa et al. and ΔH_{met} of Dermoun et al.) values obtained with the data of Ishikawa et al. and Dermoun et al. are −329 and −286 kcal, respectively, both values being considerably different from the theoretical values calculated by either set of investigators. But which methods of theoretical calculation are most correct? We do not know for sure. The most accurate calculation cannot be shown to be valid unless it gives a value that can be produced experimentally, and if the calorimetric measurement is too low or too high, this validity will never be demonstrated.

Ishikawa et al. defined an interesting index to show the extent of oxidation of the carbon source as a function of heat of combustion values.

$$\begin{aligned} H_r &= \frac{\text{assimilated energy in cell mass} + \text{heat evolved}}{\text{total energy available}} \\ &= \frac{\text{heat of combustion of cells} + \Delta H_m}{\text{heat of combustion of glucose} + \text{ammonia}} \end{aligned} \qquad (9.42)$$

In Eq. 9.42 ΔH_m represents the observed heat of growth. Thus, for their experiment,

$$H_r = \frac{2.99(-138.4) - 257}{-673 + 0.777(-91.46)} = 0.901$$

for which they concluded that since 90% of the initially available energy was recovered as cells or as dissipated heat, 10% of this energy remained unmetabolized in solution in the medium. On the other hand, it may be that the measurement of the observed heat of growth was too low. If the value of 300 kcal obtained by Dermoun et al. is used instead, the value for H_r rises to 96%, and if that of −329 kcal calculated by means of conservation theory is used, H_r rises to 99.8%. This last would be expected if the cells were to oxidize all of the organic material resulting from the metabolism of glucose. However, for whatever reason this was not determined experimentally.

The experiments of Ishikawa et al. are important in that, just as with the experiments of Dermoun et al., a serious attempt was made to work with an accurate microcalorimeter using a thermodynamically definable biological system. That the results of the two investigations did not agree particularly well is simply an indication that more critical work needs to be done.

9.6. BATTLEY'S ATTEMPT TO RELATE THEORETICALLY THE FREE ENERGY AND ENTHALPY CHANGES ACCOMPANYING MICROBIAL GROWTH

Of the three thermodynamic quantities that are important to consider with respect to microbial growth, the only one that is convenient to determine is the enthalpy change. However, it has long been known that it is the free-energy change which is the true measure of the extent to which a reaction or process can proceed. In the case of reactions involving the complete biological oxidation of organic substances to CO_2, H_2O, and NH_3, the accompanying $\Delta G^{\circ\prime}$ and $\Delta H^{\circ\prime}$ values are usually very close, so that by determining the latter one can often estimate the former to within 5% and often much better. This is *not* true with respect to fermentations, where the $\Delta G^{\circ\prime}$ and $\Delta H^{\circ\prime}$ values are usually greatly different. There would be a considerable advantage if some means could be devised whereby a parameter related to a free-energy change could be obtained by making thermal measurements.

In 1979 E. H. Battley[157] attempted to relate the energy changes accompanying the biological oxidation of organic substances to their percent reduction. The idea of a percent reduction with respect to organic substances seems to have originated in 1949 with Spoehr and Milner[73] in their studies of *Chlorella*. The carbon in CH_4 was considered to be 100% reduced, and that in CO_2 0% reduced, the percent reduction of the carbon in all other carbon-containing substances falling somewhere in between these two values. In the case of substances containing only carbon, hydrogen, and oxygen, the percent reduction of the carbon, R, can be calculated by means of the following equation.

$$R = \frac{(\%\text{C} \times 2.664) + (\%\text{H} \times 7.936) - \%\text{O}}{398.982} \tag{9.43}$$

In Eq. 9.43 %C, %H, and %O represent the percentages by weight of these elements in an organic substance. As an example of the calculation of an R value, acetic acid contains 40.00% C, 6.71% H, and 53.29% O, and is 26.71% reduced. Some substances contain nitrogen, and Spoehr and Milner reasoned that since this exists largely as amino nitrogen which is oxidized to nitrogen gas during combustion analysis, it could be regarded as being inert. The principal contribution this element makes to the percent reduction of the carbon in a substance is that it reduces the percentages of the other elements. Thus, glycine contains 32.00% C, 6.71% H, 42.63% O, and 18.66% N and, using only the percentages for carbon, hydrogen, and oxygen is calculated by means of Eq. 9.43 to be 24.03% reduced as compared with 26.71% for acetic acid.

Because of its simplicity, Battley preferred to refer to the percent reduction of the carbon in an organic substance by the term "degree of reduction," which is effectively the same as the term "reductance degree" used by Minkevich and Eroshin (see Section 9.4). The concepts behind the use of the two terms is, however, quite different.

Battley was able to show that the degree of reduction depended upon how it was calculated, and concluded that the most appropriate method of calculation was one taking into consideration the energy content per gram of substance. The reductance degree is extremely useful for calculating the number of electrons transferred during a reaction. However, not all electrons are equivalent with respect to the energy which can be obtained from them. For example, although the reductance degree of formaldehyde (CH_2O), acetic acid (CH_3COOH), and glucose ($C_6H_{12}O_6$ is 4.000 for all three substances, the $\Delta G°$ values per gram atom of carbon accompanying their oxidation by $O_2(g)$ to $CO_2(g)$ and $H_2O(l)$ are -135.36, -104.50, and -111.56 kcal, respectively. Thus, for a more accurate expression of reduction, energy contents would be better to work with than numbers of electrons.

It was Battley's opinion that the most appropriate method of calculating a value for the degree of reduction, R, was to take into consideration the quantity of energy per gram of substance that could be obtained if the substance were to be oxidized to CO_2, H_2O, and NH_3 under the conditions in which this would occur biologically, that is, in aqueous solution. Also, there are different values for degrees of reduction (R values) depending on the kinds of energy that are considered. Thus, $R_{G°'}$ and $R_{H°'}$ represent R values which are obtained using standard free energies or enthalpies of formation, respectively, for their calculation. The equations Battley used to calculate these R values are as follows.

$$R_{G°'} = \frac{\begin{array}{c}100 \times \Delta G°' \text{ in kcal accompanying the oxidation by} \\ O_2(\text{aq}) \text{ of 1 g of organic substance to } CO_2(\text{aq}), \\ H_2O(\text{l}), \text{ and } NH_3(\text{aq}) \text{ at 298.16°K and 1 atm}\end{array}}{\begin{array}{c}\Delta G°' \text{ in kcal accompanying the oxidation by} \\ O_2(\text{aq}) \text{ of 1 g of } CH_4(\text{aq}) \text{ to } CO_2(\text{aq}) \text{ and } H_2O(\text{l}) \\ \text{at 298.16°K and 1 atm}\end{array}}$$

$$R_{H°'} = \frac{\begin{array}{c}100 \times \Delta H°' \text{ in kcal accompanying the oxidation by} \\ O_2(\text{aq}) \text{ of 1 g of organic substance to } CO_2(\text{aq}), \\ H_2O(\text{l}), \text{ and } NH_3(\text{aq}) \text{ at 298.16°K and 1 atm}\end{array}}{\begin{array}{c}\Delta H°' \text{ in kcal accompanying the oxidation by} \\ O_2(\text{aq}) \text{ of 1 g of } CH_4(\text{aq}) \text{ and } H_2O(\text{l}) \\ \text{at 298.16°K and 1 atm}\end{array}}$$

If $R_{G^{\circ\prime}}$ or $R_{H^{\circ\prime}}$ values for various substances are plotted against $\Delta G^{\circ\prime}\ g^{-1}$ or $\Delta H^{\circ\prime}\ g^{-1}$ of the same substance, straight lines are obtained (by construction). The relationship between the two lines is

$$\frac{\Delta G^{\circ\prime}}{g} = \frac{-12.791}{-13.005} \times \frac{\Delta H^{\circ\prime}}{g}$$
$$= 0.9835\ \Delta H^{\circ\prime}\ g^{-1} \qquad (9.44)$$

Since Battley[157] had previously determined the enthalpy changes accompanying the growth of *S. cerevisiae* on glucose anaerobically and aerobically and on ethanol, he was able to use his results to calculate the $\Delta H_f'$ for a unit-carbon formula weight (UCFW) of cells. Inserting this value plus the other appropriate values for $\Delta H_f^{\circ\prime}$ into an equation representing the combustion in the aqueous state (biological conditions) of this quantity of cells to $CO_2(aq)$, $H_2O(l)$, and $NH_3(aq)$, a value for $\Delta H'\ (UCFW)^{-1}$ and from this for $\Delta H'\ g^{-1}$ could be calculated. Inserting this latter value into Eq. 9.44 gave a value for $\Delta G'\ g^{-1}$ of cells and from this a value for $\Delta G'\ (UCFW)^{-1}$ could be found. Inserting this value plus the other appropriate values for $\Delta G_f^{\circ\prime}$ in the combustion equation gave a value for $\Delta G_f'\ (UCFW)^{-1}$ of cells. This could then be inserted into an equation representing the growth of *S. cerevisiae* on a given substrate, anaerobically or aerobically, to find the free-energy change accompanying the growth process. In general, Battley[156] found such calculated values to be in good agreement with those calculated using the assumption that the free-energy change accompanying cellular synthesis was zero.

Calculations such as those attempted by Battley could be useful in calculating $\Delta G_f'\ (UCFW)^{-1}$ values for complex substances such as malt extract, peptone, and so on, on which microorganisms are often commercially grown, or for the mixture of unknown substances remaining after a fermentation has ceased. However, one criticism that can be leveled against his methods is the use of $CH_4(aq)$ as the standard, 100% reduced organic substance against which all other organic substances are compared. The C–H bonds in methane are all energetically equivalent, but they may not represent the C–H bonds in a typical organic substance. These latter might be better represented by those in a methylene group. Thus, if a value for $\Delta G_f'$ of $-CH_2(aq)-$ could be found, it might represent a better standard for 100% reduced biological carbon.

9.7. SUMMARY

Of the many experimental studies described in this chapter, only a few are directly applicable to the calculation of changes in free energy, enthalpy, and entropy which accompany microbial growth. For these it is necessary to be able to write a balanced equation representing the conversion of reactants to products, included among which are the cells, which occurs during growth.

Without such an equation we are unable to account for whatever heat exchange occurs when microbial cells are grown in a microcalorimeter. It will be the ability through the use of these equations to predict the exact amount of heat produced during growth that will prove the correctness of any theories pertaining to the energetics of microbial growth. Such a proof will involve not only the accurate determination of equations representing growth, but the accurate measurement of the heat of growth as well. Errors in either will completely frustrate the proof of such theories, and it is clear that it is necessary to carry out a greater number of critically designed and executed experiments.

REFERENCES

1. H. J. Morowitz, Some order–disorder considerations in living systems, *Bull. Math. Biophys.* **17,** 81–86 (1955).
2. H. J. Morowitz, Cells: Living; Molecular Physics, in O. Glasser, Ed., *Medical Physics*, Vol. III, The Yearbook Publishers, Chicago, 1960, pp. 111–113.
3. H. J. Morowitz, Some consequences of the application of the second law of thermodynamics to cellular systems, *Biochim. Biophys. Acta* **40,** 340–345 (1960).
4. A. Passynsky, On the limitations in biological role of entropy, Abstracts of Contributed Papers, International Biophysics Congress, Stockholm, Proceedings, p. 53 (1961).
5. P. Chambadal, *Evolution et Application du Concept d'Entropie*, Dunod, Paris, 1963.
6. H. J. Morowitz, *Energy Flow in Biology*, Academic Press, New York, 1968.
7. A. Lwoff, *Biological Order*, M.I.T. Press, Cambridge, MA, 1962.
8. A. Lwoff, L'Ordre Biologique, Latond, Paris, 1969.
9. P. Morrison, A thermodynamic characterization of self-reproduction, *Rev. Mod. Phys.* **36,** 517–524 (1964).
10. F. Gorski, *Plant Growth and Entropy Production*, Polish Academy of Sciences, Krakow, 1966.
11. F. Gorski, Entropie structurale des organismes, *Acta Biol. Cracov., ser. botan.* **11,** 39–57 (1968).
12. F. Gorski, Entropy of living matter and the problem of reductionism, *Bull. Acad. Polon. Sci., ser. biol.* **18,** 121–126 (1968).
13. F. Gorski, Structural entropy of organisms, *Folia Biol.* (*Krakow*) **26,** 263–275 (1978).
14. F. Gorski, Entropia strukturalna, *Problemy, Zesz. nr.* **1,** 2–8, Warszawa (1980) (in Polish).
15. F. Gorski, Entropia żywego organizmu, *Kosmos ser. A., Biol.* (*Warszawa*), *nr.* **1,** 49–72 (1981) (in Polish).
16. J. Monod, *Recherches sur la Croissance des Cultures Bacteriennes*, Actualites Scientifiques et Industrielles, No. 911, Hermann et Cie., Paris, 1958.
17. T. Bauchop, Observations on some molar growth yields of bacteria and yeasts, *J. Gen. Microbiol.* **18,** vii (1958).

18. T. Bauchop and S. R. Elsden, The growth of microorganisms in relation to their energy supply, *J. Gen. Microbiol.* **23,** 457–469 (1960).

19. J. C. Senez, Some considerations on the energetics of bacterial growth, *Bacteriol. Rev.* **26,** 95–107 (1962).

20. O. Meyerhof, Untersuchungen über den Atmungsvorgang nitrifizierender Bakterien, *Pfleugers Archiv Gesamte Physiol. Menschen Tiere* **164,** 353–427 (1916); **165,** 229–284 (1916); **166,** 240–280 (1917).

21. L. G. N. Baas-Becking and G. S. Parks, Energy relations in the metabolism of autotrophic bacteria, *Physiol. Rev.* **7,** 85–106 (1927).

22. H. Lees, *Biochemistry of Autotrophic Bacteria*, Butterworth & Co., Ltd., London, 1955.

23. E. H. Battley, On the calculation of the thermodynamic changes accompanying the growth of microorganisms, *Int. Biophys. Congr., Stockholm, Abstr. Cont. Papers BI*, p. 55 (1961).

24. R. E. Hungate, Polysaccharide storage and growth efficiency in *Ruminococcus albus*, *J. Bact.* **86,** 848–854 (1963).

25. P. N. Hopson, Continuous culture of some anaerobic and facultatively anaerobic rumen bacteria, *J. Gen. Microbiol.* **38,** 167–180 (1965).

26. E. Hernandez and M. J. Johnson, Anaerobic growth yields of *Aerobacter cloacae* and *Escherichia coli*, *J. Bact.* **94,** 991–995 (1967).

27. A. H. Stouthamer, Determination and significance of molar growth yields, in J. R. Norris and D. W. Ribbons, Eds., *Methods in Microbiology*, Vol. 1, Academic Press, New York, 1969, pp. 629–663.

28. W. J. Payne, Energy yields and the growth of heterotrophs, *Ann. Rev. Microbiol.* **24,** 17–52 (1970).

29. K. Decker, K. Jungermann, and R. K. Thauer, Energy production in anaerobic organisms, *Angew. Chem. Int. Ed.* **9,** 138–158 (1970).

30. W. W. Forrest and D. J. Walker, The generation and utilization of energy during growth, *Adv. Microb. Physiol.* **5,** 213–274 (1971).

31. A. H. Stouthamer and C. Bettenhausen, Utilization of energy for growth and maintenance in continuous cultures, *Biochim. Biophys. Acta* **301,** 53–70 (1973).

32. A. H. Stouthamer, *Yield Studies in Microorganisms*, Patterns of Progress, Meadowfield Press, Ltd., 1976.

33. D. J. McGill and E. A. Dawes, Glucose and fructose metabolism in *Zymomonas anaerobia*, *Biochem. J.* **125,** 1059–1068 (1971).

34. W. de Vries, W. M. C. Kapteijn, E. G. van der Beek, and A. H. Stouthamer, Molar growth yields and fermentation balances of *Lactobacillus casei* L3 in batch cultures and in continuous cultures, *J. Gen. Microbiol.* **63,** 333–345 (1970).

35. M. S. Karasch and B. Sher, The electronic conception of valence and heats of combustion of organic compounds, *J. Phys. Chem.* **29,** 625–658 (1925).

36. M. S. Karasch, Heats of combustion of organic compounds, *Bur. Stand. U.S. J. Res.* **2,** 359–430 (1929).

37. W. R. Mayberry, G. J. Prochazka, and W. J. Payne, Growth yields of bacteria on selected organic compounds, *Appl. Microbiol.* **15,** 1332–1338 (1967).

38. I. MacKechnie and E. A. Dawes, An evaluation of the pathways of metabolism

of glucose, gluconate, and 2-oxogluconate by *Pseudomonas aeruginosa* by measurement of growth yields, *J. Gen. Microbiol.* **55,** 341–349 (1969).

39. W. R. Mayberry, G. J. Prochazka, and W. J. Payne, Factors derived from studies of aerobic growth in minimal media, *J. Bacteriol.* **96,** 1424–1426 (1968).
40. G. J. Prochazka, W. J. Payne, and W. R. Mayberry, Calorific contents of certain bacteria and fungi, *J. Bacteriol.* **104,** 646–649 (1970).
41. G. J. Prochazka, W. J. Payne, and W. R. Mayberry, Calorific contents of microorganisms, *Biotech. Bioeng.* **15,** 1007–1010 (1973).
42. R. H. Mennett and T. O. M. Nakayama, Influence of temperature on substrate and energy conversion in *Pseudomonas fluorescens, Appl. Microbiol.* **22,** 772–776 (1971).
43. D. W. Tempest and J. R. Hunter, The influence of temperature and pH value on the macromolecular composition of Mg-limited and glycerol limited *Aerobacter aerogenes* growing in a chemostat, *J. Gen. Microbiol.* **41,** 267–273 (1965).
44. C. M. Brown and A. H. Rose, Effects of temperature on composition and cell volume of *Candida utilis*, *J. Bacteriol.* **97,** 261–272 (1969).
45. K. P. Ho and W. J. Payne, Assimilation efficiency and energy contents of prototrophic bacteria, *Biotech. Bioeng.* **21,** 787–802 (1979).
46. S. Yamagata, Über die elementare Zusammensetzung des Schimmelkorpers, *Acta Phytochim.* **8,** 107–116 (1934).
47. I. G. Minkevich and V. K. Eroshin, Productivity and heat generation of fermentation under oxygen limitation, *Folia Microbiol.* **18,** 376–385 (1973).
48. W. J. Payne and M. L. Williams, Carbon assimilation from simple and complex media by prototrophic heterotrophic bacteria, *Biotech. Bioeng.* **18,** 1653–1655 (1976).
49. W. J. Payne and W. J. Wiebe, Growth yield and efficiency in chemosynthetic microorganisms, *Ann. Rev. Microbiol.* **32,** 155–183 (1978).
50. E. H. Battley, A calculation of the free-energy changes accompanying the growth of *Saccharomyces cerevisiae* on several substrates, *First European Biophys. Congr., Proc.* **4,** 299–305 (1971).
51. G. H. Bell, Yield factors and their significance, *Process Biochem.* **7,** 21–25 and 34 (1972).
52. P. L. McCarty, Discussion of Ref. 15, *J. Sanit. Eng. Div., Am. Soc. Chem. Eng.* **89,** 65–68 (1963).
53. H. Tamiya, Zur Theorie des respiratorischen Quotienten nebst einer Bemerkung über den Einfluss der oxydoreduktiven Zellvorgänge auf den Gaswechsel der Zellen. Beiträge zur Atmungsphysiologie der Schimmelpilze. I, *Acta Phytochim.* **6,** 227–263 (1932).
54. O. R. Placak and C. G. Ruchoft, Studies of sewage purification. XVII. The utilization of organic substrates by activated sludge, *Sewage Works J.* **19,** 423–440 (1947).
55. E. N. Helmers, J. D. Frame, A. E. Greenberg, and C. N. Sawyer, Nutritional requirements in the biological stabilization of industrial wastes. II. Treatment with domestic sewage, *Sewage and Ind. Wastes J.* **23,** 884–899 (1951).
56. H. A. Barker, The oxidative metabolism of the colorless alga, *Prototheca zopfii*, *J. Cell. Comp. Physiol.* **8,** 231–250 (1936).

57. R. E. McKinney, *Microbiology for Sanitary Engineers*, McGraw-Hill, New York, 1962.

58. R. E. McKinney, Mathematics of complete-mixing activated sludge, *J. Sanit. Eng. Div. Proc. Am. Soc. Civ. Eng.* **88,** 87–113 (1962).

59. E. H. Battley, Growth-reaction equations for *Saccharomyces cerevisiae*, *Physiol. Plant.* **13,** 192–203 (1960).

60. C. E. Burkhead and R. E. McKinney, Application of complete-mixing activated sludge design equations to industrial wastes, *J. Water Pollut. Control Fed.* **40,** 557–570 (1968).

61. J. A. Servizi and R. H. Bogan, Free-energy as a parameter in biological treatment., *J. Sanit. Eng. Div. Proc. Am. Soc. Civ. Eng.* **89,** 17–40 (1963).

62. B. V. Siegel and C. E. Clifton, Oxidative assimilation of glucose by *E. coli.*, *J. Bact.* **60,** 113–118 (1950).

63. R. E. McKinney, H. D. Tomlinson, and R. L. Wilcox, Metabolism of aromatic compounds by activated sludge, *Sewage Ind. Wastes J.* **28,** 547–557 (1956).

64. S. R. Hoover and N. Porges, Assimilation of dairy wastes by activated sludge. II. The equation of synthesis and rate of oxygen utilization, *Sewage Ind. Wastes J.* **24,** 306–312 (1952).

65. J. M. Symons and R. E. McKinney, The biochemistry of nitrogen in the synthesis of activated sludge, *Sewage Ind. Wastes J.* **30,** 874–890 (1958).

66. J. A. Servizi and R. H. Bogan, Thermodynamic aspects of biological oxidation and synthesis, *J. Water Pollut. Control Fed.* **36,** 607–618 (1964).

67. P. L. McCarty, Thermodynamics of biological synthesis and growth, *2nd International Conference on Water Pollution Research, Tokyo, Japan*, Pergamon Press, New York, 1965, pp. 169–187.

68. K. Burton, Energy of adenosine triphosphate, *Nature* **181,** 1594–1595 (1958).

69. A. L. Lehninger, *Bioenergetics*, W. A. Benjamin, Inc., New York, 1965, 258 pp.

70. P. L. McCarty, Energetics and bacterial growth, in S. D. Faust and J. V. Hunger, Eds., *Organic Compounds in Aquatic Environments*, Marcel Dekker, New York, 1971, pp. 495–531.

71. E. D. Schroeder and A. W. Busch, Validity of energy change as a growth parameter, *J. Sanit. Eng. Div. Proc. Am. Soc. Civ. Eng.* **94,** 193–200 (1968).

72. C. E. Burkhead and R. E. McKinney, Energy concepts of aerobic microbial metabolism, *J. Sanit. Eng. Div. Am. Soc. Civ. Eng.* **95,** 253–268 (1969).

73. H. A. Spoehr and H. W. Milner, The chemical composition of *Chlorella*; effect of environmental conditions, *Plant. Physiol.* **24,** 120–149 (1949).

74. G. J. Prochazka, W. R. Mayberry, and W. J. Payne, Model systems for studying energy yields from synthetic organic compounds, *Develop. Ind. Microbiol.* **8,** 167–178 (1967).

75. W. J. Payne, G. J. Prochazka, and W. R. Mayberry, Measurement of yields from cultures growing on synthetic organic compounds, *J. Gen. Microbiol.* **45,** 6–7 (1967).

76. I. G. Minkevich and V. K. Eroshin, Oxygen and water expense during the growth of microorganisms, *Izv. Akad. Nauk. SSSR.*, *Ser. Biol.* **37,** 245–254 (1972) (in Russian).

77. I. G. Minkevich, G. P. Popkov, and V. K. Eroshin, Theoretical evaluation of the elemental balance during microbial cultivation on N-alkanes and N-alcohols, *Prikl. Biokh. Microbiol.* **8,** 160–166 (1972) (in Russian).
78. I. G. Minkevich and V. K. Eroshin, Oxygen consumption and heat generation during microbial growth, *Izv. Akad. Nauk. SSSR., Ser. Biol.* **3,** 414–415 (1973) (in Russian).
79. I. G. Minkevich and V. K. Eroshin, Theoretical calculation of mass balance during the cultivation of microorganisms, in B. Sikyta, A. Prokop, and M. Novak, Eds., *Biotechnology and Bioengineering Symposium* No. 4, 21–25 (1973).
80. I. G. Minkevich and V. K. Eroshin, Energetic indices of microbial growth, *Studia Biophys. Berlin* **49,** 43–52 (1975).
81. V. K. Eroshin, Biomass production and efficiency of substrate utilization, *Biotechnology and Bioengineering Symposium* No. 4, 995–1001 (1974).
82. I. G. Minkevich and V. K. Eroshin, Intracellular growth balance and its influence on growth characteristics of microbial cultures, *Studia Biophys. Berlin* **59,** 67–75 (1976).
83. I. G. Minkevich and V. K. Eroshin, Regulation of the intercellular material-energetic balance in microbial growth, *Uspekhi Sovremennoi Biologii* **82,** 103–116 (1976).
84. V. K. Eroshin, Material-energetic balance of microorganisms growth, *Process Biochem.* **12,** 29–31 (1977).
85. I. G. Minkevich, Relationship between energetics and reduction level of organic compounds as an imporant factor of energy balance in metabolism, *Biofizika* **25,** 675–679 (1980).
86. V. K. Eroshin and I. G. Minkevich, On the upper limit of mass yield of an organic product from an organic substrate, *Biotechnol. Bioeng.* **24,** 2263–2266 (1982).
87. I. G. Minkevich, Physico-chemical properties of organic compounds and the energetics of metabolism, *J. Theor. Biol.* **95,** 569–590 (1982).
88. I. G. Minkevich and V. K. Eroshin, Material and energy balance of product biosynthesis by microorganisms, *Mikorbiologicheskii Ghurnal* **44,** 6–11 (1982).
89. I. G. Minkevich, Mass energy balance for microbial product synthesis. Biochemical and cultural aspects, *Biotechnol. Bioeng.* **25,** 1267–1294 (1983).
90. M. Sobotka, J. Votruba, I. Havelik, and I. G. Minkevich, The mass energy balance of anaerobic methane production, *Folia Microbiol.* **28,** 195–204 (1983).
91. S. B. Mutafov, I. G. Minkevich, and V. K. Eroshin, A study of the effectiveness indices for the growth of yeast on different substrates, *Acta Biotchnol.* **4,** 125–136 (1984).
92. I. G. Minkevich, Estimation of available efficiency of microbial growth on methanol and ethanol, *Biotechnol. Bioeng.* **27,** 792–799 (1985).
93. L. E. Erickson, I. G. Minkevich, and V. K. Eroshin, Application of mass and energy balance regularities in fermentation, *Biotechnol. Bioeng.* **20,** 1595–1621 (1978).
94. L. E. Erickson, S. E. Selga, and U. E. Viesturs, Application of mass and energy balance irregularities to product formation, *Biotechnol. Bioeng.* **20,** 1623–1638 (1978).

95. L. E. Erickson, I. G. Minkevich, and V. K. Eroshin, Utilization of mass-energy balance regularities in the analysis of continuous culture data, *Biotechnol. Bioeng.* **21,** 575–591 (1979).

96. L. E. Erickson, Application of mass and energy balance regularities in fermentation processes, *Ann. N.Y. Acad. Sci.* **326,** 73–85 (1979).

97. L. E. Erickson, Energetic efficiency of biomass and product formation, *Biotechnol. Bioeng.* **21,** 725–743 (1979).

98. A. Ferrer and L. E. Erickson, Evaluation of data consistency and estimation of yield parameters in hydrocarbon fermentations, *Biotechnol. Bioeng.* **21,** 2203–2233 (1979).

99. L. E. Erickson, Growth and product energetic yields of *Rhodopseudomonas sphaeroides* in dark and aerobic chemostat cultures, *J. Ferment. Technol.* **58,** 53–60 (1980).

100. S. A. Patel and L. E. Erickson, Estimation of heats of combustion of biomass from elemental analysis using available electron concepts, *Biotechnol. Bioeng.* **23,** 53–60 (1980).

101. L. E. Erickson and J. L. Hess, Analysis of growth and polysaccharide yields in chemostat cultures of *Rhizobium trifolii*, *Ann. N.Y. Acad. Sci.* **369,** 81–89 (1981).

102. L. E. Erickson, Energetic yields associated with hydrocarbon fermentations, *Biotechnol. Bioeng.* **23,** 793–803 (1981).

103. B. O. Solomon, L. E. Erickson, and J. L. Hess, Application of data consistency tests and new parameter estimation methods to microbial growth on corn dust in batch culture, *Biotechnol. Bioeng.* **23,** 2333–2360 (1981).

104. B. O. Solomon, L. E. Erickson, J. L. Hess, and S. S. Yang, Maximum likelihood estimation of growth yields, *Biotechnol. Bioeng.* **24,** 633–649 (1982).

105. M. D. Oner, L. E. Erickson, and S. S. Yang, Estimation of true growth and product yields in aerobic cultures, *Biotechnol. Bioeng.* **24,** 633–649 (1982).

106. L. E. Erickson and M. D. Oner, Available electron and energetic yields in fermentation processes, *Ann. N.Y. Acad. Sci.* **413,** 99–113 (1983).

107. H. Y. Lee, L. E. Erickson, and S. S. Yang, The estimation of growth yield and maintenance parameters for photoautotrophic growth, *Biotechnol. Bioeng.* **26,** 926–935 (1984).

108. H. Y. Lee and L. E. Erickson, Characterization of algae using regularities, *Biotechnol. Bioeng.* **26,** 758–763 (1984).

109. M. D. Oner, L. E. Erickson, and S. S. Yang, Estimation of yield, maintenance, and product formation kinetic parameters in anaerobic fermentations, *Biotechnol. Bioeng.* **26,** 1436–1444 (1984).

110. E. H. Battley, Growth-reaction equations for *Saccharomyces cerevisiae*, *Physiol. Plant.* **13,** 192–203 (1960).

111. W. W. Forrest, Calorimeter for the continuous study of the heat production of microbial systems, *J. Sci. Inst.* 143–145 (1961).

112. W. W. Forrest, D. J. Walker, and M. F. Hopgood, Enthalpy changes associated with the lactic fermentation of glucose, *J. Bact.* **82,** 685–690 (1961).

113. R. J. Winzler and J. P. Baumberger, The degradation of energy in the metabolism of yeast cells, *J. Cell. Comp. Physiol.* **12,** 183–211 (1938).

114. E. H. Battley, Enthalpy changes accompanying the growth of *Saccharomyces cerevisiae* (Hansen), *Physiol. Plant.* **13,** 628–640 (1960).

115. W. W. Forrest and D. J. Walker, Calorimetric measurements of energy of maintenance in *Streptococcus faecalis*, *Biochem. Biophys. Res. Commun.* **13,** 217–222 (1963).

116. W. W. Forrest, Bacterial Calorimetry, in H. D. Brown, Ed., *Biochemical Microcalorimetry*, Academic Press, New York, 1969, pp. 165–180.

117. W. W. Forrest, Microcalorimetry, in D. W. Ribbons and J. R. Norris, Eds., *Methods of Microbiology 6B*, Academic Press, London, 1972, pp. 285–318.

118. W. W. Forrest, Energetic aspects of microbial growth, *Symp. Soc. Gen. Microbiol.* **19,** 65–86 (1969).

119. I. Prigogine, *Thermodynamics of Irreversible Processes*, Interscience, New York, 2nd ed., 1961.

120. P. Boivinet, *Utilisation du microcalorimètre en microbiologie.*, Thèse, Université d'Aix-Marseille, 1964.

121. J. P. Belaich and J. C. Senez, Growth yields of *Zymomonas mobilis.*, J. Bacteriol. **89,** 1195–1200 (1965).

122. J. C. Senez and J. P. Belaich, Etude micalorimétrique de la croissance bactérienne et du contrôle de l'activité métabolique par le phosphate, *Colloq. Int. Cent. Natl. Rech. Sci.* **124,** 357–363 (1963).

123. A. Grangetto, Thermogénèse de croissance aerobie d'*Aerobacter aerogenes*, Thèse, Universite d'Aix-Marseille, 1963.

124. J. Mandelstam and K. McQuillen, *Biochemistry of Microbial Growth*, Blackwell Scientific Publications, Oxford, 1968.

125. I. C. Gunsalus and C. W. Schuster, Energy-yielding metabolism in bacteria, in I. C. Gunsalus and R. Y. Stanier, Eds., *The Bacteria.*, Vol. II, Academic Press, New York, 1962, pp. 1–58.

126. E. Calvet and H. Prat, *Recent Progress in Microcalorimetry*, Pergamon Press, Ltd., Oxford, 1963.

127. E. Calvet and H. Prat, *Microcalorimétrie: Applications Physico-chimiques et Biologiques*, Masson Ed., Paris, 1956.

128. E. Calvet and H. Prat, *Récent Progrès en Microcalorimétrie.*, Dunod Ed., Paris, 1958.

129. T. H. Kitzinger and C. Benzinger, Microcalorimetry of some simple biochemical systems, *Fed. Proc.* **13,** 11 (1954).

130. C. Kitzinger and T. H. Benzinger, Principle and method of heat-burst microcalorimetry and the determination of free-energy, enthalpy, and entropy changes, in D. Glick, Ed., *Methods of Biochemical Analysis*, Vol. VIII, 1960, pp. 43–60.

131. T. H. Benzinger and C. Kitzinger, Microcalorimetry, new methods and objectives, in J. D. Hardy, Ed., *Temperature—Its Measurement and Control in Science and Industry.*, Vol. 3, pt. 3, Reinhold, New York, 1963.

132. T. H. Benzinger, Ultrasensitive Reaction Calorimetry, in P. Alexander and H. P. Lundgren, Eds., *A Laboratory Manual of Analytical Methods of Protein Chemistry*, Pergamon Press, New York, 1969.

133. I. Wadsö, Design and testing of a micro reaction calorimeter, *Acta Chem. Scand.* **22,** 927–937 (1968).

134. P. Monk and I. Wadsö, A flow micro reaction calorimeter, *Acta Chem. Scand.* **22,** 1842–1852 (1968).

135. P. Monk and I. Wadsö, Flow microcalorimetry as an analytical tool in biochemistry and related areas, *Acta Chem. Scand.* **23,** 29–36 (1969).

136. S. Delin, P. Monk, and I. Wadsö, Flow microcalorimetry as an analytical tool in microbiology, *Sci. Tools* **16,** 22 (1969).

137. I. Wadsö, Microcalorimeters, *Quart. Rev. Biophys.* **3,** 383–427 (1970).

138. I. Wadsö, A microcalorimeter for biological analysis, *Sci. Tools* **21,** 18–21 (1974).

139. J. P. Belaich, Thermogénèse et croissance de *Pseudomonas lindneri* en glucose limitant, *Compt. Rend. Col. Biol.* **157,** 316–322 (1963).

140. J. P. Belaich and J. C. Senez, Influence of aeration and of pantothenate on growth yields of *Zymomonas mobilis*, *J. Bact.* **89,** 1195–1200 (1964).

141. J. P. Belaich and J. C.Senez, Les Développements récents de la microcalorimétrie et de la thermogénèse, *Colloq. Int. Cent. Natl. Rech. Sci.* **156,** 381–394 (1965).

142. J. P. Belaich, J. C. Senez, and M. Murgier, Microcalorimetric study of glucose permeation in microbial cells, *J. Bacteriol.* **95,** 1750–1757 (1968).

143. M. Murgier and J. P. Belaich, Microcalorimetric determination of the affinity of *Saccharomyces cerevisiae* for some carbohydrate growth substrates, *J. Bacteriol.* **105,** 573–579 (1971).

144. A. Belaich and J. P. Belaich, Microcalorimetric study of the anaerobic growth of *Escherichia coli:* growth thermograms in a synthetic medium, *J. Bacteriol.* **125,** 14–18 (1976).

145. A. Belaich and J. P. Belaich, Microcalorimetric study of the anaerobic growth of *Escherichia coli*: Measurement of the affinity of whole cells for various substrates, *J. Bacteriol.* **125,** 19–24 (1976).

146. J. L. Stokes, Fermentation of glucose by suspensions of *Escherichia coli, J. Bacteriol.* **57,** 147–158 (1949).

147. J. P. Belaich, Growth and Metabolism in Bacteria. J. P. Belaich and A. E. Beezer, Eds., *Biological Calorimetry*, Academic Press, New York, 1980, pp. 1–42.

148. A. S. Traore, C. E. Hatchikian, J. P. Belaich, and J. Le Gall, Microcalorimetric studies of the growth of sulfate-reducing bacteria: Energetics of *Desulfovibrio vulgaris* growth, *J. Bacteriol.* **145,** 191–199 (1981).

149. I. Sorokin Yu, Role of carbon dioxide and acetate in the biosynthesis of sulfate-reducing bacteria, *Nature* (*London*) **210,** 551–552 (1966).

150. J. P. Belaich, Thèse, Université d'Aix-Marseille, 1967.

151. A. S. Traore, C. E. Hatchikian, J. Le Gall, and J. P. Belaich, Microcalorimetric studies of the growth of sulfate-reducing bacteria: Comparison of the growth parameters of some *Desulfovibrio* species, *J. Bacteriol.* **149,** 606–611 (1982).

152. Z. Dermoun and J. P. Belaich, Microcalorimetric study of *Escherichia coli* aerobic growth: Kinetics and experimental enthalpy associated with growth on succinic acid, *J. Bacteriol.* **140,** 377–380 (1979).

153. Z. Dermoun, R. Boussand, D. Cotten, and J. P. Belaich, A new batch calorimeter for aerobic growth studies, *Biotech Bioeng.* **27,** 996–1004 (1985).

154. Z. Dermoun and J. P. Belaich, Microcalorimetric study of *Escherichia coli* aerobic growth: Theoretical aspects of growth on succinic acid, *J. Bacteriol.* **143,** 742–747 (1980).

155. Y. Ishikawa, M. Shoda, and H. Maruyama, Design and performance of a new microcalorimetric system for aerobic cultivation of microorganisms, *Biotech. Bioeng.* **23,** 2629–2640 (1981).

156. Y. Ishikawa, Y. Nonoyama, and M. Shoda, Microcalorimetric study of aerobic growth of *Escherichia coli* in batch culture, *Biotech. Bioeng.* **23,** 2825–2836 (1981).

157. E. H. Battley, Alternate method of calculating the free-energy change accompanying the growth of *Saccharomyces cerevisiae* (Hansen) or three substrates, *Biotech. Bioeng.* **21,** 1929–1961 (1979).

10

THE ENERGETICS OF MICROBIAL GROWTH

Essentia non sunt multiplicanda praeter necessitatum.

WILLIAM OF OCKHAM

As so often happens in science, ideas or things that have taken a long time to develop often turn out to be really quite simple. This is certainly the case with microbial energetics in the sense that now, if only a very few facts are gathered with respect to the growth of a microbial culture, it is possible to describe completely the energetics of microbial growth, even to the absolute entropy of a unit mass of living cells. In Chapter 10 the most recent methods for calculating the free-energy, enthalpy, and entropy changes accompanying microbial growth will be summarized. Examples of the use of these methods will then be given, insofar as data are available to make this possible. In general, these examples provide for a pattern of interpretation of the available data. However, there is still a need for a greater quantity of critical data with which to test the correctness of the methods used to calculate the changes in thermodynamic quantities which accompany microbial growth.

10.1 GENERAL COMMENTS

If properly carried out, a study of the energetics of growth of a particular microorganism *ought* to be as exact a study as it is possible to make with respect to biological phenomena. The identity, composition, and quantity of the reactants comprising the initial state of a system involving microbial growth can be experimentally selected. With the remarkably accurate and precise techniques which are available today for chemical analysis there is no reason why the identity, composition, and quantity of the products of such a system cannot be determined within a few parts per thousand. The real difficulty lies in the inability of many investigators to appreciate the characteristics of the microorganism being grown and the fact that *it* is not a simple chemical. Thus, if inadequate attention is paid to a given microorganism's requirements for such things as temperature, pH, O–R potential, aeration, osmolality, light, darkness, trace elements, vitamins, or to limiting factors in general, even the most accurate analyses may not yield results that can be interpreted without recourse to elastic hypotheses. A good thermodynamic study ought to give results which are clean and clear. If this is not the case, something is probably wrong with the experimentation somewhere.

One of the modifications to past nomenclature which will be employed in this last chapter is to change the term "growth reaction," as originally used by Battley[1] to the term "growth process" to describe more accurately what takes place. Growth is not a simple reaction. It is really an extremely complex set of inter-relationships which would better be called a process.

Unfortunately, one of the disappointments of this last chapter will be a lack of data from which to draw firm conclusions. Very few investigators have concerned themselves with establishing well-defined growth-process systems with which to work, and of these even fewer have established in addition that is was *only* growth and not a combination of growth plus anaerobic or aerobic assimilation that was taking place. Although methods of constructing growth-process equations have become very much simplified, good growth-process equations are very scarce. By this is meant that even though such equations can be written from the data available, an inspection of the experimental conditions under which these data were obtained may indicate that these latter are highly questionable. The most obvious example of this is the brilliant work of Tamiya (see Chapter 6) and his group. Tamiya was the first to conceive of representing microbial growth as an equation and to develop a theory which related the degree of reduction of the cells to that of the substrate and the amount of growth which could be obtained from the latter. This, in turn, enabled him to determine how much of the substrate was used for growth, and how much was used for respiration. Ideally, the series of growth experiments he carried out with *Aspergillus oryzae* ought to provide perfect data with which to make calculations of energetics. However, *A. oryzae* was grown in these experiments on 0.5 *M* concentrations of substrate (a lot!!), about two orders of magnitude higher than the concentration of 0.002 *M* which is likely to give the best data. This is not only because of the difficulty of determining what values of the thermodynamic properties to use, but also because it is almost impossible not to believe that at such high substrate concentrations there was not carbohydrate storage going on within the *A. oryzae* hyphae at the same time that they were growing. Thus, whereas we can very easily construct growth-process equations using Tamiya's data and calculate the energy changes that accompany the growth of this mold, these calculations may not be meaningful so long as we are unsure that only growth is taking place. The only way to do this is to use limiting concentrations of substrate and to establish by carrying out a series of experiments from zero to higher substrate concentrations that nothing is limiting the quantity of growth except the quantity of substrate, as emphasized by Battley.[1] Ideally, the environmental conditions should be such that a given microorganism will be growing at its maximum specific growth rate (μ_{max}). With these precautions one has a chance of getting interpretable data. What happens with respect to conditions that are less than optimal for microbial growth can then be studied and compared at a future time.

10.2. GENERAL PROCEDURES FOR WRITING GROWTH-PROCESS EQUATIONS

Because of the developments of the past 50 years the construction of equations representing the growth of microorganisms no longer presents the difficulties that it once did, the greatest difficulty having been eliminated being the necessity of measuring the gas exchanges that take place during growth. In the simplest cases the only information necessary to construct equations representing the process of microbial growth is the following:

1. The elemental composition and the quantity of substrate used as the source of carbon and energy for growth. For studies on the energetics of microbial growth this should be chosen to be a single substance the thermodynamic properties of which are known.
2. The elemental composition and quantity of any organic products of the growth process. These will include the cells, which may be the only organic products of aerobic growth. The elemental composition of the cells should consist of the ash-free proportions of C, H, O, and N and is best expressed as a unit-carbon formula (UCF). This formula has the form $CHn_HOn_ONn_N$, where n_H, n_O, and n_N represent the relative numbers of hydrogen, oxygen, and nitrogen atoms, respectively, when the number of carbon atoms is taken as one. The quantities of identified organic products may be expressed in moles, but that of the cells should be in unit-carbon formula weights (UCFW) of ash-free, dry substance. The quantities of unidentified organic products may be expressed in the same manner.
3. The elemental composition of the nitrogen source. In most cases this can be chosen to be either ammonia or nitrate.

From the above data the following may be calculated.

10.2.1. The Molar Yield Coefficient for Each of the Organic Products of a Growth Process (MYC_p)

Generally, the symbol MYC_p is replaced by a more specific one, such as MYC_c for the quantity of cells produced during growth, MYC_{eth} for the quantity of ethanol produced during fermentative growth on glucose, and so on. An MYC_p has the dimensions of quantity of organic product per mole of substrate consumed. In the case of cells MYC_c has the dimensions of unit-carbon formula weights of cells produced per mole of substrate consumed (UCFW $mole^{-1}$). In the case of identified organic products, MYC_p has the dimensions of moles of product per mole of substrate consumed (moles $mole^{-1}$). If an organic product is unidentified it may be treated in the same manner as the cells.

To calculate a molar yield coefficient for the cells produced during growth it is necessary to know the molar growth yield, Y, which is the number of grams dry weight of cells produced during growth on 1 mole of substrate. Knowing this quantity and also the percentage of ash, the dry weight in grams of the organic material in the cells can be calculated. This is designated Y_{af}, representing the ash-free molar growth yield. Knowing also the unit-carbon formula weight, a value for MYC_c can be calculated as follows:

$$MYC_c = \frac{Y_{af}}{UCFW} \tag{10.1}$$

To calculate a molar yield coefficient for identified organic products of a growth-equation Y takes on the dimensions of grams of substance produced during growth on 1 mole of substrate. Equation 10.1 then becomes

$$MYC_p = \frac{Y}{MW} \tag{10.2}$$

where MW represents the molecular weight of the substance. The calculations of MYC_p for unidentified organic products may be made using Eq. 10.1.

10.2.2. The Quantity of Substrate Required for the Synthesis of a Given Organic Product (S_p)

Once a value for a given MYC_p has been obtained it can be used to calculate the quantity of substrate used in the formation of the product, thanks to the recognition by Minkevitch and Eroshin (see Section 9.4) that the available electrons in a product could only have come from the substrate. The quantity of available electrons (AE) in most organic substances can be calculated by means of the following equation:

$$AE = 4n_C + n_H - 2n_O - 3n_N \tag{10.3}$$

where n_C, n_H, n_O, and n_N represent the relative numbers of carbon, hydrogen, oxygen, and nitrogen atoms, respectively, in the empirical or molecular formula for the substance.

It should be pointed out that although the term "available electron," or AE, as first defined in 1967 by Mayberry et al.[2] has historical precedence,

Minkevich[3] has recently used the term "redoxon," or RO, to represent the same thing, the redoxon now being called the "unit of reductivity." Although "redoxon" is somewhat easier to say than "available electron," the use of the former seems to represent an unnecessary complication, especially in that the word "elcctron," which is what the definition is all about, is not present. Thus, for reasons of historical precedent and because there is no advantage to the use of the term "redoxon," the term "available electron" will continue to be used throughout this book. To do otherwise would be to ignore the historical contribution of the past.

The quantity of available electrons in a product can be represented by the symbol AE_p. As with the symbol MYC_p, this latter can be replaced by a more specific representation such as AE_c for the number of available electrons in the cells, AE_{eth} for this number in 1 mole of ethanol, and so on. The symbol AE_S is used to represent the number of available electrons in 1 mole of substrate. If the symbol S_p is used to represent the quantity of substrate which is required to form a given organic product during growth on 1 mole of substrate S, then,

$$S_p = \frac{MYC_p \text{ mole} \times AE_p \text{ mole}^{-1}}{AE_S \text{ mole}^{-1}} = \text{moles} \tag{10.4}$$

It is apparent that when using Eq. 10.4 and calculating S_c as the quantity of substrate necessary to form one unit-carbon formula weight of cells, the term UCFW can be considered formally equivalent to 1 "mole" of cells, even though this has no real chemical meaning because of its variability.

When S_p for a given product has been determined it becomes possible to write an equation representing its formation during the process of microbial growth. Such an equation can have two forms, depending on the relative degrees of reduction per unit-carbon atom of the substrate and the product. The degree of reduction of an organic substance can be expressed in several ways, but for the purposes of writing these equations the method of Minkevitch and Eroshin is the best. Here the degree of reduction is expressed as available electrons per unit carbon atom of a substance on a scale of eight, representing the number of electrons available from methane, the most reduced form of carbon. Thus, the appropriate equation for determining the degree of reduction is

$$D_r = \frac{4n_C + n_H - 2n_O - 3n_N}{n_C} \tag{10.5}$$

where D_r represents the degree of reduction, and n_C, n_H, n_O, and n_N have the same significance as for Eq. 10.3.

As originally recognized in the early 1930s by Tamiya (see Chapter 6), but using the simpler and more recent method of Minkevich and Eroshin (see Section 9.4), if the value of D_r for the substrate is less than that for an organic product of a growth process, carbon dioxide will be a product in the equation representing its formation from the substrate. This is represented as follows:

$$\left.\begin{array}{l} S_p\mathrm{Cn_CHn_HOn_O(aq)} \\ +\ n_\mathrm{N}\mathrm{MYC}_p\mathrm{NH_3(aq)} \end{array}\right\} \longrightarrow \begin{array}{l} \mathrm{MYC}_p\mathrm{C}n_\mathrm{C}\mathrm{H}n_\mathrm{H}\mathrm{O}n_\mathrm{O}\mathrm{N}n_\mathrm{N}\mathrm{(aq)} \\ +\ (\mathrm{n_C}S_p\ -\ n_\mathrm{C}\mathrm{MYC}_p)\mathrm{CO_2(aq)} \\ +\ 0.5(\mathrm{n_H}S_p\ +\ 3n_\mathrm{N}\mathrm{MYC}_p\ -\ n_\mathrm{H}\mathrm{MYC}_p)\mathrm{H_2O(l)} \end{array} \tag{10.6}$$

where $\mathrm{n_C}$, $\mathrm{n_H}$, $\mathrm{n_O}$, and $\mathrm{n_N}$ represent the relative numbers of carbon, hydrogen, oxygen, and nitrogen atoms, respectively, in the substrate, n_C, n_H, n_O, and n_N represent the relative numbers of these atoms in the product, respectively, and S_p and MYC_p have been previously defined. As recognized by Battley (see Chapter 8), if the value of D_r for the substrate is greater than that for a product, carbon dioxide will be a reactant in the equation representing its formation from the substrate. Thus,

$$\left.\begin{array}{l} S_p\mathrm{Cn_CHn_HOn_O(aq)} \\ +\ n_\mathrm{N}\mathrm{MYC}_p\mathrm{NH_3(aq)} \\ +(n_\mathrm{C}\mathrm{MYC}_p\ -\ \mathrm{n_C}S_p)\mathrm{CO_2(aq)} \end{array}\right\} \longrightarrow$$

$$\longrightarrow \begin{array}{l} \mathrm{MYC}_p\mathrm{Cn_C}\mathrm{H}n_\mathrm{H}\mathrm{O}n_\mathrm{O}\mathrm{N}n_\mathrm{N}\mathrm{(aq)} \\ +\ 0.5(\mathrm{n_H}S_p\ +\ 3n_\mathrm{N}\mathrm{MYC}_p\ -\ n_\mathrm{H}\mathrm{MYC}_p)\mathrm{H_2O(l)} \end{array} \tag{10.7}$$

10.2.3. The Quantity of Substrate Required for Respiration ($1 - \Sigma S_p$)

Because the yield coefficients for the organic products of a growth process are expressed in terms of unit-carbon formula weights, or moles, both per mole of substrate consumed, either (1) all of the substrate will be used in the formation of such products, as occurs during fermentative growth, or (2) that quantity of substrate not so used will be respired, as occurs during aerobic growth. This latter quantity is therefore represented by the term $(1 - \Sigma S_p)$ and an equation representing the respiration that takes place during aerobic growth can be written in the following manner.

$$\left.\begin{array}{l} (1\ -\ \Sigma S_p)\mathrm{Cn_CHn_HOn_O(aq)} \\ +\ 0.5(1\ -\ \Sigma S_p)(2\mathrm{n_C}\ +\ 0.5\mathrm{n_H}\ -\ \mathrm{n_O})\mathrm{(aq)} \end{array}\right\} \longrightarrow$$

$$\longrightarrow \begin{array}{l} (1\ -\ \Sigma S_p)\mathrm{n_C}\mathrm{CO_2(aq)} \\ +\ 0.5(1\ -\ \Sigma S_p)\mathrm{n_H}\mathrm{H_2O(l)} \end{array} \tag{10.8}$$

With respect to fermentative growth the addition of the equations representing the formation of all the products, either in the form of Eq. 10.6 or 10.7, will give the complete growth-process equation (see Section 10.5.1). With respect to aerobic growth the addition of the equations representing reactions during which organic products are formed (often only the cells) either in the form of

Eq. 10.6 or 10.7, plus Eq. 10.8 will give the complete growth-process equation. Obviously, when a term does not exist it is simply left out of one or other of the above equations.

If nitrate is the nitrogen source for a growth process the equation representing the construction of the cells will still take the form of Eq. 10.6 or 10.7. This is because it is assumed that ammonia is the primary nitrogen source for cellular synthesis and that other reactions reduce nitrate to ammonia for this purpose. It requires eight electrons to reduce nitrate to ammonia; therefore, the quantity of substrate needed to reduce a given quantity of nitrate to ammonia is

$$S_N = \frac{8e \text{ mole}^{-1} \text{ nitrate} \times N \text{ moles nitrate}}{\text{AE}_S \text{ mole}^{-1}} = \text{moles} \tag{10.9}$$

where S_N represents the quantity in moles of substrate necessary to reduce N moles of nitrate to ammonia. Once this quantity is found, it can be inserted into the following equation.

$$\left.\begin{array}{l} S_N\text{Cn}_\text{C}\text{Hn}_\text{H}\text{On}_\text{O}(\text{aq}) \\ +\ n_\text{N}\text{MYC}_c\text{HNO}_3(\text{aq}) \end{array}\right\} \longrightarrow \begin{array}{l} n_\text{N}\text{MYC}_c\text{NH}_3(\text{aq}) + \text{n}_\text{C}S_N\text{CO}_2(\text{aq}) \\ +\ 0.5(\text{n}_\text{H}S_N - 2n_\text{N}\text{MYC}_c)\text{H}_2\text{O(l)} \end{array} \tag{10.10}$$

In general, the substrate represented in Eq. 10.10 will not contain nitrogen. If the nitrogen source happens to be nitrate rather than ammonia, the value for MYC_c will be smaller, due to the necessity of utilizing some of the substrate for the reduction of the nitrate to ammonia. The quantity of substrate utilized for respiration then becomes $(1 - \Sigma S_p - S_N)$, this term becoming inserted into Eq. 10.8 as follows.

$$\left.\begin{array}{l} (1 - \Sigma S_p - S_N)\text{Cn}_\text{C}\text{Hn}_\text{H}\text{On}_\text{O}(\text{aq}) \\ +\ 0.5(1 - \Sigma S_p - S_N)(2\text{n}_\text{C} + 0.5\text{n}_\text{H} - \text{n}_\text{O})\text{O}_2(\text{aq}) \end{array}\right\} \longrightarrow$$

$$\longrightarrow \begin{array}{l} (1 - \Sigma S_p - S_N)\text{n}_\text{C}\text{CO}_2(\text{aq}) \\ +\ 0.5(1 - \Sigma S_p - S_N)\text{n}_\text{H}\text{H}_2\text{O(l)} \end{array} \tag{10.11}$$

The addition of Eqs. 10.6 or 10.7, 10.10, and 10.11 will give the complete growth-process equation.

The procedures for writing equations representing aerobic growth are summarized in Tables 10.1 through 10.4. For equations representing fermentative growth all that is required is the addition of a sufficient number of 10.6 equations to account for all the organic products of the growth process, since in most cases the substrate will be a carbohydrate which is less reduced than the cellular material or the other products of the fermentation.

TABLE 10.1. Construction of growth-process equations when the substrate is less reduced than the cells, when these latter are the only organic products of the process, and when the nitrogen source is ammonia

Anabolism (growth)

$$S_c \mathrm{C}n_\mathrm{C}\mathrm{H}n_\mathrm{H}\mathrm{O}n_\mathrm{O}(aq) + n_\mathrm{N}\mathrm{MYC}_c\mathrm{NH_3}(aq) \longrightarrow \mathrm{MYC}_c\mathrm{CH}n_\mathrm{H}\mathrm{O}n_\mathrm{O}\mathrm{N}n_\mathrm{N}(\text{cells}) + (n_\mathrm{C}S_c - \mathrm{MYC}_c)\mathrm{CO_2}(aq) + 0.5(n_\mathrm{H}S_c + 3n_\mathrm{N}\mathrm{MYC}_c - n_\mathrm{H}\mathrm{MYC}_c)\mathrm{H_2O}(l)$$

Catabolism (respiration)

$$(1 - S_c)\mathrm{C}n_\mathrm{C}\mathrm{H}n_\mathrm{H}\mathrm{O}n_\mathrm{O}(aq) + 0.5(1 - S_c)(2n_\mathrm{C} + 0.5\,n_\mathrm{H} - n_\mathrm{O})\mathrm{O_2}(aq) \longrightarrow (1 - S_c)n_\mathrm{C}\mathrm{CO_2}(aq) + 0.5(1 - S_c)n_\mathrm{H}\mathrm{H_2O}(l)$$

Anabolism + catabolism = growth process

$$\left.\begin{array}{l}\mathrm{C}n_\mathrm{C}\mathrm{H}n_\mathrm{H}\mathrm{O}n_\mathrm{O}(aq) + n_\mathrm{N}\mathrm{MYC}_c\mathrm{NH_3}(aq) \\ + 0.5(1 - S_c)(2n_\mathrm{C} + 0.5n_\mathrm{H} - n_\mathrm{O})\mathrm{O_2}(aq)\end{array}\right\} \longrightarrow \begin{array}{l}\mathrm{MYC}_c\mathrm{CH}n_\mathrm{H}\mathrm{O}n_\mathrm{O}\mathrm{N}n_\mathrm{N}(\text{cells}) + (n_\mathrm{C} - \mathrm{MYC}_c)\mathrm{CO_2}(aq) \\ + 0.5(n_\mathrm{H} + 3n_\mathrm{N}\mathrm{MYC}_c - n_\mathrm{H}\mathrm{MYC}_c)\mathrm{H_2O}(l)\end{array}$$

TABLE 10.2. Construction of growth-process equations when the substrate is more reduced than the cells, when these latter are the only organic products of the process, and when the nitrogen source is ammonia

Anabolism (growth)

$$S_c\mathrm{C}n_\mathrm{C}\mathrm{H}n_\mathrm{H}\mathrm{O}n_\mathrm{O}(aq) + n_\mathrm{N}\mathrm{MYC}_c\mathrm{NH_3}(aq) + (\mathrm{MYC}_c - n_\mathrm{C}S_c)\mathrm{CO_2}(aq) \longrightarrow \mathrm{MYC}_c\mathrm{CH}n_\mathrm{H}\mathrm{O}n_\mathrm{O}\mathrm{N}n_\mathrm{N}(\text{cells}) + 0.5(n_\mathrm{H}S_c + 3n_\mathrm{N}\mathrm{MYC}_c - n_\mathrm{H}\mathrm{MYC}_c)\mathrm{H_2O}(l)$$

Catabolism (respiration)

$$(1 - S_c)\mathrm{C}n_\mathrm{C}\mathrm{H}n_\mathrm{H}\mathrm{O}n_\mathrm{O}(aq) + 0.5(1 - S_c)(2n_\mathrm{C} + 0.5n_\mathrm{H} - n_\mathrm{O})\mathrm{O_2}(aq) \longrightarrow (1 - S_c)n_\mathrm{C}\mathrm{CO_2}(aq) + 0.5(1 - S_c)n_\mathrm{H}\mathrm{H_2O}(l)$$

Anabolism + catabolism = growth process

$$\left.\begin{array}{l}\mathrm{C}n_\mathrm{C}\mathrm{H}n_\mathrm{H}\mathrm{O}n_\mathrm{O}(aq) + n_\mathrm{N}\mathrm{MYC}_c\mathrm{NH_3}(aq)\\ + 0.5(1 - S_c)(2n_\mathrm{C} + 0.5n_\mathrm{H} - n_\mathrm{O})\mathrm{O_2}(aq)\end{array}\right\} \longrightarrow \begin{array}{l}\mathrm{MYC}_c\mathrm{CH}n_\mathrm{H}\mathrm{O}n_\mathrm{O}\mathrm{N}n_\mathrm{N}(\text{cells}) + (n_\mathrm{C} - \mathrm{MYC}_c)\mathrm{CO_2}(aq)\\ + 0.5(n_\mathrm{H} + 3n_\mathrm{N}\mathrm{MYC}_c - n_\mathrm{H}\mathrm{MYC}_c)\mathrm{H_2O}(l)\end{array}$$

TABLE 10.3. Construction of growth-process equations when the substrate is less reduced than the cells, when these latter are the only organic products of the process, and when the nitrogen source is nitrate

Anabolism (growth)

$$S_c\mathrm{Cn_CHn_HOn_O(aq)} + n_\mathrm{N}\mathrm{MYC}_c\mathrm{NH_3(aq)} \longrightarrow \mathrm{MYC}_c\mathrm{CH}n_\mathrm{H}\mathrm{O}n_\mathrm{O}\mathrm{N}n_\mathrm{N}\mathrm{(cells)} + (\mathrm{n_C}S_c - \mathrm{MYC}_c)\mathrm{CO_2(aq)} + 0.5(\mathrm{n_H}S_c + 3n_\mathrm{N}\mathrm{MYC}_c - n_\mathrm{H}\mathrm{MYC}_c)\mathrm{H_2O(l)}$$

Catabolism (nitrate reduction)

$$S_N\mathrm{Cn_CHn_HOn_O(aq)} + n_\mathrm{N}\mathrm{MYC}_c\mathrm{HNO_3(aq)} \longrightarrow n_\mathrm{N}\mathrm{MYC}_c\mathrm{NH_3(aq)} + \mathrm{n_C}S_N\mathrm{CO_2(aq)} + 0.5(\mathrm{n_H}S_N - 2n_\mathrm{N}\mathrm{MYC}_c)\mathrm{H_2O(l)}$$

Catabolism (respiration)

$$\left.\begin{array}{l}(1 - S_c - S_N)\mathrm{Cn_CHn_HOn_O(aq)} \\ + 0.5(1 - S_c - S_N)(2\mathrm{n_C} + 0.5\mathrm{n_H} - \mathrm{n_O})\mathrm{O_2(aq)}\end{array}\right\} \longrightarrow (1 - S_c - S_N)\mathrm{n_C}\mathrm{CO_2(aq)} + 0.5(1 - S_c - S_N)\mathrm{n_H}\mathrm{H_2O(l)}$$

Anabolism + catabolism = growth process

$$\left.\begin{array}{l}\mathrm{Cn_CHn_HOn_O(aq)} + n_\mathrm{N}\mathrm{MYC}_c\mathrm{HNO_3(aq)} \\ + 0.5(1 - S_c - S_N)(2\mathrm{n_C} + 0.5\ \mathrm{n_H} - \mathrm{n_O})\mathrm{O_2(aq)}\end{array}\right\} \longrightarrow \begin{array}{l}\mathrm{MYC}_c\mathrm{CH}n_\mathrm{H}\mathrm{O}n_\mathrm{O}\mathrm{N}n_\mathrm{N}\mathrm{(cells)} + (\mathrm{n_C} - \mathrm{MYC}_c)\mathrm{CO_2(aq)} \\ + 0.5(\mathrm{n_H} + 3n_\mathrm{N}\mathrm{MYC}_c - n_\mathrm{H}\mathrm{MYC}_c) - n_\mathrm{N}\mathrm{MYC}_c\mathrm{H_2O(l)}\end{array}$$

TABLE 10.4. Construction of growth-process equations when the substrate is more reduced than the cells, when these latter are the only organic products of the process, and when the nitrogen source is nitrate

Anabolism (growth)

$$S_c Cn_C Hn_H On_O(aq) + n_N MYC_c NH_3(aq) + (MYC_c - n_C S_c)CO_2(aq) \longrightarrow MYC_c CHn_H On_O Nn_N(\text{cells}) + 0.5(n_H S_c + 3n_N MYC_c - n_H MYC_c)H_2O(l)$$

Catabolism (nitrate reduction)

$$S_N Cn_C Hn_H On_O(aq) + n_N MYC_c HNO_3(aq) \longrightarrow n_N MYC_c NH_3(aq) + n_C S_N CO_2(aq) + 0.5(n_H S_N - 2n_N MYC_c)H_2O(l)$$

Catabolism (respiration)

$$(1 - S_c - S_N)Cn_C Hn_H On_O(aq) + 0.5(1 - S_c - S_C)(2n_C + 0.5n_H - n_O)O_2(aq) \longrightarrow (1 - S_c - S_N)n_C CO_2(aq) + 0.5(1 - S_c - S_N)n_H H_2O(l)$$

Anabolism + catabolism = growth process

$$\left.\begin{matrix} Cn_C Hn_H On_O(aq) + n_N MYC_c HNO_3(aq) \\ + 0.5(1 - S_c - S_N)(2n_C + 0.5n_H - n_O)O_2(aq) \end{matrix}\right\} \longrightarrow \begin{matrix} MYC_c CHn_H On_O Nn_N(\text{cells}) + (n_C - MYC_c)CO_2(aq) \\ + 0.5(n_H + 3n_N MYC_c - n_H MYC_c) - n_N MYC_c H_2O(l) \end{matrix}$$

10.3. GENERAL PROCEDURES FOR DETERMINING CHANGES IN THERMODYNAMIC PROPERTIES ACCOMPANYING MICROBIAL GROWTH

Once growth-process equations have been written as described in Section 10.2, it becomes possible to utilize these to calculate the energy changes accompanying the growth processes they represent. In general, using the appropriate values for the thermodynamic properties of the substances listed in Appendixes 1 and 2 and one or other of the several methods described in Chapter 1, values for $\Delta G'$, $\Delta H'$, $T\Delta S'$, and $\Delta S'$ can be calculated for any equation having to do with growth processes except those which include a term representing the cells. The energy changes accompanying the reactions represented by the former can all be combined into the category of "catabolism," whereas the latter represent "growth," or "anabolism." Thus, energy changes can usually be categorized as $\Delta G'_{\text{cat}}$, $\Delta H'_{\text{cat}}$, and $\Delta S'_{\text{cat}}$, and $\Delta G'_{\text{an}}$, $\Delta H'_{\text{an}}$, and $\Delta S'_{\text{an}}$. Catabolism is represented by Eqs. 10.6 and 10.7 when the product can be identified or represented by a unit-carbon formula and does not represent cells, and by Eqs. 10.8, 10.10, and 10.11. Anabolism is represented by Eqs. 10.6 and 10.7 when the product represents cellular material. Addition of the respective energy changes accompanying catabolism and anabolism gives the energy changes for the complete growth process, as follows:

$$\Delta G'_C = \Delta G'_{\text{cat}} + \Delta G'_{\text{an}}$$

$$\Delta H'_C = \Delta H'_{\text{cat}} + \Delta H'_{\text{an}}$$

$$\Delta S'_C = \Delta S'_{\text{cat}} + \Delta S'_{\text{an}}$$

where $\Delta G'_C$, $\Delta H'_C$, and $\Delta S'_C$ represent the free-energy, enthalpy, and entropy changes, respectively, which accompany the conservative (growth) process. Occasionally, when dealing with fermentative growth, it may be difficult to distinguish whether a particular equation represents anabolism or catabolism. This is because usually one or more organic products must be produced other than the cells in order to compensate materially and electronically for cell production, and there is probably a limited number of biochemical methods by which this can be done. The result must be that, if the cells and the organic products of the fermentation other than those produced by the nonconservative reaction are combined, they can be theoretically reacted to give the products of the nonconservative, fermentative reaction. This is illustrated in Section 10.5.1.

The most satisfactory method of determining $\Delta G'_{\text{an}}$ is to utilize Battley's hypothesis that with respect to heterotrophic microbial growth this quantity is equal to zero (see Section 8.12). This particular hypothesis, however, does not apply to $\Delta H'_{\text{an}}$ determinations, which must be made by the completely different method using Battley's hypothesis involving the concepts of nonconservative and conservative reactions (see Section 8.10). In this latter method it is necessary to establish the chemical nature of the nonconservative reaction which is relative to the growth process (which is conservative) that is being considered. This is usually not difficult except in the cases of complex fermentations, and of these latter only a few have been carefully studied. To be useful, a given nonconservative reaction equation must be balanced and must represent the interaction of substances the thermochemical properties of which are known. The free-energy, enthalpy, and entropy changes which accompany these reactions are the maximal of which a given growth-process system is capable, and for energy to be conserved within this latter these energy changes must obviously be less than maximal. A comparison of the energy changes accompanying a given nonconservative reaction and its relative conservative growth process makes it possible to determine how much energy has been conserved. This is done by difference using the following equations.

$$\Delta G'_{\text{con}} = \Delta G'_{\text{NC}} - \Sigma\, \Delta G'_{\text{cat}} \tag{10.12}$$

$$\Delta H'_{\text{con}} = \Delta H'_{\text{NC}} - \Sigma\, \Delta H'_{\text{cat}} \tag{10.13}$$

$$\Delta S'_{\text{con}} = \Delta S'_{\text{NC}} - \Sigma\, \Delta S'_{\text{cat}} \tag{10.14}$$

where $\Delta G'_{\text{NC}}$, $\Delta H'_{\text{NC}}$, and $\Delta S'_{\text{NC}}$ represent the free-energy, enthalpy, and entropy changes, respectively, accompanying a given nonconservative reaction; $\Sigma\, \Delta G'_{\text{cat}}$, $\Sigma\, \Delta H'_{\text{cat}}$, and $\Sigma\, \Delta S'_{\text{cat}}$ represent the sum of the free-energy, enthalpy, and entropy changes, respectively, accompanying the catabolism that occurs during growth; and $\Delta G'_{\text{con}}$, $\Delta H'_{\text{con}}$, and $\Delta S'_{\text{con}}$ represent the free energy, enthalpy, and entropy, respectively, which have been conserved during growth. These latter quantities would be exchanged with the environment *if* the substances within which these energies are conserved were to be reacted in such a manner that they become converted into the products of the relative nonconservative reaction. This latter is easiest to understand with respect to most aerobic growth processes in which the only organic product of growth is the cells. Here, the energy conserved within the substance of the cells can be exchanged with the environment by theoretically combusting these cells to form the carbon dioxide and water which are the products of the relative nonconservative reaction. The only other product of this theoretical combustion is ammonia. Thus, if the equation representing this combustion were added to that representing the growth process during which these cells were synthesized, the sum would be the nonconservative reaction equation. If the nitrogen source is nitrate, then an equation representing the oxidation to nitrate of the ammonia formed from the oxidation of the cellular material must be added to the others.

10.3.1. Appropriate Values to Use for the Thermodynamic Properties of Substances Participating in Growth Processes

What is the thermodynamic state of the reactants and products of a growth process? Obviously, the answer to this question is extremely important with respect to calculating energy changes which can be interpreted in a meaningful way. This is because the $\Delta G'$, $\Delta H'$, and $\Delta S'$ values that are obtained for a given growth process depend on the assumptions that are made and the methods of calculation that are used. Two aspects of making these calculations are immediately apparent. One is that the anabolism and catabolism comprising growth should be considered to be irreversible processes. This means that the free-energy, enthalpy, and entropy changes which accompany these processes must be calculated through the use of the free energies, enthalpies, and entropies of formation of the reactants and products. The second aspect is that these processes effectively take place in aqueous solution and therefore thermodynamic properties relative to the gaseous or solid state are inappropriate. The decision must then be made as to what values relative to the aqueous state apply to substances in solution as they actually interact with the cells. It is clear that in a natural environment microbial cells seldom interact with substances at the 1 m concentration characteristic of the conventional aqueous standard state. For this reason, values for the thermodynamic properties should be modified to take into account any effects of dilution from the standard concentration to an "average" concentration which is more likely to exist in a natural environment.

10.3.2 Free-Energy Changes

When calculating the free-energy changes accompanying reactions that take place in aqueous solution as the difference between the sum of the free energies of formation of the products and that of the reactants, all in the aqueous standard state, the assumption is made that the concentration of each reactant and product always remains 1 m. This does not occur in a batch culture of microorganisms. As pointed out in principle by Fulmer and Liefson[4] in 1928, if a microbial culture is allowed to proceed until the substrate becomes exhausted, this latter will effectively become diluted by being consumed. The free-energy change accompanying the process of dilution of a solute from the aqueous standard state to a lower concentration ($\Delta G'_{\text{dil}}$) can be calculated by means of the following equation.

$$\Delta G'_{\text{dil}} = \frac{-RT}{1000} \ln \frac{\text{conc. in std. st.} \times \text{activity coeff. in std. st.}}{\text{lower conc.} \times \text{activity coeff. at lower conc.}} \qquad (10.15)$$

In the hypothetical aqueous standard state the concentration is 1 m and the activity coefficient is considered to be equal to one. The numerator inside the logarithm in Eq. 10.15 thus equals unit activity. If the concentration of the solute in the denominator of Eq. 10.15 is sufficiently low, its activity coefficient also tends to approach one, and the activity can be taken equal to concentration. Then,

$$\Delta G'_{\text{dil}} = -0.592 \ln \frac{1}{a_c} \qquad (10.16)$$

where a_c represents the activity of the solute at a concentration, c, which is less than 1 m. The free energy of formation of a solute under these conditions, $\Delta G'_f$, then becomes

$$\Delta G'_f = \Delta G^{\circ\prime}_f + \left(-0.592 \ln \frac{1}{a_c}\right) \qquad (10.17)$$

It is apparent that if a_c becomes sufficiently small, $\Delta G'_{\text{dil}}$ will approach minus infinity. Therefore, as a substrate solute becomes consumed the value of its free energy of formation will become successively more negative until it approaches minus infinity. The reverse is true of the products of a growth reaction at the beginning of growth, at which time they have a concentration of zero. In the case either of reactants or of products, the true value for $\Delta G'_f$ can only be found by integration of the successive $\Delta G'_f$ values as a function of time, the graph of either of which deals with minus infinity. Obviously, this will not work. Fulmer and Liefson attempted to solve this problem by assuming small, finite values for the lowest concentrations of reactants or products, and by accepting an average of the $\Delta G'_f$ values with respect to the lowest and highest concentrations of any given substance. Thus, the initial concentration of carbon dioxide as a product in alcoholic fermentation was not taken as zero, but was assumed to be that in equilibrium with atmospheric carbon dioxide. Similarly, a small initial concentration of alcohol was assumed to be introduced to the system with the yeast inoculum used to carry out an alcoholic fermentation. The values adopted could be either measured or estimated. Using this method, they arrived at a value for the $\Delta G'$ accompanying alcoholic sucrose fermentation of -116.6 kcal mole^{-1}, in very close agreement with that of -114.55 kcal mole^{-1} obtained with presently accepted $\Delta G^{\circ\prime}_f$ values.

Another solution to the problem of dealing with zero concentrations of solutes is simply to adopt a new "practical" standard state more appropriate to the study of the energetics of microbial growth. In the second edition of Lewis and Randall's *Thermodynamics* as revised by K. S. Pitzer and L. Brewer,[5] these latter authors show (pp. 239–240) that, at a concentration of about 0.001 m, solute molecules or ions are sufficiently isolated from each other that further dilution has little effect on intermolecule, or interion interaction. This could well be the basis for a "microbiological" aqueous standard state. A substrate concentration of 0.001 m (to all intents and purposes the same as 0.001 M) is usually completely sufficient for growing microorganisms for the purposes of studying microbial energetics, and if this were taken as an "average," then concentrations of 0.002 m could also be used. At a concentration of 0.001 m $\Delta G'_f$ valves for all solutes will be more negative than $\Delta G^{\circ\prime}_f$ values by -4.089 kcal mole^{-1} (Eq. 10.16). What effect will this have on the free-energy change accompanying any given reaction compared to that calculated using $\Delta G^{\circ\prime}_f$ values? This depends on the difference between the sums of the coefficients on the right and left sides of an equation *exclusive of the coefficient for water* (since the concentration of water in a dilute solution can be regarded as constant, and not subject to further dilution). $\Delta G'$ values calculated using 0.001 m as the standard concentration will differ from $\Delta G^{\circ\prime}$ values according to the following equation,

$$\Delta G' = \Delta G^{\circ\prime} - 4.089\ (\Sigma\text{coeff}_r - \Sigma\text{coeff}_l) \tag{10.18}$$

where Σcoeff_l and Σcoeff_r represent the sums of the coefficients on the left and right sides of an equation, respectively, exclusive of the coefficient for water. The biological oxidation of glucose provides an example of the application of Eq. 10.18. The equation representing this reaction is

$$C_6H_{12}O_6(aq) + 6O_2(aq) \longrightarrow 6CO_2(aq) + 6H_2O(l) \tag{10.19}$$

Using $\Delta G_f^{\circ\prime}$ values for the reactants and products the $\Delta G^{\circ\prime}$ for the reaction represented by Eq. 10.19 is -698.51 kcal mole^{-1}. Were the free energies of formation of the reactants and products to be relative to an aqueous standard state of a real 0.001 m concentration at unit activity. The $\Delta G'$ for the reaction represented by Eq. 10.19 would become

$$\begin{aligned}\Delta G' &= -698.51 \text{ kcal mole}^{-1} - 4.089(6 - 7) \text{ kcal mole}^{-1} \\ &= -694.42 \text{ kcal mole}^{-1}\end{aligned} \tag{10.20}$$

The same answer could be obtained by using individual values of $\Delta G_f'$ for the reactants and products in solution at a real concentration of 0.001 m at unit activity. It is apparent that at least with respect to reactions of the type represented by Eq. 10.19 in which the sum of the coefficients is unequal on either side of the equation, $\Delta G'$ as calculated using Eq. 10.20 can be considerably different from that of $\Delta G^{\circ\prime}$. On the other hand, if the sums of the coefficients, exclusive of that for water, are equal on either side of the equation $\Delta G^{\circ\prime}$ would not change if concentrations of 0.001 m were used.

Why are we justified in using a new "standard state" consisting of a real 0.001 m concentration at unit activity in the calculation of $\Delta G_f'$ values? Hopefully, by doing so we will obtain $\Delta G_f'$ values that are more appropriate for growing microorganisms in natural environments. A good example of this is the value of $+3.9$ kcal mole^{-1} representing the $\Delta G_f^{\circ\prime}$ of $O_2(aq)$. The value of $\Delta G_f^{\circ\prime}$ for most substances in aqueous solution is negative, but it is positive for $O_2(aq)$ because of the insolubility of oxygen gas. In saturated solution at 20°C oxygen has a solubility of 0.0014 mole liter^{-1}, so that to arrive at a concentration of hypothetical 1 m at unit activity (the aqueous standard state) free energy has to be put into the system to concentrate the oxygen molecules. This is certainly not a natural condition, which is actually that of oxygen gas at a partial pressure of 0.2 atm in equilibrium with oxygen gas in solution under this partial pressure. The quantity in solution under these natural conditions would then become 0.2×0.0014 mole liter^{-1}, or about 0.0003 mole liter^{-1}. Assuming molarities and molalities to be indistinguishable and that activities are unity at this concentration, the $\Delta G_f'$ for O_2(aq) using Eq. 10.17 then becomes

$$\begin{aligned}\Delta G_f' &= \Delta G_f^{\circ\prime} - 0.592 \ln \frac{1}{0.0003} \\ &= 3.9 - 4.802 \\ &= -0.902 \text{ kcal mole}^{-1}\end{aligned}$$

This value is considerably more negative than that for the aqueous standard state, and by using the latter incorrect calculations will be made. Under natural environmental conditions $\Delta G_f'$ for oxygen gas will, theoretically, never be more positive than about -0.9 kcal mole^{-1}. The same reasoning applies to other substances entering into growth processes. As far as they truly represent naturally occurring conditions, all $\Delta G_f'$ values must be considered as estimates, but the closer these estimates come to the real values, the more accurate will be our interpretations. If we adopt the opinion of Pitzer and Brewer[5] (p. 240) that molecules in aqueous solution are in general not appreciably affected by one another at concentrations below 0.001 *m* and accept this as fact, by using this concentration as a practical value we avoid the mathematical problem of ($\ln a_c^{-1}$) approaching infinity. By so doing we also adopt a representative substrate concentration which is within the range of those likely to be encountered naturally, but which may also be slightly low. Obviously, the choice of an appropriate "biological standard state" involves a compromise. A table of $\Delta G_f'$ values for various substances at a 0.001 *m* concentration at and an assumed unit activity is given in Appendix 2.

An additional question can be asked with respect to calculating free-energy changes accompanying growth processes. This is whether or not the kinetic energy of a molecule in solution has anything to do with the energetics of microbial growth. It is, of course, this energy which moves a molecule from some location within a solution up to the surface of a cell where it may or may not enter into some sort of relationship with the cell. In the case of substrate molecules, to be useful in cellular metabolism such a molecule must become attached to another molecule with which it enters into a transport mechanism that then passes it into the cell for further processing; or, it must become attached to an enzyme to form an enzyme-substrate complex, perhaps after first passing into the cell by diffusion. In any case, does the kinetic energy of the substrate molecule become lost when it becomes attached to some other, much larger and essentially fixed molecule and simply become dissipated as heat? It seems likely that this would happen. If so, then the free energy of formation of the substrate molecule that enters into a growth process should not be expected to include its translational energy. It should only include energy inherent within the structure of an immobilized substrate molecule at a given physiological temperature, but still in an aqueous environment equivalent to that of protoplasm. The calculation of $\Delta G_f'$ values relative to this kind of a "standard state" with respect to substrate molecules has not been addressed. For the moment, the use of a standard state represented by a 0.001 *m* concentration (which includes translational energies) seems to be the most appropriate.

When calculating values for the free energy of formation of one unit-carbon formula weight of cells it should be kept in mind that these can be expected to be not greatly different from those of unit-carbon formula weights of organic substances having similar compositions, the thermodynamic properties of which are known. Thus, $\Delta G_f'$ values for cells would be expected always to be negative, but not more negative than about -50 kcal UCFW^{-1}. If positive values are obtained it is likely that something is wrong with the analysis of the growth-process system.

10.3.3. Enthalpy Changes

Calculations of enthalpy changes do not present the problems encountered in calculating free-energy changes with respect to an appropriate standard state. For enthalpy calculations the standard state can still be taken as a hypothetical 1 *m* solution obeying the limiting law corresponding to Henry's Law, no matter what the concentration. This is because the value of the partial molal enthalpy in the standard state is always equal to that at infinite dilution. Thus, the infinitely dilute solution may be used as the reference state in all enthalpy calculations. The value for $\Delta H_f^{\circ\prime}$ is not different at lower concentrations than it is at hypothetical 1 *m*.

10.3.4. Entropy Changes

Calculations of entropy changes can always be made by means of the Gibbs free-energy equation. One of the most interesting aspects of this possibility is the prospect of being able to calculate values for the absolute entropy of a unit mass of living cells. The entropy of formation of a unit mass of substance can be calculated by means of the following modification of the Gibbs free-energy equation.

$$\Delta S_f' = \frac{\Delta H_f' - \Delta G_f'}{T} \tag{10.21}$$

from which

$$S = \Delta S_f' + 1.372n_{\text{C}} + 15.615n_{\text{H}} + 24.501n_{\text{O}} + 22.885n_{\text{N}} \tag{10.22}$$

where the constants represent the entropy of solid graphite and one-half the entropies of hydrogen, oxygen, and nitrogen, respectively, at 298.15°K and 1 atm, and where n_{C}, n_{H}, n_{O}, and n_{N} represent the relative numbers of carbon, hydrogen, oxygen, and nitrogen atoms, respectively, in the substance. Absolute entropies of organic substances *must* have positive values. If a calculation of the entropy of one unit-carbon formula weight of cells gives a negative value something is wrong somewhere.

10.4. METHODS OF DETERMINING THE EFFICIENCY OF MICROBIAL GROWTH

The efficiency with which energy is conserved during growth can be expressed in two ways. One of these is effectively the "machine" efficiency in that it represents energy conserved (energy out)/energy consumed (energy in). This will be designated as "efficiency I," which can be calculated by means of the following equation.

$$\text{percent free-energy eff. I} = \frac{\Delta G'_{NC} - \Delta G'_C}{\Delta G'_{NC}} \times 100 \qquad (10.23)$$

where $\Delta G'_{NC}$ and $\Delta G'_C$ represent the free-energy changes accompanying a nonconservative reaction and its relative conservative growth process, respectively. It is apparent that the same equation can be used to calculate the percent enthalpy and entropy efficiencies if $\Delta H'_{NC}$ and $\Delta H'_C$, or $\Delta S'_{NC}$ and $\Delta S'_C$ are substituted for $\Delta G'_{NC}$ and $\Delta G'_C$, respectively. The second type of efficiency represents energy conserved/energy expended. This will be designated as "efficiency II," which can be calculated as follows.

$$\text{percent free-energy eff. II} = \frac{\Delta G'_{NC} - \Delta G'_C}{\Delta G'_C} \times 100 \qquad (10.24)$$

One of the presently unanswered questions with respect to growth is whether more energy can be conserved than is expended in bringing about this conservation. Equation 10.24 is useful for making this calculation.

10.5. EXAMPLES OF STUDIES OF GROWTH-PROCESS SYSTEMS

Examples are usually needed to demonstrate methods fully. Unfortunately, very few examples exist of satisfactory studies of growth-process systems. As emphasized by Battley,[1] it is essential to grow the microorganisms for these studies on low concentrations of substrate so as to avoid fermentative or oxidative assimilation, and to establish experimentally that nothing is limiting the quantity of growth except the quantity of substrate. This can be done only by plotting the quantity of growth, by whatever criterion, against substrate concentration, establishing the range of concentrations within which a linear relationship exists, and by carrying out all subsequent experiments on the system under identical physiological conditions within this concentration range.

In the examples of studies on growth-process systems which follow, all the thermodynamic calculations were made using values for the thermodynamic properties which are listed in Appendix 2.

10.5.1. Anaerobic Growth When Ammonia is the Nitrogen Source. The Growth of *Saccharomyces cerevisiae* Fermentatively on Glucose

In almost all cases in which fermentative growth takes place the substrate will be a carbohydrate, the exceptions being the fermentation of amino acids. The data of Battley[1] provide the following information. The substrate is glucose ($C_6H_{12}O_6$) and the nitrogen source is ammonia (NH_3). There are three organic products of the growth process. These are 15.349 g dry weight of cells, 0.432 mole of glycerol, and 1.300 moles of ethanol, all per mole of glucose consumed. The composition of the cells is 46.16% C, 6.73% H, 27.75% O, 10.76% N, and 8.6% ash. Multiplying these percentages of C, H, O, and N by 100%/(100% − 8.6%) = 1.094 gives the relative percentages of ash-free organic material in the cells, which then become 50.50% C, 7.36% H, 30.37% O, and 11.77% N. Dividing the latter percentages of these elements by their respective atomic weights gives the empirical formula $C_{4.205}H_{7.302}O_{1.898}N_{0.840}$ and dividing these proportions through by the subscript for carbon gives the unit carbon formula $CH_{1.737}O_{0.451}N_{0.200}$. The unit carbon formula weight (UCFW) for the cells thus becomes 23.778 g. The ash-free dry weight of cells produced during fermentative growth of glucose = $Y_{af} = Y \times (100\% - \text{ash }\%)/100$ = 15.349 g × 0.914 = 14.029 g. The molar yield coefficient, MYC_c, = Y_{af}/UCFW = 14.029 g/23.778 g = 0.590, and the fermentation analyses indicated that $MYC_{glycerol}$ and $MYC_{ethanol}$ are 0.432 and 1.300 moles, respectively, all per mole of glucose fermented. The available electrons from the cells, ethanol, glucose, and glycerol can be designated as AE_c, AE_{eth}, AE_{glu}, and AE_{gly}, and have values of 4.235, 12, 24, and 14, respectively. The degree of reduction of the substrate (glucose) per unit carbon atom = 24AE/6C = 4.000. This is less than that of cells (4.235), ethanol (6.000) or glycerol (4.666). As Tamiya first clearly pointed out through the use of balanced equations (see Section 6.2), under these circumstances carbon dioxide will be produced during the construction of the cells, and by inference this will also occur during the production of other products which are more reduced than the substrate.

Using Eq. 10.4, the quantity of substrate required for the construction of the cells = S_c = (0.590 × 4.235)/24 = 0.104 mole. Inserting this value and that for MYC_c into Eq. 10.6 and using the unit-carbon formula to represent the cells gives the equation representing anabolism.

$$\left.\begin{array}{l}0.104C_6H_{12}O_6(aq)\\ +\ 0.118NH_3(aq)\end{array}\right\} \longrightarrow \begin{array}{l}0.590CH_{1.737}O_{0.451}N_{0.200}(\text{cells})\\ +\ 0.034CO_2(aq) + 0.288H_2O(l)\end{array} \quad (10.25)$$

An equation representing the formation of glycerol from glucose can be written in a manner similar to Eq. 10.25 except that no nitrogen is involved. Here,

$MYC_{gly} = 0.432$ and $S_{gly} = (0.432 \times 14)/24 = 0.252$ mole. Because the degree of reduction per unit-carbon atom of glycerol (D_r) = 14AE/3C = 4.666, carbon dioxide will also be a product in the reaction during which glycerol is produced from glucose. Equation 10.6 should thus again be used, except that in this case the coefficient for water will turn out to be negative, making it necessary to transfer this term to the left side of the resulting equation, as follows.

$$\left.\begin{array}{l}0.252C_6H_{12}O_6(aq)\\ +\ 0.216H_2O(l)\end{array}\right\} \longrightarrow 0.432C_3H_8O_3(aq) + 0.216CO_2(aq) \qquad (10.26)$$

The formation of glycerol as represented by Eq. 10.26 appears to serve the purpose of balancing the production of the cells, both materially and electronically. The reason for this is that the glycerol and the cells, in the quantities in which they are produced, can be combined theoretically to yield the carbon dioxide and ethanol which are not produced during the growth reaction but which would be produced were the relative nonconservative reaction to go to completion. This is illustrated by Eq. 10.30. Nevertheless, the production of glycerol as represented by Eq. 10.26 is best considered a catabolic reaction by definition, since no cells are formed during this particular reaction.

An equation representing the formation of ethanol from glucose can be written in the same way as for Eq. 10.26. In this case $MYC_{eth} = 1.300$ moles and $S_{eth} = (1.300 \times 12)/24 = 0.650$ mole. The degree of reduction of ethanol per unit carbon is 12AE/2C = 6.000; thus, carbon dioxide will also be produced from glucose along with ethanol. Again using Eq. 10.6,

$$0.650C_6H_{12}O_6(aq) \longrightarrow 1.300CH_3CH_2OH(aq) + 1.300CO_2(aq) \qquad (10.27)$$

Equation 10.27 represents the fermentative catabolic reaction which provides energy for bringing about the anabolic reaction.

The addition of Eqs. 10.25, 10.26, and 10.27 gives the whole growth reaction.

$$\left.\begin{array}{l}1.006C_6H_{12}O_6(aq)\\ +\ 0.118NH_3(aq)\end{array}\right\} \longrightarrow \begin{array}{l}0.590CH_{1.737}O_{0.451}N_{0.200}(\text{cells})\\ +\ 1.300CH_3CH_2OH(aq) + 0.432C_3H_8O_3(aq)\\ +\ 1.550CO_2(aq) + 0.072H_2O(l)\end{array} \qquad (10.28)$$

Equation 10.28 is not exactly the same as Eq. 1 in Table 8.1 for the construction of which the gas exchange was measured experimentally, but it is certainly very nearly identical.

The nonconservative reaction relative to this particular growth process is that of alcoholic fermentation.

$$C_6H_{12}O_6(aq) \longrightarrow 2CO_2(aq) + 2CH_3CH_2OH(aq) \qquad (10.29)$$

Using the values for $\Delta G'_f$ and $\Delta H'_f$ found in Appendix 2, the values for $\Delta G'_{NC}$, $\Delta H'_{NC}$, and $T\ \Delta S'_{NC}$ (from the Gibbs free-energy equation) are calculated to be -64.71, -33.12, and 31.59 kcal mole^{-1} of glucose fermented, respectively. According to conservation theory (see Section 8.10), the energy changes accompanying a growth reaction are proportional to the extent to which the relative nonconservative reaction has proceeded to completion.

With respect to enthalpy changes, the $\Delta H'_{cat}$ accompanying the catabolic reaction represented by Eq. 10.26 is -0.39 kcal. That accompanying the catabolic reaction represented by Eq. 10.27 is -21.53 kcal. The value for $\Sigma\ \Delta H'_{cat}$ is then $-0.39 + (-21.53) = -21.92$ kcal. The enthalpic energy represented by the difference between $\Delta H'_{NC}$ and $\Delta H'_{cat}$ (see Eq. 10.13), $\Delta H'_{con}$, must therefore be conserved in the substance of the organic products of the growth reaction which are different from those of the relative nonconservative reaction. These products are the cells, glycerol, and a small amount of water which, if they were to be theoretically *fermented* to carbon dioxide and alcohol would yield the remaining energy having a value of $-33.12 - (-21.92) = -11.20$ kcal. This fermentation is represented by the following equation.

$$\left.\begin{array}{l} 0.590CH_{1.737}O_{0.451}N_{0.200}(\text{cells}) \\ +\ 0.432C_3H_8O_3(aq) + 0.072H_2O(l) \end{array}\right\} \longrightarrow \begin{array}{l} 0.462CO_2(aq) + 0.118NH_3(aq) \\ +\ 0.712CH_3CH_2OH(aq) \end{array} \tag{10.30}$$

The addition of Eqs. 10.28 and 10.30 should equal Eq. 10.29, and does so with an accuracy of about 0.6%. Since the $\Delta H'$ for the reaction represented by Eq. 10.30 is -11.20 kcal, the $\Delta H'_f$ for the cellular material formed during growth is calculated to be -18.35 kcal UCFW^{-1}. This value can then be inserted into Eq. 10.25 so that $\Delta H'_{an}$ for the reaction represented by this equation can be calculated to be -0.20 kcal. The enthalpy change, $\Delta H'_C$, for the complete growth reaction is then the sum of the $\Delta H'$ values for the reactions represented by Eqs. 10.25, 10.26, and 10.27, which is $-0.20 + (-0.39) + (-21.53) = -22.12$ kcal mole^{-1} of glucose consumed. This value is in good agreement with the average of the values (-22.5) obtained experimentally by Dubrunfaut[6] (-22.7), Bouffard[7] (-23.1), A. J. Brown[8] (-21.4), H. T. Brown[9] (-22.5), and Battley[10] (-23), with growing cultures of *S. cerevisiae*. Rubner's[11] value of -25.6 kcal mole^{-1} of glucose fermented should probably be rejected as being too high.

The free-energy change accompanying the anaerobic growth of *S. cerevisiae* on glucose is easier to calculate than the enthalpy change since, in accord with Battley's hypothesis (see Section 10.12), $\Delta G'_{an}$ is assumed to be equal to zero, from which $\Delta G'_f$ for the cells produced during the reaction represented by Eq. 10.25 can be calculated to be -8.11 kcal UCFW^{-1}. The $\Delta G'_{cat}$ value for the catabolic reaction resulting in the production of glycerol, using Eq. 10.26, is -5.59 kcal. Using Eq. 10.27, $\Delta G'_{cat}$ is found to be -42.06 kcal. The free-energy change, $\Delta G'_C$, for the complete growth process is then the sum of the $\Delta G'$ values for the reactions represented by Eqs. 10.25, 10.26, and 10.27, which is 0.000 + (−5.59) + (−42.06) = −47.65 kcal mole^{-1} of glucose consumed. Nearly the same free-energy change can be calculated using the method described previously for calculating $\Delta H'$ values. Here, the $\Delta G'_{cat}$ accompanying the catabolic reaction represented by Eq. 10.26 is −5.59 kcal and that represented by Eq. 10.27 is −42.06 kcal. The value for $\Sigma\ \Delta G'_{cat}$ then equals −5.59 + (−42.06) = −47.65 kcal. The difference between the $\Delta G'$ value for the nonconservative reaction represented by Eq. 10.29 and $\Delta G'_{cat}$ is then −64.71 − (−47.65) = −17.06 kcal. This is the free-energy change accompanying the reaction represented by Eq. 10.30, from which $\Delta G'_f$ for the cellular material formed during growth is calculated to be −8.78 kcal UCFW^{-1}. When this is substituted into Eq. 10.25 together with the $\Delta G'_f$ values for the other reactants and products, $\Delta G'_{an}$ is found to equal −0.37 kcal. Considering all the potentialities for experimental error, this is close enough to zero to confirm Battley's hypothesis. $\Delta G'_C$ using this second method has a value of −0.37 + (−5.59) + (−42.06) = −48.02 kcal mole^{-1} of glucose consumed, as compared with that of −47.65 kcal mole^{-1} obtained by the first method using Battley's hypothesis. The agreement of the two methods is within about 0.8%.

Using the Gibbs free-energy equation $T\Delta S'_C$ is found to be $(\Delta H'_C - \Delta G'_C) = -22.12 - (-47.65) = 25.53$ kcal mole^{-1} of glucose consumed, from which $\Delta S'_C = 25.53/298.15 = 85.6$ cal mole^{-1} of glucose consumed deg^{-1}. Similarly, $T\,\Delta S'_{an} = -0.20 - 0.000 = -0.20$ kcal, from which $\Delta S'_{an} = -0.20/298.15 = 0.6$ cal deg^{-1} for the conversion of the reactants in Eq. 10.25 into the products, or 1.0 cal UCFW^{-1} deg^{-1}. This is a negligible entropy change, and indicates that the entropy of the cells plus the other products of this reaction is the same as that of the quantity of substrate used to synthesize the cells plus the nitrogen source. Another method of comparison is to calculate the entropy of formation of one unit-carbon formula weight of *S. cerevisiae* cells. Using Eq. 10.21,

$$\Delta S'_f(\text{cells}) = \frac{\Delta H_f(\text{cells}) - \Delta G_f(\text{cells})}{T}$$

$$= \frac{-18.35 \text{ kcal UCFW}^{-1} - (-8.11) \text{ kcal UCFW}^{-1}}{298.15}$$

$$= -34.3 \text{ cal UCFW}^{-1} \text{ deg}^{-1}$$

The absolute entropy of one unit-carbon formula weight of *S. cerevisiae* cells can then be calculated by means of Eq. 10.22.

$$S = \begin{array}{l} -34.3 + 1.372 + (15.615 \times 1.737) \\ + (24.501 \times 0.451) + (22.885 \times 0.200) \end{array}$$

$$= 9.8 \text{ cal UCFW}^{-1} \text{ deg}^{-1}$$

Eff. I with respect to free-energy is calculated using Eq. 10.23 to be 26.4% and eff. II using Eq. 10.24 to be 35.9%.

10.5.2. Aerobic Growth When the Substrate is Less Reduced Than the Cellular Material and the Nitrogen Source Is Ammonia. The Growth of *S. cerevisiae* Aerobically on Glucose

Here, the data of Battley[1] will again be used. Glucose ($C_6H_{12}O_6$) is the sole source of carbon and energy and the nitrogen source is ammonia (NH_3). Some of the glucose is used for the construction of the cells, the rest being oxidized to carbon dioxide and water. The only organic product of the growth reaction is 49.250 g dry weight of cells per mole of glucose consumed. The composition of the cells is 46.68% C, 6.67% H, 28.55% O, 9.30% N, and 8.8% ash. Carrying out the same operation as for Section 10.5.1 gives a unit-carbon formula of $CH_{1.703}O_{0.459}N_{0.171}$ and a UCFW of 23.467 g. Y_{af} = 44.916 and MYC_c = 44.916/23.467 g = 1.914. AE_c = 4.272 and AE_{glu} = 24. S_c is therefore (1.914 × 4.272)/24 = 0.341 mole of glucose. Because the substrate has 4.000 AE/unit-carbon atom it is less reduced than the cellular material, and carbon dioxide will be a product of the reaction whereby cells are constructed. By inserting the appropriate values into Eq. 10.6 the following equation results.

$$\left.\begin{array}{l}0.341C_6H_{12}O_6(aq)\\ +\ 0.327NH_3(aq)\end{array}\right\} \longrightarrow \begin{array}{l}1.914CH_{1.703}O_{0.459}N_{0.171}(\text{cells})\\ +\ 0.132CO_2(aq) + 0.907H_2O(l)\end{array} \tag{10.31}$$

Since there are no organic products of the growth reaction other than the cells, what glucose is not used for construction of the cells becomes respired. This quantity, which is represented by the term $(1 - \Sigma S_p)$, can be calculated by means of Eq. 10.8. If cells are the only organic products of a growth reaction, $(1 - \Sigma S_p) = (1 - S_c)$. In the case of the aerobic growth of *S. cerevisiae* on glucose $(1 - S_c) = (1 - 0.341) = 0.659$ mole. Since this quantity of glucose becomes completely respired to carbon dioxide and water, the equation representing the catabolic reaction is then

$$0.659C_6H_{12}O_6(aq) + 3.954O_2(aq) \longrightarrow \begin{array}{l}3.954CO_2(aq)\\ +\ 3.954H_2O(l)\end{array} \tag{10.32}$$

The addition of Eqs. 10.31 and 10.32 gives the complete growth-process equation.

$$\left.\begin{array}{l}C_6H_{12}O_6(aq) + 0.327NH_3(aq)\\ +\ 3.954O_2(aq)\end{array}\right\} \longrightarrow \begin{array}{l}1.914CH_{1.703}O_{0.459}N_{0.171}(\text{cells})\\ +\ 4.086CO_2(aq) + 4.861H_2O(l)\end{array} \tag{10.33}$$

This equation should be compared with Eq. 2 in Table 8.1, for the construction of which the gas exchanges were determined experimentally.

The nonconservative reaction relative to this particular growth process is that of the complete oxidation of glucose to carbon dioxide and water,

$$C_6H_{12}O_6(aq) + 6O_2(aq) \longrightarrow 6CO_2(aq) + 6H_2O(l) \qquad (10.34)$$

for which $\Delta G'_{NC}$, $\Delta H'_{NC}$, and $T\,\Delta S'_{NC}$ are calculated to be -694.42, -684.07, and 10.35 kcal mole^{-1} of glucose oxidized, respectively. According to conservation theory (see Section 8.10), the energy changes accompanying aerobic growth processes are also proportional to the extent to which the relative nonconservative reaction has proceeded to completion.

With respect to enthalpy changes $\Delta H'_{cat}$ can be calculated using Eq. 10.32 to be $0.659 \times -684.07 = -450.80$ kcal. The difference between $\Delta H'_{NC}$ and $\Delta H'_{cat}$ represents the enthalpic energy conserved in the products of the growth process which are different from those of the relative nonconservative reaction. In this particular example the only products are the cells which, if they were to be oxidized to carbon dioxide, water, and ammonia would yield the conserved energy, $\Delta H'_{con}$, having a value of $-684.07 - (-450.80) = -233.27$ kcal. This oxidation is represented by the following equation.

$$\left.\begin{array}{l} 1.914CH_{1.703}O_{0.459}N_{0.171}(\text{cells}) \\ +\ 2.044O_2(aq) \end{array}\right\} \longrightarrow \begin{array}{l} 1.914CO_2(aq) + 1.139H_2O(l) \\ +\ 0.327NH_3(aq) \end{array} \qquad (10.35)$$

The addition of Eqs. 10.33 and 10.35 should equal Eq. 10.34, and does. Since the $\Delta H'$ for the reaction represented by Eq. 10.35 is -233.27 kcal, the $\Delta H'_f$ for the cellular material formed during the growth reaction is calculated to be -17.87 kcal UCFW^{-1} value which is close to that of -18.35 kcal UCFW^{-1} for the same cells grown anaerobically on the same substrate. When this value is inserted into Eq. 10.31 together with the other appropriate $\Delta H'_f$ values, $\Delta H'_{an}$ is calculated to be 0.004 kcal. The sum of $\Delta H'_{an}$ plus $\Delta H'_{cat}$ gives $0.004 + (-450.80) = -450.80$ kcal mole^{-1} of glucose consumed as the value of $\Delta H'_C$ for the complete growth process, in only fair agreement with the average value of -479 kcal mole^{-1} of glucose consumed obtained experimentally by Battley.[8]

With respect to free-energy changes, using Eq. 10.32 the value for $\Delta G'_{cat}$ is $0.659 \times -694.41 = -457.62$ kcal. This is also the value for $\Delta G'_C$ since $\Delta G'_{an}$ for the reaction represented by Eq. 10.31 is assumed according to Battley's hypothesis to be zero. Thus, $\Delta G'_{an} + \Delta G'_{cat} = 0.000 + (-457.62) = -457.62$ kcal mole^{-1} of glucose consumed. Because $\Delta G'_{an}$ is taken to be zero, the value for $\Delta G'_f$(cells) using Eq. 10.31 is found to be -7.95 kcal, which is nearly identical with the value of -8.15 kcal UCFW^{-1} for the same cells grown anaerobically on the same substrate.

Using the Gibbs free-energy equation $T\Delta S'_C$ is found to be $(\Delta H'_C - \Delta G'_C) = -450.80 - (-457.62) = 6.82$ kcal, from which $\Delta S'_C = 6.82/298.15 = 22.8$ cal mole^{-1} of glucose consumed deg^{-1}. Similarly $T\Delta S_f(\text{cells}) = \Delta H_f(\text{cells}) -$

ΔG_f(cells) = $-17.87 - (-7.95) = -9.92$ kcal, from which $\Delta S_f'$(cells) = $-9.92/298.15 = -33.3$ cal UCFW^{-1}. Using Eq. 10.22 the value of the absolute entropy, S, for the cells is 10.1 cal UCFW^{-1} deg^{-1}, in close agreement with the value of 9.8 cal UCFW^{-1} deg^{-1} obtained with the same cells grown anaerobically on glucose.

Eff. I with respect to free-energy is calculated using Eq. 10.23 to be 34.1% and eff. II using Eq. 10.24 to be 51.7%.

The fact that the values for $\Delta G_f'$ UCFW^{-1}, ΔH_f UCFW^{-1}, and $\Delta S_f'$ UCFW^{-1} are nearly identical with respect to *S. cerevisiae* cells grown either anaerobically or aerobically on glucose must surely indicate that the thermodynamic properties of microbial cells are completely independent of the nature of the catabolic reactions.

10.5.3. Aerobic Growth When the Substrate Is More Reduced Than the Cellular Material and the Nitrogen Source Is Ammonia. The Growth of *S. cerevisiae* on Ethanol

Using once more the data of Battley,[1] the only organic product of this growth process is 25.096 g dry weight of cells per mole of ethanol consumed, the other products being carbon dioxide and water. The composition of the cells is 49.30% C, 7.05% H, 26.81% O, 8.54% N, and 8.3% ash from which the unit-carbon formula becomes $CH_{1.704}O_{0.408}N_{0.149}$, giving a UCFW of 22.343 g. Y_{af} = 23.013 and MYC$_c$ = 23.013 g/22.343 g = 1.030 UCFWs. AE$_c$ = 4.441 and AE$_S$ = 12. Using Eq. 10.4, $S_c = (1.030 \times 4.441)/12 = 0.381$ mole of ethanol. Because the substrate has 6.000 AE per unit-carbon atom it is considerably more reduced than the cellular material. For this reason Eq. 10.6 *cannot* be used, and it becomes necessary to follow the suggestion of Battley[12–13] (see also Section 8.12) and to include carbon dioxide as a reactant rather than a product in the equation representing anabolism, as represented by Eq. 10.7. Inserting the appropriate values into this equation then gives

$$\left.\begin{array}{l} 0.381CH_3CH_2OH(aq) \\ +\ 0.153NH_3(aq) + 0.268CO_2(aq) \end{array}\right\} \longrightarrow \begin{array}{l} 1.030CH_{1.704}O_{0.408}N_{0.149}(cells) \\ +\ 0.495H_2O(l) \end{array} \quad (10.36)$$

The quantity of substrate utilized in respiration per mole of ethanol consumed during growth is $(1 - S_c) = (1 - 0.381) = 0.619$. The equation representing catabolism is therefore

$$0.619CH_3CH_2OH(aq) + 1.857O_2(aq) \longrightarrow 1.238CO_2(aq) + 1.857H_2O(l) \quad (10.37)$$

The addition of Eqs. 10.36 and 10.37 gives the complete growth-process equation.

$$\left.\begin{array}{l} CH_3CH_2OH(aq) + 0.153NH_3(aq) \\ + 1.857O_2(aq) \end{array}\right\} \longrightarrow \begin{array}{l} 1.030CH_{1.704}O_{0.408}N_{0.149}(\text{cells}) \\ + 0.970CO_2(aq) + 2.352H_2O(l) \end{array} \tag{10.38}$$

This should be compared with Eq. 3 in Table 8.1, for the construction of which the gas exchanges were measured experimentally.

The nonconservative reaction relative to this particular growth process is that of the complete oxidation of ethanol to carbon dioxide and water

$$CH_3CH_2OH(aq) + 3O_2(aq) \longrightarrow 2CO_2(aq) + 3H_2O(l) \tag{10.39}$$

for which $\Delta G'_{NC}$, $\Delta H'_{NC}$, and $T\,\Delta S'_{NC}$ are calculated to be -314.85, -325.47, and -10.62 kcal mole^{-1}, respectively.

With respect to enthalpy changes $\Delta H'_{NC}$ is calculated by means of Eq. 10.37 to be $0.619 \times -325.47 = -201.47$ kcal. The enthalpic energy conserved in the products of the growth reaction which are different from those of the relative nonconservative reaction is represented by $\Delta H'_{NC} - \Delta H'_{cat}$. This quantity, $\Delta H'_{con}$, has a value of $-325.47 - (-201.47) = -124.00$ and represents the heat energy that would be released if the cells, being the only organic products of the growth process, were to be oxidized to carbon dioxide, water, and ammonia as represented by the following equation.

$$\left.\begin{array}{l} 1.030CH_{1.704}O_{0.408}N_{0.149}(\text{cells}) \\ + 1.144O_2(aq) \end{array}\right\} \longrightarrow \begin{array}{l} 1.030CO_2(aq) + 0.648H_2O(l) \\ + 0.153NH_3(aq) \end{array} \tag{10.40}$$

The addition of Eqs. 10.38 and 10.40 should equal Eq. 10.39, and does. Since $\Delta H'$ for the reaction represented by Eq. 10.40 is -124.00 kcal, $\Delta H'_f$(cells) is calculated to be -21.13 kcal UCFW^{-1}. This value can then be inserted into Eq. 10.36, so that $\Delta H'_{an}$ is calculated to be -0.002 kcal. The sum of $\Delta H'_{an}$ + $\Delta H'_{cat}$ gives $0.002 + (-201.47) = -201.47$ kcal mole^{-1} of ethanol consumed as the value for $\Delta H'_C$ for the complete growth process, in very good agreement with the value of -204 kcal mole^{-1} determined experimentally by Battley[8] for the same growth process.

With respect to free-energy changes, using Eq. 10.37 $\Delta G'_{cat}$ is calculated to be $0.619 \times -314.85 = -194.89$ kcal. This is also the value of $\Delta G'_C$ since $\Delta G'_{an}$ for the reaction represented by Eq. 10.36 is equal to zero. Using this latter value and the same equation, $\Delta G'_f$(cells) is calculated to have a value of -16.89 kcal UCFW^{-1}.

Using the Gibbs free-energy equation $T\,\Delta S'_C$ is found to be $(\Delta H'_C - \Delta G'_C) = -201.47 - (-194.89) = -6.58$ kcal, from which $\Delta S'_C = -6.58/$

298.15 = -22.1 cal mole^{-1} of ethanol consumed degree deg^{-1}. It is worth noting that the value for $\Delta S'_C$ in this case is negative, whereas for growth on glucose it is positive. It is thus apparent that $\Delta S'_C$ can have a value which is positive, negative, or zero. Similarly, $T\,\Delta S'_f$(cells) = $\Delta H'_f$(cells) − $\Delta G'_f$(cells) = $-21.13 - (-16.89) = -4.24$ kcal, from which $\Delta S'_f$(cells) = $-4.24/298.15$ = -14.2 cal UCFW^{-1}deg^{-1}. By means of Eq. 10.22 the absolute entropy, S, for the cells is calculated to be 27.15 cal UCFW^{-1} deg^{-1}.

Eff. I with respect to free energy is calculated by means of Eq. 10.23 to be 38.1% and eff. II to be 61.5%.

10.5.4. Aerobic Growth When the Substrate Is Less Reduced Than the Cellular Material and the Nitrogen Source Is Nitrate. The Growth of *Aspergillus oryzae* on Glucose

Apart from the experiments of Battley[1] there are no others from which information can be extracted with which to write growth-process equations in which it was determined that nothing was limiting the amount of growth except the quantity of substrate, and that therefore oxidative assimilation could be presumed not to be occurring. Thus, the data of Tamiya[14] will be used here for an example in spite of the fact that he grew *A. oryzae* on 0.5 M concentrations of substrates, which is two orders of magnitude higher than the ideal concentration (0.002 M) which would better be used for studies on microbial energetics. At this concentration, especially with glucose as a substrate, there was almost surely some oxidative assimilation which took place along with growth, so that the value for MYC_c was probably higher than it would have been had only growth occurred. We do not really know. For this reason and because of the high concentration of substrate, the values for energy changes calculated here should be interpreted with caution.

Tamiya's data give $C_{415}H_{728}O_{237}N_{35}$ as the empirical formula multiplied by 100 for *A. oryzae* cells grown aerobically on glucose with nitrate as the nitrogen source. The RQ for growth under these conditions was determined to be 1.30. There were no organic products of the growth reaction other than the cells. These data are all that is necessary to construct a growth-reaction equation (were the data available the use of the unit-carbon formula and the molar growth yield would be easier). Here AE_S = 24 and AE_c = 1809, from which $S_p = S_c = 1809/24 = 75.37$ moles of glucose. Because there is no molar yield coefficient in this example, this term should be substituted by a yield coefficient, YC_c, having a value of unity. Thus, inserting the appropriate values into Eq. 10.6 gives the following equation representing anabolism.

$$75.37C_6H_{12}O_6(aq) + 35NH_3(aq) \longrightarrow \begin{array}{l} C_{415}H_{728}O_{237}N_{35}(\text{cells}) \\ + 37.22CO_2(aq) + 140.72H_2O(l) \end{array} \quad (10.41)$$

The quantity of glucose required to reduce an amount of nitrate equivalent to the amount of ammonia in Eq. 10.41 is found using Eq. 10.9, whereby $S_N = (8 \times 35)/24 = 11.66$ moles. The equation representing this reduction is Eq. 10.10. Inserting the appropriate values into this equation gives

$$11.66C_6H_{12}O_6 + 35HNO_3(aq) \longrightarrow \begin{array}{l} 35NH_3(aq) + 69.96CO_2(aq) \\ + 35H_2O(l) \end{array} \tag{10.42}$$

Using Eq. 6.34 and the data pertaining to Eq. 6.35 in Chapter 6, the quantity of oxygen needed to oxidize sufficient glucose to provide an RQ of 1.30 is 359.8 moles, from which Eq. 6.35 representing catabolism results.

$$60C_6H_{12}O_6(aq) + 359.8O_2(aq) \longrightarrow 359.8CO_2(aq) + 359.8H_2O(l) \tag{6.35}$$

Addition of the coefficients for the substrate in Eqs. 10.41, 10.42, and 6.35 gives a total value of 147.03 moles. Dividing these three equations through by this value and converting the formula for the cells into a unit-carbon formula gives the following three equations.

$$\left.\begin{array}{l} 0.513C_6H_{12}O_6(aq) \\ + 0.237NH_3(aq) \end{array}\right\} \longrightarrow \begin{array}{l} 2.822CH_{1.754}O_{0.571}N_{0.084}(\text{cells}) \\ + 0.256CO_2(aq) + 0.958H_2O(l) \end{array} \tag{10.43}$$

$$\left.\begin{array}{l} 0.079C_6H_{12}O_6(aq) \\ + 0.237HNO_3(aq) \end{array}\right\} \longrightarrow \begin{array}{l} 0.237NH_3(aq) + 0.474CO_2(aq) \\ + 0.237H_2O(l) \end{array} \tag{10.44}$$

$$\left.\begin{array}{l} 0.408C_6H_{12}O_6(aq) \\ + 2.448O_2(aq) \end{array}\right\} \longrightarrow 2.448CO_2(aq) + 2.448H_2O(l) \tag{10.45}$$

The addition of Eqs. 10.43, 10.44, and 10.45 gives the complete growth-process equation in terms of growth on 1 mole of substrate.

$$\left.\begin{array}{l} C_6H_{12}O_6(aq) + 0.237HNO_3(aq) \\ + 2.448O_2(aq) \end{array}\right\} \longrightarrow \begin{array}{l} 2.822CH_{1.754}O_{0.571}N_{0.084}(\text{cells}) \\ + 3.178CO_2(aq) + 3.643H_2O(l) \end{array} \tag{10.46}$$

With respect to Eq. 10.46, the RQ $= 3.178/2.448 = 1.30$, in accord with the data given by Tamiya.

The nonconservative reaction relative to this particular growth process is that of the complete oxidation of glucose to carbon dioxide and water, as represented by Eq. 10.34:

$$C_6H_{12}O_6(aq) + 6O_2(aq) \longrightarrow 6CO_2(aq) + 6H_2O(l) \tag{10.34}$$

for which $\Delta G'_{NC}$, $\Delta H'_{NC}$, and $T\,\Delta S'_{NC}$ are calculated to be -694.42, -684.07, and 10.35 kcal mole^{-1}, respectively.

With respect to enthalpy changes, the reactions represented by Eqs. 10.44

and 10.45 should both be considered catabolic, since no cells are formed during either reaction and since both result in the oxidation of glucose to carbon dioxide and water. Equation 10.44 represents the reduction of nitrate to ammonia. The enthalpy change, $\Delta H'_{NR}$ (NR = nitrate reduction), accompanying this reaction has a value of -32.03 kcal. The catabolic reaction represented by Eq. 10.45 has an accompanying enthalpy change, $\Delta H'_{cat}$, of -279.10 kcal. The value for $\Delta H'_{con}$ is thus $\Delta H'_{NC} - \Sigma\,\Delta H'_{cat} = \Delta H'_{NC} - \Delta H'_{NR} - \Delta H'_{cat} = -684.07 - (-32.03) - (-279.10) = -372.94$ kcal. This is the enthalpic energy which would be released as heat were the cells produced during this growth process to be oxidized to carbon dioxide, water, and in this case, nitrate (since this is the original nitrogen source) as follows.

$$\left.\begin{array}{l} 2.822CH_{1.754}O_{0.571}N_{0.084}(\text{cells}) \\ +\ 3.550O_2(aq) \end{array}\right\} \longrightarrow \begin{array}{l} 2.822CO_2(aq) + 2.356H_2O(l) \\ +\ 0.237HNO_3(aq) \end{array} \qquad (10.47)$$

Using Eq. 10.47, $\Delta H'_f$(cells) is calculated to be -24.30 kcal UCFW^{-1}. This value can then be inserted in Eq. 10.43 to give a value of 0.147 kcal for $\Delta H'_{an}$. The enthalpy change for the complete growth process, $\Delta H'_C$, is then the sum of $\Delta H'_{an} + \Delta H'_{NR} + \Delta H'_{cat} = 0.147 + (-32.03) + (-279.1) = -311.0$ kcal mole^{-1} of glucose consumed. The addition of Eqs. 10.46 and 10.47 should equal Eq. 10.34, and does. Similarly, the addition of the enthalpy changes accompanying the reactions or processes represented by these equations should equal that of the relative nonconservative reaction, and also does.

With respect to free-energy changes, $\Delta G'_{NR}$ and $\Delta G'_{cat}$ for the reactions represented by Eqs. 10.44 ands 10.45 have values of -36.71 and -283.32 kcal, respectively. $\Delta G'_{an}$ as represented by Eq. 10.43 is zero, from which $\Delta G'_f$(cells) is calculated to be -13.39 kcal. The free-energy change for the complete growth process, $\Delta G'_C$, is then $\Delta G'_{an} + \Delta G'_{NR} + \Delta G'_{cat} = 0 + (-36.71) + (-283.32) = -320.03$ kcal mole^{-1} of glucose consumed.

Using the Gibbs free-energy equation $T\,\Delta S'_C = (\Delta H'_C - \Delta G'_C) - 311.0 - (-320.0) = 9.00$ kcal, from which $\Delta S'_C = 9.00/298.15 = 30.2$ cal deg^{-1}. Similarly, $T\,\Delta S'_f(\text{cells}) = \Delta H'_f(\text{cells}) - \Delta G'_f(\text{cells}) = -24.30 - (-13.39) = -10.90$ kcal, from which $\Delta S'_f(\text{cells}) = -36.6$ cal UCFW^{-1} deg^{-1}. The absolute entropy, S, for the cells is then calculated using Eq. 10.22 to be 8.1 cal UCFW^{-1} deg^{-1}.

Eff. I and eff. II with respect to free energy are calculated using Eqs. 10.23 and 10.24 to be 53.9% and 117.0%, respectively.

10.5.5. Aerobic Growth When the Substrate Is More Reduced Than the Cellular Material and the Nitrogen Source Is Nitrate. The Growth of *A. oryzae* on Ethanol

Once again the data of Tamiya[14] will be used in spite of the fact that his cultures were grown on such high concentrations of substrate, simply because his data are some of the few available. Tamiya gives $C_{411}H_{752}O_{227}N_{48}$ as the empirical formula multiplied by 100 for *A. oryzae* cells grown aerobically on ethanol with nitrate as the nitrogen source. There were no organic products of the growth process other than the cells. The RQ for growth under these conditions was found to be 0.65. AE_S = 12 and AE_c = 1798, from which $S_p = S_c$ = 1798/12 = 149.83 moles of ethanol. Again because there is no molar yield coefficient in this example, this term should be substituted by a yield coefficient, YC_c, having a value of unity. Inserting the appropriate values into Eq. 10.7 gives the following equation representing anabolism.

$$\left.\begin{array}{l} 149.83CH_3CH_2OH(aq) \\ +\ 48NH_3(aq) + 111.34CO_2(aq) \end{array}\right\} \longrightarrow \begin{array}{l} C_{411}H_{752}O_{227}N_{48}(cells) \\ +\ 145.49H_2O(l) \end{array} \qquad (10.48)$$

The quantity of ethanol required to reduce the amount of nitrate equivalent to the amount of ammonia in Eq. 10.48 is found using Eq. 10.9, whereby $S_N = 8 \times 48/12 = 32$ moles. Inserting the appropriate data into Eq. 10.10, which represents this reduction, gives

$$32CH_3CH_2OH(aq) + 48HNO_3(aq) \longrightarrow \begin{array}{l} 48NH_3(aq) + 64CO_2(aq) \\ +\ 48H_2O(l) \end{array} \qquad (10.49)$$

Using Eq. 6.37 and the data pertaining to Eq. 6.38 in Chapter 6, the quantity of carbon dioxide produced during the oxidation of sufficient ethanol to produce an RQ of 0.65 is 1811 moles, from which Eq. 10.50 results.

$$950CH_3CH_2OH(aq) + 2850O_2(aq) \longrightarrow 1900CO_2(aq) + 2850H_2O(l) \qquad (10.50)$$

Addition of the coefficients for the substrate in Eqs. 10.48, 10.49, and 10.50 gives a total value of 1132 moles. Dividing these three equations through by this value and converting the formula for the cells into a unit-carbon formula gives the following three equations.

$$\left.\begin{array}{l} 0.132CH_3CH_2OH(aq) \\ +\ 0.042NH_3(aq) \\ +\ 0.099CO_2(aq) \end{array}\right\} \longrightarrow \begin{array}{l} 0.363CH_{1.830}O_{0.552}N_{0.117}(cells) \\ +\ 0.128HO(l) \end{array} \qquad (10.51)$$

$$\left.\begin{array}{l} 0.028CH_2CH_2OH(aq) \\ +\ 0.042HNO_3(aq) \end{array}\right\} \longrightarrow \begin{array}{l} 0.042NH_3(aq) + 0.056CO_2(aq) \\ +\ 0.042H_2O(l) \end{array} \qquad (10.52)$$

$$0.840CH_3CH_2OH(aq) + 2.520O_2(aq) \longrightarrow \begin{array}{l} 1.680CO_2(aq) \\ +\ 2.520H_2O(l) \end{array} \qquad (10.53)$$

The addition of Eqs. 10.51, 10.52, and 10.53 gives the complete growth-process equation in terms of 1 mole of substrate.

$$\left.\begin{array}{l} CH_3CH_2OH(aq) \\ +\ 0.042HNO_3(aq) \\ +\ 2.520O_2(aq) \end{array}\right\} \longrightarrow \begin{array}{l} 0.363CH_{1.830}O_{0.552}N_{0.117}(\text{cells}) \\ +\ 1.637CO_2(aq) + 2.690H_2O(l) \end{array} \quad (10.54)$$

With respect to Eq. 10.54, the RQ = 1.637/2.520 = 0.65, which is what was reported by Tamiya.

The nonconservative reaction relative to this particular growth process is that of the complete oxidation of ethanol to carbon dioxide and water, as represented by Eq. 10.39.

$$CH_3CH_2OH(aq) + 3O_2(aq) \longrightarrow 2O_2(aq) + 3H_2O(l) \quad (10.39)$$

for which $\Delta G'_{NC}$, $\Delta H'_{NC}$, and $T\,\Delta S'_{NC}$ are calculated to be -314.85, -325.47, and -10.62 kcal mole^{-1}, respectively.

The enthalpy changes accompanying the catabolic reactions represented by Eqs. 10.52 and 10.53, $\Delta H'_{NR}$ and $\Delta H'_{cat}$, are calculated to be -5.20 and -273.42 kcal, respectively. The conserved enthalpic energy, $\Delta H'_{con}$, then equals $\Delta H'_{NC} - \Sigma\,\Delta H'_{cat} = \Delta H'_{NC} - \Delta H'_{NR} - \Delta H'_{cat} = -325.47 - (-5.20) - (-273.42) = -46.85$ kcal mole^{-1} of ethanol consumed. This is the heat energy that would be released were the cells produced during this growth reaction to be oxidized to carbon dioxide, water, and in this case nitrate (since this is the original nitrogen source), as follows:

$$\left.\begin{array}{l} 0.363CH_{1.830}O_{0.552}N_{0.117}(\text{cells}) \\ +\ 0.481O_2(aq) \end{array}\right\} \longrightarrow \begin{array}{l} 0.363CO_2(aq) + 0.311H_2O(l) \\ +\ 0.042HNO_3(aq) \end{array} \quad (10.55)$$

Using Eq. 10.55, $\Delta H'_f$(cells) is calculated to be -30.27 kcal UCFW^{-1}. This value can then be inserted into Eq. 10.51 to give a value of -0.148 kcal for the enthalpy change, $\Delta H'_{an}$, accompanying anabolism. The enthalpy change for the complete growth process, $\Delta H'_C$, is then $\Delta H'_{an} + \Sigma\,\Delta H'_{cat} = -0.15 + (-278.62) = -278.77$ kcal mole^{-1} of ethanol consumed. The addition of Eqs. 10.54 and 10.55 should equal Eq. 10.39, and does. Similarly, the addition of the enthalpy changes accompanying the reactions represented by these equations should equal that of the nonconservative reaction, and also does.

Using Eqs. 10.51, 10.52, and 10.53 $\Delta G'_{an}$, $\Delta G'_{NR}$, and $\Delta G'_{cat}$ are calculated to be 0.000, -5.60, and -264.43 kcal, respectively, the addition of which gives a value for $\Delta G'_C$ of -270.03 kcal mole^{-1} of ethanol consumed. By means of Eq. 10.51 $\Delta G'_f$(cells) is calculated to be -24.55 kcal UCFW^{-1}.

Using the Gibbs free-energy equation $T\,\Delta S'_C = (\Delta H'_C - \Delta G'_C) = -278.74 - (-270.03) = -8.71$ kcal, from which $\Delta S'_C = -8.71/298.15 = -29.2$ kcal mole^{-1} of ethanol consumed deg^{-1}. Similarly, $T\,\Delta S'_f$(cells) $= \Delta H'_f$(cells) $-$

$\Delta G'_f$(cells) = $-30.27 - (-24.55) = -5.72$ kcal, from which $\Delta S'_f$(cells) = $-5.72/298.15 = -19.2$ cal UCFW^{-1} deg^{-1}. The absolute entropy, S, for the cells is then calculated by means of Eq. 10.22 to be 27.1 cal UCFW^{-1} deg^{-1}.

Eff. I and eff. II with respect to free energy are calculated using Eqs. 10.23 and 10.24 to be 14.2% and 16.6%, respectively.

10.6. WHAT IS INDICATED BY THE AVAILABLE DATA

Appendixes 3 through 13 contain most of the available data which could be put together to give some idea of the energetic changes which take place during the process of microbial growth. These data must be interpreted with caution since, as has been continually emphasized throughout the last five chapters of this book, it must be established that it is only growth that is being studied and not growth plus fermentative or oxidative assimilation. The only investigator so far to have done this is Battley, from whose experiments the data in Appendixes 3 through 5 have been compiled. The data in the other tables may or may not be equally valid, and this will be discussed subsequently.

10.6.1. The Relationship Between the Quantity of Energy Initially Available and That Conserved During Growth

Ideally, when studying this relationship, experiments should be carried out with a given microorganism grown under optimal physiological conditions on a series of substrates having different energy contents. With these data one can then look for "regularities" of energetic behavior. These need not necessarily exist at all, but if they *are* found to exist with one microorganism it is reasonable to postulate that similar regularities ought to exist with others, and if this is not found to be the case the experimental conditions for the nonconforming microorganisms might well be reconsidered. The data in Appendix 4 with respect to *S. cerevisiae* can be used to illustrate this. Figures 10.1 and 10.2 show, respectively, a linear relationship between ΔG_C and ΔG_{NC}, and between the machine efficiency of growth (eff. I) and the quantity of energy per gram substrate. These examples represent certain regularities in microbial energetics. On the other hand, as is shown in Figs. 10.3 and 10.4, using data with respect to *A. oryzae* and *Pseudomonas saccharophila* taken from Appendixes 7 and 10, respectively, with these microorganisms no such regularities seem to exist. Does this mean that regularities do not naturally occur, or that something was wrong with the experiments? This is difficult to answer. Because of the precautions taken by Battley to assure that nothing was limiting the quantity of growth except the quantity of substrate, we can assume, at least for the sake of argument, that indeed, only growth and no oxidative assimilation was occurring with respect to his experiments. However, *suppose* that the conditions of his experiments were such that when yeast was grown aerobically on glucose, 20% of the glucose was oxidatively assimilated and the rest was used

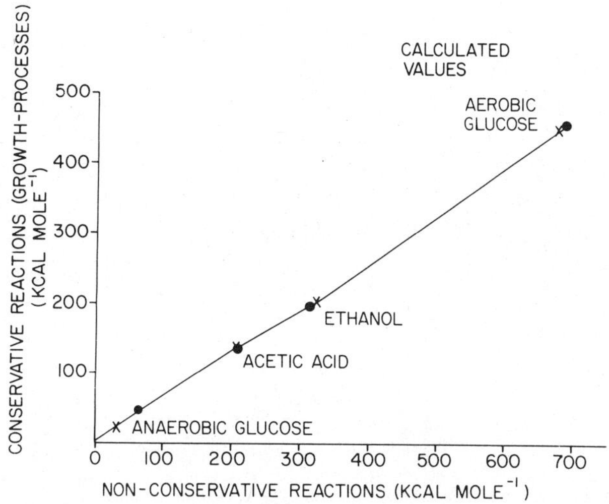

FIGURE 10.1. Graph of a calculated energy changes accompanying the conservative reactions (growth processes) during which *Saccharomyces cerevisiae* is grown on different substrates plotted against the energy changes of the relative nonconservative reactions. (—●—●—), free-energy changes; (—X—X—), enthalpy changes. Source: Data in Appendix 4.

for the growth process. Under these circumstances what would be expected to take place can be represented by the following equations.

80% growth

$$\left.\begin{array}{l} 0.800C_6H_{12}O_6(aq) \\ +\ 0.262NH_3(aq) + 3.165O_2(aq) \end{array}\right\} \longrightarrow \begin{array}{l} 1.531CH_{1.703}O_{0.459}N_{0.171}(\text{cells}) \\ +\ 3.269CO_2(aq) + 3.888H_2O(l) \end{array}$$

20% oxidative assimilation

$$0.200C_6H_{12}O_6(aq) \longrightarrow 0.200(C_6H_{10}O_5)_n + 0.200H_2O(l)$$

During this process the ash-free dry weight of the cellular material produced (including the assimilated glucose) would be 68.326 g instead of 44.916 g, an increase in weight of 52.1% over that which would occur if growth were the only process taking place! At the same time, the free-energy change accompanying growth would be $0.8 \times -457.6 = -366.1$ kcal and that accompanying the storage of glycogen would be 1.5 kcal, giving an *apparent* $\Delta G'_C$ value of -364.6 kcal mole^{-1} of glucose consumed. This is 20.3% less than would occur were growth alone to take place, and has the result of increasing eff. I from 34.1 to 47.3%. *It is thus evident that only a small amount of oxidative or fermentative assimilation can have a considerable effect on the yield of cellular material, the free-energy change of the growth process, and on the apparent efficiency of growth.* And, since efficiencies of oxidative assimilation have been

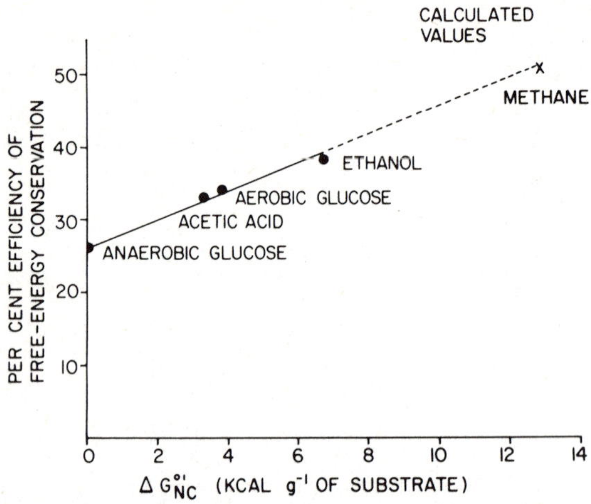

FIGURE 10.2. Graph of the percent efficiency of free-energy conservation when *Saccharomyces cerevisiae* is grown on different substrates plotted against the free-energy change accompanying the relative nonconservative reactions per unit carbon atom of the substrate. The graph is extrapolated to the most reduced form of carbon, methane. Source: Data in Appendix 4.

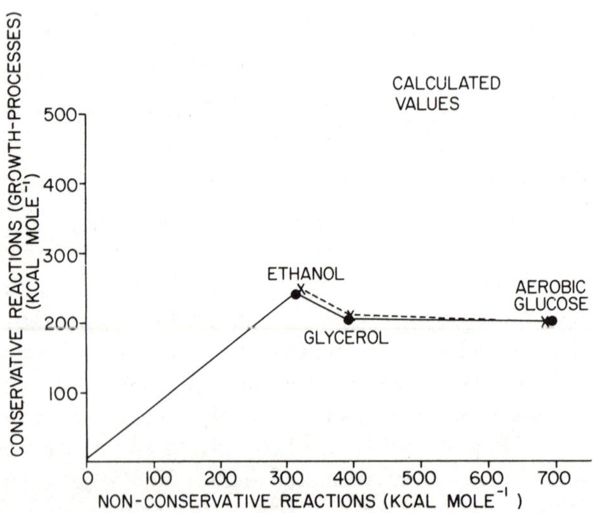

FIGURE 10.3. Graph of calculated energy changes accompanying the conservative reactions (growth processes) during which *Aspergillus oryzae* is grown on different substrates plotted against the energy changes of the relative nonconservative reactions. (—●—●—), free energy changes; (—X—X—), enthalpy changes. The nitrogen source is ammonia. Source: Data in Appendix 7.

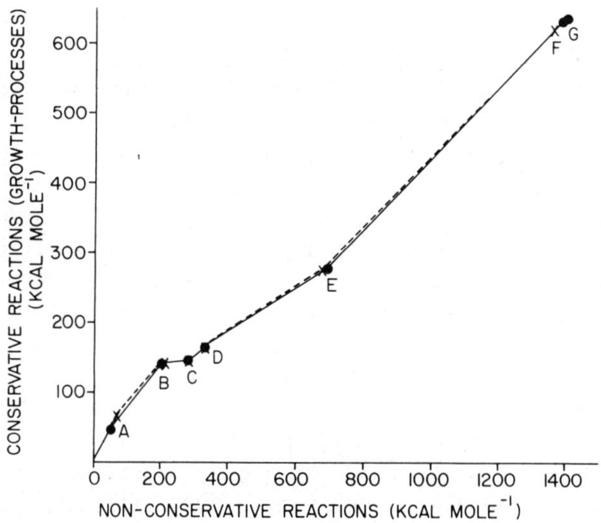

FIGURE 10.4. Graph of calculated energy changes accompanying the conservative reactions (growth processes) during which *Pseudomonas saccharophila* is grown on different substrates plotted against the energy changes of the relative nonconservative reactions. A, hydrogen gas; B, acetic acid; C, pyruvic acid; D, lactic acid; E, glucose; F, maltose; G, sucrose. (—●—●—), free energy changes; (—X—X—), enthalpy changes. Source: Data in Appendix 10.

reported as high as 66% (see Chapter 5), the necessity of viewing "growth" processes having low $\Delta G'_C$ values and high efficiencies with caution may be well founded. On the other hand, unduly high ΔG_C values may be due simply to the catabolism of the substrate with a resulting low yield of cells, due to suboptimal conditions. What, then, represents the true energetics of growth? The answer should probably be that if regularities such as can be demonstrated with *S. cerevisiae* are seen, the experimental results are likely to indicate a real energetic relationship. If such regularities cannot be demonstrated, it is impossible at the present time to determine if the energetics are truly representative of the growth process for that particular microorganism, or represent some artifact of experimentation, or an improper or incomplete definition of the system. Why, for example, is the efficiency of growth of *A. oryzae* in Appendix 7 so much less when grown on ethanol than when it is grown on glycerol or on glucose, even when considered on an energy-per-unit-carbon basis? Or, when $\Delta G'_{NC}/nC$ for acetate (-102.5 kcal) is not greatly different from that for glucose (-115.7 kcal), why is eff. I in Appendix 10 for the growth of *P. saccharophila* on acetate (31.0%) only half of that for the growth of the same microorganism under presumably identical conditions on glucose (59.9%). Somehow this does not seem to make sense. Questions such as these point out uncertainties, or irregularities which cause one to doubt the data. However, with the data on *S. cerevisiae* one can ask if the $\Delta G'_C$ for the growth of this organism anaerobically on 1 mole of glucose (-47.6 kcal) can be added to that from aerobic growth on the ethanol produced during the anaerobic growth reaction ($1.300 \times -194.9 = -253.37$ kcal) and to the $\Delta G'$ accompanying the complete oxidation of the glycerol produced during the anaerobic reaction ($0.432 \times -392.14 = -169.41$ kcal) to equal $\Delta G'_C$ accompanying its aerobic growth on glucose (-457.6 kcal mole^{-1}), and find that it very nearly can. This summation $= -47.6 + (-253.4) + (-169.4) = -470.5$ kcal, representing an agreement which is only 2.8% too high with respect to the aerobic glucose growth reaction. This agreement is excellent considering the potentialities for experimental error in that four equations involving experimentally determined fractional coefficients, seven molecules each having experimentally determined thermodynamic properties compiled from the work of many investigators, and three experimentally determined unit-carbon formulas were all put together to make this comparison. Thus, the *S. cerevisiae* set of growth processes has an internal consistency which inspires considerable confidence in the calculated energetics. These data may then quite reasonably be used as a basis from which to make various generalizations.

The graphs in Figs. 10.1 and 10.2 are good evidence that, under the appropriate experimental conditions, a definite relationship exists between the quantity of energy initially available and that conserved during a growth process. The relationship in Fig. 10.1 is described by the equation:

$$\Delta G'_C = K \, \Delta G'_{NC}$$

where K was designated by Battley[12] as a "thermodynamic growth constant," but it might better be defined as an "energetic growth constant" since in the strict sense it is not a "constant" at all and the adjective "energetic" would be more appropriate considering the variety of meanings energetics has come to have in biology. The energetic growth constant, K, can be considered a constant only in its application to a given microorganism. Although it *can* have the same value for other microorganisms, this ought not to be expected. This is because what seems to determine the energetics of growth is the composition of the cells, and it is abundantly evident that this is different for different kinds of microorganism and even, to a lesser extent, for the same microorganism grown on different substrates. Thus, the slight scatter of the points on the graph in Figure 10.1 may represent experimental error *or*, accurate data resulting from the growth of *S. cerevisiae* on different substrates with a consequent slightly different empirical cellular composition. In this latter case the "constant" would represent only a close approximation. Nevertheless, the idea of an energetic growth constant implies a degree of predictability such that, were a value for the constant to be determined by growing a microorganism under appropriate experimental conditions on one substrate, the amount of growth which would be obtained from any other substrate on which it was capable of growing could be closely estimated provided the experimental conditions were the same. It is unfortunate that Battley's data are at present the only data available with which to test the concept of an energetic growth constant which establishes a close relationship between $\Delta G'_C$ and $\Delta G'_{NC}$.

10.6.2. The Free Energy and Enthalpy of Formation of Cellular Material

The unique aspect of growth processes as opposed to more conventional chemical reactions is that one of the products of the former is cells, which are extremely complex assemblages of all sorts of substances. To what extent can we regard cells as having any kind of constancy of composition resulting in a constancy with respect to thermodynamic properties? For example, at 298.15°K and 1 atm pressure 1 mole of glucose in aqueous solution at a concentration of 0.001 *m* has an accurately known and constant free energy, enthalpy, and entropy of formation. These thermodynamic properties are the same for glucose at this temperature, pressure, and concentration no matter how this substance participates as a reactant or a product in any aqueous reaction. Can the same be said of *S. cerevisiae* cells grown under optimal physiological conditions but using different substrates? The data do not indicate a definite answer. Some investigators[15,16] have apparently assumed that microbial cells of the same kind have the same composition. Others[1,14,17] have not made this assumption and have found experimentally that the empirical composition of microbial cells does, indeed, differ somewhat depending upon the substrate on which it is grown. However, it should be remembered that techniques for preparing microbial cells for elemental analysis have never been standardized, nor have there even been many such analyses done. Even worse, where these analyses *have* been done there has seldom been any attempt to perform multiple determinations in conformity with good, chemical analytical practice, so that some degree of confidence could be placed in the data obtained. To what extent, then, are the differences in the empirical composition of cells grown on different substrates as shown in Appendixes 3, 6, and 12 real, or are they artifacts of preparation or analysis? We do not know. Nevertheless, accepting the "regularities" with respect to the relationship between $\Delta G'_C$ and $\Delta G'_{NC}$ in the growth of *S. cerevisiae* discussed in the previous section, the data in Appendix 5 are of considerable interest. These indicate that the thermodynamic properties of *S. cerevisiae* cells are different, depending on the substrate, but that when grown on the same substrate anaerobically or aerobically they are nearly the same. Thus, considering the differences both in the empirical composition and in the thermodynamic properties of one unit-carbon formula weight of cells grown on different substrates, it must be concluded that *S. cerevisiae* cells are not the same under all conditions of *growth* (in which only the fabric of the cells is being synthesized, and not this plus various storage products). The same conclusion can be reached with respect to *A. oryzae* cells if we disregard the fact that growth regularities are absent and that efficiencies of free-energy conservation are possibly high. Here, the data in Appendix 8 show that values for the free energy of formation of one unit-carbon formula weight of cells are remarkably similar when this microorganism is grown on a given substrate on either ammonia or nitrate as the nitrogen source. This is understandable if it is remembered that nitrate becomes reduced to ammonia, which then becomes the nitrogen source which is actually used. On the other hand, such similarities are not so apparent with respect to enthalpies of formation. In Appendix 12 very similar values for $\Delta G_f'$(cells) are found when *Chlorella pyrenoidosa* cells are grown heterotrophically on glucose, using either ammonia or nitrate as the nitrogen source, the same being true for values of $\Delta H_f'$(cells) under the two conditions. The data in Appendix 13 for *Rhizobium meliloti* are as good as those for *Chlorella*. This leaves the data in Appendix 11 on *Pseudomonas saccharophila*, which was calculated in the same manner as those in the other appendixes. For whatever reason the free energy and enthalpy of formation values are wide ranging and much lower than those pertaining to the other microorganisms previously discussed. One value for $\Delta H_f'$ is even positive, which is mostly unlikely. *P. saccharophila* is an extremely difficult bacterium from which to get consistent growth yields, and this may be reflected in the data in Appendixes 9–11.

We can conclude, then, that the value of $\Delta G_f'$(cells) for a given type of microbial cell is for all practical purposes constant when it is grown on a given substrate, and that the same conclusion applies to values of $\Delta H_f'$(cells). However, the values of both of these thermodynamic properties may be different with respect to growth on different substrates. The range of these differences for the microorganisms considered seems to be relatively narrow. The values for $\Delta G_f'$(cells) range from -6.73 kcal for *R. meliloti* (a bacterium) to -24.89 kcal for *A. oryzae* (a fungus) with values for *S cerevisiae* (a yeast) and *C. pyrenoidosa* (an alga) being intermediate. Values for $\Delta H_f'$(cells) range from -17.80 kcal for *S. cerevisiae* and *R. meliloti* to -30.27 kcal for *A. oryzae*. The data in these comparisons have not included those for *P. saccharophila*, which, while seemingly erratic, may indicate that the above values are in general lower for bacteria than they are for molds, yeasts, or algae. In all cases values for $\Delta H_f'$(cells) are 3–11 kcal more negative than the corresponding values for $\Delta G_f'$(cells). In this respect cells appear to exhibit much the same thermochemical behavior as do simple organic molecules.

10.6.3. The Free-Energy Change Accompanying Anabolism

The anabolism of (or growth on) substrate substances is clearly different from the catabolism of these same substances, as shown from the equations representing anabolism and catabolism in the examples given in Section 10.5. *Anabolism does not require elemental oxygen*, which should not be surprising since anabolic processes were presumably going on long before this became available through the evolutionary development of green-plant photosynthesis and the accompanying photolysis of water. Anabolism does require a source of reduced carbon in the form of an organic molecule, and of ammonium or amino nitrogen, in addition to the very small amounts of other things such as minerals, trace elements, and vitamins. If the source of reduced carbon, or substrate, is less reduced than the average degree of reduction of the substances comprising the cells, anabolism takes on the form of what is effectively an internal dismutation, whereby a part of the substrate becomes reduced to form the cells while the other part becomes oxidized to form carbon dioxide, during which process the nitrogen source becomes incorporated. This type of anabolism can be called anabolism I. It is in some respects analogous to the types of fermentations that also produce carbon dioxide and a product that is more reduced than the substrate, an example of which is alcoholic fermentation. In common with such fermentations, the sum of the available electrons of the reactants is equal to the sum of the available electrons of the products. If the substrate is more reduced than the average degree of reduction of the substances comprising the cells anabolism includes what is effectively a carbon dioxide fixation, during which carbon dioxide becomes incorporated along with the nitrogen source so that the average level of reduction of the substrate substances becomes that of the cells which are produced. This type of anabolism can be called anabolism II. Just as with anabolism I, the sum of the available electrons of the reactants is equal to that of the products. It is interesting to speculate that during evolution anabolism I may have preceded anabolism II. This is because the substrates for anabolism I must be carbohydrate-like substances capable of undergoing internal dismutations, or fermentations, resulting in the production of carbon dioxide and organic products which are more reduced than the substrate such as cells, ethanol, propionic acid, butanol, and so on. Simple molecules such as these latter cannot be fermented, and would tend to accumulate in the "primordial soup." Conceivably anabolism II evolved so as to take advantage of these more highly reduced sources of energy in the absence of oxygen, as in bacterial photosynthesis.

Since in both anabolism I and II the number of available electrons of the reactants is equal to that of the products, it is immediately apparent that the free energy changes accompanying such reactions are going to be small and may even be zero. The types of regularities shown in Fig. 10.1 and 10.2 are the results of calculations made using this hypothesis. A further test of this hypothesis would be a similar result using a different method of calculation. The results of this test are shown in Appendixes 4, 7, 8, 12, and 13, in which values for $\Delta G'_{an}$ are given based on Battley's hypothesis cited above, and on his conservation theory, which was discussed in Section 8.11. Using this latter method, the amount of free energy, enthalpy, or entropy conserved within the substance of the cells can be calculated, and from these values those for $\Delta G'_f$, $\Delta H'_f$ and $\Delta S'_f$, all per unit-carbon formula weight of cells, can be determined. These values can then be inserted into the anabolic reaction equations along with the other appropriate energies of formation for the purpose of calculating $\Delta G'_{an}$, $\Delta H'_{an}$, and $\Delta S'_{an}$. In the case of the anaerobic growth of *S. cerevisiae* on glucose (Appendix 4) $\Delta G'_{an}$ has a value of -0.37 kcal rather than zero. The aerobic growth of *P. saccharophila* on glucose and lactate gave values of -0.29 and 0.23 kcal, respectively. All the other values for $\Delta G'_{an}$ in

Appendixes 4, 7, 10, 12, and 13 are less than 0.20 kcal, and 11 are 0.00 kcal. These calculations represent a strong confirmation of the hypothesis that $\Delta G'_{\text{an}}$ values are in fact zero, or are at least *extremely* small.

Battley's[3] alternative method (see Section 9.6) for calculating values for $\Delta G'_f$(cells) can also be used to calculate values for $\Delta G'_{\text{an}}$ but is theoretically less suitable than the method using conservation theory. Nevertheless, using this former method values for $\Delta G'_{\text{an}}$ are also close to zero, strengthening the original hypothesis even further. In fact, the validity of this hypothesis can now reasonably be considered to be established.

10.6.4. The Enthalpy Change Accompanying Anabolism

The hypothesis that $\Delta G'_{\text{an}}$ equals zero cannot be directly applied with respect to enthalpy changes. For this reason, the enthalpic energy conserved within the substance of the cells must be calculated and a value for $\Delta H'_f$(cells) obtained. This value is then inserted into an anabolic reaction equation together with the appropriate $\Delta H'_f$ values for the other products and for the reactants to calculate $\Delta H'_{\text{an}}$. As the data in Appendixes 4, 7, 10, 12, and 13 show, these values, interestingly, are also very low, in fact, within about the same ranges as the calculated $\Delta G'_{\text{an}}$ values. In some cases, however, values as high as 14.64 (Appendix 12) and 30.03 (Appendix 13) kcal are obtained. Whether these latter are artifacts or real is uncertain. In general the $\Delta H'_{\text{an}}$ values are under 0.20 kcal, just as are the calculated $\Delta G'_{\text{an}}$ values and these former might well be considered also equal to zero. This would be in accord with the previous conclusion of Molliard[18] (see Section 3.10) as far back as 1922; the observations of Senez and Belaich,[19] Grangetto,[20] Boivinet,[21] and Belaich and Senez;[22] and the conclusion of Forrest [23] that the heat production accompanying microbial growth is entirely due to catabolic reactions.

10.6.5. The Entropy Change Accompanying Anabolism

In general, entropy changes are calculated by means of the Gibbs free-energy equation. Thus, in the case of anabolic reactions

$$\Delta G'_{\text{an}} = \Delta H'_{\text{an}} - T\,\Delta S'_{\text{an}}$$

from which

$$\Delta S'_{\text{an}} = \frac{\Delta H'_{\text{an}} - \Delta G'_{\text{an}}}{T}$$

Because both $\Delta G'_{\text{an}}$ and $\Delta H'_{\text{an}}$ have in general very small values, $\Delta S'_{\text{an}}$ is going to be equally small, and in those examples where both $\Delta G'_{\text{an}}$ and $\Delta H'_{\text{an}}$ have values of less than 0.200 kcal; $\Delta S'_{\text{an}}$ will be less than 1 cal UCFW^{-1} deg^{-1}.

There seems to be little point in trying to do an analysis of the data involved since such small values involve large relative errors. In addition, except for Battley's data (see Fig. 8.1) there is no reasonable certainty that it was only growth that was being studied. Thus, we can only generalize. What the data do indicate is that in spite of these uncertainties the entropy change for an anabolic reaction is close to zero, and the conclusion can be drawn that the sum of the $\Delta S_f'$ values for the products of such a reaction is very nearly equal to that of the reactants. This is an extremely interesting observation in light of the connection biologists make between entropy and information. Cells are such complex substances and perform such intricate biochemical operations that we instinctively feel that a great deal of information, and therefore negative entropy, must be incorporated within them. However, if either no, or very little, entropic change takes place when cells are synthesized from simple molecules such as glucose and ammonia, this connection between information and entropy must be meaningless, at least between information as the average biologist thinks of it and physical entropy, which is a thermal measurement. The word "information" has significance only within a certain context in which a meaning can be ascribed to it. For example, a protein molecule is made up of a sequence of amino acids to form a chain which may subsequently fold into a secondary or tertiary structure having an enzymatic activity that can be (but need not be in some cases) quite specific. Any sequence (random or otherwise) of the same number and type of amino acids will contain information, all of which will be meaningless except for the particular sequence that creates an active enzyme. However, except for what must be extremely small differences in entropy which may (or may not) be involved in the formation of secondary or tertiary structures, the physical entropy of the meaningful and meaningless sequences would be expected to be exactly the same. Thus the use of the word "information" is unfortunate. The word ought to apply only to things that have a meaning for the observer. Thus, a protein sequence has meaning, or information, *only* if it has an activity which can be observed and which conveys information to us. It will either have this activity or it will not. But it will *always* have an entropy, and given the same number and type of amino acids making up the protein, this entropy will be the same whether or not it yields a given kind of information. This same argument can be applied to the higher levels of organization of the DNA molecules of the cell. We know that a cell is capable of a great number of activities, and therefore, at one time or other during its life cycle, contains a quantity of information. Suppose, however, that we irradiate a living cell briefly with ultraviolet light or X-rays so that it is killed. The energetic changes that take place in the cell as a result of this treatment are probably so small as to be impossible to measure. However, the cell has suddenly lost all the information it possessed relative to the continuation of its life processes. Does this mean that its physical entropy has changed? The answer can only be that if it has changed, this change is so infinitesimally small that it would be difficult if not impossible to compare the entropy of the cell before and after irradiation. Thus, although a great deal of information with respect to the cell has been lost, its entropy to all intents and purposes remains the same.

What do the above comments have to do with the entropy change accompanying metabolism, which seems to be very close to zero? The initial impetus for equating information with entropy apparently originated in statistical mechanics whereby it becomes possible, by knowing something about the specific mechanical properties of molecules, to use statistical laws to calculate values for entropy as a thermodynamic property. Thus, the most general definition of statistical mechanical entropy is given by the following equation:[24]

$$S = -k\Sigma p \ln p \tag{10.56}$$

where k is the Boltzmann constant and where the p values are the probabilities of the different microstates of a system. The concept of probability involves information, which is dimensionless. The information that is useful in statistical mechanics acquires a physical meaning through the introduction of the Boltzmann constant, giving S the dimensions of cal deg^{-1}. Statistical mechanics works well with simple molecules for which translational, vibrational, rotational, and internal rotational (all of which are mechanical) entropies can be determined. Cells are effectively precipitates, and therefore the contribution of translational entropy to the total entropy of a unit mass of cells is presumably small. Alternatively, the formation of cells leads to a loss of translational entropy of the substrate molecules. It is largely the internal entropy, all types of entropy other than translational, with which we are dealing, and of these it is probably vibrational entropy which is the most important. It is entirely conceivable that the vibrational entropy between the atoms in an amino acid in solution is not very different from that in a protein incorporated in a cell, and that the internal entropies of a unit mass of small molecular weight molecules such as those comprising substrate substances for the growth of microbial cells are not greatly different from those for the same unit mass of cellular fabric. There is presently little evidence one way or the other for this. The efforts of Linschitz[25,26] and of Morowitz[27,28] to use statistical mechanics to calculate the entropy of a living cell were by their own admission at best educated guesses. On the other hand, the entropy change accompanying an anabolic reaction would be expected to be very small, which is what is observed (see, however, Section 10.6.7).

10.6.6. A Comparison of the Physical Entropy of Cells and of Simple Organic Substances

The ability to calculate values for $\Delta G_f'$(cells) and $\Delta H_f'$(cells), both per unit-carbon formula weight, makes it possible to calculate a value for $\Delta S_f'$(cells) and from this and the unit-carbon formula to use Eq. 10.22 to calculate a value for the absolute entropy of one unit-carbon formula weight of cells. Dividing this value by the unit-carbon formula weight gives the absolute entropy per gram dry weight of cells. Exactly what such values mean is conjectural. We think of the cells in terms of unit-carbon formula weights, or grams dry weight of substance, but this does not represent living cells. Were we actually to put a unit quantity of dried cells into a low-temperature calorimeter and determine thermally the absolute entropy of this material (and this has never been done), we would likely find this value to be different from the same value calculated in the above manner. This is because in dealing with microbial growth in terms of growth-process equations, we are using the unit-carbon formulas for the cells solely to obtain a mass balance. No matter how the cells are represented the implication is that data obtained for them apply to hydrated cells in the living state and not to cells that have been dried. The thermodynamic changes that would accompany drying the cells for an entropy measurement are not presently known, but these changes would have to be added to this latter to obtain the absolute entropy of the living cells. The use of growth-process equations is presently our only method of gaining some insight into cellular entropy.

It is of interest to compare the thermodynamic properties of microbial cells with those of amino acids on a unit-carbon basis. This is done in Appendix 14. The data are merely suggestive. It is apparent that for glycine, alanine, valine, and leucine, all of which have only one carboxyl group, the values for $\Delta G_f^{o\prime}$, $\Delta H_f^{o\prime}$, and $S^{o\prime}$ all decrease with an increasingly long carbon skeleton.The same trend holds for dicarboxylic L-aspartic and L-glutamic acids, except that values for the thermodynamic properties are all higher than those for monocarboxylic amino acids of the same chain length. Valine and leucine have AE/nC values of 5.000 and 4.800, respectively, both of which are reasonably close to the AE_c values for a unit-carbon formula weight of cells. On this same basis of comparison, the $\Delta G_f^{0\prime}$ and $\Delta H_f^{0\prime}$ values for these amino acids fall within the range of the same values calculated for the cells. In general, the calculated values of S for the cells are higher than those for these two amino acids except for *S. cerevisiae* grown anaerobically or aerobically on glucose (Appendix 5) and for *C. pyrenoidosa* and *R. meliloti* (Appendixes 12 and 13, respectively). These data suggest that microbial cells may have thermodynamic properties on a unit-carbon (or other) basis which are not markedly different from those of the simple molecules of the substrates on which they are grown, but they are far too limited to enable us to arrive at any firm conclusions. Other data are more conclusive. *If* the number of available electrons in the substrate (carbon and energy source plus the nitrogen source) is the same as that in the cells plus the other products of anabolism (which it is), and *if* the free-energy change accompanying anabolism is zero (which it seems to be), and *if* the enthalpy change accompanying anabolism is very small (which it demonstrably is, since the heat produced during growth is, within experimental error, equal to that of catabolism), *then* the entropy change accompanying anabolism must be *extremely* small, and $\Sigma\ \Delta S_f'$ (cells plus products of anabolism) can be considered equal to $\Sigma\ \Delta S$ (substrate plus nitrogen source).

10.6.7. Cells and Information

The data and considerations in Section 10.6.6 appear to be contradictory to common thought with respect to how we regard the complexity of cells. Surely, the experimental act of growing microbial cells from simple substrate molecules in solution the information content of which would seem to be small by comparison with the cells, would result in a considerable decrease in entropy. In fact, there is apparently no change in entropy, at least with respect to the process of anabolism, so that, as stated previously, $\Sigma\ \Delta S_f'$ (cells plus products) equals $\Sigma\ \Delta S_f'$ (substrate plus nitrogen source). But this tells us nothing directly with respect to "information," since the concept of this latter is based on absolute entropy and its relation to the probabilities of the microstates of a system and not on a similar relationship with respect to the entropy of formation.

There are two separate probabilities associated with anabolism. One is the probability of creating a cell, and the other is the probability of reproducing it. It is only with this latter that we must be concerned. The reason it is not necessary to consider the former may be the same that Duclaux had in mind in 1896 with respect to the Gay-Lussac equation for alcoholic fermentation when he wrote:

> About 1850, anyone wishing to obtain some idea of the extent to which the equation for alcoholic fermentation had become accepted would have had the right to be completely sceptical on this subject, above all, if he had asked why all the chemists who were themselves occupied with the question [of alcoholic fermentation] passed obstinately in silence over the yeast that Lavoisier had been obliged to add in order to ferment his sugar, and without which it was impossible to obtain a fermentation. Why had this yeast, that was so necessary for the experimentation, disappeared from his interpretation? [Duclaux,[24] p. 78.]

When one is concerned with a particular aspect of microbial growth that is reflected in the final state of a growth-process system, it is sometimes easy to forget that it was initially necessary to inoculate the microorganism. The improbable act of the formation of the cells has already taken place *before* the experiment is begun and the cells forming the inoculum are placed in a nutrient solution. If we quite reasonably accept a single, "average" microbial cell (averaging out the information content during a life cycle) as being the basic unit of "information," then the information content of a cell serving as an inoculum would not be expected to be different from that of an "average" cell that had been grown in the culture medium. This being so, the number of microstates associated with such cells and therefore their "probabilities" would be expected to be the same both before and after an experiment involving growth takes place, provided, of course, that healthy cells are inoculated. Effectively, the probabilities of inoculum cells and grown cells are the same (whatever they are), and there is therefore no difference in entropy between them. This is not an unreasonable conclusion. On the other hand, cells may be regarded as very complex, self-reproducing catalysts which are also precipitates, and there might well be expected to be a change in entropy associated with reproduction of inoculum cells, during which substances in solution become condensed into cellular fabric. Because of this latter, less random (supposedly) condition, anabolism would be expected to result in a cellular biomass having a lower absolute entropy than that of the substrate molecules from which it was synthesized. This can be demonstrated using Eq. 10.31 representing the formation of yeast cells from glucose and ammonia, repeated below for ease of reference.

$$\left.\begin{array}{l} 0.341C_6H_{12}O_6(aq) \\ +\ 0.327NH_3(aq) \end{array}\right\} \longrightarrow \begin{array}{l} 1.914CH_{1.703}O_{0.459}N_{0.171}(\text{cells}) \\ +\ 0.132CO_2(aq) + 0.907H_2O(1) \end{array} \qquad (10.31)$$

Because the entropy change accompanying the above process is zero, it becomes possible to calculate a $\Delta S_f'$ value for the cells (biomass). This is found from Appendix 5 to be -33.3 cal UCFW^{-1} deg^{-1}. This value can be used with Eq. 10.22 to calculate that the absolute entropy of the same mass of cells is 10.10 cal UCFW^{-1} deg^{-1} which, when multiplied by 1.914 gives 19.33 cal (g total biomass)$^{-1}$ deg^{-1}. Using the value from Appendix 2 of -265.67 cal mole^{-1} deg^{-1} for $\Delta S_f'$ of α-D-glucose and inserting this into Eq. 10.22 gives 76.95 cal mole^{-1} deg^{-1} as its absolute entropy in the "biological" 0.001 *m* standard state. Multiplying this value by 0.341 gives 26.24 cal (g total substrate utilized)$^{-1}$ deg^{-1} as the absolute entropy for the quantity of glucose consumed in the anabolic process represented by Eq. 10.31. It is apparent that there is less entropy associated with the biomass than is associated with the substrate from which it is formed, which is a satisfying result if we feel that a lesser entropy is associated with the higher degree of organization we ascribe to cells. However, such a comparison may be more apparent than real, since the mass of the substrate (61.38 g) is not equal to that of the biomass (44.91 g), being more oxidized and not containing nitrogen. A more appropriate comparison could be made on the basis of unit masses of substrate and biomass, that is, on an absolute entropy per gram basis, whereupon 19.33 cal (44.91 g biomass)$^{-1}$ deg^{-1} becomes 0.43 cal (g biomass)$^{-1}$ deg^{-1}, and 26.24 cal (61.38 g substrate)$^{-1}$ deg^{-1} becomes 0.43 cal (g substrate)$^{-1}$ deg^{-1}. Thus, the absolute entropies per unit mass of substrate and of biomass are exactly the same, and if we equate entropy in any way with information, it can only be concluded that per unit mass, the information content of the cells is exactly the same as that of the substrate from which they are synthesized. The identity of the absolute entropy per unit mass of cells and of substrate may be correlated with the identity of the number of available electrons in the cells and in the quantity of substrate from which the cells are formed, but if so, this relationship is presently conjectural.

The function of the cell in anabolism is apparently that of an extremely complex, self-reproducing catalyst which serves as a template for the construction of another cell. The process by which this takes place involves no net change in free-energy, and results in a product which is the same in its information content per unit mass as that of the substrate from which it is formed.

10.6.8. Cells and Negative Entropy

From what has been discussed previously, it appears that the changes in thermodynamic properties accompanying anabolism are all zero, including entropy. It is thus difficult to believe that cells and negative entropy are in any way related, at least with respect to anabolism.

10.6.9. The Free-Energy Changes of Growth Processes

From an inspection of the examples in Section 10.5 and from the foregoing discussions it is apparent that very little, if any, energy change accompanies anabolic reactions as they are represented, and that virtually all of the free-energy conversion into enthalpy and entropy that occurs during a growth process results from catabolic reactions, whether they are anaerobic and fermentative or aerobic and oxidative. This conversion is accompanied by a conservative process involving the transient participation of ATP or similar substances, the energy of which becomes utilized, but not incorporated, in constructing the fabric of the microbial cells which are being grown. An anabolic reaction cannot proceed by itself because $\Delta G'_{\text{an}}$ is zero. These are therefore not spontaneous reactions. On the other hand, they do not require free energy to be put into them, that is, they are not endergonic. What they do require is for work (in the sense of something acting on something else) to be done such that substances serving as substrates become incorporated into cells with the result that the number of available electrons in the latter is the same as in the former, and that $\Sigma\ \Delta G'_f$ (products) is equal to $\Sigma\ \Delta G'_f$ (reactants). The energy for this work is provided by ATP, as described in the following section.

10.6.10. ATP and Growth

It must be remembered that the treatment of the energetics of microbial growth in this book involves the use of growth-process equations that can be partitioned into anabolic and catabolic reaction equations, all of which represent *net processes.* None of these represents an actual reaction. Thus, the treatment is more conceptual than real. The reason it can be used is because we can apply Hess's Law in adding up the energy changes which accompany these processes as they are represented.

It has long been known that most substrate molecules are unsuitable for direct incorporation into cellular substance, and that substrates or parts of them must be activated by the addition of other substances such as phosphate groups or coenzymes. These activated substances then interact with others in reactions whereby syntheses take place, the free energy of activation then becoming lost as heat and entropy. In 1960 Bauchop and Elsden[30] came to the conclusion that the amount of microbial growth was proportional to the amount of ATP that could be generated from the degradation of the substrate. This led to the concept of the Y_{ATP}, representing the yield in grams dry weight of microorganism produced per mole of ATP generated. They originally felt that the Y_{ATP} was a constant, having a value of about 10. However, subsequent work by a great many investigators showed that Y_{ATP} values were different for different microorganisms, and that they can even vary with respect to a given organism depending on the conditions of growth (see the monograph by Stouthamer[31]). The difficulty with Y_{ATP} studies is the determination of how many moles of ATP are generated per mole of substrate catabolized. This is generally calculated on the basis of the biochemical pathways involved in the catabolic degradation of the substrate. For anaerobic growth, the catabolic pathways always involve substrate-level phosphorylation and the number of moles of ATP generated per mole of substrate catabolized can often be determined with reasonable certainty. Similar values for aerobic growth reactions, where considerable uncertainties may exist with respect to the number of oxidative phosphorylations resulting in the production of ATP molecules, are correspondingly uncertain.

It has been well established that alcoholic fermentation results in the net production of 2 moles of ATP per mole of glucose fermented. This does not mean, however, that if the molar growth yield is 15.349 g dry weight of *S. cerevisiae* cells grown anaerobically on glucose (from Section 10.5.1), that 2 moles of ATP have been generated. Some of the substrate has to have been utilized for the construction of the cells. What, then, is the energetics of ATP utilization in this particular growth process? This is represented in Appendix 15. It is apparent that the free-energy changes associated with ATP synthesis and utilization cancel out. The addition of the $\Delta G'$ values for catabolism coupled to ATP synthesis, for anabolism coupled to ATP utilization, and for the production of glycerol gives $-25.81 + (-16.25) + (-5.59) = -47.65$ kcal mole^{-1} as the $\Delta G'_C$ for the complete process. The calculated Y_{ATP} value with respect to anaerobic growth on glucose is 15.349 g (not ash-free)/1.3 = 11.8. This value is consistent with published Y_{ATP} values for *Saccharomyces* sp.

The ATP utilization energetics for the growth of *S. cerevisiae* aerobically on glucose is shown in Appendix 16. It is again apparent that the free-energy changes associated with ATP synthesis and utilization cancel out. The addition of the $\Delta G'$ values for catabolism coupled to ATP synthesis and anabolism coupled to ATP utilization gives $-226.99 + (-230.62) = -457.61$ kcal mole^{-1} as the $\Delta G'_C$ for the complete growth process. Assuming a P/O ratio of 2, the quantity of ATP formed during growth on 1 mole of glucose is 18.45 moles. The calculated Y_{ATP} value with respect to aerobic growth on glucose is therefore 49.250 g (not ash-free)/18.45 = 2.67 g mole^{-1} of ATP. The reason for the characteristically lower Y_{ATP} values obtained for aerobic growth using

P/O ratios of 1, 2, or 3 has never been satisfactorily explained. This may be because the concept of a Y_{ATP}, interesting and plausible though it seemed at its inception, is no longer tenable as it was originally conceived.

There appear to be two forms of energy conservation taking place during a microbial growth process. One is the conservation of free energy in ATP. The other is the conservation of free energy in the cells or other products of the growth process. To what extent is one related to the other? As represented in Appendixes 15 and 16 the free energy from ATP conserved during catabolism is utilized to incorporate substrate substances into cells during anabolism. The free energy in the substrate substances is *materially separated* from that in the ATP (i.e., the substrate substances are molecules that are different from ATP molecules). The free energy in these latter, upon being utilized, becomes converted into heat and entropy. This does *not* happen in the former, since the free energy accompanying anabolism is zero. The question must then be raised as to whether more energy can be conserved during anabolism than is expended (i.e., converted into heat and entropy) as a result of coupling anabolism to the hydrolysis of ATP, without which anabolism could not occur. Or, put another way, is the quantity of free energy conserved during anabolism more than, equal to, or less than the quantity of free energy conserved in ATP during catabolism? Without some understanding of this the whole concept of a Y_{ATP} begins to lose meaning, since it is not only the number of ATP molecules involved during growth per unit mass of cells which are grown, but the energies conserved in each and how these are related.

Accepting Burton's[32] value of -12.5 kcal mole^{-1} for the hydrolysis of ATP to ADP and P_i under physiological conditions, the quantity of energy conserved in ATP during the anaerobic growth of *S. cerevisiae* on glucose using alcoholic fermentation as the catabolic reaction could not be more than $(1.3 \times -12.5 \times 100)/-64.7 = 25.1\%$. The quantity of free energy conserved during this particular growth process is calculated to be 26.4%. All things considered, this is good agreement, and leads to the hypothesis that as an upper limit, *as much* free energy can be conserved in the cells as has been conserved in ATP. Suppose that the hypothesis is made that the amount of energy conserved during growth cannot be *more* than is utilized from ATP in bringing about this conservation. Whatever the *in vivo* value for ATP hydrolysis, the amount of energy conserved in this growth-process system is $0.264 \times 64.7 = -17.08$ kcal mole^{-1} of glucose consumed. Since 1.3 moles of ATP are generated during catabolism and *if* more free energy cannot be conserved during anabolism than is utilized from ATP hydrolysis in causing anabolism to take place, *then* the maximum value of $\Delta G'$ for *in vivo* hydrolysis becomes -17.08 kcal conserved/1.3 moles ATP utilized $= -13.1$ kcal mole^{-1} of ATP hydrolyzed. This is not an unlikely figure considering the low concentrations of reactants and products of this reaction as they are thought to exist in microbial cells, and quite close to that of -12.5 kcal mole^{-1} calculated by Burton. With respect to aerobic growth on glucose, using the data in Appendix 16 the quantity of free energy conserved is $-694.4 - (-457.6) = -236.8$ kcal mole^{-1} of glucose consumed which, when divided by 18.45 (assuming a P/O ratio of 2) moles of ATP gives −12.8 kcal as the maximum value for the free-energy change accompanying ATP hydrolysis if the quantity of free energy conserved during anabolism cannot be more than that conserved in ATP during catabolism. Considering the uncertainties involved, this value is in good agreement with that of −13.1 kcal $mole^{-1}$ obtained by a similar method of calculation with respect to anaerobic growth on glucose. Similarly, when calculations are made using the same P/O ratio with respect to the growth of *S. cerevisiae* on 1 mole of ethanol or acetate, the values obtained are −12.9 and −12.7 kcal $mole^{-1}$ of ATP hydrolyzed, respectively. This represents an excellent agreement, indicating that the values for *in vivo* ATP hydrolysis are constant for the same microorganism. It therefore must be concluded that the efficiency with which ATP energy is utilized during anabolism has no direct relationship to the efficiency with which free energy is conserved during the same process. This is understandable if, as was suggested by Forrest and Walker (see Section 9.4.1), ". . . only a small fraction of the energy available from catabolism is actually incorporated into the organisms as energies of chemical bonds." (Forrest and Walker,[33] p. 266.) It might be safer to state that *no net ATP energy becomes incorporated at all into cellular fabric.* Such incorporation cannot occur if the free-energy change accompanying anabolism is zero. This tends to confirm the idea that the two forms of energy conservation are quite separate.

The preceding discussion notwithstanding, the concept of a Y_{ATP} must still be valid, even though there does not appear to be a constant relationship between the total amount of ATP energy conserved during catabolism and the amount of microbial growth. The energy of ATP or other similar molecules is certainly required to bring about the growth of microorganisms. Also, it has been clearly demonstrated that the fabric of a given microorganism has nearly, although not necessarily exactly, the same empirical composition no matter on what substrate it is grown. It is thus reasonable to assume that a nearly constant quantity of ATP must be required for the synthesis of a unit mass of a given microorganism no matter on what substrate it is grown, even though the energy of this ATP does not become incorporated into the substance of cellular fabric. If this is true, then the quantity of ATP required for microbial synthesis may be less than that generated through catabolism. Recognizing this, a differentiation can be made between the "true" Y_{ATP} represented by the former and the "catabolic" Y_{ATP} represented by the latter, which is what has always been studied. What the "true" Y_{ATP} is for given microorganisms has not yet been determined; however, since ATP energy is essentially activation energy the hypothesis can be made that the former is related to the latter. Such a hypothesis is testable, the most simple relationship being represented by the following equation,

$$\text{"true" } Y_{ATP} = \frac{Y_{af} \times (\Delta G' \text{ for ATP hydrolysis})}{E}$$

where E is the activation energy for the growth of a given microorganism as determined by an Arrhenius plot. Although Y has been traditionally determined by simply washing and drying the cells to constant weight, it might be more accurate to express Y in terms of g ash-free dry wt/moles of substrate consumed.

10.6.11. Efficiencies of Microbial Growth Processes

With respect to the data in Appendix 4 it is apparent that no eff. I value is greater than 38% or less than 26%. This is not true for most of the other efficiency values in Appendixes 7, 10, 12 and 13. The difference *may* be due to the data in Appendix 4 being from experiments in which there was a reasonable certainty that it was only growth that was taking place, and not growth plus primary fermentative or oxidative assimilation to produce storage products. This was not established for the experiments from which the data in these other appendixes were taken. The only data presently available having any predictive value are those in Fig. 10.2, from which it is evident that the percent efficiency of free-energy conservation during growth is directly proportional to $\Delta G'_{NC}\ g^{-1}$ of substrate, or to the overall degree of reduction of the carbon atoms in the substrate. According to Fig. 10.2, if the graph is extrapolated to $\Delta G'_{NC}\ g^{-1}$ of methane, which is the most reduced form of carbon, the maximal machine efficiency of free-energy conservation cannot be greater than about 50%. This has the satisfying aspect about it that more energy cannot be conserved in the fabric of the cells than is expended during catabolism to bring about the process of anabolism. But, this may be illusory. Whether this kind of limitation of efficiency also applies to other kinds of microorganisms is not presently known. Bacteria may turn out to be more efficient.

10.6.12. Maintenance Energy

With respect to microorganisms which are grown *exponentially* in an environment which is *nonlimiting* with respect to everything except the quantity of substrate, it appears that the free-energy change accompanying anabolism is zero and that all the catabolized free-energy becomes converted into heat and entropy. Is any of this later used for maintenance? The answer is that it is probably not. This is a satisfying answer in that intuitively it ought not to be necessary to maintain a structure that was in the process of being built. But, what happens when the source of energy for cells which have been exponentially grown under nonlimiting conditions becomes exhausted? There are several indications that under these conditions no storage products are produced that could enable the cell to maintain itself. Thus, there is no further oxygen uptake, no further increase in optical density, no further heat output, and so on. What appears to happen is that if the environmental conditions are not hostile, and if autolysis does not occur, the synthetic machinery of the cell simply comes to a stop. This has been demonstrated by the experiments of many investigators. Among these are two investigations which are notable in that the microorganisms were not lyophilized or dried, but were preserved for extended periods of time in aqueous suspension under conditions closely approaching those of the natural environment.

In 1961 J. E. DeVay and W. C. Schnathorst[34] isolated more than 100 strains of *Pseudomonas syringae* from soil and from diseased plant tissue. Loopfuls of cells from single colonies were then transferred to 5 ml of sterile, distilled water in screw-capped tubes, dispersed in water, and stored at 10°C. In 1965 an identical procedure was used with *Agrobacterium tumefaciens.* These same suspensions were subcultured in 1985 by Iacobellis and DeVay[35] with the result that 92% of the *P. syringae* isolates and 90% of the *A. tumefaciens* isolates were shown to be viable after 24 and 20 years, respectively. This must surely represent an "infinity" of time relative to that of the normal life cycles of these microorganisms. The simplest conclusion can only be that they must have used up in a short time (weeks to months) whatever materials were available within their protoplasts as sources of energy, after which no energy for maintenance was available, or required, to maintain the viability of 90–92% of the cells in the suspensions. An important aspect of this investigation is that these suspensions must have been continually in contact with dissolved oxygen in the tubes, indicating that at least for the microorganisms used a reducing atmosphere is not required for their preservation over long periods of time.

Of even greater interest is an unpublished investigation that was carried out at the Laboratorium voor Microbiologie, Technische Hoogeschool in Delft, Holland by J. van der Toorn, who has kindly given his permission[36] to comment on his results. In 1924–1925 a series of glass ampoules containing 3% NaCl, 10% $CaCO_3$, and either fish extract water or peptone water was inoculated with luminescent bacteria by either C. B. Van Niel or L. E. den Dooren de Jong, who were at that time assistants in the laboratory under Professor A. J. Kluyver. After 2 days of aerobic growth the ampoules were hermetically sealed with a flame, after which they were kept at room temperature, something that fluctuated considerably from winter to summer in older Dutch laboratories prior to the 1960s. The author (EHB) remembers having been shown these tubes when he was working at the laboratory in 1956, at which time they were more or less regarded as museum pieces. When van der Toorn became curator of the culture collection at the laboratory he inherited the tubes, many of which he opened in 1979 and inoculated into "photobacterium broth" (Difco) to determine if any of the cultures were still viable. No growth was obtained from any of the cultures that were originally inoculated into fish extract water medium. However, one culture isolated by M. W. Beijerinck in 1914 and called *Photobacterium splendidum,* which was inoculated in 1925 into ampoules containing peptone water rather than fish extract water, did grow after 54 years in aqueous suspension at room temperature. This is most remarkable. What seems to have happened is that is that the bacteria continued to grow in the hermetically sealed tubes until the oxygen became depleted, after which their aerobic source of energy became exhausted. However, according to van der Toorn, Beijerinck's *Photobacterium splendidum* is now recognized as being a

Vibrio sp., and as such possibly capable of facultative growth. Thus, it could conceivably have grown anaerobically for a short time by utilizing whatever fermentable substances happened to be in the peptone water. In either case, these activities could not be expected to have gone on for long (again, weeks to months), and the condition would eventually arise wherein the quantity of energy available for maintenance would have fallen asymptotically to zero, including any stored within the cells in whatever form it could be used for this purpose. Although undoubtedly some cells must have died during the following 54 years, for those that survived the most simple conclusion is, again, that no energy for maintenance was available or required. The protoplasmic machinery in the cells evidently slows down under these conditions until it comes into some state of energetic equilibrium with the environment inside the tubes, and then it stops, because the $\Delta G'$ for all the reactions within each cell drops to zero. Upon subculture the cellular machinery slowly begins to move, and eventually the cells once more return to a metabolically active state during which they can grow and reproduce.

It should be pointed out that with respect to the two investigations cited above none of the bacteria used were capable of forming spores, so that they must have survived suspension for extremely long periods of time in an aqueous medium in the absence of an energy source as vegetative cells.

The example of the luminescent bacteria which has just been described, and undoubtedly many other similar examples in which microorganisms have been shown to survive in liquid culture long after the substrate has become exhausted, represent special cases in which the environment was not hostile. What this demonstrates is that *the maintenance of the fabric of microbial cells is not intrinsically necessary*. On the other hand, under conditions in which the environment is hostile, such as too acid or alkaline an aqueous medium, too high a temperature, too high or low an osmotic pressure, and so on, it is common knowledge that after the source of energy runs out microbial cells do not long survive. Obviously, in these cases, energy in addition to that required to bring about growth is necessary to enable the microorganism to protect itself against its environment. Thus, microbial cells must be provided with that quantity of activation energy necessary for the synthesis of a given mass of cells plus what maintenance energy is required depending on the vicissitudes of the environment. This can be represented under any given environmental conditions by the following equation.

$$\Delta G'_{\text{cat}} = \Delta G'_A + \Delta G'_M \tag{10.57}$$

where $\Delta G'_{\text{cat}}$ represents the total amount of substrate free energy utilized during catabolism, $\Delta G'_A$ the activation energy required for anabolism, and $\Delta G'_M$ the energy of maintenance, all in kcal mole^{-1} of substrate utilized during metabolism. Included within $\Delta G'_A$ is all the free energy required to transport substrate materials into the cells from the environment and to incorporate these within the fabric of the cells, together with all the ancillary processes required for this to occur. The inclusion of transport energy in $\Delta G'_A$ is analogous to including the cost of transporting building materials to a construction site as a part of the total cost of constructing a building. Also included is the energy required to assemble the cellular fabric, which is analogous to the mechanical and manual labor required to construct a building. Other analogies can also be imagined, but do not include an energy for maintenance while the building is under construction provided that the construction is not limited in any way by an adverse environment. A subtropical climate would emulate these conditions. On the other hand, if the construction had to be done in a temperate climate in the middle of winter some maintenance energy would be required at the same time that a building was being constructed. Parts of the building would have to be enclosed as they were framed in so as to provide a heated and lighted space for workmen to function more efficiently. Such a condition is analogous to microbial growth under limiting conditions, when maintenance energy is required as well as activation energy. In the building trade if only a certain amount of money is available for construction, the necessity for using some of the money for maintenance will result in the construction of a smaller building. The same applies with respect to the energetics of microbial growth. Cost overrides are not allowed in nature. Included in $\Delta G'_M$ is all the energy required to maintain the cells in a viable, reproductive state when the environmental conditions are limiting, that is, when the specific growth rate, μ, is less than the maximum specific growth rate, μ_{max}, which occurs when the environmental conditions are nonlimiting in any way. Under these latter conditions $\mu = \mu_{max}$, and no maintenance energy is necessary. At least, so it would seem.

Assuming a constant "true" Y_{ATP}, the molar growth yield is related to the molar activation energy for growth in the following manner.

$$Y \propto \Delta G'_A$$

from which

$$Y = k\,\Delta G'_A \tag{10.58}$$

where k is a constant having the dimensions of g mole kcal^{-1}. From Eq. 10.58, $\Delta G'_A = Y/k$. Inserting this into Eq. 10.57 gives

$$\Delta G'_{cat} = \frac{Y}{k} + \Delta G'_M \tag{10.59}$$

From Eq. 10.59 it is apparent that with respect to any given quantity of catabolic energy that is available, the more energy that is required for maintenance, the less can be used for growth. This is what is actually observed on either side of any given optimum physiological condition for growth. For example, at the optimum pH, all other conditions being optimal, a given microorganism will grow at its maximum intrinsic growth rate. On either side of this pH optimum the growth rate will decrease until at the pH limits for growth, no growth will occur at all, and all the catabolic energy will be diverted to maintenance. The entirely plausible hypothesis can be made that when a microorganism is growing at its maximum intrinsic growth rate, μ_{max}, no maintenance energy is necessary and therefore a maximum amount of catabolic energy is available for use as activation energy. Under these optimum, nonlimiting environmental conditions, whether all of the energy from catabolism is utilized as activation energy or only part of it, the rest simply being wasted, is not clear at present. One representation of the ways in which free energy can be utilized during the growth process as a function of the hostility of the environmental conditions is given in Fig. 10.5. The percent "hostility" in Fig. 10.5 is impossible to determine directly since it has no dimensions. It is used to represent possibly several measurable parameters which vary with the degree of hostility, or the degree of limitation of the environment. One of these is the percent molar growth yield with 100% = Y_{max}. Another is the percent specific growth rate, with 100% = μ_{max}. According to Fig. 10.5,

$$\Delta G'_{NC} = \Delta G'_{con} + \Delta G'_A + \Delta G'_M$$

and

$$\% \ \Delta G'_{con} + \% \ \Delta G'_A + \% \ \Delta G'_M = 100\% \ \Delta G'_{NC} \qquad (10.60)$$

where $\Delta G'_{con}$ represents the free energy conserved during anabolism and

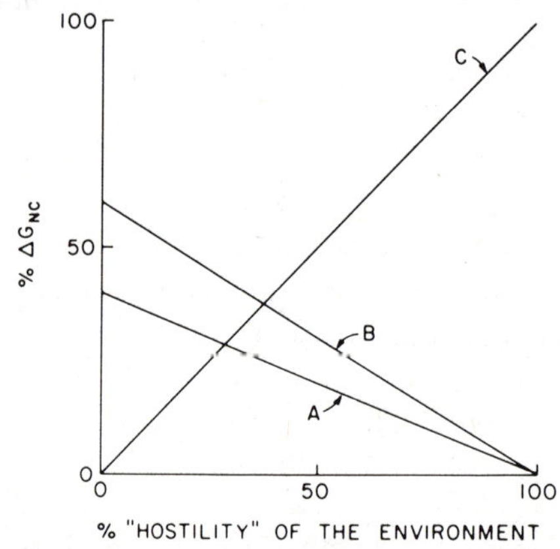

FIGURE 10.5. Graph of the percent free energy available from a given nonconservative reaction plotted against the percent "hostility" of the environment. A, free energy conserved during the growth process; B, free energy utilized for activation energy during the growth process; C, free energy utilized for maintenance.

$\Delta G'_{\mathrm{NC}}$ the free-energy change accompanying the relative nonconservative reaction. Rearranging Eq. 10.60 results in the following equation.

$$\% \ \Delta G'_M = 100\% \ \Delta G'_{\mathrm{NC}} - \% \ \Delta G'_{\mathrm{con}} - \% \ \Delta G'_A \qquad (10.61)$$

but,

$$\frac{\% \ \Delta G'_{\mathrm{con}}}{\% \ \Delta G'_A} = k = \frac{\text{eff. II}}{100} \qquad (10.62)$$

where k is a dimensionless constant which is numerically equal to eff. II/100 (see Section 10.4), that is, under nonhostile, nonlimiting conditions the maintenance energy = 0 (see Fig. 10.5) and $\Delta G'_A = \Delta G'_{\mathrm{cat}}$. From Eq. 10.62 $\% \ \Delta G'_A = \% \ \Delta G'_{\mathrm{con}}/k$, and substituting this value in Eq. 10.61 gives

$$\% \ \Delta G'_M = 100\% \ \Delta G'_{\mathrm{NC}} - \% \ \Delta G'_{\mathrm{con}} - \frac{\% \ \Delta G'_{\mathrm{con}}}{k}$$

$\Delta G^{\circ\prime}_{\mathrm{NC}}$ and $\Delta G'_{\mathrm{con}}$ are values that can always be determined for any given, definable growth-process system whether or not the environmental conditions for growth are optimum. Thus, provided that a value for constant k in Eq. 10.62 can be determined under optimum, nonlimiting environmental conditions, a value for $\Delta G'_M$ can also be determined for growth under environmental conditions that are not optimum and therefore limiting and hostile. It should be emphasized that Fig. 10.5 may well represent an oversimplification in that A, B, and C may not be linear. This is because as environmental conditions change from optimum to hostile, the metabolism of a microorganism may also change and with it the system being studied. This is not likely to happen with aerobic organisms which catabolize their substrates completely to carbon dioxide and water, but may well happen with fermentations. These latter cannot be represented simply and would require careful analyses. In addition, the value for the ratio $\Delta G'_{\mathrm{con}}/\Delta G'_A$ under optimum (nonlimiting) conditions, or k in Eq. 10.62, might well be expected to be constant for a given microorganism but ought not to be expected to be identical for different microorganisms. In any case, the sum of $\Delta G'_{\mathrm{con}}$, $\Delta G'_A$, and $\Delta G'_M$ would always be expected to equal $\Delta G'_{\mathrm{NC}}$. The main point Fig. 10.5 is intended to illustrate is that the proportions of total free energy conserved during growth and needed for maintenance are not constant, but vary with the condition of the environment.

The concepts of maintenance energy presented here tend to fly in the face of conventional wisdom in that they are different from those introduced by Pirt[37] in 1965 and modified by him[38] in 1982. The presently described approach to calculating maintenance energy is essentially energetic (thermodynamic), whereas that of Pirt is essentially kinetic. Since it is not the purpose of this book to describe in detail the energetics of cell maintenance, this subject will not be treated further. Suffice it to state that if the two methods of calculating maintenance energy are equally valid, they should both give the same answers.

10.6.13. The Energetics of Microbial Growth and Evolution

In Section 1.20 the concept of the classical, thermodynamic "reversible" process is described as one by which a system passes from one thermodynamic state to another while still remaining in equilibrium with any tendency which may exist to oppose it, such that if this tendency is infinitesimally increased the process will be reversed, and vice versa. Such systems are characterized by a $\Delta G^{\circ\prime}$ which is equal to zero. Surely something very close to this condition of reversibility must have existed during the primitive conditions on this planet envisioned by Oparin[39] when chemical structural entities were forming, eventually resulting in the synthesis of highly complex, self-reproducing, organic catalysts that might be called the origin of Life. This was the "primeval soup," consisting of many different types and quantities of organic molecules in aqueous solution, all ultimately synthesized from inorganic substances under the influence of heat and radiation. Once formed, these must have begun to collide and interact, forming larger molecules and coalescences. It is conceivable that these latter were formed over long periods of time as a consequence of a long series of chemical events, each of which required only an extremely small amount of energy. Under these conditions, each reaction could well be considered to be "reversible," and accompanied by a $\Delta G'$ equal to zero. A summation of these at any given time would also be equal to zero, so that, by the time primitive life-forms emerged, their actual synthesis over whatever time it took might be considered the result of a thermodynamically reversible sequence of equilibrium reactions all taking place in the primeval soup. It is likely that the potentiality for such equilibrium reactions diminished as the components of the soup began to be used up faster than they were formed, and about this time it must have been necessary for these self-reproducing, organic catalysts to become separated from their environment by some sort of limiting boundary within which their chemical reproductive mechanisms could be contained. The cytosol of present-day prokaryotes may be what remains of the primeval soup. As the earth's crust cooled, the thermal activation energy that had been utilized previously in these reversible, biosynthetic reactions slowly became unavailable, and was replaced equally slowly by chemical activation energy. This may have been when the use of activating molecules such as ATP first began. What is important here is that systems for the provision of chemical activation energy evolved separately from and subsequent to the evolution of a self-reproducing, cellular fabric. The former results in the utilization of chemical free energy during the process of activation, after which the activation energy is recovered quantitatively as heat and entropy. The utilization of activation energy therefore involves a loss of free energy to the environment, and is exergonic. The formation of the cellular fabric, on the other hand, can be conceived to have been a reversible process under the primitive conditions which we imagine to have existed in the primeval soup. The formation of cellular fabric under these conditions must therefore have occurred by means of a process of growth in which the free-energy change was neither positive nor negative, but zero. The amazing thing is that it still seems to be so. The reproduction of a primitive form of life is equivalent to what we call anabolism as it occurs within presently existing microbial cells. From the evidence presented in this book it appears that anabolism is also accompanied by a zero change in free energy. The process of microbial growth (anabolism) must thus be considered the most conservative of all biological processes, since it apparently has not changed energetically since the primitive life-forms became closed off from their environment. This is not true with respect to catabolism, which represents the provision of activation energy for growth and chemical energy for maintenance, both of which act through the generation of ATP or similar substances and both of which are exergonic. Catabolic processes have undergone a considerable evolution. The first general catabolic process was probably that of the fermentation of carbohydrates in the primeval soup. In the reducing environment that existed prior to the evolution of the photolysis of water, fermentations represented an uncomplicated method of producing ATP. Supporting this conjecture is the fact that the enzymes catalyzing fermentative processes are in general considered to be soluble, thereby being in effect separate from the cellular fabric. This may represent an initial, physical separation of both the structures and activities of anabolism and catabolism. Later, as carbohydrates in the primitive environment became limiting, parts of the cellular fabric became utilized in that nonsoluble, membranous structures acquired catabolic, enzymatic activity and the physical separation between the anabolic and the catabolic systems became indistinct. This did not mean, however, that the two systems did not remain chemically distinct. In fact, they seem to have done so, since a variety of physical and chemical mechanisms have evolved for the purpose of providing catabolic energy, including the various types of bacterial photosyntheses, anaerobic respiration, green-plant photosynthesis, and aerobic respiration. Thus, although we can conceive of a rational series of evolutionary steps involving the energetics of catabolic processes, during the same time period the process of growth (anabolism) apparently remained energetically unchanged.

10.6.14. The Energetics of Growth of Higher Organisms

The use of microorganisms for the study of the energetics of growth has the distinct advantage that most of them take what is needed for growth from the environment in the form of small molecular weight, soluble substances the thermodynamic properties of many of which are known. The same can be said for the products of their metabolism. Thus, it is possible to devise thermodynamically definable growth-process systems. The same is not true for most higher, heterotrophic organisms which ingest from the environment particulate food the thermodynamic properties of which are most obscure. In addition, higher heterotrophs need to store sources of energy with which to move from one place to another in search of food, and most of their total food intake is used for mechanical work which is either internal or external. Very little food is used for growth. Thus, about all that can be studied conveniently is the enthalpy change accompanying growth or metabolism which was first so brilliantly carried out by Rubner. Nevertheless, *if* a thermodynamically definable system involving a higher organism could be found, there is every reason to think that the energetics of growth that can be described with respect to bacteria as prokaryotes, and yeasts, filamentous fungi, and algae as eukaryotes should also be applicable to protozoa, coelenterates, nematodes, and so on. Carrying out the required experiments will, however, present a considerable challenge.

10.6.15. A Possible Difference Between the Use of Batch Cultures and Continuous Cultures with Respect to Studying the Energetics of Microbial Growth

As discussed previously, the use of batch cultures of microorganisms grown exponentially under nonlimiting experimental conditions, when $\mu = \mu_{max}$, has the advantage that no storage products are accumulated within the cells, and it is only the fabric of the cells which is synthesized. This may not be true of continuous cultures, in which the rate of growth is always limited by the experimental conditions and can vary from slightly less than μ_{max} down to zero. Although growth under these conditions is still theoretically considered to be exponential, the physical state of the cells *may* be quite different from that of cells grown under nonlimiting conditions. For example, it is conceivable that cells grown under environmental conditions of substrate limitation may slowly accumulate storage material to the point where the quantity of stored energy is sufficient to carry them through one division, which then takes place. There is then a prolonged time while more storage material is accumulated until another division can occur. If this is so, then a greater or lesser proportion of the cells will contain storage products at any given time, and an analysis of the cells will give an empirical formula and an apparent yield which will be different from that when the cells are grown under nonlimiting conditions. Thus, although the energetics of microbial growth would be expected to be the same for both batch and continuous cultures *if* the same system were being studied, it may be in fact that the two methods of culture do not provide the same system.

10.7. SUMMARY

1. It is possible to represent the growth of microorganisms as a balanced equation in which the cells are represented by a unit-carbon formula.

2. The equation in (1) can be divided into two balanced equations representing anabolism, or growth proper, and catabolism, respectively.

3. Gaseous oxygen does not enter into anabolism. It only functions as an electron acceptor during aerobic catabolism.

4. Using conservation theory $\Delta G'_{\text{an}}$, $\Delta H'_{\text{an}}$, and $\Delta S'_{\text{an}}$ with respect to heterotrophically grown microorganisms are all calculated to be zero.

5. From (4), if $\Delta G'_{\text{an}}$ is zero, then no *net* free energy from catabolic processes can become incorporated into the fabric of the cells. The free energy from ATP and similar molecules is therefore utilized strictly as activation energy which is recovered quantitatively as heat and entropy. This means that a Y_{ATP} cannot be calculated on the basis of the number of ATP molecules generated during catabolism. Exactly how much of the available energy from ATP is used as activation energy for growth is uncertain.

6. From (4), if $\Delta H'_{\text{an}}$ is zero, then all of the heat produced during microbial growth must come from catabolism. This has been demonstrated experimentally to be true.

7. From (4), if $\Delta S'_{\text{an}}$ is zero, it becomes possible to calculate $\Delta S'_f$ UCFW^{-1} of cells formed during anabolism, and from this the absolute entropy of this mass of cells in the living state. With this datum and that of the absolute entropy of the substrate, it can be calculated that S g^{-1} of cells is equal to S g^{-1} of substrate from which they have been synthesized, and that if information is in any way related to absolute entropy the quantity of information per gram of cells is equal to the quantity of information per gram of substrate.

8. If the early forms of life in the "primordial soup" are postulated to have occurred as the result of thermodynamically reversible reactions, then anabolism can be said not to have changed energetically since the time these early life-forms closed off from the environment. Instead, what has evolved is the various forms of catabolism in response to the sequential elimination of various carbon and energy sources.

9. It is not necessary to utilize methods designed to study the thermodynamics of irreversible processes to calculate changes in energy accompanying microbial growth. The use of classical thermodynamics is quite sufficient. On the other hand, if the use of the two methods is equally valid, it should be expected that the same results would be obtained.

10.8. CONCLUSION

In 1875 Pasteur wrote:

> I have full confidence that the theory of fermentation is founded. It will be established mathematically the day that science will be sufficiently advanced to put into a ratio the quantity of heat that the life of the yeast releases during the

> decomposition of sugar in the absence of air, with the quantity of heat furnished by combustions caused by free oxygen gas when the life of the yeast causes this to happen under conditions where this gas is furnished in more or less great abundance. [Pasteur,[40] p. 456.]

It is apparent that we have come this far and considerably farther in our comprehension of various aspects of the energetics of microbial growth. What is needed is further work in this field, and it is to be hoped that other investigators will be stimulated to challenge some of the ideas and concepts presented here, and to look deeper into the energy exchanges accompanying the process of growth in microbes and also in higher organisms to the extent that this latter is possible.

REFERENCES

1. E. H. Battley, Growth-reaction equations for *Saccharomyces cerevisiae, Physiol. Plant.* **13,** 192–203 (1960).
2. W. R. Mayberry, G. J. Prochazka, and W. J. Payne, Growth yields of bacteria on selected organic compounds, *Appl. Microbiol.* **15,** 1332–1338 (1967).
3. I. G. Minkevich, Estimation of available efficiency of microbial growth on methanol and ethanol, *Biotech. Bioeng.* **27,** 792–799 (1985).
4. E. I. Fulmer and E. Leifson, The free-energy decrease of alcoholic fermentation, *Iowa St. J. Sci.* **2,** 159–177 (1928).
5. G. N. Lewis and M. Randall, *Thermodynamics;* revised by K. S. Pitzer and L. Brewer, 2nd ed., McGraw-Hill, New York, 1961.
6. M. Dubrunfaut, Note sur la chaleur et le travail méchanique produit par la fermentation vineuse, *C. R. Hebd. Séances Acad. Sci.* **42,** 945–948 (1856).
7. A. Bouffard, Détermination de la chaleur dégagée dans la fermentation alcoolique, *C. R. Hebd. Séances Acad. Sci.* **121,** 357–360 (1895).
8. A. J. Brown, The heat of fermentation, *Inst. Brew. London J.* **7,** 93–103 (1903).
9. H. T. Brown, Some studies on yeast, *Ann. Bot.* **28,** 197–226 (1914).
10. E. H. Battley, Enthalpy changes accompanying the growth of *Saccharomyces cerevisiae* (Hansen), *Physiol. Plant.* **13,** 628–640 (1960).
11. M. Rubner, Die Umsetzungswärme bei der Alkoholgärung, *Arch. Hyg.* 355–418 (1904).
12. E. H. Battley, A calculation of the free-energy changes accompanying the growth of *Saccharomyces cerevisiae* on several substrates, *Proc. First European Biophys. Congr., V.IV, Baden, Austria,* pp. 299–305 (1971).
13. E. H. Battley, Alternate method of calculating the free-energy change accompanying the growth of *Saccharomyces cerevisiae* (Hansen) on three substrates, *Biotech. Bioeng.* **21,** 1929–1961 (1979).
14. H. Tamiya, *Le Bilan Matériel et l'Energétique des Synthèses Biologiques,* Actualités Scientifiques et Industrielles No. 214, Série 1935, Hermann & Cie., Editeurs, Paris.

15. R. Whelton and M. Doudoroff, Assimilation of glucose and related compounds by growing cultures of *Pseudomonas saccharophila, J. Bact.* **49,** 177–186 (1945).
16. S. R. Hoover and F. E. Allison, The growth metabolism of *Rhizobium,* with evidence on the interrelations between respiration and synthesis, *J. Biol. Chem.* **134,** 181–192 (1940).
17. H. Samejima and J. Meyers, On the heterotrophic growth of *Chlorella pyrenoidosa, J. Gen. Microbiol.* **18,** 107–117 (1958).
18. M. Molliard, Recherches calorimétriques sur l'utilisation de l'énergie au cours du development d'une culture de *Sterigmatocystis nigra, C. R. Séances Soc. Biol. Fil.* **87,** 219–221 (1922).
19. J. C. Senez and J. P. Belaich, Etude microcalorimétrique de la croissance bactérienne et du contrôle de l'activité métabolique par le phosphate, *Colloq. Int. Cent. Natl. Rech. Sci.* **124,** 357–363 (1963).
20. A. Grangetto, *Thermogénèse de croissance aérobie d'Aerobacter aerogenes,* Thèse, Université d'Aix-Marseille, 1963.
21. P. Boivinet, *Utilisation du microcalorimètre en microbiologie,* Thèse, Université d'Aix-Marseille, 1964.
22. J. P. Belaich and J. C. Senez, Growth yields of *Zymomonas mobilis, J. Bacteriol.* **89,** 1195–1200 (1965).
23. W. W. Forrest, Energetic aspects of microbial growth, *Symp. Soc. Gen. Microbiol.* **19,** 65–86 (1969).
24. J. D. Fast, *Entropy,* 2nd ed., MacMillan, London, 1970.
25. H. Linschitz, Information and physical entropy., in H. Quastler, Ed., *Information Theory in Biology,* Univ. of Illinois Press, Urbana, 1953, pp. 14–15.
26. H. Linschitz, The information content of a bacterial cell., in H. Quastler, Ed., *Information Theory in Biology,* Univ. of Illinois Press, Urbana, 1953, pp. 251–262.
27. H. J. Morowitz, Some order-disorder considerations in living systems, *Bull. Math. Biophysics* **17,** 81–86 (1955).
28. H. J. Morowitz, Some consequences of the application of the second law of thermodynamics to cellular systems, *Biochim. Biophys. Acta* **40,** 340–345 (1960).
29. E. Duclaux, *Pasteur, Histoire d'un Esprit,* Scaux, Parix, 1896.
30. T. Bauchop and S. R. Elsden, On the growth of microorganisms in relation to their energy supply, *J. Gen. Microbiol.* **23,** 457–469 (1960).
31. A. H. Stouthamer, *Yield Studies in Microorganisms,* Meadowfield Press, Ltd., Durham, 1976.
32. K. Burton, Energy of adenosine triphosphate, *Nature* **181,** 1594–1595 (1958).
33. W. W. Forrest and D. J. Walker, The generation and utilization of energy during growth, *Adv. Microb. Physiol.* **5,** 213–274 (1971).
34. J. E. DeVay and W. C. Schnathorst, Single-cell isolation and preservation of bacterial cultures, *Nature* (*Lond*), **199,** 775–777 (1963).
35. N. S. Iacobellis and J. E. Devay, Long-term storage of plant-pathogenic bacteria in sterile distilled water, *Appl. Environ. Microbiol.* **52,** 338–339 (1986).
36. J. van der Toorn, personal communication, 1986.

37. S. J. Pirt, The maintenance energy of bacteria growing in cultures, *Proc. R. Soc. Lond. B Biol. Sci.* **163,** 224–231 (1965).
38. S. J. Pirt, Maintenance energy: a general model for energy-limited and energy sufficient growth, *Arch. Microbiol.* **133,** 300–302 (1982).
39. A. I. Oparin, *The Origin of Life,* Macmillan, New York 1938.
40. L. Pasteur, Nouvelles observations sur la nature de la fermentation alcoolique (Réponse à *MM.* Brefeld et Traube)., *C. R. Hebd. Séances Acad. Sci.* **80,** 452–457 (1875).

APPENDIXES

APPENDIX 1. Standard values for chemical thermodynamic properties of selected substances functioning as convenient substrates or as products in microbial growth processes[a]

Substance	Formula	ΔG_f°	$\Delta G_f^{\circ\prime}$	ΔH_f°	$\Delta H_f^{\circ\prime}$	ΔS_f°	$\Delta S_f^{\circ\prime}$	Reference
Acetic acid	$C_2H_4O_2$	−93.08	−96.58	−115.73	−115.98	−75.97	−65.07	2
Acetic acid (eq. buf.)			−99.79		−116.06		−54.57	2
Acetate ion	$C_2H_3O_2-$		−90.08		−116.06		−87.14	2
Ammonia	NH_3	−3.94	−6.35	−11.02	−19.19	−23.75	−43.06	1
Ammonia (eq. buf.)			−9.58		−31.67		−74.09	2
Ammonium ion	NH_4^+		−18.97		−31.67		−42.60	1
Bicarbonate ion	HCO_3^-		−140.26		−165.39		−84.28	1
Carbon dioxide	CO_2	−94.25	−92.26	−94.05	−98.90	0.06	−2.23	1
Carbon dioxide (eq. buf.)			−93.37		−97.49		−13.82	1
Carbonic acid	H_2CO_3		−148.94		−167.22		−61.31	1
Ethanol	C_2H_6O	−41.63	−43.25	−66.20	−68.60	−82.40	−85.02	2
Formic acid	CH_2O_2	−86.38	−89.0	−101.51	−101.68	−50.74	−42.53	1
Formate ion	CHO_2^-		−83.9		−101.71		−59.73	1
α-D-Glucose	$C_6H_{12}O_6$	−217.63	−218.58	−304.60	−301.88	−29.17	−27.94	2
α,β-D-Glucose			−219.16		−302.05		−278.01	2
Glycerol	$C_3H_8O_3$	−114.6	−118.9	−160.3	−161.7	−153.28	−143.55	2
Glycine	$C_2H_5O_2N$	−88.09	−88.62	−126.22	−122.85	−127.89	−114.81	1
Glycine (eq. buf.)			−90.80		−125.0		−114.71	2
Glycinate ion	$C_2H_4O_2N^-$		−75.28		−112.28		−124.10	1
Hydrogen	H_2	0.00	4.2	0.00	−1.0	0.00	−17.44	1

Hydrogen ion	H^+		0.00		0.00		0.00	1
Hydroxyl ion	OH^-		−37.59		−54.97		−58.29	1
L(+)-Lactic acid	$C_3H_6O_3$	−125.06	−128.77	−165.88	−164.01	−136.91	−118.19	2
L(+)-Lactic acid (eq. buf.)			−133.21		−164.11		−103.64	2
β-Lactose	$C_{12}H_{22}O_{11}$	−374.5	−374.32	−534.6	−533.82	−536.97	−534.96	2
Methane	CH_4	−12.13	−8.22	−17.88	−21.28	−19.28	−43.80	1
Nitrogen	N_2	0.00	4.32	0.00	−2.52	0.00	−22.94	2
Nitric acid	HNO_3	−19.31	−26.61	−41.61	−49.56	−74.79	−76.97	1
Nitrate ion	NO_3^-		−26.61		−49.56		−76.97	1
Oxygen	O_2	0.00	3.9	0.00	−2.89		−22.77	2
Sucrose	$C_{12}H_{22}O_{11}$	−369.1	−370.8	−531.0	−529.60	−543.0	−532.62	2
Succinic acid	$C_4H_6O_4$	−178.62	−178.45	−224.86	−218.02	−155.08	−132.72	2
Succinic acid (eq. buf.)			−184.75		−217.19		−108.80	2
Water	H_2O	−56.69		−68.31		−38.97		1

[a]The symbols ΔG_f°, ΔH_f°, and ΔS_f° represent standard Gibbs free-energy, enthalpy, and entropy values, respectively, at 298.15°K and 1 atm. The symbols $\Delta G_f^{\circ\prime}$, $\Delta H_f^{\circ\prime}$, and $\Delta S_f^{\circ\prime}$ represent standard Gibbs free-energy, enthalpy, and entropy values, respectively, relative to the aqueous standard state at 298.15°K and 1 atm. ΔG_f°, $\Delta G_f^{\circ\prime}$, ΔH_f°, and $\Delta H_f^{\circ\prime}$ values are in kcal mole^{-1}. ΔS_f° and $\Delta S_f^{\circ\prime}$ values are in cal mole^{-1} deg^{-1}. The designation (eq. buf.) indicates an equilibrium mixture of species of the substance in dilute solution at a pH of 7.

References

1. D. D. Wagman, W. H. Evans, V. B. Barker, I. Halow, S. M. Bailey and R. H. Schumm, NBS Technical Note 270-3. *Selected Values of Chemical Thermodynamic Properties. Tables for the first thirty-four elements in the standard order of arrangement*, National Bureau of Standards, Washington, D.C., 1968.
2. R. C. Wilhoit, Selected values of thermodynamic properties, in *Biochemical Microcalorimetry*, ed. by H. D. Brown, Academic Press, New York, 1969, pp. 305–317. (There are a great many more values of chemical interest included here, other than those listed above.)

APPENDIX 2. Practical values for chemical thermodynamic properties of selected substances functioning as convenient substrates or as products in microbial growth processes[a]

Substance	Formula	$\Delta G_f'$	$\Delta H_f^{\circ\prime}$	$\Delta S_f'$
Acetic acid	$C_2H_4O_2$	−100.67	−115.98	−51.35
Acetic acid (eq. buf.)		−103.88	−116.06	−40.85
Acetate ion	$C_2H_3O_2^-$	−94.17	−116.06	−73.42
Ammonia	NH_3	−10.44	−19.19	−29.35
Ammonia (eq. buf.)		−13.67	−31.67	−60.37
Ammonium ion	NH_4^+	−23.06	−31.67	−28.88
Bicarbonate ion	HCO_3^-	−144.35	−165.39	−20.57
Carbon dioxide	CO_2	−96.35	−98.90	−8.55
Carbon dioxide (eq. buf.)		−97.46	−97.49	0.00
Carbonic acid	H_2CO_3	−153.03	−167.22	−47.59
Ethanol	C_2H_6O	−47.36	−68.60	−71.24
Formic acid	CH_2O_2	−93.09	−101.68	−28.81
Formate ion	CHO_2^-	−87.99	−101.71	−46.02
α-D-Glucose	$C_6H_{12}O_6$	−222.67	−301.88	−265.67
α,β-D-Glucose		−223.25	−302.05	−264.30
Glycerol	$C_3H_8O_3$	−122.99	−161.7	−129.83
Glycine	$C_2H_5O_2N$	−92.71	−122.85	−101.09
Glycine (eq. buf.)		−94.89	−125.0	−100.99
Glycinate ion	$C_2H_4O_2N^-$	−79.37	−112.25	−110.28
Hydrogen	H_2	0.11	−1.0	−1.11
Hydrogen ion	H^+	0.00	0.00	0.00
Hydroxyl ion	OH^-	−41.68	−54.97	−44.57
L(+)-Lactic acid	$C_3H_8O_3$	−132.86	−164.11	−104.81
L(+)-Lactic acid (eq. buf.)		−137.30	−164.11	−104.81
α-Lactose	$C_{12}H_{22}O_{11}$	−378.41	−533.82	−521.25
Methane	CH_4	−12.31	−21.28	−30.08
Nitrogen	N_2	0.23	−2.52	−7.68
Nitric acid	HNO_3	−30.70	−49.56	−63.25
Nitrate ion	NO_3^-	−30.70	−49.56	−63.25
Oxygen	O_2	−0.19[b]	−2.89	−9.05
Sucrose	$C_{12}H_{22}O_{11}$	−374.89	−529.60	−518.90
Succinic acid	$C_4H_6O_4$	−182.54	−218.02	−119.00
Succinic acid (eq. buf.)		−188.84	−217.19	−95.09

[a]The symbol $\Delta G_f'$ refers to the Gibbs free energy of formation of substances in aqueous solution at a concentration of 0.001 *m*, $\Delta H_f^{\circ\prime}$ to the enthalpy of formation of substances in the aqueous standard state (which is also that for 0.001 *m*), and $\Delta S_f'$ to the entropy of formation as calculated from $\Delta G_f'$ and $\Delta H_f^{\circ\prime}$ using the Gibbs free-energy equation, all at 298.15°K and 1 atm. $\Delta G_f'$ and $\Delta H_f^{\circ\prime}$ values are in kcal mole^{-1}. $\Delta S_f'$ values are in cal mole^{-1} deg^{-1}. Eq. buf. indicates the mixture of species in solution at pH 7.0.

[b]This value for oxygen is calculated for a 0.001 *m* aqueous solution, rather than for the 0.0003 *m* solution that exists in equilibrium with oxygen in the atmosphere, in view of the opinion of Pitzer and Brewer (Chapter 10, Ref. 3, p. 240) that below a mole fraction of 0.001 there will be very little further effect of dilution on the thermodynamic properties of solute molecules.

Source: $\Delta G_f'$ values calculated from analogous $\Delta G_f^{\circ\prime}$ values in Appendix 1. $\Delta H_f'$ values taken directly from Appendix 1.

APPENDIX 3. Equations representing the growth at 298.15°K and 1 atm of *Saccharomyces cerevisiae* anaerobically on glucose and aerobically on glucose, ethanol, and acetic acid, using ammonia as the nitrogen source

1. Anaerobic growth on glucose.

$$\left.\begin{array}{l} C_6H_{12}O_6(aq) \\ +\ 0.118NH_3(aq) \end{array}\right\} \longrightarrow \begin{array}{l} 0.590CH_{1.737}O_{0.451}N_{0.200}(\text{cells}) + 0.072H_2O(l) \\ +\ 1.300CH_3CH_2OH(aq) + 0.432C_3H_8O_3(aq) \end{array}$$

2. Aerobic growth on glucose.

$$\left.\begin{array}{l} C_6H_{12}O_6(aq) + 0.327NH_3(aq) \\ +\ 3.956O_2(aq) \end{array}\right\} \longrightarrow \begin{array}{l} 1.914CH_{1.703}O_{0.459}N_{0.171}(\text{cells}) \\ +\ 4.086CO_2(aq) + 4.860H_2O(l) \end{array}$$

3. Growth on ethanol.

$$\left.\begin{array}{l} CH_3CH_2OH(aq) + 1.153NH_3(aq) \\ +\ 1.851O_2(aq) \end{array}\right\} \longrightarrow \begin{array}{l} 1.030CH_{1.704}O_{0.408}N_{0.149}(\text{cells}) \\ +\ 0.970CO_2(aq) + 2.346H_2(l) \end{array}$$

4. Growth on acetate.

$$\left.\begin{array}{l} CH_3COOH(aq) + 0.094NH_3(aq) \\ +\ 1.336O_2(aq) \end{array}\right\} \longrightarrow \begin{array}{l} 0.620CH_{1.619}O_{0.442}N_{0.151}(\text{cells}) \\ +\ 1.380CO_2(aq) + 1.639H_2O(l) \end{array}$$

Source: References 1 and 12 in Chapter 10.

APPENDIX 4. Thermodynamic changes associated with the growth at 298.15°K and 1 atm of *Saccharomyces cerevisiae* anaerobically on glucose, and aerobically on glucose, ethanol, and acetate, using ammonia as the nitrogen source[a]

Substrate	$\Delta G'_{NC}$	$\Delta G'_C$	$\Delta G'_{an}$	$\Delta H'_{NC}$	$\Delta H'_C$	$\Delta H'_{an}$	$\Delta S'_{NC}$	$\Delta S'_C$	Eff. I	Eff. II
Glucose	−64.7	−47.6	0	−33.1	−22.1	−0.20	105.9	85.5	26.4	35.9
(anaerobic)		(−48.0)	(−0.37)					(86.9)	(25.8)	(34.8)
Glucose	−694.4	−457.6	0	−684.1	−450.8	0.13	34.5	22.8	34.1	51.7
(aerobic)		(−457.6)	(−0.00)					(22.8)	(34.1)	(51.7)
Ethanol	−314.8	−194.9	0	−325.5	−201.4	0.12	−35.6	−21.8	38.1	61.5
		(−194.9)	(0.00)					(−21.8)	(38.1)	(61.5)
Acetate	−205.0	−137.0	0	−212.7	−124.1	0.01	−25.8	−17.1	33.2	49.6
		(−137.0)	(0.00)					(−17.1)	(33.2)	(49.6)

[a]The dimensions of $\Delta G'_{NC}$, $\Delta G'_C$, $\Delta H'_{NC}$, and $\Delta H'_C$ are kcal $mole^{-1}$ of substrate consumed; those of $\Delta G'_{an}$ and $\Delta H'_{an}$ are kcal $(MYC_c \times UCFW)^{-1}$ of cells produced; and those of $\Delta S'_{NC}$ and $\Delta S'$ are cal $mole^{-1}$ of substrate consumed deg^{-1}. Eff. I and eff. II have the dimensions of percent free energy conserved. The thermodynamic properties used to obtain these calculated values were taken from Appendix 2. $\Delta G'_{an}$ was calculated using two methods; the values not in parentheses being in accord with Battley's hypothesis that $\Delta G'_{an} = 0$, and those in parentheses being calculated in the same manner as $\Delta H'_{an}$ on the basis of Battley's hypothesis with respect to energy conservation. The values in parentheses then enter into the calculations of $\Delta S'_C$, eff. I, and eff. II.

Source: The growth-process equations in Appendix 3.

APPENDIX 5. The thermodynamic properties at 298.15°K and 1 atm of *Saccharomyces cerevisiae* cells grown anaerobically on glucose, and aerobically on glucose, ethanol, and acetic acid[a]

Substrate	ΔG_f	ΔH_f	ΔS_f	S	S g^{-1}
Glucose	−8.11	−18.35	−34.3	9.8	0.412
(anaerobic)	(−8.77)		(−32.1)	(−12.0)	(0.505)
Glucose	−7.95	−17.87	−33.3	10.1	0.430
(aerobic)	(−7.95)		(−33.3)	(10.1)	(0.430)
Ethanol	−16.89	−21.13	−14.2	27.1	1.177
	(−16.89)		(−14.2)	(27.1)	(1.177)
Acetate	−21.02	−24.68	−12.3	28.6	1.253
	(−20.99)		(−12.4)	(28.5)	(1.248)

[a]The dimensions of ΔG_f and ΔH_f are kcal UCFW^{-1}, and those of ΔS_f and S are cal UCFW^{-1}deg^{-1}. S g^{-1} has the dimensions of cal g^{-} dry weight of cells deg^{-1}. The values of ΔG_f not in parentheses were calculated using Battley's hypothesis that ΔG equals zero; those in parentheses being calculated in the same manner as ΔH_f on the basis of Battley's hypothesis with respect to energy conservation. The values of ΔS_f, S, and S g^{-1} were obtained using ΔH_f and either value of ΔG_f.
Source: The growth-process equations in Appendix 3.

APPENDIX 6. Equations representing the growth of *Aspergillus oryzae* on glucose, ethanol, and glycerol, using either ammonia or nitrate as a nitrogen source

1. Growth on glucose and ammonia.
$$C_6H_{12}O_6(aq) + 0.446NH_3(aq) + 1.734O_2(aq) \longrightarrow 3.990CH_{1.753}O_{0.753}O_{0.570}N_{0.112}(cells) + 2.010CO_2(aq) + 3.173H_2O(l)$$
2. Growth on glucose and nitrate.
$$C_6H_{12}O_6(aq) + 0.237HNO_3(aq) + 2.488O_2(aq) \longrightarrow 2.822CH_{1.754}O_{0.571}N_{0.141}(cells) + 3.178CO_2(aq) + 3.645H_2O(l)$$
3. Growth on ethanol and ammonia.
$$CH_3CH_2OH(aq) + 0.089NH_3(aq) + 2.316O_2(aq) \longrightarrow 0.634CH_{1.803}O_{0.533}N_{0.141}(cells) + 1.366CO_2(aq) + 2.562H_2O(l)$$
4. Growth on ethanol and nitrate.
$$CH_3CH_2OH(aq) + 0.042HNO_3(aq) + 2.520O_2(aq) \longrightarrow 0.363CH_{1.830}O_{0.552}N_{0.117}(cells) + 1.637CO_2(aq) + 2.688H_2O(l)$$
5. Growth on glycerol and ammonia.
$$C_3H_8O_3(aq) + 0.235NH_3(aq) + 1.830O_2(aq) \longrightarrow 1.592CH_{1.695}O_{0.528}N_{0.148}(cells) + 1.408CO_2(aq) + 3.003H_2O(l)$$
6. Growth on glycerol and nitrate.
$$C_3H_8O_3(aq) + 0.130HNO_3(aq) + 1.799O_2(aq) \longrightarrow 1.314CH_{1.802}O_{0.556}N_{0.099}(cells) + 1.686CO_2(aq) + 2.881H_2O(l)$$

Source: The empirical formula and RQ data for constructing these equations came from Reference 14 in Chapter 10.

APPENDIX 7. Thermodynamic changes associated with the growth at 298.15°K and 1 atm of *Aspergillus oryzae* on glucose, ethanol, and glycerol, using either ammonia or nitrate as the nitrogen source[a]

Substrate and Nitrogen Source	$\Delta G'_{NC}$	$\Delta G'_{C}$	$\Delta G'_{an}$	$\Delta H'_{NC}$	$\Delta H'_{C}$	$\Delta H'_{an}$	$\Delta S'_{NC}$	$\Delta S'_{C}$	Eff. I	Eff. II
Glucose (NH_3)	−694.4	−200.7	0	−684.1	−197.7	0.003	34.7	10.0	71.1	246.0
		(−200.7)	(−0.00)					(10.0)	(71.1)	(246.0)
Glucose (HNO_3)	−694.4	−320.0	0	−684.1	−311.0	0.171	34.7	30.2	53.9	117.0
		(−320.0)	(−0.11)					(30.2)	(53.9)	(117.0)
Ethanol (NH_3)	−314.8	−243.1	0	−325.5	−249.3	1.914	−35.6	−21.0	22.8	29.5
		(−242.9)	(0.17)					(−21.6)	(22.8)	(29.5)
Ethanol (HNO_3)	−314.8	−270.1	0	−325.5	−278.7	−0.147	−35.6	−29.2	14.2	16.6
		(−270.2)	(−0.12)					(−29.2)	(14.1)	(16.4)
Glycerol (NH_3)	−392.1	205.1	0	−398.1	−208.2	0.002	−20.1	−10.4	47.7	91.2
		(−221.3)	(−0.02)					(−10.0)	(47.7)	(91.1)
Glycerol (HNO_3)	−392.1	221.3	0	−398.1	−192.5	0.001	−20.1	9.6	45.6	77.2
		(−221.3)	(−0.02)					(9.6)	(45.6)	(77.2)

[a]The dimensions of the symbols are the same as for those in Appendix 4.

Source: The growth-process equations in Appendix 6.

APPENDIX 8. The thermodynamic properties at 298.15°K and 1 atm of *Aspergillus oryzae* cells grown on glucose, ethanol, and glycerol, using either ammonia or nitrate as the nitrogen source[a]

Substrate and Nitrogen Source	ΔG_f	ΔH_f	ΔS_f	S	$S\ g^{-1}$
Glucose, NH_3	13.74	−24.46	−35.3	9.9	0.405
	(−13.74)		(−35.9)	(9.4)	(0.384)
Glucose, HNO_3	−13.39	−24.30	−36.6	8.1	0.336
	(−13.35)		(−36.7)	(8.1)	(0.336)
Ethanol, NH_3	−23.81	−25.93	−7.1	38.7	1.590
	(−23.54)		(−7.9)	(37.9)	(1.558)
Ethanol, HNO_3	−24.55	−30.27	−19.2	27.1	1.114
	(−24.89)		(−17.9)	(28.2)	(1.159)
Glycerol, NH_3	−15.70	−22.19	−21.8	22.4	0.924
	(−15.81)		(−21.4)	(22.8)	(0.941)
Glycerol, HNO_3	−16.88	−24.21	−24.6	20.8	0.863
	(−16.90)		(−24.5)	(20.8)	(0.863)

[a]The dimensions of the symbols are the same as for those in Appendix 5.
Source: The growth-process equations in Appendix 6.

APPENDIX 9. Equations representing the growth at 298.15°K and 1 atm of *Pseudomonas saccharophila* on several substrates using ammonia as the nitrogen source

1. Growth on acetate.
$$\left.\begin{array}{l}CH_3COOH(aq) + 0.097NH_3(aq)\\ + 1.380O_2(aq)\end{array}\right\} \longrightarrow \begin{array}{l}0.560CH_{1.618}O_{0.327}N_{0.174}(\text{cells})\\ + 1.440CO_2(aq) + 1.693H_2O(l)\end{array}$$
2. Growth on glucose.
$$\left.\begin{array}{l}C_6H_{12}O_6(aq) + 0.564NH_3(aq)\\ + 2.406O_2(aq)\end{array}\right\} \longrightarrow \begin{array}{l}3.240CH_{1.618}O_{0.327}N_{0.174}(\text{cells})\\ + 2.760CO_2(aq) + 4.230H_2O(l)\end{array}$$
3. Growth on hydrogen.
$$\left.\begin{array}{l}H_2(aq) + 0.086CO_2(aq)\\ + 0.015NH_3(aq) + 0.405O_2(aq)\end{array}\right\} \longrightarrow \begin{array}{l}0.086CH_{1.618}O_{0.327}N_{0.174}(\text{cells})\\ + 0.953H_2O(l)\end{array}$$
4. Growth on lactate.
$$\left.\begin{array}{l}C_3H_6O_3(aq) + 0.230NH_3(aq)\\ + 1.533O_2(aq)\end{array}\right\} \longrightarrow \begin{array}{l}1.320CH_{1.618}O_{0.327}N_{0.174}(\text{cells})\\ + 1.680CO_2(aq) + 2.275H_2O(l)\end{array}$$
5. Growth on maltose.
$$\left.\begin{array}{l}C_{12}H_{22}O_{11}(aq) + 1.023NH_3(aq)\\ + 5.472O_2(aq)\end{array}\right\} \longrightarrow \begin{array}{l}5.880CH_{1.618}O_{0.327}N_{0.174}(\text{cells})\\ + 6.120CO_2(aq) + 7.778H_2O(l)\end{array}$$
6. Growth on pyruvate.
$$\left.\begin{array}{l}C_3H_4O_3(aq) + 1.888NH_3(aq)\\ + 1.300O_2(aq)\end{array}\right\} \longrightarrow \begin{array}{l}1.080CH_{1.618}O_{0.327}N_{0.174}(\text{cells})\\ + 1.920CO_2(aq) + 1.408H_2O(l)\end{array}$$
7. Growth on sucrose.
$$\left.\begin{array}{l}C_{12}H_{22}O_{11}(aq) + 1.023NH_3(aq)\\ + 5.472O_2(aq)\end{array}\right\} \longrightarrow \begin{array}{l}5.880CH_{1.618}O_{0.327}N_{0.174}(\text{cells})\\ + 6.120CO_2(aq) + 7.778H_2O(l)\end{array}$$

Source: The empirical formula and the numerical data for constructing these growth-process equations were taken from Reference 15 in Chapter 10.

APPENDIX 10. Thermodynamic changes associated with the growth at 298.15°K and 1 atm of *Pseudomonas saccharophila* on several substrates using ammonia as the nitrogen source[a]

Substrate	$\Delta G'_{NC}$	$\Delta G'_{C}$	$\Delta G'_{an}$	$\Delta H'_{NC}$	$\Delta H'_{C}$	$\Delta H'_{an}$	$\Delta S'_{NC}$	$\Delta S'_{C}$	Eff. I	Eff. II
Acetate	−205.0	−141.5	0	−212.7	−146.8	−0.11	25.6	−17.9	31.0	44.9
		(−141.5)	(0.00)					(−17.9)	(31.0)	(44.9)
Glucose	−694.4	−278.5	0	−684.5	−274.5	−0.07	33.2	13.4	59.9	149.3
		(−278.8)	(−0.29)					(14.1)	(59.9)	(149.3)
Hydrogen	−56.7	−45.9	0	−65.8	−53.28	0.00	−30.7	−7.4	19.1	23.5
		(−45.9)	(0.00)					(−7.4)	(19.1)	(23.5)
Lactate	−325.7	−166.4	0	−328.9	−167.8	0.27	−11.0	−4.7	48.9	95.7
		(−166.2)	(0.23)					(−5.3)	(49.0)	(95.9)
Maltose	−1397.0	−637.0	0	−1368.7	−626.5	0.01	95.1	35.2	54.4	119.3
		(−637.0)	(0.02)					(35.2)	(54.4)	(119.3)
Pyruvate	−281.6	−146.4	0	−280.9	−146.1	0.01	2.1	1.0	48.0	92.3
		(−146.4)	(0.00)					(1.0)	(48.0)	(92.3)
Sucrose	−1402.6	−639.6	0	−1374.0	−626.5	0.06	95.9	43.9	54.4	119.3
		(−639.6)	(0.01)					(43.9)	(54.4)	(119.3)

[a]The dimensions of the symbols are the same as for those in Appendix 4.

Source: The growth-process equations in Appendix 9.

APPENDIX 11. The thermodynamic properties at 298.15°K and 1 atm of *Pseudomonas saccharophila* cells grown on several substrates using ammonia as the nitrogen source[a]

Substrate	ΔG_f	ΔH_f	ΔS_f	S	S g^{-1}
Acetate	15.53	−19.00	−11.6	27.0	1.27
	(−15.59)		(−11.4)	(27.2)	(1.27)
Glucose	−0.54	−9.91	−31.4	7.2	0.34
	(−0.63)		(−31.1)	(7.5)	(0.35)
Hydrogen	−3.67	9.88?	45.4?	84.0?	3.94?
	(−3.66)		(45.4)?	(84.0)?	(3.94)?
Lactate	−8.44	−14.48	−20.3	18.3	0.86
	(−8.27)		(−20.8)	(17.8)	(0.84)
Maltose	−0.23	−9.84	−33.7	4.9	0.23
	(−0.23)		(−33.8)	(4.8)	(0.22)
Pyruvate	3.89	−11.63	−26.0	12.6	0.59
	(−3.89)		(−26.0)	(12.6)	(0.59)
Sucrose	0.74	−9.34	−33.8	4.8	0.22
	(−0.74)		(−33.8)	4.8	(0.22)

[a]The dimensions of the symbols are the same as for those in Appendix 5.

Source: The growth-process equations in Appendix 9.

APPENDIX 12. Equations representing the growth at 298.15°K and 1 atm of *Chlorella pyrenoidosa* heterotrophically on glucose using either ammonia or nitrate as the nitrogen source, the thermodynamic changes associated with the growth processes represented by these equations, and the thermodynamic properties of the cells formed during these processes[a]

1. Growth on glucose and ammonia.
   $C_6H_{12}O_6(aq) + 0.345NH_3(aq) + 2.598O_2(aq) \longrightarrow 3.083CH_{1.740}O_{0.526}N_{0.112}(cells) + 2.917CO_2(aq) + 3.835H_2O(l)$
2. Growth on glucose and nitrate.
   $C_6H_{12}O_6(aq) + 0.400HNO_3(aq) + 2.316O_2(aq) \longrightarrow 2.683CH_{1.750}O_{0.500}N_{0.149}(cells) + 3.317CO_2(aq) + 3.852H_2O(l)$

Substrate and Nitrogen Source	$\Delta G'_{NC}$	$\Delta G'_C$	$\Delta G'_{an}$	$\Delta H'_{NC}$	$\Delta H'_C$	$\Delta H'_{an}$	$\Delta S'_{NC}$	$\Delta S'_C$	Eff. I	Eff. II
Glucose, NH_3	−694.4	−300.7	0	−684.5	−281.7	14.65	33.2	63.6	56.7	130.9
		(−300.7)	(0.00)					(63.6)	(56.7)	(130.9)
Glucose, HNO_3	−694.4	−331.8	0	−684.5	−315.9	0.06	33.2	53.2	52.2	109.3
		(−331.8)	(0.00)					(53.2)	(52.2)	(109.3)

	ΔG_f	ΔH_f	ΔS_f	S	$S\ g^{-1}$
Glucose, NH_3	−9.41	−19.94	−35.3	7.9	0.340
	(−9.41)		(−35.3)	(7.9)	0.340
Glucose, HNO_3	−10.14	−20.50	−34.7	9.7	0.406
	(−10.14)		(−34.7)	(9.7)	(0.406)

[a]The dimensions of the symbols are the same as for those in Appendixes 4 and 5.

Source: The empirical formula and the numerical data for constructing the growth-process equations were taken from Reference 17 in Chapter 10.

APPENDIX 13. Equations representing the growth at 298.16°K and 1 atm of *Rhizobium meliloti* on sucrose, using either ammonia or nitrate as the nitrogen source, the thermodynamic changes associated with the growth processes represented by these equations, and the thermodynamic properties of the cells formed during these processes[a]

1. Growth on sucrose and ammonia.
   $C_{12}H_{22}O_{11}(aq) + 0.772NH_3(aq) + 3.876O_2(aq) \longrightarrow 7.354CH_{1.684}O_{0.473}N_{0.105}(cells) + 4.646CO_2(aq) + 5.966H_2O(l)$
2. Growth on sucrose and nitrate.
   $C_{12}H_{22}O_{11}(aq) + 0.748HNO_3(aq) + 2.616O_2(aq) \longrightarrow 7.126CH_{1.684}O_{0.473}N_{0.105}(cells) + 4.874CO_2(aq) + 5.374H_2O(l)$

Substrate and Nitrogen Source	$\Delta G'_{NC}$	$\Delta G'_C$	$\Delta G'_{an}$	$\Delta H'_{NC}$	$\Delta H'_C$	$\Delta H'_{an}$	$\Delta S'_{NC}$	$\Delta S'_C$	Eff. I	Eff. II
Sucrose, NH_3	−1402.6	−453.0	0	−1374.0	−413.7	30.04	95.9	131.8	67.7	209.6
		(−453.0)	(0.00)					(131.8)	(67.7)	(209.6)
Sucrose, HNO_3	−1402.6	−427.6	0	−1374.0	−401.8	−0.01	95.9	86.5	69.5	228.0
		(−427.6)	(0.00)					(86.5)	(69.5)	(228.0)

	ΔG_f	ΔH_f	ΔS_f	S	$S\ g^{-1}$
Glucose, NH_3	−6.92	−17.99	−37.1	4.6	0.200
	(−6.92)		(−37.1)	(4.6)	(0.200)
Glucose, HNO_3	−6.73	−17.80	−37.2	4.5	0.200
	(−6.73)		(−37.2)	(4.5)	(0.200)

[a]The dimensions of the symbols are the same as for those in Appendixes 4 and 5.

Source: The empirical formula and the numerical data for constructing the growth-process equations were taken from Reference 16 in Chapter 10.

APPENDIX 14. The thermodynamic properties of some small molecular weight organic substances on a unit-carbon basis[a]

Substrate	nC	$\Delta G_f^{\circ\prime}$	$\Delta H_f^{\circ\prime}$	$S^{\circ\prime}$
Glycine	2	−45.40	−62.50	18.95
L-Alanine	3	−29.57	−44.2	12.67
L-Valine	5	−17.14	−29.54	8.46
L-Leucine	6	−14.03	−25.63	8.23
L-Aspartic acid	4	−44.01	−56.37	17.42
L-Glutamic acid	5	−34.50	−46.94	11.90
L-Glutamine	5	−25.24	−38.48	12.00
Glycyl-glycine	4	−31.09	−43.87	?

[a]The dimensions of $\Delta G_f^{\circ\prime}$ and $\Delta H_f^{\circ\prime}$ are kcal mole^{-1} and that of $S^{\circ\prime}$ is cal mole^{-1}deg^{-1}. The data are for an equilibrium mixture of species formed from a 1 *m* solution of the species listed in a solution buffered at pH 7.

Source: R. C. Wilhoit, Selected values of thermodynamic properties, in *Biochemical Microcalorimetry,* ed. by H. D. Brown, Academic Press, New York, 1969, pp. 305–317.

APPENDIX 15. Equations representing ATP utilization during the anaerobic growth of *Saccharomyces cerevisiae* on glucose[a]

Catabolism:

$$0.650C_6H_{12}O_6(aq) \longrightarrow 1.300CH_3CH_2OH(aq) + 1.300CO_2(aq) \qquad \Delta G'_{cat} = -42.06$$

ATP synthesis:

$$1.30ADP(aq) + 1.30P_i(aq) \longrightarrow 1.30ATP(aq) + 1.30H_2O(aq) \qquad \Delta G' = +16.25$$

I. *Catabolism coupled to* ATP *synthesis*

$$0.650C_6H_{12}O_6(aq) + 1.30ADP(aq) + 1.30P_i(aq) \longrightarrow \begin{matrix} 1.300CH_3CH_2OH(aq) + 1.30H_2O(l) + 1.300CO_2(aq) \\ + 1.30ATP(aq) \end{matrix} \qquad \Delta G' = -25.81$$

Anabolism:

$$0.104C_6H_{12}O_6(aq) + 0.118NH_3(aq)\} \longrightarrow 0.590CH_{1.737}O_{0.451}N_{0.200}(cells) + 0.034CO_2(aq) + 0.288H_2O(l) \qquad \Delta G' = 0.00$$

ATP utilization:

$$1.30ATP(aq) + 1.30H_2O(l) \longrightarrow 1.30ADP(aq) + 1.30P_i(aq) \qquad \Delta G' = -16.25$$

II. *Anabolism coupled to* ATP *utilization*

$$\left.\begin{matrix} 0.104C_6H_{12}O_6(aq) + 0.118NH_3(aq) \\ + 1.30ATP(aq) + 1.30H_2O(l) \end{matrix}\right\} \longrightarrow \begin{matrix} 0.590CH_{1.737}O_{0.451}N_{0.200}(cells) + 0.034CO_2(aq) \\ + 0.288H_2O(l) + 1.30ADP(aq) + 1.30P_i(aq) \end{matrix} \qquad \Delta G' = -16.25$$

III. *Glycerol production* (possibly not enough energy here for ATP production)

$$0.252C_6H_{12}O_6(aq) + 0.216H_2O(l)\} \longrightarrow 0.432C_3H_8O_3(aq) + 0.216CO_2(aq) \qquad \Delta G' = -5.59$$

Equations I + II + III above:

$$1.006C_6H_{12}O_6(aq) + 0.118HN_3(aq) \longrightarrow \begin{matrix} 0.590CH_{1.737}O_{0.451}N_{0.200}(cells) + 1.300CH_3CH_2OH(aq) \\ + 0.432C_3H_8O_3(aq) + 1.550CO_2(aq) + 0.072H_2O(l) \end{matrix} \qquad \Delta G'_C = -47.65$$

[a] The value of $\Delta G'$ for the reaction ATP $\rightarrow$ ADP + P_i has been taken as -12.5 kcal (Reference 32 in Chapter 10). The other data are from Section 10.5.1. $\Delta G'$ values are in kcal.

APPENDIX 16. Equations representing ATP utilization during the aerobic growth of *Saccharomyces cerevisiae* on glucose[a]

Catabolism:

$$0.659C_6H_{12}O_6(aq) + 3.954O_2(aq) \longrightarrow 3.954CO_2(aq) + 3.954H_2O(l) \qquad \Delta G_{cat} = -457.61$$

ATP synthesis:[b]

$$18.45ADP(aq) + 18.45P_i(aq) \longrightarrow 18.45ATP(aq) \qquad \Delta G' = +230.62$$

I. *Catabolism coupled to* ATP *synthesis*

$$\left.\begin{array}{l}0.659C_6H_{12}O_6(aq) + 3.954O_2(aq)\\ + 18.45ADP(aq) + 18.45P_i(aq)\end{array}\right\} \longrightarrow 3.954CO_2(aq) + 3.954H_2O(l) + 18.45ATP(aq) \qquad \Delta G' = -226.99$$

Anabolism:

$$0.341C_6H_{12}O_6(aq) + 0.327NH_3(aq)\} \longrightarrow 1.914CH_{1.703}O_{0.459}N_{0.171}(cells) + 0.132CO_2(aq) + 0.907H_2O(l) \qquad \Delta G'_{an} = 0.00$$

ATP utilization:

$$18.45ATP(aq) \longrightarrow 18.45ADP(aq) + 18.45P_i(aq) \qquad \Delta G' = -230.62$$

II. *Anabolism coupled to* ATP *utilization*

$$C_6H_{12}O_6(aq) + 3.954O_2(aq) \longrightarrow \begin{array}{l}1.914CH_{1.703}O_{0.459}N_{0.171}(cells) + 4.086CO_2(aq)\\ + 4.860H_2O(l) + 18.45ADP(aq) + 18.45P_i(aq)\end{array} \qquad \Delta G' = -230.62$$

Reactions I plus II above:

$$C_6H_{12}O_6(aq) + 0.327NH_3(aq) + 3.954O_2(aq)\} \longrightarrow 1.914CH_{1.703}O_{0.459}N_{0.171}(cells) + 4.086CO_2(aq) + 4.860H_2O(l) \qquad \Delta G'_C = -457.61$$

[a]The value of $\Delta G'$ for the reaction ATP → ADP + P_i has been taken as −12.5 kcal (Reference 32 in Chapter 10). The other data are from Section 10.5.2. $\Delta G'$ values are in kcal. [b]Assuming a P/O ratio of 2.

APPENDIX 17. Definitions of symbols used in Chapter 10

a_c	The activity of a solute in aqueous solution at concentration c.
AE, AE_c, AE_p, AE_S	Available electrons, available electrons per UCFW of cells, available electrons per UCFW of product, available electrons per mole of substrate, respectively.
(aq)	The postscript representing a substance in aqueous solution under unspecified conditions.
c	Cells, or referring to cells.
cal	Small calories.
(cells)	The postscript representing the "state" of living cells.
$CHn_HOn_ONn_N$	The unit-carbon formula representing products of a growth process (these could be cells or other products), where n_H, n_O, and n_N represent the relative numbers of H, O, and N atoms, respectively, when the number of carbon atoms is taken as 1.
$Cn_CHn_HOn_ONn_N$	The formula for an organic substrate, where n_C, n_H, n_O, and n_N represent the relative numbers of C, H, O, and N atoms, respectively, in the substrate molecule.
D_r	The degree of reduction of an organic substance on a scale of 8 = the number of available electrons per unit-carbon atom of a substance = $(4n_C + n_H - 2n_O - 3n_N) \times n_C^{-1}$.
eff. I	The "machine efficiency" of a growth process = $[(\Delta G'_{NC} - \Delta G'_C) \times 100] \times \Delta G'^{-1}_{NC}$, or similar equation with respect to $\Delta H'$ or $\Delta S'$ changes.
eff. II	$[(\Delta G'_{NC} - \Delta G'_C) \times 100] \times \Delta G'^{-1}_C$, or similar equation with respect to $\Delta H'$ or $\Delta S'$ changes.
$\Delta G^{\circ\prime}$, $\Delta H^{\circ\prime}$, $\Delta S^{\circ\prime}$	The standard Gibbs free-energy, enthalpy, and entropy changes, respectively, accompanying reactions in aqueous solution at 298.15°K and 1 atm at pH 7.
$\Delta G'$, $\Delta H'$, $\Delta S'$	Nonstandard Gibbs free-energy, enthalpy, and entropy changes, respectively, accompanying reactions in aqueous solution.
$\Delta G'_{an}$, $\Delta H'_{an}$, $\Delta S'_{an}$	The nonstandard Gibbs free-energy, enthalpy, and entropy changes, respectively, accompanying anabolism.
$\Delta G'_C$, $\Delta H'_C$, $\Delta S'_C$	The nonstandard Gibbs free-energy, enthalpy, and entropy changes, respectively, accompanying a conservative reaction or process.
$\Delta G'_{cat}$, $\Delta H'_{cat}$, $\Delta S'_{cat}$	The nonstandard Gibbs free-energy, enthalpy, and entropy changes, respectively, accompanying catabolism.
$\Delta G'_{con}$, $\Delta H'_{con}$, $\Delta S'_{con}$	The nonstandard Gibbs free-energy, enthalpy, and entropy conserved during a conservative reaction or a growth process. With respect to the latter, these represent the energy changes that would occur were the products of the growth process to be reacted in such a way as to form the products of the relative nonconservative reaction.
$\Delta G'_{dil}$	The free-energy change accompanying the dilution of a solute in aqueous solution.
$\Delta G^{\circ\prime}_f$, $\Delta H^{\circ\prime}_f$, $\Delta S^{\circ\prime}_f$	The standard Gibbs free-energy, enthalpy, and entropy changes, respectively, accompanying the formation from the elements of 1 mole of substance in aqueous solution at 298.15°K and 1 atm, at pH 7.

APPENDIX 17. (*Continued*)

$\Delta G'_f$, $\Delta H'_f$, $\Delta S'_f$	The nonstandard Gibbs free-energy, enthalpy, and entropy changes, respectively, accompanying the formation from the elements of 1 mole or of 1 unit wt of substance in aqueous solution or in a hydrated state (as found in living cells), respectively.
$\Delta G'_{NC}$, $\Delta H'_{NC}$, $\Delta S'_{NC}$	The nonstandard Gibbs free-energy, enthalpy, and entropy changes, respectively, accompanying a nonconservative reaction in aqueous solution.
kcal	Kilocalories.
MYC_c, MYC_p	The molar yield coefficient for 1 UCFW of cells, or of 1 mole or 1 UCFW of product, respectively.
nC, nH, nO, nN	The numbers of C, H, O, and N atoms, respectively, in substrate molecules.
*n*C, *n*H, *n*O, *n*N	The numbers of C, H, O, and N atoms, respectively, in product substances.
p	Product, or referring to a product.
R	The gas constant, here having a value of 1.987 cal $mole^{-1}deg^{-1}$.
(s), (l), (g)	Postscripts designating the state of a substance as being solid, liquid, or gas, respectively.
S_c	The quantity of substrate in moles required to construct the mass of cells grown on 1 mole of substrate = $[(MYC_c \times UCFW) \times (AE_c \times UCFW^{-1})] \times (AE_S \times mole^{-1})^{-1}$.
S_p	The quantity of substrate in moles required to construct the quantity of a noncellular product formed during growth on 1 mole of substrate = $[(MYC_p \times$ UCFW or molecular wt$) \times (AE_p \times$ UCFW or molecular wt$)] \times (AE_S \times mole^{-1})^{-1}$.
S_N	The quantity of substrate in moles required to reduce *N* moles of nitrate to ammonia = $(8e \times mole^{-1} \times n$ moles nitrate$) \times (AE_S \times mole^{-1})^{-1}$.
T	The temperature in °K.
UCF	Unit-carbon formula.
UCFW	Unit-carbon formula weight.
Y	The molar growth yield (not ash-free) = g dry wt of cells × $moles^{-1}$ of substrate.
Y_{af}	Ash-free molar growth yield = g ash-free dry wt of cells × $moles^{-1}$ of substrate.

INDEX